Flaxseed in Human Nutrition
Second Edition

Editors

Lilian U. Thompson
Department of Nutritional Sciences
Faculty of Medicine
University of Toronto

Stephen C. Cunnane
Department of Nutritional Sciences
Faculty of Medicine
University of Toronto

Champaign, Illinois

Library of Congress Cataloging-in-Publication Data
Flaxseed in human nutrition / editors, Lilian U. Thompson, Stephen C. Cunnane.-- 2nd ed.
    p. cm.
Includes bibliographical references and index.
  ISBN 1-893997-38-3 (alk. paper)
 1. Flaxseed in human nutrition. 2. Flaxseed. I. Thompson, Lilian U.
II. Cunnane, Stephen C.

  QP144.O44F56 2003
  613.2'8--dc21

2003011329

Printed in the United States of America with vegetable oil-based inks.

5 4 3 2 1

Preface

The first edition of *Flaxseed in Human Nutrition* was published by AOCS Press in 1995. The objective then was to describe in a single multiauthored volume the state of the art regarding properties of flaxseed intended for human consumption and its potential uses in human nutrition and health. Consumer and food industry interest in flaxseed as a beneficial component in the human diet has continued to grow as the scientific literature on this subject has expanded over the past decade. Hence, in 2002 we agreed to edit this, the second edition of *Flaxseed in Human Nutrition*, with the goal of providing the current status of the knowledge about the analysis and composition of flaxseed, the metabolism and bioavailability of its major components (α-linolenic acid and lignans), the effect of flaxseed on development and disease, processing of flaxseed, and availability of flaxseed products. Some of the research in these areas was just emerging in the early to mid-1990s and was incompletely or not described when the first edition was published but is much better substantiated now.

For instance, in the past few years, the cardioprotective attributes of α-linolenic acid were given a firm boost by the Lyon Heart Study and by reports from several epidemiological studies. Affirmation of this important component of flaxseed is welcome at the start of the twenty-first century when cardiovascular disease continues to be a major killer in both developed and developing countries. Fish oils are highly efficacious in moderating heart disease risk, but dietary preferences are changing and fish stocks are becoming endangered. People are therefore seeking healthy alternatives, and flaxseed is one of them, not only for its α-linolenic acid but for its lignans, soluble fiber, and probably protein.

The book is divided into four sections. Section 1 provides updated and comprehensive information on the structure and composition of flaxseed and the biosynthesis of lignans. Section 2 describes the metabolism and bioavailability of α-linolenic acid and lignans and the effect of flaxseed on sex hormone metabolism, nutrition, and hematology. Section 3 covers the effect of flaxseed and its major components (α-linolenic acid, lignan, and fiber) on brain function and infant development, cancer, cardiovascular disease, diabetes, kidney disease, bone metabolism, menopause, inflammatory disease, immune function, reproductive function, and fetal development. Both the potential health and adverse effects are included. Section 4 describes the recent advances in the processing of flaxseed to different fractions, availability and labeling of flaxseed food products and supplements, and the use of flaxseed in poultry and pig diets to increase n-3 fatty acids in products derived from them.

Recent studies investigating how flaxseed is potentially beneficial in many chronic diseases have identified the major components responsible for the effect of flaxseed and provided some insights on the mechanisms of their actions. Lignans have been identified as the components partly responsible for the beneficial health effects of flaxseed. Hence, they have been a focus of much research activity since the publication of the first edition. Studies on their effect on cancer have advanced from the animal model to clinical trials to epidemiology. Knowledge of their effect on other diseases continues to evolve. Research on their bioavailability and analysis is providing guidance on the appropriate level of flaxseed and lignan intake. Their safety has been assessed from their effect on reproductive and fetal development. Methods of isolating or concentrating lignans from flaxseed have been developed, and some have been commercialized.

Progress in finding out how "functional foods" like flaxseed work is an important step in encouraging their consumption. It is equally rewarding to see that the food industry is attuned to the scientific evidence and to consumer demand, and is capable of producing flaxseed-enriched products aimed at those seeking healthier food alternatives. Canada should therefore maintain its international prominence as a flaxseed producer and as a leader in research into the nutritional and health attributes of flaxseed.

Many people have helped make the second edition of *Flaxseed in Human Nutrition* possible. Thanks go to the Books and Special Publications Committee of AOCS Press, who identified the need for an update, and the authors who have taken the time and effort from busy schedules and respected the near-impossible deadline we set. Mary Lane and her staff at AOCS Press, including Melissa Blankenship, Connie Winslow, and Brian Moore, have tirelessly and patiently compiled the manuscripts and have been a beacon of encouragement. We are proud to have edited the second edition of *Flaxseed in Human Nutrition* and hope it will be an asset in helping to place flaxseed in the mainstream of human nutrition.

Lilian U. Thompson and Stephen C. Cunnane
Department of Nutritional Sciences
Faculty of Medicine
University of Toronto

Contents

Chapter 1

Structure, Composition, and Variety Development of Flaxseed

James K. Daun[a], Véronique J. Barthet[a], Tricia L. Chornick[a], and Scott Duguid[b]

[a]Canadian Grain Commission Grain Research Laboratory, 1404-303 Main Street, Winnipeg, Manitoba R3C 3G8, Canada, and [b]Agriculture and Agri-Food Canada, Morden, Manitoba, Canada

Introduction

Production and Utilization of Flaxseed

World flaxseed production has averaged about 2.5 million tonnes over the past five years, less than 1% of the total world oilseed production (1; Table 1.1). Canada is by far the world's largest flaxseed producer, accounting for about one-third of the total world production and most of the export trade.

Although flaxseed has been a human food for thousands of years, industrial uses of flax oil (more commonly known as linseed oil) have predominated since the industrial revolution. The rapid oxidation properties of linseed oil make it especially useful for the production of linoleum, paints, stains, and surface coatings as well as in the production of oleo chemicals based on α-linolenic acid (ALA). Flaxseed fiber also has unique properties that make it useful in the production of fiber products such as cloth (linen) or paper. In Europe especially, flax has been cultivated in long-stem form for linen production, primarily as a fiber crop with the seeds being processed as a by-product. Flaxseed is the predominant term for oilseed flax in North America, whereas the term linseed is used in Europe.

The use of flaxseed as a functional food has been traditional in many parts of the world, especially in northern Europe where incorporation of flaxseed into bread has been traditional for many years. Recently, however, the functional properties of flax have received considerable attention in North America. The health-related reasons for this increase are the topic of many other chapters in this book. Flaxseed has been incorporated into foods such as breads, muffins, cereals, crackers, energy bars, baking mixes, snacks, soups, and waffles. The use of whole or milled flaxseed in home baking or even as an adduct to fruit juices has increased significantly.

The use of flaxseed in pet food has also seen a dramatic increase in recent years. A search (GOOGLE™ Canada) of the internet with the key words "flax pet food" produced 775 hits, mostly to different companies offering flax-based products.

In addition, there is considerable interest in the use of flaxseed as an animal feed ingredient. Research has demonstrated benefits to animal health from incor-

TABLE 1.1
World Flaxseed Supply and Disposition (1)

	Local marketing year				
	1998–1999	1999–2000	2000–2001	2001–2002[a]	2002–2003[b]
Harvested area (Mha)	3.52	3.50	3.09	2.93	3.06
Average yield (t/ha)	0.77	0.79	0.73	0.71	0.69
	(1000 tonnes)				
Carryin stocks	170	310	560	410	270
Production					
Canada[c]	1081	1022	693	715	679
China	523	404	520	420	465
United States[d]	170	200	273	291	335
India	280	290	220	260	245
EU-15	312	565	211	135	125
Other	330	293	333	269	251
Total production	2696	2774	2250	2090	2100
Total supply	2866	3084	2810	2500	2370
Crush	2219	2130	2015	1910	1850
Other	337	394	385	320	300
Total use	2556	2524	2400	2230	2150
Carryout stocks	310	560	410	270	220
Trade	923	710	793	774	750

[a]Estimate, *Oil World*, September 2002.
[b]Forecast, AAFC, December 2002.
Source: Oil World, except [c]Statistics Canada and [d]USDA.

poration of flaxseed into the diet of animals. As discussed elsewhere in this book, this has resulted not only in improved health and performance of the animals but also in increased deposition of ALA in tissues including eggs, pork fat, and beef fat.

A conservative estimate of the amount of flaxseed entering the health market including both human and pet consumption exceeds 100,000 tonnes/y (1). When combined with the amount of flaxseed entering animal feed, world estimates for total feed and food use of flaxseed have been greater than 400,000 tonnes annually.

Scope of Chapter

Despite the growing use of flaxseed as a nutritional source, there has been a relative scarcity of information regarding the composition of the seed. This paucity of information is further complicated by some erroneous information present in the scientific and technical literature regarding such basic components as oil content and ALA composition. Since the first edition of this book in 1995, especially over the past five years, there has been an increasing awareness of the need for accurate information on the composition of flaxseed. Because flaxseed destined for nutritional purposes is often sold in small parcels, it is also important to provide information on the variability of composition due to both variety and environment. For example, the Flax Council of Canada publishes a pamphlet (2) that lists the general composition of flaxseed. The pamphlet has a link to the Canadian Grain Commission's quality data summaries of flaxseed (available at http://www.grainscanada.gc.ca/Quality/grlreports/Flax/flax-menu-e.htm). This site provides information on the composition of the major components of flaxseed from different varieties and years.

Along with variety and environment, analytical methodology is another source of variability in compositional data for flaxseed. Although this may come as a surprise to many who are used to numbers being absolute, it must be recognized that all analytical methods are subject to variability due to random factors. In addition, some methods may be more accurate than others. For example, the discussion on oil content below will point out that a common method for determining the fat content in food is inaccurate when applied to flaxseed and other soft oilseeds.

This chapter will focus on providing information, supplemental to that provided in the first edition of the book, on the chemical composition of flaxseed; how that composition varies with variety and environment; and what are the best analytical methods available to measure specific components. The chapter will also identify areas where the authors feel that more information or better analytical methodology is needed in order to provide a complete picture of the composition of flaxseed used for human and animal nutrition.

Where necessary, data derived from studies on meal or oil presented in this chapter have been recalculated to reflect the contents in whole flaxseed. Average values of 40% oil were used for the recalculation.

General Structure and Composition of Flaxseed

Seed Structure

Flaxseed is composed of an embryo or germ, a thin endosperm, and two cotyledons encased in a seed coat (or hull) called a spermoderm (3; Fig. 1.1). The seeds are 4–6 mm in length, and the seed coat, which may range in color from deep brown to yellow or be variegated, is usually smooth and shiny. Seed weight is about 5 ± 1 g/1000, and there is a positive correlation between seed weight and oil content (4). Test weight for flaxseed ranges from about 55–70 kg/hL with no relationship between test weight and oil content.

Hand-dissected flaxseed is made up of 55% cotyledon, 36% seed coat and endosperm (hull), and 4% embryo. The cotyledons are the major oil storage tissue, containing about three-quarters of the seed oil. About 22% of the oil is found in the seed coat and 3% in the embryo. Oil is found mainly as triacylglycerols in oil bodies having an average diameter of 1.3 μm. The organelles contain about 98% neutral lipid, 1.3% encasing protein (oleosin), 0.9% phospholipids, and 0.1% free fatty acids. Most of the phospholipids are made up of phosphatidyl choline and phosphatidyl serine (5).

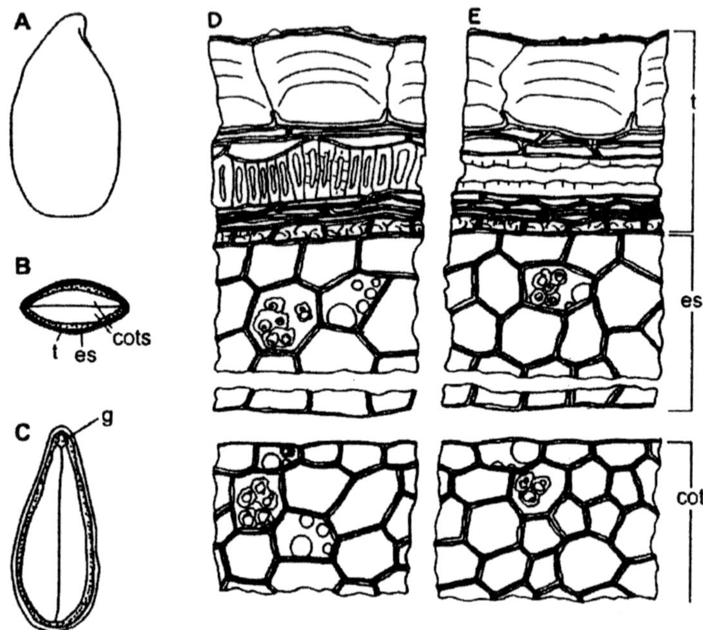

Fig. 1.1. Structure of flaxseed. A–C, three views of a flaxseed (approx. 7×): A, seed; B, transverse section; C, longitudinal section. D, transverse section and (×400); E, longitudinal section (×400), showing testa endosperm and cotyledon cells. *Abbreviations:* cots, cotyledons; es, endosperm; g, germ; t, testa. *Source:* Based on Vaughan, J.G., *The Structure and Utilization of Oil Seeds*, Chapman and Hall, London, 1970, p. 140.

Several different commercial processes for separating the hull and cotyledon from flaxseed have resulted in hull and cotyledon yields of about 40% each, with the residue made up of nonseparated material (6–8). Although previous studies have shown that dehulling of flaxseed can serve to concentrate components of interest, no commercial use of this process has yet occurred. Recent findings that the nutraceutical lignans are concentrated in the seed coat fraction (9) may result in an economically feasible process.

Seed coat color in flaxseed can range from brown to light yellow. In Canada, the light yellow seed coat characteristic has been reserved for solin (low ALA flax) varieties, but in the United States, several varieties, with yellow seed coats are popular for food use. There has been a suggestion that the seed coat in the yellow-seeded flax is thinner (10), but Dribnenki, a Canadian breeder of yellow-seeded solin, reported that although the yellow seed produced by the recessive yellow gene had less seed coat than seeds produced by the dominant brown gene, there was no difference in seed coat weight. He noted that some yellow-seeded strains had a greater propensity to have seed coats with a faulty seam holding the two halves of the seed coats together.

Proximate Composition

Flaxseed is recognized as having about 40% oil, 30% dietary fiber, 20% protein, 4% ash, and 6% moisture (Table 1.2). Estimates of proximate composition vary considerably according to the samples involved and the methodology employed to measure each of the components. No data on proximate composition, or of any other component measured in flaxseed, should be accepted unless it is accompanied by details giving the source of the samples tested as well as the analytical procedures involved in the testing. Both sample source and analytical methodology can be responsible for large differences in results. Variation due to sample source cannot be controlled. Variation due to analytical methodology can be controlled, however, by choosing only analytical methods that have been shown to give accurate measurements of the component in question and where the sources of error have been documented.

Chemical Components of Flaxseed

Oil (or Fat) Content

Oilseed flax is characterized as being made up of about 45% oil and 55% meal on a dry basis. Nutritional and labeling guidelines use the term fat to refer to neutral lipids from any source, and the nutritional and labeling definition of fat refers to the amount of triacylglycerol that would be made up from the fatty acids derived from all of the acyl lipids in a food. For oilseeds used for processing to provide oil and meal, the term oil content refers to those neutral lipids obtained by exhaustive extraction with a nonpolar lipid solvent, usually hexane or a related mixture of hydrocarbons. This mixture is composed almost exclusively of triacylglycerols (11) and provides a good estimate of the nutritional fat content although it fails to account for a small amount of polar lipid, mainly from phospholipids which might make up as much as 2% of the total mass of flax seeds (12).

TABLE 1.2

Proximate Analysis of Flaxseed (Various Estimates)

Source	Fat	Fiber[a]	Fiber[b]	Protein	Moisture	Ash	Energy (cal)
			(g/100 g)				
University of Saskatchewan (55)	41	28	—	26	0	4	ND
USDA (17)	34	27.9	—	19.5	8.74	3.5	492
Flax Council of Canada (35)	41	28	—	20	7	4	450
Canadian Grain Commission (35)	39.8–45.6	—	30.5–36.8	17.4–24.1	4.2–4.9	ND	ND

[a]Total dietary fiber.
[b]Sum of soluble and insoluble fiber.
Abbreviation: ND, no data.

                                    J.K. Daun et al.

Canada, the major producer of flaxseed, also has developed the most comprehensive set of published data on the oil contents of commercially grown flaxseed (13). Over the past five years, Canadian flaxseed contained, on average, 44% oil. Oil content on individual farm samples can vary by as much as 15%, with a range of 35–50% being reported for farm-grown samples tested over the five-year period of 1998–2002. As the seed from individual farm samples moves through the handling system it is combined, and the range of oil content decreases so that the range in Canadian exports of flaxseed has been only 3.5%, from 42.3–45.8% (Fig. 1.2).

Effect of Analytical Methods on Oil Content Results. The increased utilization of flaxseed in food has resulted in some controversy over oil or fat content values. The

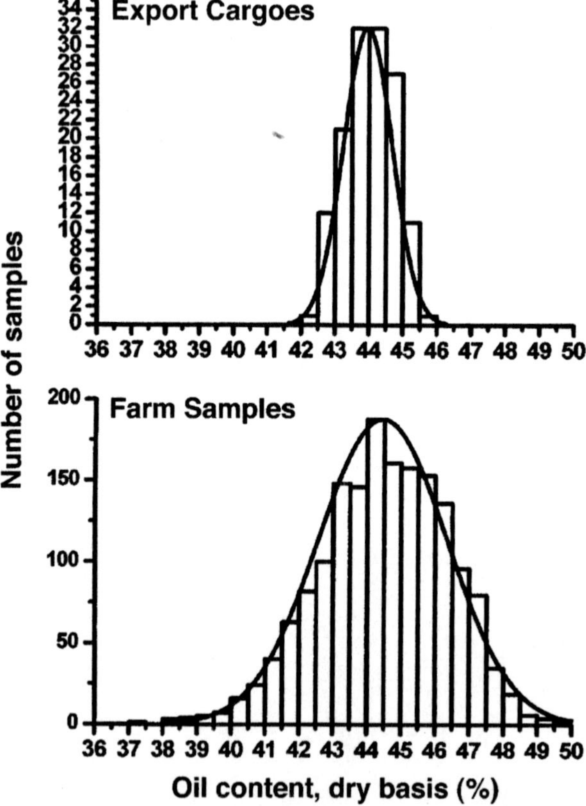

Fig. 1.2. Difference in range of oil contents in flax samples between farm samples and export cargoes based on data from Canadian Grain Commission surveys from 1998–2002. Export oil contents ranged from 42.3–45.8%, and farm samples ranged from 35.7–50%. The decrease in range between exported and farm flax oil content is due to the averaging effect of the grain handling system.

controversy rests in the methodology used for the analysis. The physical and chemical structure of flaxseed makes it relatively difficult to remove the oil from the seed. For this reason, the methods most often utilized for determining fat or oil in food, including those methods most often used for food labeling purposes, give inaccurately low values when applied to flaxseed. These values can be inaccurate by as much as 5% of the oil content (11,14). The best methods for oil content determination rely on exhaustive extraction of the ground seed using neutral solvents. Fineness of grind is very important. It is necessary to regrind and extract the sample three times in order to ensure that all of the neutral lipid is removed. If a nutritional fat value is desired, this neutral lipid can be transesterified and the methyl ester content determined by GC using internal standards (15). For completeness, the polar lipids in the residual meal should be treated to a hydrolysis/ etherification method (16; AOAC 996.06).

This difference caused by methods can be extremely important economically if the payment is based on oil content or a factor, such as ALA, that is dependent on oil content. It can be even more important nutritionally. For example, the USDA Nutrient database shows flaxseed as having only 34% fat (17), a value that is likely low by as much as 6%. This would result in a significant error in the calculated energy content and, because the analysis is based on GC determination of fatty acid methyl esters, would also result in a significant underestimation of the different fatty acids (e.g., ALA) present in a portion.

Some research on oil and fat content of flaxseed remains to be carried out. In particular, work is required to detail an accurate and precise method for determining the fat content of flaxseed according to the nutritional labeling definition. This method would include estimation of both the neutral lipid, now determined by exhaustive extraction, and the polar lipids. In addition, rapid methods for accurately determining and preparing the neutral lipid contents are required. These methods must give values that correspond to the values from the current exhaustive extraction methods. The lipids isolated by these methods should be unchanged so that other quality factors such as acidity and peroxide value can be determined. It is also important that indirect methods for oil content determination such as NIR be calibrated against the exhaustive extraction method.

Lipid Composition

Lipid Classes. Exhaustive extraction with hexane does not remove all the lipids from flaxseed. Like other biological materials, flaxseed contains complex polar lipids that are not soluble in hexane. To remove these lipids it is necessary to extract with a more polar solvent. The most common of these is a mixture of chloroform and methanol. Total lipid extractions of flaxseed (Table 1.3) have demonstrated the predominant lipid class to be acylglycerols and fatty acids, and sterol esters, glycolipids, and phospholipids make up about 5% of the total lipids. There is a need to further study the makeup of the minor components of flaxseed lipids. In particular, the phospholipid classes need further clarification. Although all studies agreed that phosphatidylcholine made up the largest group of phospholipids,

TABLE 1.3
Lipid Components in Flaxseeds

Component	Min	Max	Min	Max	Min	Max
	(g/100 g seed)		(Rel. %)		(mg/100 g seed)	
Triacylglycerols	36.0	40.7				
Monoacylglycerols	0.5	0.9				
Diacylglycerols	0.9	2.6				
Free fatty acids	0.1	3.4				
Sterols, sterol esters, and hydrocarbons	0.5	2.9				
Cholesterol			0	0.09		
Brassicasterol			0.1	0.7		
Campesterol			25	31		
Stigmasterol			6	9		
β-sitosterol			45	53		
δ-5 avenasterol			8	12		
δ-7 stigmastero			0	3		
δ-7 avenasterol			0	0.6		
Total neutral lipids	38.6	43.4				
Complex lipids	0.4	1.8				
Glycolipids	1.0	2.9				
Monogalactosyldiacylglycerol			40.5	43.0		
Digalactosyldiacylglycerol[b]			24.9	26.2		
Acylatedsterylgalactoside			15.4	16.5		
Sterylgalactoside			15.4	17.1		
Phospholipids	0.1	2.3				
Phosphatidyl inositol			6.9	23.4		
Phosphatidyl choline			29.1	67.5		*(Continued)*

there was disagreement about the relative importance of the other phospholipids. In addition, the makeup of the glycolipid fraction should be confirmed. The content and composition of sterols, sterol esters, hydrocarbons, and wax esters also require further study.

Fatty Acid Composition. Traditional flaxseed has very high levels of ALA, usually making up greater than 50% of the total fatty acids (Fig. 1.3). Other fatty acids include palmitic (about 5%), stearic (about 3%), oleic (about 18%), and linoleic (about 14%). Although most publications list only the major fatty acids in flax oil, the oil from *Linum usitatissimum* seeds also contains about 1% of other fatty acids with chain lengths as long as C_{24}. The duplication of structure of the C_{18} fatty acids in the C_{20} fatty acids is evident with significant amounts of eicosatrienoic acid (18) in flax oil but only trace amounts in solin oil (with larger amounts of eicosadienoic acid in solin oil). The identity of the minor peak between $C_{22:0}$ and $C_{24:0}$ is unknown, but analogy suggests it might be $C_{22:3}$. Flaxseed oil is one of the most unsaturated common oils, and its iodine value is usually greater than 185 Wijs units. Iodine

TABLE 1.3
(*Cont.*)

Component	Min	Max	Min	Max	Min	Max	Min	Max
	(g/100 g seed)		(Rel. %)		(mg/100 g seed)		(mg/kg seed)	
Phosphatidyl glycerol and lysophosphatidyl glycerol			2.5	19.5				
Phosphatidyl ethanolamine and phosphatidyl serine			3.1	35.9				
Cardiolipin			19.5	20.3				
Tocopherols								
α-tocopherol					0.0	1.2		
β-tocopherol						2.4		
γ-tocopherol					8.5	39.5		
δ-tocopherol					0.2	1.1		
Plastochromanol-8					11.9	13.9		
Total					8.8	57.2		
Pigments								
β-carotene								1.8
trans β-carotene							0.3	1.1
cis β-carotene							0.1	0.5
Lutein								27.7
Total carotenoids[b]							8.0	46.1
Chlorophyll							0.7	2.0

[a]Results calculated to reflect a flaxseed with about 40% oil content. Lipid class analysis (5,31,32,132,133) by chloroform methanol extraction; sterol composition (40); glycolipids (31,34); phospholipids (5,31–33) by various chromatographic techniques; and tocopherols by spectrophotometry (39) or by various HPLC methods (35,37,134–136). Pigments by spectrophotometric (137,138) or HPLC (43,44).
[b]Based on analysis in the authors' laboratory and published results (137).

value, a traditional measurement of the total degree of unsaturation in oils, has been shown to be strongly correlated with levels of ALA and negatively correlated with levels of oleic, linoleic, and saturated fatty acids (19).

The level of unsaturation varies with both variety and environment. The level of ALA in individual farm samples of Canadian flaxseed harvested in 2002 was found to range from 52–63%. Environmental studies have demonstrated that flaxseed grown under cool conditions has a higher degree of unsaturation (20–22). Between 1952–1994, the average iodine value of Canadian flaxseed was strongly correlated with the mean summer temperature (Fig. 1.4). Location and varietal effects have also been demonstrated to play a significant role in determining the degree of unsaturation in flax (23). In addition, the degree of unsaturation has been shown to be affected by agronomic practices including seeding rate, time of swathing, and dessication (24,25).

Plant breeders, using mutation selection processes, have developed lines of *L. usitatissimum* with low levels of ALA (< 3%) (26; Fig. 1.3). These lines have been given the common name of solin to differentiate them from flaxseed. The ALA con-

　　　　　　　　　　　　　　　J.K. Daun et al.

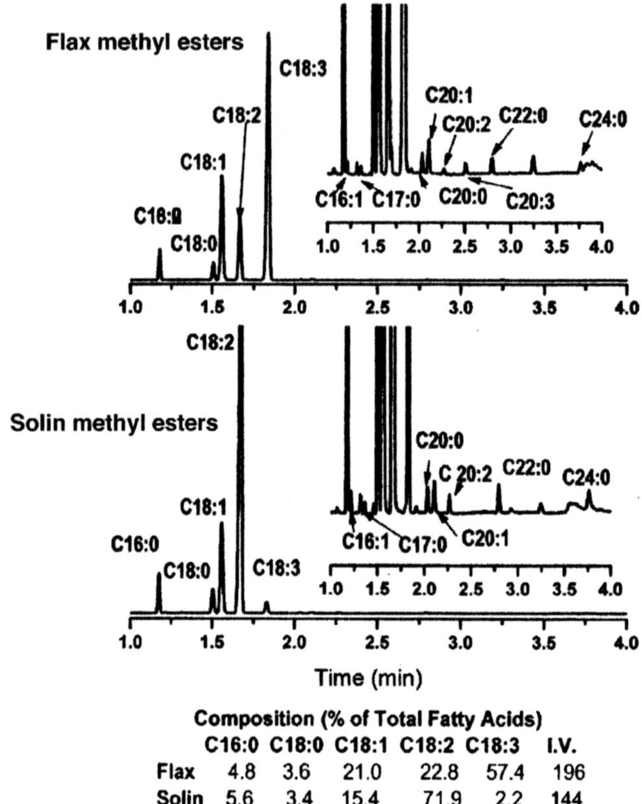

Composition (% of Total Fatty Acids)					
	C16:0	C18:0	C18:1	C18:2	C18:3	I.V.
Flax	4.8	3.6	21.0	22.8	57.4	196
Solin	5.6	3.4	15.4	71.9	2.2	144

Fig. 1.3. Fatty acid composition of flax and solin oils. GC traces from methyl ester preparations of hexane extracts from seed. Chromatographic conditions: Samples injected (split 30:1) into a 15 m × 0.32 mm column with a 0.25 µm Supelcowax 10 coating; carrier gas H_2 at linear flow 35 cm/s; temperature program, initial 220°C for 1 min; 10°C/min to 245°C; hold for 2.5 min.

tent of flaxseed has a strong genetic component (27). Using different breeding techniques, including mutation selection, varieties of flaxseed with different levels of fatty acid have been developed (Table 1.4).

The fatty acid composition expressed in Figure 1.3 reflects mainly the composition of the triacylglcyerols. Although unsaturated fatty acids have been shown to predominate at the 2 position of vegetable fats, Brockerhoff and Yurkowski (28), in their classic paper on stereospecific analysis of the structure of vegetable oils, noted that ALA was found in nearly equivalent amounts in all three positions on the glycerol molecule. Linoleic acid predominated in the 2 position, in a proportion similar to other vegetable oils. Palmitic acid was preferentially inserted in the 1 position. The high proportion of ALA in the 1 and 3 positions was confirmed (29). In a study of

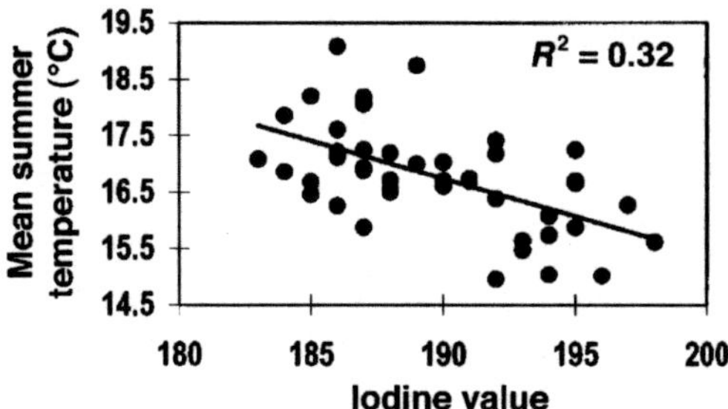

Fig. 1.4. Effect of temperature on unsaturation in flaxseed. Data from mean iodine values from the Canadian Grain Commission's harvest surveys. Weather data from Environment Canada.

triacylglycerol species, the highly unsaturated species linoleyoldi-α-linolenyl (13.8%) and tri-α-linolenyl (20.9%) were the predominant species in flax oil (30). Both glycolipids and phospholipids were found to have lower amounts of ALA and much higher amounts of saturated fatty acids than triacylglcyerols (31,32). Kulkarni *et al.* (33) noted that there was no difference in the fatty acid composition between phospholipids and seed oils from Indian linseed lines and that although there were larger amounts of saturated fatty acid in glycolipids (34), ALA was still the predominant fatty acid. This work highlights the need for further studies on the lipid components in flaxseed.

The fatty acid composition was found to vary according to the position in the seed, with the embryo axis containing much less ALA and more saturated fatty acids (3). It was noted that although the embryo axis contained only 4.5% of the seed oil, its unique composition contributed greatly to the overall fatty acid composition of the seed.

Tocopherols and Tocotrienols

Flaxseed, like most other oilseeds, contains a group of vitamin E compounds including both tocopherols and at least one tocotrienol. The major tocopherol in flaxseed is γ-tocopherol present at about 30 mg/100 g (Table 1.3). Flax also contains about 12

TABLE 1.4

Fatty Acid Variation in *Linum usitatissimum*

Sample	$C_{16:0}$	$C_{18:0}$	$C_{18:1}$	$C_{18:2}$	$C_{18:3}$
			(%)		
Flax[a]	4.8	3.6	21.0	22.8	57.4
Hi-saturates (139)	27.8	2.8	11.3	6.6	44.0
Solin[a]	5.6	3.4	15.4	71.9	2.2
Hi-oleic (139)	15.3	4.0	49.3	21.6	9.3
High ALA[a]	4.5	2.3	10.0	10.7	71.8

[a]Analysis of samples in the authors' laboratory.

mg/100 g of a tocotrienol, plastochromanol-8 (14,35,36). This compound has been reported to have better activity than α-tocopherol in preventing fat oxidation in a model system (14). Unfortunately, this component has not been included in many studies. Analytical methodology has also played a large role in the values reported for tocopherols. In particular, the reports by Oomah *et al.* (37) and Buden *et al.* (38) have much lower values than those reported by other researchers. This discrepancy is likely due to analytical methodology—especially calibration and extraction techniques—but also to their neglecting to saponify those tocopherols that are esterified.

Tocopherol levels have been reported to have significant interactions with both variety and environment (37,39). There has been some active interest in breeding samples with higher levels of tocopherols.

There is a need to provide better reference methodology for determination of tocopherols in seeds. The common reference methods, both involving HPLC, are AOCS Ce8-89 (40) and ISO 9936 (41). These methods are based on commercial fats and oils and do not provide instructions for complete extraction of tocopherols from seeds prior to determination. This is a common problem for many other determinations of seed lipid components. The data in Table 1.3 is based on extractions with hexane, commercially extracted crude oils, or extractions with solvents more polar than hexane (diethyl ether or hexane with 5% isopropanol).

Other Lipid Components (Sterols, Pigments)

Flaxseed sterols are characterized by major amounts of stigmasterol, campesterol, and δ-5 avenasterol (Table 1.3). Much of the information on flax sterols is comparatively old, and sterols are expressed as a relative percentage of the total. It would be useful to carry out a study using modern analytical methods to determine the content of each sterol and its variation due to environment and variety.

The major carotenoid pigment in flaxseed is lutein, although violaxanthin has been reported in developing embryos (42). Recent interest in the vitamin and antioxidant activity of carotene has resulted in estimates of this component (43,44; Table 1.3), but only small amounts are present amounting to about 100 retinol activity equivalents per kilogram of seed. Flaxseed is a chloroembryophyta, plants with green embryos. During maturation, however, all of the chlorophyll and other photosynthetic pigments in flax embryos disappear. Measurements in the authors' laboratory showed <1 mg/kg chlorophyll in top-quality Canadian flaxseed, and only about 2 mg/kg chlorophyll in grade No. 3 Canada.

Flax unsaponifiables were found to contain 11% hydrocarbons and 22% triterpene alcohols including β-amirina, cycloartenol, and 24-methylenecycloartanol (18).

Protein Content

Flaxseed meal is considered a high protein meal for animal feed supplementation. Proteins are complex polymers of amino acids. Many thousands of different proteins with molecular weights ranging from several thousand to several million make up much of the structure and all of the enzymic activity sites of living organisms. Because of the complex nature of proteins, it is impossible to provide a truly accurate measure of their content in foods. Instead, estimation is made based on a measure of the nitrogen content of the food.

Crude and True Protein. On average, nitrogen makes up about 16% of amino acids, although the actual amount varies from 7.7 (tyrosine)–32% (arginine). Early studies showed that proteins from animal sources contained about 16% nitrogen, and from these values the nitrogen to protein conversion factor of 6.25 was adopted to give an estimate of "crude protein." It was generally accepted that this value would likely be high because it would not account for the sources of nonprotein nitrogen in a sample. In proteins from plant sources, particularly, the value of 6.25 as a conversion factor is significantly high because plants not only have significant amounts of nonprotein nitrogen but also have different ratios of amino acids to the animal proteins from which the factor was derived. It is extremely difficult to obtain accurate values for calculating the "true protein" of plant sources (45). An estimate based on amino acid analyses of flax meal range is 5.41 (46). Recent work (47) has shown that there was little variation in this factor due to environment or variety.

Using the true protein factor will lower crude protein results by 15%, a significant amount. The "true protein" values derived from amino acid analyses do not account for the nonprotein nitrogen contained in the sample. For flaxseed these compounds would include some vitamins, sinapine, choline, and cyanogenic glycosides. An estimate of the corrected true protein value can be made from the recovery of nitrogen from amino acid analysis. This corrected value for true protein would be in the range of 4.9, giving protein contents 22% lower than crude protein.

Although it is important that researchers know the difference between crude protein and true protein, the significant differences between them do not imply that previous nutritional studies are in doubt. It is hoped that both human and animal nutritionists will make increasing use of true protein values in their studies. The use of true protein conversion factors is even more important in protein isolation studies.

Over the past five years, Canadian flaxseed has been found to contain about 23% crude protein (or 20% true protein or 18% true protein corrected for nonprotein nitrogen). On individual farm samples, crude protein ranged from 17.4–29.2%, whereas on exported flaxseed the range was 20.9–25.1%. As with other oilseeds, there is an inverse relationship between oil and protein in flaxseed (Fig. 1.5). The relationship for flaxseed (correlation coefficient $r \approx -0.50$), however, is not as strong as for other oilseeds such as canola ($r \approx -.85$), and unlike canola, the meal protein content is not related to the oil content.

Proteins in flaxseed were found to be made up of about 20% albumins (1.6S and 2S) and 80% legumin-like proteins (11S and 12S) (48). Flaxseed proteins were found to be structurally more lipophilic than soybean proteins (49). The legumin-like fraction was found to be rich in sulphur amino acids, but the 1.6S fraction was reported to act as a sulphur reserve for seed germination.

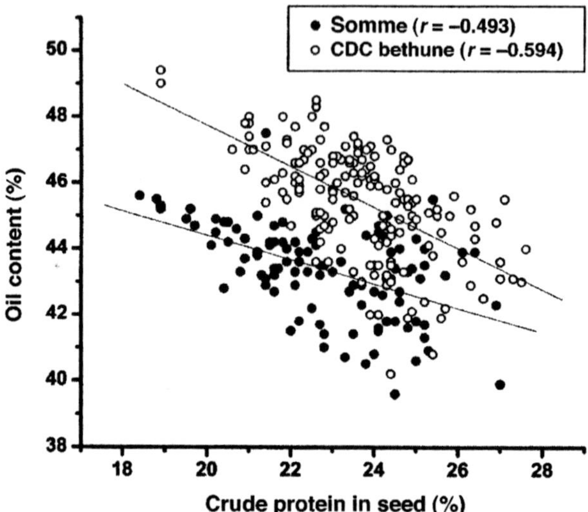

Fig. 1.5. Relationship between oil and crude protein in two varieties of flaxseed. Data from samples collected during Canadian Grain Commission harvest surveys, 1998–2002.

Methodology for Protein Determination. Total protein has conventionally been measured by determining nitrogen as discussed above and then applying a conversion factor. For many years, the most common method to determine nitrogen in biological materials was the Kjeldahl method. This method involves dissolution of the sample in concentrated sulphuric acid during which organic nitrogen is converted to ammonia. The ammonia is distilled from the mixture into a boric acid solution followed by an acidimetric titration. Although the Kjeldahl method was well established, it had some drawbacks, including a tendency to lower recoveries of nitrogen from oilseed samples unless special precautions were taken. In addition, the method generates large amounts of chemical waste making it environmentally undesirable.

More recently, the Dumas combustion method for total nitrogen has supplanted the Kjeldahl method as the method of choice for determining total nitrogen. This method involves burning a sample in a stream of oxygen, converting all of the organic nitrogen to nitrogen oxides. After catalytic conversion to known oxides, the nitrogen content of a known volume of combustion gas is determined. This method has been demonstrated to be more robust than the Kjeldahl method, to give equivalent precision and per sample cost, and to be much more environmentally friendly. Although many reports have shown that the results for the Kjeldahl and Dumas nitrogen methods are equivalent, some have shown that the Dumas method gave higher (and more correct) results for oilseeds (50).

In biochemical studies, protein may be determined using a colorimetric method, usually that described by Lowry *et al.* (51). These methods require calibration against a pure protein, and although relative results within a study may be acceptable, care should be taken when considering absolute results. For example, in Sammour's characterization of flaxseed protein (48), determination of total protein with the Lowry method calibrated against bovine serum albumin gave a total protein content of flaxseed meal of 49%. This result is considerably higher than results obtained from nitrogen determinations on flaxseed or meal, even using the crude protein factor of 6.25.

Amino Acid Composition

Amino acid composition is commonly reported as either g/100 g crude protein (i.e., g/16 g nitrogen) or as g/100 g nitrogen. Given the uncertainty of what is meant by protein (as discussed above), the latter method is probably preferred, and this is given in Table 1.5. Amino acids are routinely determined by hydrolyzing the protein in acid solution followed by determination of the amino acids either by ion exchange chromatography followed by detection with ninhydrin solution (41) or by HPLC after derivatization (52). Hydrolysis of the protein remains the critical step. In order to obtain a complete profile, the sample may require two separate hydrolyses and a reaction with performic acid to determine sulfur amino acids.

The effects of methodological differences can be seen if the results from Bhatty and Cherdkiatgumchai (53) are compared with those from Daun and Przybylski (35) and the classic paper by Tkachuk (54). The first limiting amino acid in flaxseed is lysine (55) followed by methionine and cystine. Significant environmental differences were observed for arginine, glutamate, methionine, serine, and tyrosine whereas aspartate and isoleucine showed varietal variation (35).

Carbohydrates

Carbohydrates can be classified into two groups, those that can be digested by human enzymes and those that cannot. Digestible carbohydrates generally include simple sugars and starch, but flaxseed contains little of these. Flaxseed may contain a small percentage of soluble sugars, but a definitive amount has not been reported and it is likely to be less than 1–2% (53,56). The majority of the carbohydrates present in flaxseed are of the group that are resistant to the action of human digestive enzymes, namely dietary fiber. Flaxseed stands out from other whole grains because it is an excellent source of both soluble and insoluble dietary fiber, and in total it accounts for approximately 28% of the dry seed weight (57). The ratio of soluble to insoluble fiber ranges from 20:80–40:60 depending on the methods of extraction and chemical analysis used (56).

Dietary Fiber of Flaxseed. The seed coat (hull) of flax consists of four layers (Fig. 1.1), and the outermost layer (epidermal layer) of the seed contains the soluble fiber portion also known as mucilage (58). Mucilage makes up approximately 8% of the seed weight, but its yield depends greatly on the method used for extraction (59) and varies among cultivars (59,60). Flaxseed mucilage has potential to be used as a food

TABLE 1.5
Amino Acid Composition of Flaxseed Expressed as g/100 g N or g/100 g Flaxseed

	Daun and Przybylski (35)		Tkachuk (54)	Bhatty and Cherdkiatgumchai (53)		Daun and Przybylski (35)		Tkachuk (54)	Bhatty and Cherdkiatgumchai (53)	
	Min	Max		Min	Max	Min	Max		Min	Max
Amino acid			(g/100 g N)					(g/100 g flaxseed)		
Alanine	26.9	28.4	27.6	31.3	36.3	0.99	1.05	1.02	1.15	1.33
Ammonia	12.3	13.2	13.5			0.45	0.49	0.50		
Arginine	51.0	55.3	55.6	64.4	76.9	1.88	2.04	2.05	2.37	2.83
Aspartate	56.0	60.2	58.3	71.3	83.1	2.06	2.22	2.15	2.62	3.06
Cystine	9.8	10.8	11.0	22.5	29.4	0.36	0.40	0.40	0.83	1.08
Gluatamate	105.0	118.8	131	153.1	175.0	3.86	4.37	4.82	5.64	6.44
Glycine	30.3	37.6	37.7	40.0	46.9	1.12	1.38	1.39	1.47	1.73
Histidine	12.9	15.6	13.6	16.9	20.6	0.48	0.57	0.50	0.62	0.76
Isoleucine	26.2	28.0	28.0	27.5	33.8	0.97	1.03	1.03	1.01	1.24
Leucine	35.0	37.2	36.9	40.0	46.9	1.29	1.37	1.36	1.47	1.73
Lysine	23.7	26.1	23.1	21.9	30.0	0.87	0.96	0.85	0.81	1.10
Methionine	9.8	11.6	10.1	12.5	16.3	0.36	0.43	0.37	0.46	0.60
Phenylalanine	26.4	28.1	28.0	30	35.0	0.97	1.04	1.03	1.10	1.29
Proline	21.1	22.4	22.6	28.8	36.3	0.78	0.82	0.83	1.06	1.33
Serine	23.9	29.1	27.4	33.1	38.9	0.88	1.07	1.01	1.22	1.43
Threonine	22.4	23.9	23.9	28.1	33.9	0.82	0.88	0.88	1.04	1.24
Tryptophan	8.0	8.8	12.2	10.0	11.9	0.29	0.32	0.45	0.37	0.44
Tyrosine	13.7	14.9	13.8	16.9	20.6	0.51	0.55	0.51	0.62	0.76
Valine	31.9	34.0	36.5	31.3	36.3	1.17	1.25	1.34	1.15	1.33

gum as a result of its thickening and emulsifying properties (61–63). The functionality of a food gum is affected by many factors, some of which include molecular size, orientation and molecular association of the polymers, water binding and swelling, concentration, particle size, and degree of dispersion (64). The health benefits of dietary fiber are also well documented, and over the last few years consumers have become more aware of the important role dietary fiber plays in the human diet. Certain forms of soluble dietary fiber have been shown to promote beneficial physiological effects including lowering the cholesterol/triglyceride lipid level in the blood and/or blood glucose attenuation (65–69). Given the apparent nutritional benefits of soluble fiber and the potential for using flaxseed mucilage as a food gum, it has been receiving more attention in the research arena than the insoluble fiber portion of flaxseed. The most recent research on flaxseed mucilage has examined different extraction conditions and how they affect the composition and rheological properties of the mucilage (58,70–72). In addition, several researchers have looked at how variety affects the composition and rheological properties, but there is limited information on how environment affects these same parameters (60,62,73,74). The insoluble fiber of flaxseed consists of nonstarch polysaccharides, the structural polymers found in cell walls, namely lignin and cellulose.

Flaxseed Mucilage. Flaxseed mucilage is a water-soluble hydrocolloid, and the most common method used for its isolation is an aqueous extraction procedure followed by ethanol precipitation. The alcohol precipitation is used not only for the collection of the dissolved polysaccharides, but also for removal of cyanogenic glycosides that are likely to be extracted along with the mucilage (66). A number of researchers have reported that it is better to extract the mucilage from whole seed as opposed to ground meal in order to reduce the amount of protein contamination (71,75). In addition, other factors such as temperature, pH, and water/seed ratio have been shown to have an effect on the yield, purity, composition, and rheological properties of the extracted mucilage (58,71,74). In response to these findings, Cui and colleagues (72) found it necessary to optimize the extraction process with respect to several parameters including yield, purity, viscosity, and energy cost. Using response surface methodology, the optimum extraction conditions were determined to be as follows: temperature 85–90°C; pH 6.5–7.0; and water/seed ratio 13:1.

Composition and Functional Properties of Flaxseed Mucilage. Flaxseed mucilage is composed primarily of two polysaccharides: a neutral and an acidic polymer (71,76,77). The neutral polymer consists of a (1 → 4)-linked β-D-xylose backbone, which contains side chains of arabinose and galactose at positions 2 and/or 3. Evidence suggests that the main chain of the acidic polymer consists of (1→ 2)-linked α-L-rhamnopyranosyl and (1 → 4)-linked D-galactopyranosyluronic acid residues, with side chains of fucose and galactose. A number of researchers have set out to separate the two polysaccharides in order to study their composition and physicochemical properties individually (71,76,77).

Effect of Variety on Flaxseed Mucilage Properties. There is limited information available on how the dietary fiber content varies among flax cultivars because it is not a component of flaxseed that is commonly monitored, unlike oil, protein, and fatty acid composition. One study measured the total dietary fiber content of 11 flaxseed cultivars grown in North Dakota, and a 10% variation in the amount of total dietary fiber was found among the cultivars (78). There is also limited information available on how the soluble fiber content varies among the cultivars. Due to the viscous nature of the mucilage, dietary fiber analysis of flaxseed tends to be tedious and time consuming. As a result, measuring the insoluble and soluble dietary fiber content of a large number of samples is not always practical. Therefore, the yield of mucilage obtained from the aqueous extraction procedure has often been used as an indicator for the amount of soluble fiber present in a particular flax cultivar. Mucilage was extracted from seven flaxseed cultivars (all grown in the same location), and the mucilage yields ranged from 4.9–7.2% (70). Variability in the yields was also observed when Cui *et al.* (60) extracted mucilage from six different flaxseed cultivars (Table 1.6).

Compositional analysis of the mucilage extracts has revealed that the monosaccharide composition of the mucilage varies significantly among cultivars (59,60,62,70,74). Xylose and rhamnose are the major components of the neutral and acidic polysaccharides, respectively; therefore, the ratio of rhamnose/xylose is often used to estimate the ratio of acidic/neutral polysaccharides (59,60,70,71). The rhamnos/xylose ratios for seven flaxseed cultivars commonly grown in Canada were calculated from the monosaccharide composition results, and the ratios ranged from 0.55–1.50 (70; Table 1.7). This is in agreement with ratios reported by Oomah and colleagues (59) who studied the monosaccharide composition of mucilage extracted from 109 different flaxseed accessions and found that the ratio of rhamnose/xylose ranged from 0.3–2.2. The content of acidic and neutral polysaccharides varies greatly among flaxseed cultivars. In general, cultivars having higher levels of neutral polysaccharides tend to exhibit higher apparent viscosity and stronger gellike properties (60,62), but Chornick (70) reported that the variation in the rheological behavior observed among cultivars could not be fully accounted for by the difference in xylose content (primary component of the neu-

TABLE 1.6

Effect of Cultivar on Mucilage Extract Yields Expressed as a Percentage of Seed Weight

Cultivar (70)	Yield (%)	Cultivar (60)	Yield (%)
CDC Bethune	6.3	Norman	7.9 ± 0.5
AC Carnduff	4.9	Royal	5.5 ± 0.2
AC Emerson	6.8	Reina	6.7 ± 0.3
AC Linora	6.4	22-87-2-159	6.0 ± 0.2
AC McDuff	5.7	Verne	6.0 ± 0.3
NorLin	7.2	Atlante	6.0 ± 0.1
Vimy	7.0		

TABLE 1.7
Rhamnose/Xylose Ratios of Mucilage Extracts (70)

Cultivar	Rhamnose/xylose
CDC Bethune	0.65
AC Carnduff	0.78
AC Emerson	0.55
AC Linora	0.75
AC McDuff	1.50
NorLin	0.70
Vimy	0.87

tral polysaccharide). This is because the proportion of neutral polysaccharides may be greater in one cultivar, but these polysaccharides may be of a lower molecular weight or have a slightly different structural conformation, both being factors that have an effect on rheological characteristics. Flaxseed mucilage is a complex, polydisperse hydrocolloid, and the diverse rheological properties observed among the cultivars were caused by differences in the proportion of neutral and acidic polysaccharides as well as differences in molecular weight and structural conformation of the extracted polysaccharides.

Only limited research has been carried out on how the functional properties of flaxseed mucilage differ among cultivars. Chornick (70) examined the emulsifying ability of mucilage extracts from seven flaxseed cultivars. It was found that those cultivars that exhibited increased viscosity and higher intrinsic viscosity values also showed the greatest potential for the stabilization of oil-in-water emulsions (Fig. 1.6)

The diverse rheological properties that have been observed among different flaxseed cultivars show that genetic variability exists in the physicochemical properties of flaxseed mucilage. This genetic variability may allow plant breeders to develop cultivars that contain mucilage with rheological and functional characteristics for specific end uses. The effect that growing location and environment has on the content, composition, and physicochemical properties of flaxseed mucilage still needs to be examined.

Minor Components

Minerals. Potassium and phosphorus were the major mineral components of flaxseed (Table 1.8). Flaxseed also contained significant quantities of iron, zinc, and manganese although the maximum levels reported, calculated from data including commercial meal samples, might include some mineral components derived from the processing equipment (Table 1.9).

Flaxseed, like many other grains and oilseeds, incorporates small amounts of cadmium into its seed (79). Although no levels have yet been set for oilseeds, amounts in excess of the 0.5 ppm considered by WHO/FAO (80) do occur in flaxseed. Even lower limits have been considered by some countries such as

                              J.K. Daun et al.

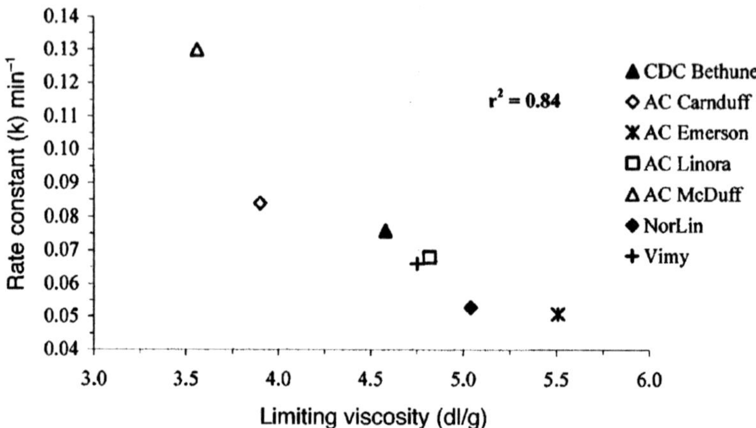

Fig. 1.6. Relationship between the rates of emulsion decay and intrinsic viscosity for the mucilage extracts (70).

Switzerland (81). The risk to health from the slightly higher levels of cadmium in flaxseed is considered considerably less than the risk from consumption of other plants containing cadmium, such as rice or wheat (82).

Cadmium accumulation in flaxseed is dependent on both the cadmium levels in the soil (83,84) and the variety of seed (85–88). There are active programs underway to develop low cadmium-accumulating varieties of flax, and some lines have been identified. The maximum levels identified in Table 1.8 represent the largest numbers in a large selection of lines from the USDA collection (87) grown on soil with high levels of cadmium. In studies of several thousand samples from Canadian harvests the levels found were much lower, in the range of 0.2–0.6 mg/kg (47). Studies carried out at the Canadian Grain Commission found only one sample above 3 mg/kg out of several thousand.

Mineral levels in flaxseed are normally determined by either atomic absorption spectroscopy (AAS) or by inductively coupled plasma linked atomic emission spectroscopy. AAS methods are available through ISO or AOAC International.

Water Soluble Vitamins. Flaxseed contains a range of B vitamins, and there has been one report of vitamin C (Table 1.8). The variation in vitamin content is large, partly due to the different methodologies. Methods using HPLC are less expensive but may not be as sensitive or accurate as biological methods. In a study of a number of flaxseed accessions grown in widely varying climatic situations, Marquard and coworkers (39) concluded that differences in location were greater than those of variety.

Phenolics and Lignans. Most oilseeds contain a range of phenolic compounds, occurring as the hydroxylated derivatives of benzoic and cinnamic acids, as

TABLE 1.8

Mineral and Water Soluble Vitamin Content of Flaxseed

Minerals (17,35,53)	Min	Max	Minerals (17,35,53)	Min	Max	Vitamins (17,39,53,131)	Min	Max
	(mg/g)			(mg/kg)			(mg/100 g)	
Potassium	5.5	10.6	Iron	36.7	164.4	Ascorbate, C		1.3
Magnesium	3.2	4.1	Zinc	38.2	93.6	Thiamine, B_1	0.03	0.6
Calcium	2.0	4.4	Manganese	13.0	42.8	Riboflavin, B_2	0.1	0.3
Sodium	0.2	0.6	Copper	7.9	17.1	Niacin, B_3	1.4	5.5
Phosphorus	4.4	7.6	Aluminum	2.2	9.5	Pantothenic acid, B_5	1.5	7
Sulphur	2.3	2.5	Nickel	0.8	2.8	Pyridoxine, B_6	0.4	10
			Cadmium (87)	0.2	3.6	Folate		278
			Cobalt	0.1	0.4	Cyanocobalamine, B_{12}	0	0.5
			Selenium		0.6			

TABLE 1.9

Cyanide and Cyanoglycoside Content of Flaxseed from Various Sources and Methodologies

Source	Method	Linustatine	Neolinustatine	Linamarin	Lautostralin	Total CN equivalent
			(g/100 g flaxseed)			(ppm or mg/kg)
Canada (123)	HPLC	0.0003	0.0004			0.40
Canada (118)	HPLC/UV	0.168–0.380	0.072–0.238	0–0.403		157–448
Canada (111)	HPLC Pyridine/pyrazole Barbituric A/pyridine					
Unknown (119)	HPLC	0.218–0.538	0.083– 0.454			197–647
Unknown (112)	HPLC/silver detection					
Canada (140)	Acid hydrolysis Picrate					
Australia (140)	Acid hydrolysis Picrate					
Meal (Canada) (141)[a]	HPLC	0.265	0.114			249
Immature seed unknown origin) (108)	TLC	0.200–0.320	0.106–0.185	0.333–0.605	0.141–0.240	712–1243

[a]Calculations made based on 40% oil content.

flavonoids and as lignans. Flax has been found to contain between 0.8–1.3/100 g of phenolic acids of which approximately 0.5/100 g was in the esterified form and 0.3–0.5/100 g was in the etherified form (89). There were significant variety and environment effects. Phenolic acids in dehulled flaxseed meal included trans-ferulic (46%), trans-sinapic (36%), p-coumaric (7.5%), and trans-caffeic (6.5%) (90). Flax is a rich source of ferulic acid. The latter compound constituted 50% of the total phenolics in a methanol extract of flax meal compared with 34.2% for chlorogenic acid (91).

Flaxseed also contains significant quantities of complex phenolics known as lignans (see Chapter 4). The lignan component in flaxseed of particular interest is secoisolariciresinol diglucoside (SDG) which appears to be present in the order of 1–10 µM/g seed depending on the seed source and methodology (92). Other lignans including isolariciresinol, pinoresinol, and matairesinol have also been isolated (93). Several other ferulic acid derivatives have also been reported to be present (94). Problems in the measurement of lignan components have been highlighted elsewhere (92,95; Chapter 4). The continued interest in commercial exploitation of these components makes the development of standardized methodology imperative.

Plant Antinutrients

Flaxseeds contain antinutrients as do other plants. Phytic acid, cyanogenic glucoside, and linatin are the main flax antinutrients. No adverse effects, including acute poisoning, have been mentioned in the literature.

Phytic Acid. In the late nineteenth century, phytic acid was identified as an important component in seeds. The chemical structure was elucidated as *myo*-inositol(1,2,3,4,5,6)-hexa*kis*phosphate. Phytic acid is a charged compound and can react with other charged molecules. It is a very strong chelator of mineral cations (mono- or divalent) such as potassium, magnesium, iron, and zinc and also binds to proteins and starch (96).

In flax seeds, the amount of phytic acid varies from approximately 0.80–1.5% of the dry seed weight (97). These levels are comparable to the amount of phytic acid found in peanuts and soybeans but lower than the usual 2–5.20% found in other oilseeds (98). As with other oilseeds, cultivar and environment affect the phytic acid content of the seeds (53,97). For some flax cultivars, such as AC Emerson and Flanders, the level of phytic acid was independent of the environmental conditions, whereas for others, such as AC Linora, Linola 947, and McGregor, the phytic acid levels were highly dependent on the environment (97). Phytic acid represents about 60–90% of the total seed phosphorus; it is the primary form of phosphorus storage in the seeds. It is involved in germination and seedling growth. Because of its mineral-chelating properties, it plays an important role in the mineral status of the seed. It has been suggested that phytic acid plays an important role in the viability and vigor of the seeds (99).

Zn forms a stable, insoluble complex with phytic acid. As a consequence, Zn bioavailability is greatly affected, potentially leading to Zn deficiency (96). As with all biological molecules, phytic acid can have positive or negative effects. Several reviews have been published on the dual role of phytic acid (100,101). It has been reported that phytic acid has a negative effect on calcium and iron absorption and protein digestibility due to its strong chelating properties. On the positive side, phytic acid was shown to decrease the blood glucose response by reducing starch digestion. More important, several studies have shown the positive role phytic acid plays in reducing the incidence of colon cancer in rats. It was speculated that by complexing with iron, phytic acid reduces the amount of hydroxyl radicals in the colon.

Most of the ingested phytic acid (99%) is hydrolyzed by lactating dairy cows (102). However, nonruminant animals do not possess the enzymes necessary for its catabolism, and the ingested phytic acid is eliminated without modification. This leads to several nutritional and environmental problems, and it is not cost effective. Phytic acid is the main phosphorus source in grains, and exogenous phosphorus must be added to avoid poor nutrition for poultry, swine, and fish (99,100). This increases the phosphorus in the waste, leading to environmental problems (103). The addition of a phytase from fungal origin (*Rhizopus* or *Aspergillus*) to grain feed (99,100) improved the mineral status of the animals (104).

Thompson (101) mentioned that the term antinutrient for phytic acid was inappropriate. Phytic acid is one of those molecules that is as good as it is bad—everything depends on the quantity. Research on the phytic acid content of flaxseed can be conducted in several directions. Oomah *et al.* (97) reported that preliminary results suggest that the stability of the phytic acid content might be genetically controlled. Some research has been conducted to try to reduce the amount of phytic acid in grain through breeding (99), but no study of this nature has yet been carried out on flaxseed. Phytic acid is only a minor component in regard to quantity; its role in the seed as well as in animal and human nutrition has not been completely elucidated. This could become an important issue when flaxseed is used more routinely as feed for nonruminants (Table 1.9).

Cyanogenic Glycosides. The cyanogenic glycosides are nitrile-containing glycosides with the nitrile group alpha to the glycosidic linkage (Fig. 1.7). They have the ability to release hydrogen cyanide upon acidic or enzymatic hydrolysis. There are about 75 known cyanogenic glycosides in more than 2500 plant species, including wheat and barley (105,106).

In flaxseed, the main cyanogenic glycosides (Fig. 1.7) are linustatin (1-cyano1-ethyl-ethyl-β-D-glucopyranosyl-β-D-glucopyranoside) and neolinustatin (1-cyano1-methyl-ethyl-β-D-glucopyranosyl-β-D-glucopyranoside) (107). Linamarin (1-cyano1-ethyl-ethyl-β-D-glucopyranoside) and lotaustralin (1-cyano1-methyl-ethyl-β-D-glucopyranoside) were also found by other researchers (43,108). Conn (109) reported that linen flax contained a mixture of 50-50 linamarin and lotaustralin. As in other cyanophoric plants, cultivar, age, development, environment, and nutritional and genetic factors affect the cyanogenic contents of flaxseeds and

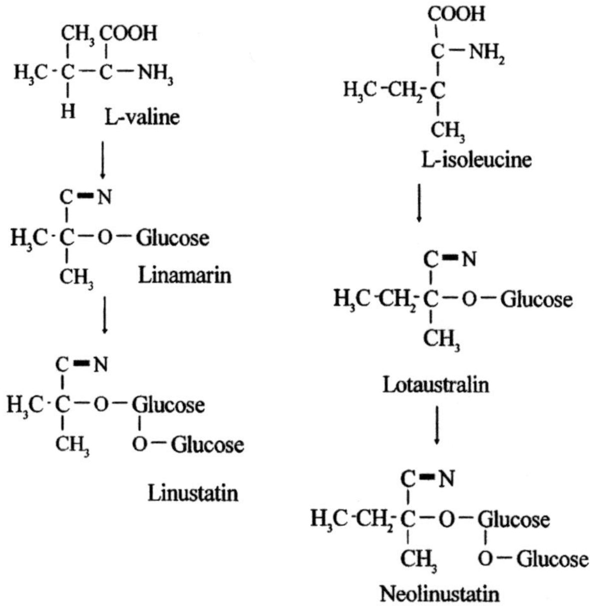

Fig. 1.7. Scheme of the synthetic pathway of flax cyanogenic glycosides (107).

flax oil. In mature flaxseed, the amount of cyanogenic glycosides was found to be around 0.1% of the dry weight of the seeds, whereas in young green seeds this level could reach up to 5% of the dry weight of the seeds (110).

Amino acids are the parent molecules of flax cyanogenic glycosides. Conn (109) reported that *L*-valine and isoleucine were the precursors of linamarin and lotaustralin, which are monoglycosides. A scheme of the synthetic pathway of flax cyanogenic glycosides is presented in Figure 1.7 (107).

Degradation of cyanogens occurs in damaged flaxseeds. Disruption of cellular structures allows the enzymes and cyanogenic substrates previously present in different cellular structures to react (105). Crude flax extracts contain two distinct β-glucosidases, which perform a stepwise removal of the glucose residues (Fig. 1.8). The first enzyme or linustatinase, is responsible for the hydrolysis of the diglycosides linustatin and neolinustatin into monocyanogenic glycosides; this enzyme is inactive on the monoglycosides linamarin and lotaustralin. Then these two monoglycosides are hydrolyzed by a second β-glucosidase, called linamarase, which is inactive toward diglycoside (111). The common characteristic of these enzymes is that they are specific toward their native substrates (105).

It has been found that during heating, ground flaxseeds were able to retain more β-glucosidase activity than whole flaxseeds (112). It was speculated that the enzymes responsible for the hydrolysis of the cyanogenic glycoside were more effectively inactivated in the whole seeds than in the ground seeds because of the slower water loss (113).

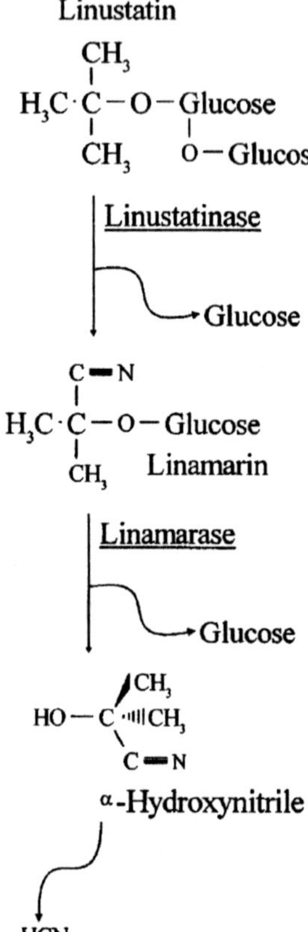

Fig. 1.8. Enzymatic hydrolysis of linustatin in flaxseeds (105).

The importance of the metabolic effect of cyanogenic glycosides depends on (i) the amount of cyanogenic glycosides ingested, (ii) the presence of other interacting components, (iii) the nutritional and health status of the consumer (protein deficiency), and (iv) the frequency of consumption (114). It is believed that an adult can detoxify from 30–100 mg of cyanide/d (115); however, when combined with a low-protein diet and low sulfur availability, the detoxification ability is greatly decreased (116,117). With flaxseed being used only as a minor ingredient in food products such as flax bread, muffins, or cereals, the cyanogenic glycosides are not really a problem for human consumption. Moreover, after cooking, no cyanide was detected in flaxseed muffins (118) or bread (112).

Studies have shown that when calculated in cyanide equivalent, the amount of cyanide may vary from 190–1000 mg HCN/kg of flaxseed. In other words, an adult can consume more than 1 kg of ground flax/d before being susceptible to exhibiting acute cyanide toxicity.

Two approaches are used to measure the cyanogenic glycoside content of flaxseed and flaxseed products. They either measure (i) the cyanogenic glycosides present in the sample using HPLC or TLC (118,119), or (ii) the HCN released from the cyanogenic glycoside hydrolysis (75,120–122).

In general, the measure of cyanogenic glycosides by measuring the cyanide content is very long, tedious, and could lead to an underestimation of the cyanoglycoside content if the cyanide, which is volatile, is not completely recovered or if the hydrolysis of the cyanoglycoside is not complete. The other major inconvenience of such methods is that the identity of the cyanogenic glycoside is unknown after analysis.

There was no statistical difference between HPLC and colorimetric results, even if the HPLC results were higher (about 10%) than the colorimetric results. There is also discrepancy regarding the nature of the cyanogenic glycosides reported in flaxseed. Some researchers, such as Oomah *et al.* (118), showed the presence of only linustatin, neolinustatin, and linamarin, whereas others reported that four cyanogenic glycosides were present (108). On the other hand, Schicher and Wilkens-Sauter (119) and Cunnane *et al.* (123) reported only two cyanogenic glycosides (linustatin and neolinustatin). Smith *et al.* (107) concluded that linamarin and lotaustralin were artifacts produced by the hydrolysis of the diglycosides linustatin and neolinustatin by a β-glucosidase.

Some research needs to be developed in several directions regarding the cyanogenic glycosides. The analytical methods to measure their content, the true role of these compounds in flax, and their usefulness against disease or for pest protection are some of the research areas that should be studied.

Linatine—An Antipyridoxine Factor. Kratzer (124) and Kratzer and Vohra (125) reported that chicks fed a diet containing linseed meal showed vitamin B_6 deficiency (loss of appetite, poor growth, nervous disorders with convulsions, and anemia). They were able to demonstrate that adding pyridoxine to the diet counteracted the poor growth of the chicken. Klosterman *et al.* (126) showed that the inhibitor was a polar compound (extracted with water and/or 70% ethanol) and probably had an amine moiety; moreover, it was concentrated in the cotyledon of the flaxseed. This antivitamin B_6 compound was named linatine. Linatine (Fig. 1.3) is formed from 1-amino-D–proline, a secondary hydrazine, bound to glutamic acid by a peptide bond engaging the γ-carbonyl of the glutamic acid. Upon hydrolysis, 1-amino-D–proline is released; it can then form a stable derivative with pyridoxal or pyridoxal phosphate leading to the vitamin B_6 deficiency. This hydrolysis can occur in the gut of the animal (Fig. 1.3). The 1-amino-D–proline inhibits enzymes such as aminotransferase and glutamic decarboxylase which are involved in amino acid metabolism. *In vitro*, linatine did not inhibit any of these enzymes because the hydrazine function was masked (127). The amount of linatine present in flax seed was 100 ppm (126). However, the method used six steps of which four were preparative thin layer chromatographies, and this suggests that some linatine might have been lost during the analysis leading to its underestimation.

Linatine effects are easily counteracted by addition of pyridoxal phosphate in the animal feed. For human consumption, the linatine is not a real problem due to the recommended guideline for flaxseed in food products. Moreover, a varied diet will provide enough vitamin B_6 to avoid deficiency.

Other Antinutritional Factors. Flax contains small amounts of trypsin inhibitor factor measured as between 20–30 units/g of seed (75,128). This is much lower than amounts found in soybeans or canola. Flax also has allergenic properties (79,82,129) but these are not any more significant than those of other grains and oilseeds (130). Limited studies in the authors' laboratory, using a commercially available diagnostic assay, suggested that flaxseed contained less than 10 mg/kg oxalate. Flaxseed showed no amylase inhibitor activity or hemaglutinating activity (128).

Variety Development in Flaxseed in Canada

Effect of Variety on Composition

Three major breeding programs develop flax and solin varieties for Canada: the Agriculture and Agri-Food Canada program located at the Morden Research Station in Morden, Manitoba; the Crop Development Centre program located at the University of Saskatchewan in Saskatoon, Saskatchewan; and the Agricore United program at the Morden Research Station. Additionally, some seed companies are introducing cultivars from other countries.

Since the early 1900s, Agriculture and Agri-Food Canada and its predecessors have been active in the development of new flax varieties for Canada and, in particular, for the Canadian prairies. The initial program at the Central Experimental Farm in Ottawa produced varieties such as Diadem, Ottawa 770B, Ottawa 829C, and Novelty. During the 1950s, this program was particularly active, releasing varieties such as Linott, Raja, and Rocket. The 1950s and 1960s also marked the beginning of an evolution and transition in flax breeding in Canada. A new program was initiated at the Indian Head Experimental Farm and the Winnipeg Cereal Breeding Laboratory, which led to the development of the variety Cree. In the 1960s, a breeding program was conducted in Alberta at the Fort Vermillion Experimental Farm and Beaverlodge Research Station, producing the variety Noralta, the predominant variety grown in northern Alberta and Saskatchewan. The breeding programs of Agriculture and Agri-Food Canada were consolidated and moved to Winnipeg in 1960; they then moved to Morden, Manitoba, where they still exist. The varieties Dufferin, McGregor, NorLin, NorMan, AC Linora, AC McDuff, AC Emerson, AC Carnduff, Lightning, Hanley, Macbeth, and Prairie Blue have been released by Agriculture and Agri-Food Canada.

A modest breeding program was carried out at the University of Saskatchewan from the 1920s through the 1960s, which produced the varieties Royal and Redwood 65. The program was enlarged in 1974 when the Crop Development Centre (CDC) initiated a flax breeding program. It has since produced eight cultivars: Vimy, Somme, Flanders, CDC Normandy, CDC Valour, CDC Arras, CDC Bethune, and CDC Mons. Other varieties produced at the CDC include Andro (tissue-culture derived) and CDC Triffid (first transgenic flax cultivar). Both of these varieties have now been deregistered and are not commercially available.

In 1987, a solin breeding program was initiated by Biotechnia Canada in cooperation with Australia's Commonwealth Scientific & Industrial Research Organization to develop low ALA flax, subsequently known as solin. In 1990, UGG Ltd. (UGG) (now known as Agricore United with the merger of United Grain Growers and Agricore) purchased Biotechnia's interest in the program and moved the program from Calgary to the Agriculture and Agri-Food Canada Research Station located at Morden, Manitoba, and the Agricore United research and evaluation farm at Rosebank, Manitoba. This breeding program has produced the solin cultivars, Linola TM 947, 989, 1084, 2047, and 2090.

All flax varieties registered in Canada are brown-seeded and have high levels of ALA fatty acid. Solin varieties, with less than 5% ALA, produce polyunsaturated edible oil similar to sunflower oil and, in Canada, must have yellow seed. Unregistered yellow-seeded flax varieties, with high levels of ALA, are grown under contract for use in edible products and for the health food market. Yet there is a growing trend back to natural fibers for industrial and textile applications. This trend will only continue as pressures increase to produce materials that are recyclable or decomposable.

Variety Development Objectives

The focus of the breeding efforts of the three Canadian breeding programs is very similar. Overall, all the programs contribute to the stability and competitiveness of flax production in Canada and to the acceptability of the product for food and industrial utilization, by developing high yielding and higher quality flax varieties that reduce risk through genetic resistance to biotic and abiotic stresses.

The competitiveness of Canadian flax on the world markets depends on the quality of the oil and meal for specific end uses. The most important quality parameter for flax is high oil content. Therefore, in terms of improving the quality of flax, the primary goal of Canadian flax breeding programs is to increase the oil content of flax varieties. As mentioned earlier, over the past five years, Canadian flax seed contained, on average, 44% oil and ranged from a low of 35 to 50%. As the seed from individual farm samples moves through the handling system, it is combined and the range of oil content decreases, so that in Canadian exports of flaxseed it has been only 3.5%, from 42–46%. The variety producing the highest oil at present is AC McDuff with an oil content of 48%.

Canadian flax varieties are registered through the Canadian Food Inspection Agency (CFIA). Information on current varieties registered may be found at the CFIA website at http://www.inspection.gc.ca/english/plaveg/variet/liste.shtml.

Another objective of Canadian breeding programs at present is the development of varieties with improved fatty acid composition. The nutritional quality of a vegetable oil for human and animal consumption or in an industrial application as well as its end-use qualities are determined by its fatty acid composition. Flaxseed contains both n-3 and n-6 polyunsaturated fatty acids. The level of ALA in individual farm samples of Canadian flaxseed harvested in 2002 was found to range from 52–63%. Varieties have been developed with higher levels of ALA using conventional breeding techniques. The cultivar with the highest level of ALA commercially available is AC Emerson, whereas Hanley has the highest level of unsaturation in its oil.

The need for various fatty acid modifications in flax is well established. Research efforts are underway to develop cultivars with various fatty acid profiles for use by the world flax industry whether as a functional food, pharmaceutical, or with improved biodegradability for industrial uses. Various sources of higher levels of oleic, linoleic, and ALA are available in flax. It would be possible to introgress this material into elite germplasm through classical breeding approaches.

One of the main reasons for the improvement and stability of the quality of flax produced in Canada is not a result of breeding for quality directly by improving oil content and fatty acid composition. Rather, it is a result of breeding varieties for improved disease resistance, earlier maturation, and resistance to lodging. Since the turn of the nineteenth century, flax breeders and pathologists have developed resistant varieties of flax to diseases such as rust (caused by *Melamspora lini*), fusarium wilt (*Fusarium oxysporium* f. sp. *lini*), powdery mildew (*Oidium lini*), and pasmo (*Septoria linicola*). Each of these diseases can cause losses in yield, and in some cases complete crop failures can result from infections of susceptible varieties. Additionally, these diseases also affect the overall quality of the seed in terms of oil content, fatty acid composition, and protein content. The quality of flax straw can also be adversely affected. Disease resistance to rust and wilt has been emphasized by all the programs in order to keep these problems under control. Thus, all registered flax and solin varieties commercially grown in Canada are resistant to rust and must have moderate resistance to fusarium wilt. The development of genetic resistance is the most efficient and cost-effective method to prevent losses to these diseases and reduce the risk to flax production. Resistance strategies will also have a positive environmental impact by reducing the need for chemical disease control. The success of this strategy is demonstrated by the fact that the last outbreak of rust occurred in 1973, and the resistance within commercially grown cultivars in Canada has continued to hold. Various sources of resistance to these diseases are available in flax, and these are being introgressed into elite germplasm through classical breeding approaches.

Historically, damage to crops due to early or late frosts, excess moisture, drought, and heat, as well as damage caused by disease pressure have resulted in yield losses along with reduced overall quality of the seed. One strategy is the development of early maturing varieties to avoid these stresses and potentially open up new areas of production for the crop such as the northern areas of the Canadian prairies or further production in the brown soil zones. Due to northern environmental factors, new cultivars may have enhanced value-added components (e.g., oil quantity and quality) that could expand the market opportunities for both the industrial and functional food marketplaces. Oil content and fatty acid composition are key components. As demands of the industrial and functional food arenas change, both nationally and internationally, further improvements in bioproduct quality will enhance marketability.

Another objective important to the sustainability of producing a high-quality flax variety particularly in the black and dark-grey soil zones is lodging resistance. Lodging occurs in areas of high soil fertility and moisture and can result in reductions in yield, grade, and quality. Severely lodged flax is also more susceptible to diseases such as pasmo and sclerotinia. In 1993, 80% of the flax acreage in Manitoba was grown with medium-early maturing varieties, predominantly NorLin, NorMan, and Somme. These varieties are more susceptible to lodging than the late maturing varieties McGregor and AC McDuff. Yield losses in flax resulting from lodging have not been well documented; however, in a susceptible variety, Vimy, yield losses are often more than 50% higher than for other varieties of similar maturity in cooperative tests grown under severe lodging conditioning.

Future Objectives and Possibilities

Although a considerable amount of time and effort has been expended in the improvement of oil content and fatty acid composition by breeders, less attention has been spent in improving the meal of flax regarding protein content or components such as mucilage and lignans, or the removal of antinutritional factors such as cyanogenic glucosides or minerals such as cadmium. These are examples of potential future directions and objectives for flax breeding programs, along with the continued development of improved oil content, fatty acid composition, and yield.

Historically, the flaxseed crush in Europe was driven by the demand for linseed oil to be used in the production of linoleum, paints, and other industrial products. Recently, however, the demand for nongenetically modified high-protein meal is beginning to drive the crush and the production of linseed oil. The linseed meal is fed to livestock, and is largely consumed in the country in which it is produced. As with other oilseeds, there is an inverse relationship between oil and protein in flaxseed. This relationship is not as strong as in other oilseeds such as canola. Various sources of higher protein content are available in flax germplasm. The potential exists that the protein content of flax could be improved in the seed and meal for use in the animal and human food market. Following this, the next step would be the improvement of amino acid composition.

There has been considerable interest in the inclusion of flaxseed in Western diets or in the development of isolated and purified compounds for food, animal feed, and the improvement of human health. Part of this interest comes from studies indicating that there are significant and beneficial effects (123). For instance, flaxseed mucilage is an excellent source of dietary fiber, and due to its highly viscous nature also shows potential for being used as a food gum. Genetic variability exists for the content (4–7%) and physiochemical properties of flaxseed mucilage. This variability would allow the development of cultivars that contain mucilage with rheological and functional characteristics for specific end uses. Another possibility of developing new varieties for a specific end use or as a general flax variety would be to improve the lignan content of flax. The level of SDG found in Canadian cultivars is significant, and an increase in this level would be beneficial. If a variety were developed with a meal

SDG content of approximately 40 mg/g, it would represent more than double the SDG content of commercial flax meal which is typically in the 1.2–1.7% range.

Characteristics that limit trade and utilization of flaxseed meal include the presence of antinutritional factors such as cyanogenic glucosides. As in the case of canola, a leading factor in international competitiveness would be the development of improved varieties low in such antinutritional factors. It would also solidify flax as a safe, high-quality food and animal feed. Hydrolysis of cyanogenic glucosides can produce hydrogen cyanide, a potent respiratory inhibitor. This limits the quantity (8–10%) of flaxseed that can be used in food products or feed rations. Reduction of the cyanogenic glucosides would reduce potential health hazards and maintain the export market of flaxseed and its associated products. Lowering of cyanogenic glucosides may also increase the demand for other value-added products in the food and pharmaceutical markets such as proteins, mucilage, gums, and lignans. The potential exists that this product could be marketed in an identity-preserved manner, which may lead to additional development of value-added products.

There is a tremendous potential for the breeding of new varieties with specific end uses and for developing a food and feed product that opens this market up further. The development of a new flax variety takes approximately 10–12 y from initial hybridization to being commercially grown. Therefore, a breeder or industry wanting to develop an identity-preserved flax must anticipate what the farmer, the end user, and the consumer will require in the future.

Theoretically, flax breeders may be able to produce a flax variety with high oil, ALA, mucilage, and lignan and with low cyanogenic glucosides and cadmium. To achieve this, a breeder must have both sufficient genetic variability available to alter the trait and efficient testing procedures that are inexpensive, reliable, and nondestructive in order to evaluate thousands of lines on a yearly basis. The potential for developing a variety possessing each of these attributes all at once is low. Compromises for one attribute or another as well as a series of steps over time may have to occur within the program to achieve this goal. For instance, if the primary goal was to obtain high mucilage and high lignans from this flax variety, it may be more beneficial for a processor to receive a low-oil, low-protein-content flax with high levels of mucilage and lignans. Along with the vision, the breeder or industry must be willing to undertake a long-term approach with resources.

Summary

Although a great deal of information has been accumulated regarding the composition of flaxseed, there are many areas in which further work is required. In particular, more work is required to elucidate the composition of minor lipids in flaxseed, particularly wax esters, sterol esters, and hydrocarbons. The reports on some other lipid classes, such as glycolipids and phospholipids, are also somewhat sparse or contradictory, and further work is required. Methodology plays a major role in determining the quality of information. Use of inappropriate methodology, for example as in the case of oil content, has been shown to have as great an effect on flax composition as the environmental and genetic effect. Methodology for some important components, such as cyanogenic glycosides and lignans, needs development of internationally recognized standard methods.

The components of flaxseed elucidated to date show that this seed has some unique properties that make it valuable as a food source. The content of minor amounts of antinutritional components, especially cyanogenic glycosides and the thiamine inhibitor linatine, must be considered when flaxseed is included in diets without heat processing.

Acknowledgment

Barry Misener and Christina Scott provided assistance in analysis of some components reported from the authors' laboratory.

References

1. Gower, D., Flaxseed, *Bi-weekly Bulletin, 15*:1–4 (2002).
2. Flax Council of Canada, Report 0598PD200, Flaxseed: A Smart Choice, Winnipeg Flax Council of Canada, 1998.
3. Dorrell, D.G., Distribution of Fatty Acids Within the Seed of Flax, *Can. J. Plant Sci. 50*:71–75 (1970).
4. Geddes, W.F., and F.H. Lehberg, Flax Studies. I. The Relation Between Weight per Measured Bushel, Weight per Thousand Kernels and Oil Content of Flaxseed, *Can. J. Res. 14*:45–47 (1936).
5. Tzen, J.T.C., Y.L.P. Cao, C. Ratnayake, and A.H.C. Huang, Lipids, Proteins and Structure of Seed Oil Bodies from Diverse Species, *Plant Physiol. 101*:267–276 (1993).
6. Sosulski, F., Fractionation of Oilseed Meals into Flour and Hull Components with Liquid Cyclones, *Prog. Food Eng.* 553–557 (1983).
7. Oomah, B.D., and G. Mazza, Effect of Dehulling on Chemical Composition and Physical Properties of Flaxseed, *Food Sci. Technol. 30*:135–140 (1997).
8. Oomah, B.D., G. Mazza, and E.O. Kenaschuk, Dehulling Characteristics of Flaxseed, *Food Sci. Technol. 29*:245–250 (1996).
9. Madhusudhan, B.J., D. Wiesenborn, J. Schwarz, K. Tostenson, and J. Gillespie, A Dry Mechanical Method for Concentrating the Lignan Secoisolariciresinol Diglucoside in Flaxseed, *Food Sci. Technol. 33*:268–275 (2000).
10. Mittapalli, O., and G. Rowland, The Effect of Seed Colour on Oil Concentration in Flax, in *Proceedings of the 59th Flax Institute of the United States*, Fargo, North Dakota, 2002, pp. 150–155.
11. Barthet, V.J., T. Chornick, and J.K. Daun, Comparison of Methods to Measure the Oil Contents in Oilseeds, *J. Oleo Sci. 51*:589–597 (2002).
12. Oomah, B.D., G. Mazza, and R. Przybylski, Comparison of Flaxseed Meal Lipids Extracted with Different Solvents, *Food Sci. Technol. 29*:654–658 (1996).
13. Daun, J.K., and D.R. DeClercq, Sixty Years of Canadian Flaxseed Quality Surveys at the Grain Research Laboratory, in *Proceedings of the 55th Meeting of the Flax Institute of the United States*, Fargo, North Dakota, 1994, pp. 192–200.
14. Olejnik, D., M. Gogolewski, and K.M. Nogala, Isolation and Some Properties of Plastochromanol-8, *Nahrung 41*:101–104 (1997).

15. House, S.D., P.A. Larson, R.R. Johnson, J.W. Devries, and D.L. Martin, Gas Chromatographic Determination of Total Fat Extracted from Food Samples Using Hydrolysis in the Presence of Antioxidant, *J. AOAC Int. 77*:960–965 (1994).

16. *Official Methods of Analysis*, AOAC International, Washington, D.C., 1999.

17. USDA, Nutrient Data Laboratory United States Department of Agriculture, available at http://www.nal.usda.gov/fnic/foodcomp/ (2002).

18. Conte, L., G. Lercker, P. Capella, and M. Catena, The Composition of Linseed Oil, *Riv. Ital. Sost. Gras. 46*:339–342 (1979).

19. Eckey, E.W., *Vegetable Fats and Oils*, Reinhold, New York, 1954, p. 541.

20. Canvin, D.T., The Effect of Temperature on the Oil Content and Fatty Acid Composition of the Oils from Several Oil Seed Crops, *Can. J. Bot. 43*:63–69 (1965).

21. Dybing, C.D., and D.C. Zimmerman, Temperature Effects on Flax (*Linum usitatissimum* L.) Growth, Seed Production, and Oil Quality in Controlled Environments, *Crop Sci. 5*:184–187 (1965).

22. Dybing, C.D., and D.C. Zimmerman, Fatty Acid Accumulation in Maturing Flaxseeds as Influenced by Environment, *Plant Physiol. 41*:1463–1470 (1966).

23. Schuster, W., H. Iran-Nejad, and R. Marquard, Ertragsleistungen und einige Qualitäsmerkmale von verschiedenen Ölleinsorten (*Linum usitatissimum* L.) aur ökologisch stark unterschiedlichen Standorten, *Fette Seifen Anstrich. 80*:133–143 (1978).

24. Gubbels, G.H., D.M. Bonner, and E.O. Kenaschuk, Effect of Time of Swathing and Desiccation on Plant Drying, Seed Color and Germination of Flax, *Can. J. Plant Sci. 73*:1001–1007 (1993).

25. Gubbels, G.H., and E.O. Kenaschuk, Effect of Seeding Rate on Plant and Seed Characteristics of New Flax Cultivars, *Can. J. Plant Sci. 69*:791–795 (1989).

26. Green, A.G., and D.R. Marshall, Isolation of Induced Mutants in Linseed (*Linum usitatissimum*) Cultivar Glenelg Having Reduced ALA Content, *Euphytica 33*:321–328 (1984).

27. Green, A.G., Genetic Control of Poly-Unsaturated Fatty Acid Biosynthesis in Flax (*Linum usitatissimum*) Seed Oil, *Theor. Appl. Genetics 72*:654–661 (1986).

28. Brockerhoff, H., and M. Yurkowski, Stereospecific Analyses of Several Vegetable Fats, *J. Lipid Res. 7*:62–64 (1966).

29. Drozdowski, B., Effect of the Unsaturated Acyl Position in Triglycerides on the Hydrogenation Rate, *J. Amer. Oil Chem. Soc. 54*:600–603 (1977).

30. Tarandjiiska, R.B., I.N. Marekov, D.B.M. Nikolova, and B.S. Amidzhin, Determination of Triacylglycerol Classes and Molecular Species in Seed Oils with High Content of Linoleic and ALA Fatty Acids, *J. Sci. Food Agric. 72*:403–410 (1996).

31. Tonnet, M.L., and A.G. Green, Characterization of the Seed and Leaf Lipids of High and Low ALA Flax Genotypes, *Arch. Biochem. Biophys. 252*:646–654 (1987).

32. Wanasundara, P.K.J.P.D., F. Shahidi, and M.E. Brosnan, Changes in Flax (*Linum usitatissimum*) Seed Nitrogenous Compounds During Germination, *Food Chem. 65*:289–295 (1999).

33. Kulkarni, S., R.R. Khotpal, H. Bhakare, and P.B. Shingwekar, Studies on Phospholipids of Some Linseed Varieties, *Ind. J. Pharmaceut. Sci. 61*:384–385 (1999).

34. Kulkarni, A.S., R.R. Khotpal, and H.A. Bhakare, Glycolipids Composition of Some Indian Linseed Varieties, *J. Food Sci. Technol. India 35*:245–246 (1998).

35. Daun, J.K., and R. Przybylski, Environmental Effects on the Composition of Four Canadian Flax Cultivars, in *Proceedings of the 58th Flax Institute of the United States*, Fargo, North Dakota, 2000, pp. 80–91.

36. Velasco, L., and G.D. Goffman, Tocopherol, Plastochromanol and Fatty Acid Patterns in the Genus *Linum*, *Plant Sys. Evol. 221*:77–88 (2000).

37. Oomah, B.D., E.O. Kenaschuk, and G. Mazza, Tocopherols in Flaxseed, *J. Agric. Food Chem. 45*:2076–2080 (1997).

38. Budin, J.T., W.M. Breene, and D.H. Putnam, Some Compositional Properties of Camelina (*Camelina sativa* L. Crantz) Seeds and Oils, *J. Amer. Oil Chem. Soc. 72*:309–315 (1995).

39. Marquard, R., W. Schuster, and H. Iran-Nejad, Untershuchungen überd Tokopherol- und Thiamingehalt in Leinsaat aus welweitem Anbau und aus dem Phytotron unter definierten Klimabedingunden, *Fette Seifen Anstrich. 79*:265–270 (1977).

40. Firestone, D.E. (ed.), *Official Methods and Recommended Practices of the AOCS, Method Ce 8-89,* AOCS Press, Champaign, Illinois, 1998.

41. International Organization for Standardization, ISO NWI 13903: Animal Feeding Stuffs—Determination of Amino Acids Content, Geneva International Organization for Standardization, Switzerland, 2003.

42. Pretova, A., and M. Vojtekova, Chlorophylls and Carotenoids in Flax Embryos During Embryogenesis, *Photosynthetica 19*:194–197 (1985).

43. Luterotti, S., M. Franko, and D. Bicanic, Fast Quality Screening of Vegetable Oils by HPLC-Thermal Lens Spectrometric Detection, *J. Amer. Oil Chem. Soc. 79*:1027–1031 (2002).

44. Luterotti, S., M. Franko, M. Sikovec, and D. Bicanic, Ultrasensitive Assays of *trans-* and *cis*-Beta-Carotenes in Vegetable Oils by High-Performance Liquid Chromatography-Thermal Lens Detection, *Anal. Chim. Acta 460*:193–200 (2002).

45. Mossé, J., J.-C. Huet, and J. Baudet, Changements de la Composition en Acides Aminés des Graines de Pois en Fonction de Leur Taux d'Azote, *Sci. Aliments 7*:301–324 (1987).

46. Tkachuk, R., Nitrogen-to-Protein Conversion Factors for Cereals and Oilseed Meals, *Cereal Chem. 46*:419–423 (1969).

47. Daun, J.K., and R. Przybylski, Effect of Environment as Determined by Oil Content and Unsaturation on the Composition of Flaxseed, *INFORM Abstracts, 91st AOCS Annual Meeting & Expo, San Diego, California, April 25–28, 2000, 11*:106 (2000).

48. Sammour, R.H., Proteins of Linseed (*Linum usitatissimum* L.): Extraction and Characterization by Electrophoresis, *Bot. Bull. Acad. Sin. 40*:121–126 (1999).

49. Oomah, B.D., and G. Mazza, Flaxseed Proteins—A Review, *Food Chem. 48*:109–114 (1993).

50. Daun, J.K., and D.R. DeClercq, Comparison of Combustion and Kjeldahl Methods for Determination of Nitrogen in Oilseeds, *J. Amer. Oil Chem. Soc. 71*:1068–1072 (1994).

51. Lowry, O.H., A.I. Ronserough, A.L. Farr, and R.J. Randell, Protein Measurement with the Folin Phenol Reagent, *J. Biol. Chem. 193*:265–275 (1951).

52. Ma, Z.L., Y.P. Wang, C.X. Wang, and F.Z. Miao, HPLC Determination of Muscle, Collagen, Wheat, Shrimp, and Soy Proteins in Mixed Food with the Aid of Chemometrics, *Amer. Lab. 29*:27–29 (1997).

53. Bhatty, R.S., and P. Cherdkiatgumchai, Compositional Analysis of Laboratory-Prepared and Commercial Samples of Linseed Meal and of Hull Isolated from Flax, *J. Amer. Oil Chem. Soc. 67*:79–84 (1990).

54. Tkachuk, R., Amino Acid Compositions of Cereals and Oilseed Meals, *Cereal Chem. 46*:206–218 (1969).

55. Bhatty, R.S., Nutrient Content of Whole Flaxseed and Flaxseed Meal, in *Flaxseed in Human Nutrition,* 1st edn., edited by S.C. Cunnane and L.U. Thompson, AOCS Press, Champaign, Illinois, 1995, pp. 22–42.

56. Vaisey-Genser, M., and D. Morris, Description and Composition of Flaxseed, in *Flaxseed Health, Nutrition and Functionality Edition*, edited by M. Vaisey-Genser and D. Morris, Flax Council of Canada, Winnipeg, Manitoba, 1997, pp. 10–17.

57. Flax Council of Canada, A Focus on Fiber, Available at www.flaxcouncil.ca/flaxnut6.htm.

58. Mazza, G., and C.G. Biliaderis, Functional Properties of Flax Seed Mucilage, *J. Food Sci. 54:*1302–1305 (1989).

59. Oomah, B.D., E.O. Kenaschuk, W. Cui, and G. Mazza, Variation in the Composition of Water-Soluble Polysaccharides in Flaxseed, *J. Agric. Food Chem. 43:*1484–1488 (1995).

60. Cui, W., E. Kenaschuk, and G. Mazza, Influence of Genotype on Chemical Compositon and Rheological Properties of Flaxseed Gums, *Food Hydrocoll. 10:*221–227 (1996).

61. Fedeniuk, R.W., Compositional Analysis and Physical Properties of Water-Soluble Polysaccharides from Linseed, M.Sc. Thesis, University of Manitoba, Winnipeg, Manitoba, Canada, 1993.

62. Wannerberger, K., T. Nylander, and M. Numan, Rheological Properties of Mucilage in Different Varieties from Linseed (*Linum usitatissimum*), *Acta Chem. Scand. 41:*311–319 (1991).

63. Bemiller, J.N., Quince Seed, Psyllium Seed, Flaxseed and Okra Gums, in *Industrial Gums*, 2nd edn., edited by R.L. Wistler and J.N. Bemiller, Academic Press, New York, 1973, pp. 331–337.

64. Ward, F., and S. Andon, The Use of Gums in Bakery Foods, *Amer. Baking Inst. Res. Dept. Tech. Bull. 15:*108 (1993).

65. DFSRC (Dietary Fiber Scientific Review Committee), The Definition of Dietary Fiber, *Cereal Foods World 46:*112–125 (2000).

66. Rickard, S.E., and L.U. Thompson, Health Effects of Flaxseed Mucilage, Lignans, *INFORM 8:*860–865 (1997).

67. Ink, S.L., and H.D. Hurt, Nutritional Implications of Gums, *Food Technol. 41:*77–82 (1987).

68. Cunnane, S.C., M.J. Hamadeh, A.C. Liede, L.U. Thompson, T.M.S. Wolever, and D.J.A. Jenkins, Nutritional Attributes of Traditional Flaxseed in Healthy Humans, *Am. J. Clin. Nutr. 61:*62–68 (1995).

69. Glicksman, M., Functional Properties of Hydrocolloids, in *Food Hydrocolloids Edition*, edited by M. Glicksman, CRC Press, Boca Raton, Florida, 1982, pp. 48–93.

70. Chornick, T.L., Effect of Cultivar and Sequential Ethanol Precipitation on the Physicochemical Properties of Flaxseed Mucilage, M.Sc. Thesis, University of Manitoba, Winnipeg, Manitoba, Canada, 2002.

71. Fedeniuk, R.W., and C.G. Biliaderis, Composition and Physicochemical Properties of Linseed (*Linum usitatissimum* L.) Mucilage, *J. Agric. Food Chem. 42:*240–247 (1994).

72. Cui, W., G. Mazza, B.D. Oomah, and C.G. Biliaderis, Optimization of an Aqueous Extraction Process for Flaxseed Gum by Response Surface Methodology, *Food Sci. Technol. 27:*363–369 (1994).

73. Chornick, T., L. Malcolmson, M. Izydorczyk, S. Duguid, and C. Taylor, Effect of Cultivar on the Physicochemical Properties of Flaxseed Mucilage, in *Proceedings of the 59th Flax Institute of the United States*, Fargo, North Dakota, 2002, pp. 7–13.

74. Cui, W., and G. Mazza, Physicochemical Characteristics of Flaxseed Gum, *Food Res. Inter. 29:*397–402 (1996).

75. Bhatty, R.S., Further Compositional Analyses of Flax—Mucilage, Trypsin Inhibitors and Hydrocyanic Acid, *J. Amer. Oil Chem. Soc. 70:*899–904 (1993).

76. Cui, W., G. Mazza, and C.G. Biliadereis, Chemical Structure, Molecular Size Distributions and Rheological Properties of Flaxseed Gum, *J. Agric. Food Chem. 42*:1891–1895 (1994).

77. Muralikrishna, G., P.V. Salimath, and R.N. Tharanathan, Structural Features of an Arabinoxylan and a Rhamnogalacturonan Derived from Linseed Mucilage, *Carbohydr. Res. 161*:265–271 (1987).

78. Hettiarachchy, N.S., G.A. Hareland, A. Ostenson, and G. Baldner-Shank, Chemical Compostion of Eleven Flax Seed Varieties Grown in North Dakota, in *Proceedings of the 53rd Flax Institute of the United States*, Fargo, North Dakota, 1990, pp. 36–40.

79. Dabrowski, K.J., and F.W. Sosulski, Composition of Free and Hydrolyzable Phenolic Acids in Defatted Flours of Ten Oilseeds, *J. Agric. Food Chem. 32*:128–130 (1984).

80. Matthys, A., NFPA International-CCFAC Report, National Food Processors Association, available at http://www.nfpa-food.org/members/international/codex_ccchreport01_03.htm (2001).

81. World Health Organization, Lead, Cadmium and Mercury in Foods, Switzerland, *Int. Dig. Health Leg. 43*:326 (1992).

82. Butler, G.W., and E.E. Conn, Biosynthesis of Cyanogenic Glucosides. I. Labeling Studies *in vivo* with *Linum usitatissimum*, *J. Biol. Chem. 239*:1674–1679 (1964).

83. Grant, C.A., and L.D. Bailey, Effects of Phosphorus and Zinc Fertiliser Management on Cadmium Accumulation in Flaxseed, *J. Sci. Food Agric. 73*:307–314 (1997).

84. Grant, C.A., J.C.P. Dribnenki, and L.D. Bailey, Cadmium and Zinc Concentrations and Ratios in Seed and Tissue of Solin (cv Linola-T-M 947) and Flax (cvs McGregor and Vimy) as Affected by Nitrogen and Phosphorus Fertiliser and Provide (*Penicillium bilaji*), *J. Sci. Food Agric. 80*:1735–1743 (2000).

85. Hocking, P.J., and M.J. McLaughlin, Genotypic Variation in Cadmium Accumulation by Seed of Linseed, and Comparison with Seeds of Some Other Crop Species, *Austral. J. Agric. Res. 51*:427–433 (2000).

86. Chaney, R.L., J.A. Ryan, Y-M. Li, and J.S. Angle, Transfer of Cadmium Through Plants to the Food Chain, in *Proceedings Workshop Environmental Cadmium in the Food Chain: Sources, Pathways, and Risks*, Scientific Committee on Problems of the Environment, Paris, 2001, 76–81.

87. Hammond, J.J., J.F. Miller, C.E. Green, and R.L. Chaney, Screening the USDA Flax Collection for Seed Cadmium, in *Proceedings of the 58th Flax Institute of the United States*, Fargo, North Dakota, 2000, pp. 108–115.

88. Chen, Y., E. Kenaschuk, and P. Dribnenki, Response of Flax Genotypes to Doubled Haploid Production, *Plant Cell Tiss. Org. Cult. 57*:195–198 (1999).

89. Oomah, B.D., E.O. Kenaschuk, and G. Mazza, Phenolic Acids in Flaxseed, *J. Agric. Food Chem. 43*:2013–2019 (1995).

90. Becher M., A. Woerner, and S. Schubert, Cd Translocation into Generative Organs of Linseed (*Linum usitatissimum* L.), *Z. Pflanzen. Bodenkunde 160*:505–510 (1997).

91. Li, Y-M., R.L. Chaney, A.A. Schneiter, J.F. Miller, E.M. Elias, and J.J. Hammond, Screening for Low Grain Cadmium Phenotypes in Sunflower, Durum Wheat and Flax, *Euphytica 94*:23–30 (1997).

92. Muir, A.D., N.K. Westcott, K. Ballantyne, and S. Northrup, Flax Ligans—Recent Developments in the Analysis of Lignans in Plant and Animal Tissues, in *Proceedings of the 58th Flax Institute of the United States*, Fargo, North Dakota, 2000, pp. 23–31.

93. Meagher, L.P. Isolation, and Characterization of the Lignans Isolariciresinol and Pinoresinol in Flaxseed Meal, in *Proceedings of the 58th Flax Institute of the United States*, Fargo, North Dakota, 2000, pp. 2–8.

94. Westcott, N.K., T.W. Hall, and A.D. Muir, Evidence for the Occurrence of Ferulic Acid Derivatives in Flaxseed Meal, in *Proceedings of the 58th Flax Institute of the United States*, Fargo, North Dakota, 2000, pp. 49–52.

95. Muir, A., N.D. Wescott, K.D. Reschny, and S.F. Northrup, Flaxseed Lignan Analysis: Methods and Strategies, in *Proceedings of the 59th Flax Institute of the United States*, Fargo, North Dakota, 2002, pp. 203–208.

96. Thompson, L.U., Nutritional and Physiological Effects of Phytic Acid, in *Food Proteins Edition*, edited by J.E. Kinsella and W.G. Soucie, AOCS Press, Champaign, Illinois, 1989, pp. 410–431.

97. Oomah, B.D., E.O. Kenaschuk, and G. Mazza, Phytic Acid Content of Flaxseed as Influenced by Cultivar, Growing Season, and Location, *J. Agric. Food Chem. 44*:2663–2666 (1996).

98. Reddy, N.R., S.K. Sathe, and D.K. Salunkhe, Phytates in Legumes and Cereals, in *Advances in Food Research Edition*, edited by C.O. Chichester, E.M. Mik, and G.F. Stewart, Academic Press, New York, 1982, pp. 1–92

99. Raboy, V., Seeds for a Better Future: 'Low Phytate' Grains Help to Overcome Malnutrition and Reduce Pollution, *Trends Plant Sci. 6*:458–462 (2001).

100. Harland, B.F., and E.R. Morris, Phytate, a Good or Bad Food Component, *Nutr. Res. 15*:733–654 (1995).

101. Thompson, L.U., Nutritional and Physiological Effects of Phytic Acid, in *Food Proteins Edition*, edited by J.E. Kinsella and W.G. Soucie, AOCS Press, Champaign, Illinois, 1989, pp. 410–431.

102. Morse, D., H.H. Head, and C.J. Wilcox, Disappearance of Phosphorus in Phytate from Concentrates *in vitro* and from Rations Fed to Lactating Dairy Cows, *J. Dairy Sci. 75*:1979–1986 (1994).

103. Van der Molen, D.T., A. Breeuwsma, and P.C.M. Boers, Agricultural Nutrient Losses to Surface Water in the Netherlands: Impact, Strategies and Perspective, *J. Environ. Qual. 27*:4–11 (1998).

104. Harland, B.F., and G. Narula, Food Phytate and Its Hydrolysis Products, *Nutr. Res. 19*:947–961 (1999).

105. Conn, E.E., Cyanogenesis—A Personal Perspective, *Acta Hort. 375*:31–43 (1994).

106. Jones, D.A., Why Are So Many Food Plants Cyanogenic? *Phytochemistry 48*:155–162 (1998).

107. Smith, C.R., D. Weisleder, and R.W. Miller, Linustatin and Neolinustatin: Cyanogenic Glycosides of Linseed Meal That Protect Animals Against Selenium Toxicity, *J. Org. Chem. 45*:507–510 (1980).

108. Niedzwiedz-Siegien, I., Cyanogenic Glucosides in *Linum usitatissimum*, *Phytochemistry 49*:59–63 (1998).

109. Conn, E.E., Cyanogenic Glycosides, *J. Agric. Food Chem. 17*:519–526 (1969).

110. Butler, G.W., and E.E. Conn, Biosynthesis of the Cyanogenic Glucosides Linamarin and Lotaustralin. I. Labeling Studies *in vivo* with *Linum usitatissimum*, *J. Biol. Chem. 239*:1674–1679 (1964).

111. Kobaisy, M., B.D. Oomah, and G. Mazza, Determination of Cyanogenic Glycosides in Flaxseed by Barbituric Acid-Pyridine, Pyridine-Pyrazolone and High-Performance Liquid Chromatography Methods, *J. Agric. Food Chem. 44*:3178–3181 (1996).

112. Chadha, R.K., J.F. Lawrence, and W.M.N. Ratnayake, Ion Chromatographic Determination of Cyanide Released from Flaxseed Under Autohydrolysis Conditions, *Food Add. Contam. 12*:527–533 (1995).

113. Poulton, J.E., Toxic Compounds in Plant Foodstuffs: Cyanogens, in *Food Proteins Edition*, edited by W.G. Soucie and J.E. Kinsella, AOCS Press, Champaign, Illinois, 1989, pp. 381–401.

114. Jackson, F.L., The Bioanthropological Impact of Chronic Exposure to Sublethal Cyanogens from Cassava in Africa, *Acta Hort. 375:*295–309 (1994).

115. Roseling, H., Measuring Effects in Humans of Dietary Cyanide Exposure from Cassava, *Acta Hort. 375:*271–283 (1994).

116. Schulz, V.L.A.a.G.T., Resorption of Hydrocyanic Acid from Linseed [Resorption von blausäure aus leinsamen], *Leber Magen Darm 13*:10–14 (1983).

117. Tewe, O.O., Serum and Tissue Thiocyanate Concentrations in Growing Pigs Fed Cassava Peel or Corn Based Diets Containing Grade Proteins Levels, *Toxicol. Lett. 23:*169–176 (1984).

118. Oomah, B.D., G. Mazza, and E.O. Kenaschuk, Cyanogenic Compounds in Flaxseed, *J. Agric. Food Chem. 40:*1346–1348 (1992).

119. Schicher, v.H., and M. Wilkens-Sauter, Quantitative Bestmmung Cyanogener Glycoside in *Linum usitatissimum* mit hilfe HPLC, *Fette. Seifen. Anstrich. 88:*287–290 (1986).

120. Rao, P., and S.K. Hahn, An Automated Enzymatic Assay for Determining the Cyanide Content of Cassava (*Manihot esculenta* Crantz) and Cassava Products, *J. Sci. Food Agric. 35:*426–436 (1984).

121. Cooke, R.D., An Enzymatic Assay for the Total Cyanide Content of Cassava (*Manihot esculenta* Crantz), *J. Sci. Food Agric. 29:*345–352 (1978).

122. Harris, J.R., G.H.J. Merson, M. Hardy, and D.J. Curtis, Determination of Cyanide in Animal Feeding Stuffs, *Analyst 105:*974–980 (1980).

123. Cunnane, S.C., S. Ganguli, C. Menard, A.C. Liede, M.J. Hamadeh, Z-Y. Chen, T.M.S. Wolever, and D.J.A. Jenkins, High α-ALA Flaxseed (*Linum usitatissimum*): Some Nutritional Properties in Humans, *Brit. J. Nutrit. 69:*443–453 (1993).

124. Kratzer, F.H., The Treatment of Linseed Meal to Improve Its Feeding Value for Chicks, *Poultry Sci. 25:*541–542 (1946).

125. Kratzer, F.H., and P. Vohra, The Use of Flaxseed as a Poultry Feedstuff, *Poultry Fact Sheet No. 21 Cooperative Extension*, University of California, Davis, Available at http://animalscience.ucdavis.edu/Avian/pfs21.htm (1996).

126. Klosterman, H.J., G.L. Lamoureux, and J.L. Parson, Isolation, Characterization, and Synthesis of Linatine: A Vitamin B$_6$ Antagonist from Flaxseed (*Linum usitatissimum*), *Biochemistry 6:*170–177 (1967).

127. Klosterman H.J., Vitamin B$_6$ Antagonists of Natural Origin, *J. Agric. Food Chem. 22:*13–16 (1974).

128. Madhusudhan, K.T., and N. Singh, Studies on Linseed Proteins, *J. Agric. Food Chem. 37:*959–963 (1983).

129. Alonso, L., M.L. Marcos, J.G. Blanco, J.A. Navarra, S. Juste, M.D.M. Garces, R. Perez, and P.J. Carretero, Anaphylaxis Caused by Linseed (Flaxseed) Intake, *J. Allergy Clin. Immun. 98:*469–470 (1996).

130. Harris, R.K., J. Greaves, D. Alexander, T. Wilson, and W.J. Haggerty, Development of Stability-Indicating Analytical Methods for Flaxseed Lignans and Their Precursors, in *Food Phytochemicals for Cancer Prevention II—Teas, Spices, and Herbs Edition*, edited by C.T. Ho, T. Osawa, and R.T. Rosen, American Chemical Society, Washington, D.C., 1974–1994. (547) p., 1994, pp. 295–305.

131. Goldberg, J.M., and K. O'Mara, Flaxseed Info, available at http://members.tripod.com/~himolocarb/flax.htm.

132. Khotpal, R.R., A.S. Kularni, and H.A. Bhakare, Studies on Lipids on Some Varieties of Linseed (*Linum usitatissimum*) of Vidarbha Region, *Ind. J. Pharmaceut. Sci. 59*:157–158 (1997).

133. El Shattory, Y., Chromatographic Column Fractionation and Fatty Acid Composition of Different Lipid Classes of Linseed Oil, *Nahrung 20*:307–311 (1976).

134. Green, A.G., and J.C.P. Dribnenki, Linola, a New Premium Polyunsaturated Oil, *Lipid Tech. March/April*:29–33 (1994).

135. Coors, U., Anwendung des Tocopheromusters zur Erkennung von Fett- und Övermischungen, *Fat Sci. Technol. 92*:519–526 (1991).

136. Syvaoja, E.L., V. Piironen, P. Varo, P. Koivistoinen, and K. Salminen, Tocopherols and Tocotrienols in Finnish Foods: Oils and Fats, *J. Amer. Oil Chem. Soc. 63*:328–329 (1986).

137. Box, J.A.G., and H.A. Boekenoogen, Vegetable Oil Pigments: Carotenoids and Pheophytines in Soybean, Rapeseed and Linseed Oils, *Fette Seifen Anstrich. 69*:724–729 (1967).

138. Daun, J.K., N. Buhr, J.T. Mills, L.L. Diosady, and T. Magin, *Grains and Oilseeds, Handling, Marketing, Processing, 4th edn.*, edited by E. Bass, Canadian International Grains Institute, Winnipeg, Manitoba, 1993, pp. 883–935.

139. Hormis, Y.A., and G.G. Roland, The Development of High Palmitic Fatty Acid Solin Cultivars, in *Proceedings of the 56th Flax Institute of the United States*, Fargo, North Dakota, 1996, pp. 172–176.

140. van Zoonen, P., R. Hoogerbrugge, S.M. Gort, H.J. van de Wiel, and H.A. van't Kooster, Some Practical Examples of Method Validation in the Analytical Laboratory, *Trends Analyt. Chem. 18*:584–593 (1999).

141. Wanasundara, P.K.J.P.D., R. Amarowicz, M.T. Kara, and F. Shahidi, Removal of Cyanogenic Glycosides of Flaxseed Meal, *Food Chem. 48*:263–266 (1993).

Chapter 2

Delineating the Metabolic Pathway(s) to Secoisolariciresinol Diglucoside Hydroxymethyl Glutarate Oligomers in Flaxseed (*Linum usitatissimum*)

Keat H. Teoh, Joshua D. Ford, Mi-Ran Kim, Laurence B. Davin, and Norman G. Lewis

Institute of Biological Chemistry, Washington State University, Pullman, Washington 99164-6340, USA

Introduction

Flaxseed (*Linum usitatissimum*) contains many biologically active phytochemicals that are beneficial to human health. One group of these phytochemicals is currently being extensively investigated for its biological roles in preventing the onset of hormone-related cancers. These are the flaxseed 8-8' linked lignans (1), a subclass of the structurally very diverse lignan natural products widely distributed throughout the plant kingdom (2,3). Typically, in different plant species, the lignans exist as phenylpropanoid pathway derived dimers and are found in all plant parts including stems, roots, leaves, flowers, seeds, oils, and exuded resins. Although much still remains to be understood about the biochemical coupling processes, the data, to date suggest that they can result via dimerization of monolignols (4–13), allylphenols (3,11), and hydroxycinnamic acids (12,13), respectively, depending upon the actual lignan skeletal subclass formed.

Lignans have had a long history of pharmacological importance. For example, podophyllotoxin **1** (bold numerals in text refer to structural formulas in Figs. 2.1–2.10, 2.12), from *Podophyllum hexandrum* or *P. peltatum*, is used for the treatment of venereal warts (14), whereas its semisynthetic derivatives, etoposide **2**, etopophos **3**, and teniposide **4**, are successfully employed in chemotherapy (often in combination with other agents) for Hodgkin's and non-Hodgkin's lymphomas, testicular/small-cell lung cancers, and acute leukemia (15–19; Fig. 2.1). Additionally, the allylphenol-derived creosote bush (*Larrea tridentata*) lignans have long been utilized by native North American Indians for a variety of ailments including digestive disorders, sores, and rheumatism (20). More recently, creosote bush lignans, such as nordihydroguaiaretic acid **5** and its derivatives, have been shown to display potent antiviral (e.g., anti-HIV) (21–23), anticancer (24), and antioxidant (25) properties. Sesame lignans, such as sesamin **6** and sesamolinol **7**, also help reduce cholesterol levels in the diet (26), whereas the lignans termilignan **8**, thannilignan **9**, and anolignan B **10**, from *Terminalia bellerica* fruit, exhibit antimalarial and antifungal activities (27). Although the physiological roles of lignans in plants remain to be fully elucidated, the functions

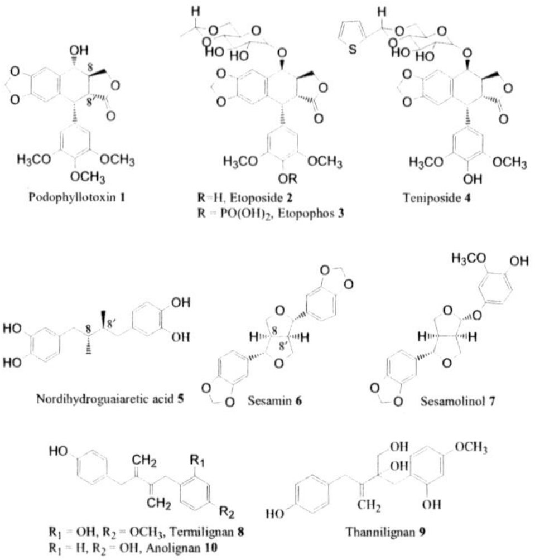

Podophyllotoxin 1

R = H, Etoposide 2
R = PO(OH)$_2$, Etopophos 3

Teniposide 4

Nordihydroguaiaretic acid 5

Sesamin 6

Sesamolinol 7

R$_1$ = OH, R$_2$ = OCH$_3$, Termilignan 8
R$_1$ = H, R$_2$ = OH, Anolignan 10

Thannilignan 9

Fig. 2.1. Selected lignan and lignan derivatives with known pharmaceutical and/or bioactive properties.

appear to be primarily defense related, such as antibacterial, antiviral, and antifungal, as well as having feeding deterrent and allelopathic properties (see 3, for a review).

From a pharmacological perspective, in 1982 the potential value of various lignans as anticancer (chemopreventive) agents was noted when high levels of "mammalian" lignans were detected in the urine of postmenopausal women fed a plant-based diet, the presence of which was coincident with a significantly lower incidence rate of breast cancers (28). These mammalian lignans had previously been identified as enterolactone **11** and enterodiol **12** (Fig. 2.2) by two independent groups in 1980, Setchell and coworkers (29) and Stitch and coworkers (30). Both metabolites were shown to be derived (at least in part) from the plant lignans matairesinol **13** and secoisolariciresinol **14** (31,32), respectively.

Formation of both enterodiol **12** and enterolactone **11** from **13–15** is considered to be via the action of gastrointestinal flora catalyzed transformations, which in turn are speculated to be engendered by *Clostridia* sp. (32). [Other reports have described the conversion of these plant lignans into mammalian lignans *in vitro* using crude isolates from human fecal flora (33,34).] It is of interest that although chiral analyses of mammalian lignan products have not been described, it has been reported that enterolactone **11** is racemic (30,35) and enterodiol **12** is probably racemic (35). This is curious given that most plant lignans are optically pure, even though the particular antipode can vary with plant species [e.g., (+)-secoisolariciresinol **14a** in flax, and its (−)-form **14b** in *Forsythia* sp. (discussed below)].

Figure 2.3 depicts possible transformations (36) catalyzed by gut microflora that give rise to the lignans enterolactone **11** and enterodiol **12**, using secoisolariciresinol **14** as an example. The conversions engendered involve enzymatic demethylations and

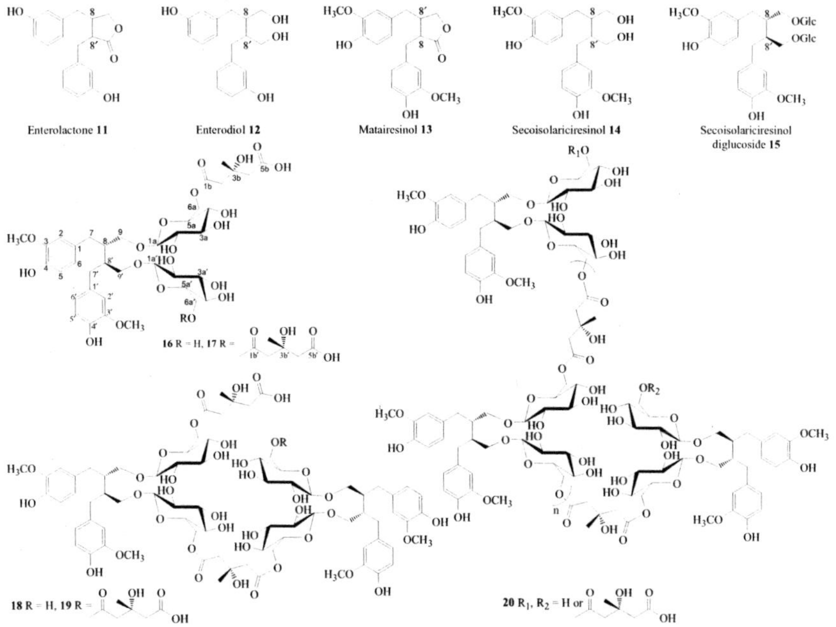

Fig. 2.2. The "mammalian" lignans enterolactone **11** and enterodiol **12**, their presumed plant lignan precursors, matairesinol **13**, secoisolariciresinol **14**, secoisolariciresinol diglucoside (SDG) **15**, and the SDG-hydroxymethyl glutaric acid (HMG) oligomers **16–20**. [SDG: all depicted as the (+)-antipode; average value of $n = 3$.]

dehydrations, and possibly even racemization, with the actual sequence of transformations involving any or all of the permutations illustrated. [In the case of matairesinol **13**, equivalent transformations could be envisaged at the lactone (rather than at the diol) level (36).] It is interesting that the proposed dehydration of the putative 1,4 dihydrophenyl intermediates **21** and **24** has a biochemical precedence, that is, in the conversion of arogenic acid **25** into phenylalanine **26** (37).

From a more general perspective, resolution of the enantiospecificity (if any) of these microbial transformations may have an important bearing on both the type of lignan to be targeted for health protection and the type of plant(s) to be considered. This is for the following reasons: First, does the metabolism of an essentially optically pure lignan, such as (+)-secoisolariciresinol diglucoside (SDG) **15**, indeed involve racemization at positions 8 and 8′ as implied, or did previous feeding studies fortuitously use only precursors containing similar amounts of both enantiomeric forms? Second, are other lignans, such as pinoresinol **27** or lariciresinol **28** (Fig. 2.4), also converted into enterodiol **12**/enterolactone **11**? Biochemically, this would require only an additional reductive step of the furan ring(s) to generate enterodiol **12**, and hence this possibility cannot be dismissed. Another question, not yet experimentally tested, is whether the lignin biopolymers present in plant foodstuffs also contribute to forma-

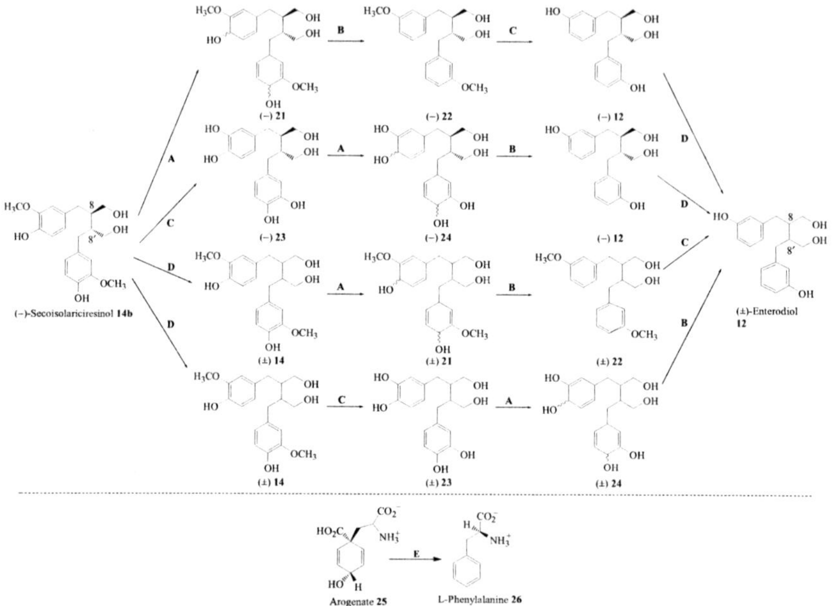

Fig. 2.3. Proposed catabolic pathway to enterodiol **12** from secoisolariciresinol **14**. (**A**) reduction, (**B**) dehydration, (**C**) demethylation, and (**D**) racemization at $C_8,C_{8'}$; (**E**) for comparative purposes, dehydratase-decarboxylase catalyzed formation of L-phenylalanine **26**.

tion of the mammalian lignans enterodiol **12** and enterolactone **11**. Lignins are formed by dehydrogenative polymerization of monolignols, such as coniferyl alcohol **29**, and thus their polymeric matrices contain substructures, which include pinoresinol **27**. Given that lignins represent the second most abundant naturally occurring substances, next to cellulose, and that their biopolymers are composed of racemic substructures, it seems important to ascertain what, if any, their contributions are in the diet to enterodiol **12**/enterolactone **11** formation.

In terms of flaxseed and health protection, the lignan SDG **15** (Fig. 2.2), which is released from its seed by mild alkali treatment (38), has received much attention over the years. This is mainly because of its established cancer preventive properties. For example, Thompson and coworkers (1,39) and Jenab and Thompson (40) demonstrated that administration of SDG **15** can significantly inhibit development of tumors in rats treated with the carcinogen dimethylbenzanthracene as well as reduce the number of tumors per rat in each experimental group (1,39). Furthermore, the relationship between mammalian lignans and the diminished risk of cancer was supported by two separate epidemiological studies, one conducted in western Australia (41) and the other in eastern Finland (42), which revealed an inverse relationship between serum and urine enterolactone **11** concentrations and breast cancer risk. Moreover, the role of SDG **15** goes beyond its anticancer properties: it has also been reported to be an effective antioxidant in mediating the effect of type 1 (43) and type 2 (44) diabetes in rats, reducing the serum cholesterol level and lipid peroxidation in rabbits (45), and improving renal function in patients suffering from lupus nephritis (46). To date, flaxseed is the richest known source of SDG **15** with levels approaching ~3% of dry weight (38). It is, however, not present in the seed as such, but exists as a number of hydroxymethyl glutarate ester oligomers (e.g., Fig. 2.2, **16–20**; 47,48).

Given the significant role of SDG **15** in promoting human health, and flaxseed as a rich source of SDG (following hydrolytic cleavage of the fascinating array of SDG-hydroxymethyl glutarate esters, e.g., **16–20**), the question of how such substances are biosynthetically formed is of considerable interest. This chapter therefore focuses upon what is currently known about lignan biosynthesis (secoisolariciresinol **14**, matairesinol **13**, and podophyllotoxin **1**), as well as the emerging pathways leading to the SDG oligomers **16–20**. As discussed below, both secoisolariciresinol **14** and matairesinol **13** are also intermediates in the biosynthesis of the antiviral/anticancer agent podophyllotoxin **1** present mainly in the rhizomes of *Podophyllum* sp. (49).

Unraveling the Biosynthetic Pathway to Secoisolariciresinol Using *Forsythia intermedia* and *Thuja plicata*

Both *Forsythia intermedia* and *Thuja plicata* contain various lignans: the former has lignans mainly composed of matairesinol **13**, arctigenin **30**, and arctiin **31** (Fig. 2.4; 50,51), whereas in *T. plicata* (western red cedar), they are primarily plicatic acid **32**-derived (52–54). Both plant species display interesting features: *Forsythia* has relatively few pathogens and the dried fruits of *F. suspensa* are used in Japan and China under the name "Rengyo" for the treatment of tonsillitis, and pharyngitis, as well as being used as an anti-inflammatory drug (55,56); in *T. plicata*, its lignans help confer color, durability, and longevity to the heartwood. Note that the plicatic acid **32** deriv-

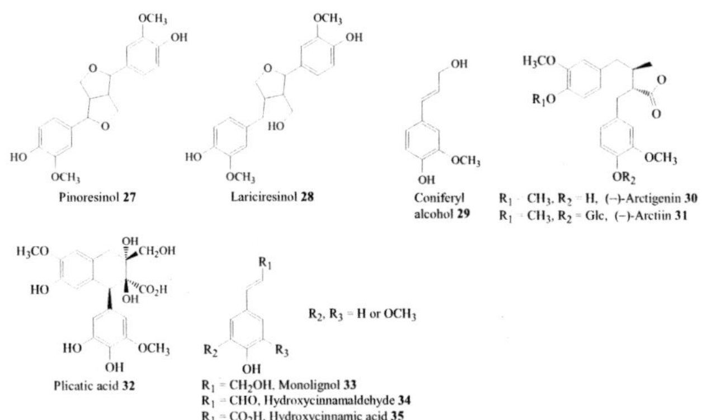

Fig. 2.4. Selected lignans and phenylpropanoids.

atives in cedar wood can also engender various allergenic responses, such as occupational asthma (57,58), upon lengthy (chronic) exposure to the heartwood during handling.

Stereoselective Coupling and Dirigent Proteins

Our initial efforts in defining the biochemical pathways to the 8-8′ linked lignans (such as matairesinol **13**) involved identifying the molecular entities undergoing initial coupling and determining whether this conversion was at the monolignol **33**, hydroxycinnamaldehyde **34**, or hydroxycinnamic acid **35** level (Fig. 2.4) or whether mixed coupling (e.g., of monolignols **33** and hydroxycinnamic acids **35**) could occur. Of additional interest was whether this coupling led to the generation of optically active moieties, inasmuch as many of the dimeric 8-8′ linked lignans are optically active (50).

Entry into the pathway was ultimately demonstrated to result from the stereoselective coupling of two molecules of *E*-coniferyl alcohol **29** to afford (+)-pinoresinol **27a** (Fig. 2.5). Initially, this conversion was detected using crude *F. intermedia* plant material after removal of readily soluble proteins (4), with the remaining proteins and enzymes subsequently solubilized (59). In this way, it was found that a tightly bound ~26 kDa glycoprotein (estimated by SDS-page), which lacked any oxidative capacity by itself, could convert *E*-coniferyl alcohol **29** into (+)-pinoresinol **27a** in the presence of an oxidase (laccase) and/or an oxidant (5,6). The term dirigent protein (Latin: *dirigere*, to guide or to align) was introduced to describe this new class of proteins, and its proposed mechanism of action is summarized in Figure 2.6.

More detailed analyses of the *F. intermedia* dirigent protein established that it is dimeric (~50 kDa), as evidenced by electrospray ionization mass spectrometry (ESI-MS), gel permeation chromatography–multiangle laser light scattering (GPC-MALLS), and matrix-assisted laser desorption/ionization–time of flight mass spec-

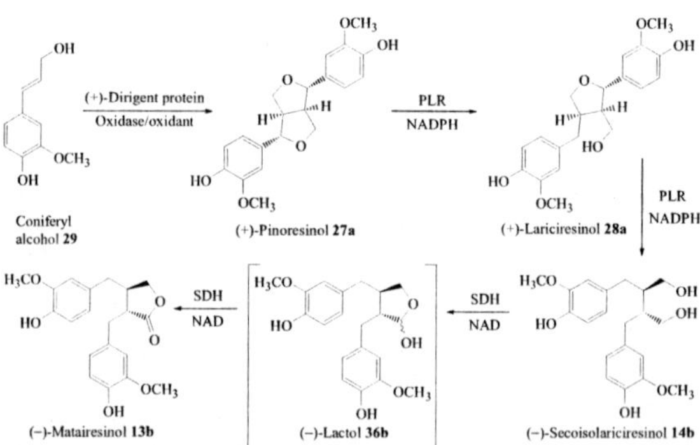

Fig. 2.5. Biosynthetic pathway to the lignans (–)-secoisolariciresinol **14b** and (–)-matairesinol **13b** in *Forsythia intermedia*.

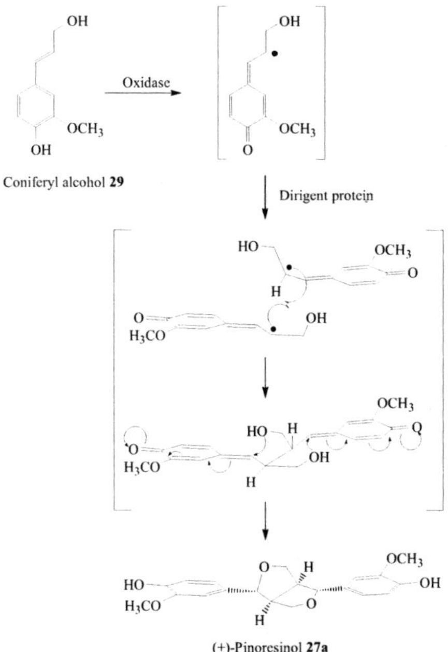

Fig. 2.6. Proposed stereoselective coupling mechanism of (+)-pinoresinol forming dirigent protein, in presence of an oxidase/oxidant. [Structures in brackets represent putative intermediates.]

trometry (MALDI-TOF MS) analyses, as well as by ultracentrifugation (8), with the latter analysis revealing a tendency of the dirigent protein to aggregate, this presumably being due to its significant β-sheet character (35–42%).

Cloning of the gene revealed that it encoded a glycosylated dirigent protein of ~18 kDa (7), with a transit peptide targeting it to the secretory system. [Subsequent glycosylation of the native dirigent (+)-pinoresinol forming protein gives the monomeric entity ~26 kDa (6,7).] It is notable that the protein displayed no significant homology to any other known protein, implying that this radical-radical coupling dirigent protein function had evolved during the transition of plants from an aquatic to a land base. *In situ* hybridization has also been carried out to establish where the (+)-pinoresinol dirigent forming protein transcripts are expressed, this primarily occurring in the cambial regions of *F. intermedia* (60).

The dirigent proteins in both *F. intermedia* and *T. plicata* were established to be part of a multigene family containing at least three and nine members, respectively. This observation led to the view that there are multidimensional networks controlling all aspects of radical-radical coupling leading to both lignans, as well as the polymeric lignins via involvement of dirigent proteins and/or proteins harboring (arrays of) dirigent sites, respectively (9,10). Since then, dirigent protein homologues have been established in the pharmacologically important plants *Linum flavum* (49), *Podophyllum peltatum* (49), and *Sesamum indicum* (61) as well as in other species such as *Arabidopsis* (16 homologues). Of these, some have been established to engender (+)-pinoresinol **27a** formation (9), whereas others are of unknown function (and are presumed to be involved in other coupling processes with other substrates).

Pinoresinol/Lariciresinol Reductase

In *F. intermedia* and *T. plicata*, it was demonstrated that pinoresinol **27** could undergo an enantiospecific conversion to afford initially lariciresinol **28** and subsequently secoisolariciresinol **14** (Fig. 2.5; 62,63). In *F. intermedia*, this bifunctional (~35 kDa) pinoresinol/ lariciresinol reductase (PLR) was purified to apparent homogeneity and established to catalyze the enantiospecific NADPH-dependent conversion of (+)-pinoresinol **27a** into (+)-lariciresinol **28a** and (–)-secoisolariciresinol **14b**, respectively; the corresponding (–)-antipode **27b** did not undergo this conversion (62). Initial velocity studies were also carried out with the purified protein, giving values for K_m of 27 ± 1.5 and 121 ± 6.0 µM and V_{max} of 16.2 ± 0.4 and 25.2 ± 0.7 µmol h^{-1} mg^{-1} protein for (+)-pinoresinol **27a** and (+)-lariciresinol **28a**, respectively (62).

Using specifically labeled [4*R*- and 4*S*-³H]NADPH (64), it was established that only the 4-pro-*R* hydrogen (hydride) was abstracted during enzymatic reduction. Additionally, using [7,7′-²H₂]pinoresinol **27** and [7,7′-²H₃]lariciresinol **28** as substrates, it was determined that the hydride transfer was >99% stereospecific, where the incoming hydride took up the pro-*R* position at C-7′ (and/or C-7) in lariciresinol **28** and secoisolariciresinol **14**, respectively, i.e., during lariciresinol **28** formation, the incoming hydride from the NADPH took up the pro-*R* position, with the initial carbon-deuterium bond geometry at C-7′ undergoing inversion to assume the pro-*S* position (Fig. 2.7).

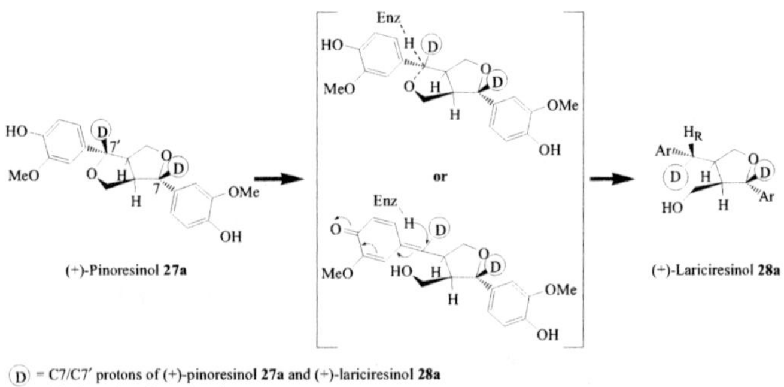

Fig. 2.7. Stereospecificity of hydride transfer by pinoresinol/lariciresinol reductase (PLR) from *Forsythia intermedia*. [D = ²H; structures in brackets represent putative intermediates.]

The corresponding gene was also obtained (*plr–Fi1*), this encoding a polypeptide of calculated molecular mass of 34.9 kDa (62). In this case, no signal peptide was present, the enzyme being cytosolic. The heterologously expressed protein in *Escherichia coli*, *plr–Fi1*, catalyzed the same stereospecific reactions as for the native *Forsythia* protein. Two reductase cDNA's, *plr–Tp1* and *plr–Tp2*, were also identified from a western red cedar cDNA library. When expressed in *E. coli*, the recombinant *plr–Tp1* and *plr–Tp2* proteins displayed *different* activities (63): *plr–Tp1* was able to catalyze the sequential conversion of (–)-pinoresinol **27b** into (–)-lariciresinol **28b** and (+)-secoisolariciresinol **14a** (see Fig. 2.8), whereas *plr–Tp2* utilized (+)-pinoresinol **27a** to give (+)-lariciresinol **28a** and then (–)-secoisolariciresinol **14b** (i.e., the same stereochemical specificity as *plr–Fi1*). Thus, in western red cedar, parallel pathways to lignan formation were established, but with enzymes with different enantiospecificities (63).

Secoisolariciresinol Dehydrogenase (SDH)

The dehydrogenation of secoisolariciresinol **14** to matairesinol **13** thus marks the final step in the biosynthesis to one of the presumed precursors of the mammalian lignans (Fig. 2.5). In *F. intermedia*, this step was demonstrated to be catalyzed by secoisolariciresinol dehydrogenase (SDH) in the presence of NAD as cofactor (65,66). However, the transformation was established to be completely enantiospecific; the (–)-antipode of secoisolariciresinol **14b** underwent dehydrogenation, whereas the (+)-antipode **14a** did not. Following its >6000-fold purification to apparent homogeneity, the ~32 kDa SDH was cloned with an open reading frame (ORF) of ~831 base pairs, with corresponding homologues subsequently obtained from *P. peltatum* using a PCR-based strategy (66). The latter was overexpressed in *E. coli* to afford a fully functional recombinant SDH which was again able to catalyze only the enantiospecific conversion of (–)-secoisolariciresinol **14b** into (–)-matairesinol **13b**. In addition, the recombinant protein was found to catalyze this conversion via the intermediary lactol **36** (see Fig. 2.5), which was not observed with the native *F. intermedia* protein. When (–)-lactol **36** was used as substrate, apparent values for K_m of 160.2 ± 0.8 µM and velocities (V_{max}) of 7.1 ± 0.02 mmol min^{-1} mg^{-1} protein were obtained (66).

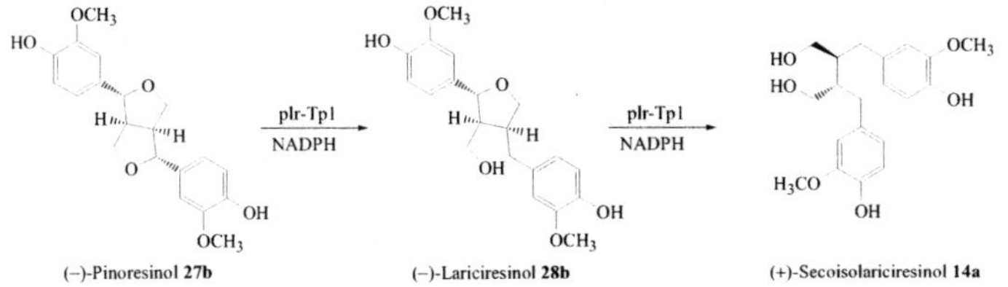

Fig. 2.8. *Thuja plicata* PLR catalyzed conversion of (–)-pinoresinol **27b** into (+)-secoisolariciresinol **14a**. *Abbreviation:* See Figure 2.7.

Biosynthesis of Podophyllotoxin in *Linum flavum* and *Podophyllum* sp.

Matairesinol **13** has been shown to be an important precursor to the antiviral/anti-cancer lignans podophyllotoxin **1** in *Podophyllum* sp. and to 5-methoxypodophyllo-toxin **37** in *L. flavum* root tissue. A gene encoding a dirigent protein has been cloned from *P. peltatum*, with the protein heterologously expressed in *Drosophila melanogaster* cells, and crossreacting with *F. intermedia* dirigent polyclonal antibod-ies (49). A partially purified pinoresinol/lariciresinol reductase from *L. flavum* was also obtained which catalyzed the enantiospecific conversion of (+)-pinoresinol **27a** into (+)-lariciresinol **28a** and (–)-secoisolariciresinol **14b**, respectively (49). The resulting (–)-secoisolariciresinol **14b** was then enantiospecifically dehydrogenated into (–)-matairesinol **13b**, as shown by conversion of both stable- and radioisotopical-ly labeled secoisolariciresinol **14** into matairesinol **13** (49). An additional emphasis of this study then demonstrated that in *L. flavum* matairesinol **13** could be hydroxylated at the 7′ position to afford 7′-hydroxymatairesinol **38** (49), which in turn was metabo-lized into 5-methoxypodophyllotoxin **37** through a series of steps involving arylte-trahydronaphthalene ring formation, additional hydroxylations, *O*-methylations, and methylenedioxy bridge formation (Fig. 2.9; 49,67).

SDG and SDG-Hydroxymethyl Glutaric Acid (HMG) Ester-Linked Oligomers in Flaxseed

In 1954, Klosterman and Clagett (68) showed that treatment of a (defatted) flaxseed extract with base (NaOMe) results in release of soluble SDG **15**, this in turn being attributed to cleavage of a readily hydrolyzable, soluble, "polymeric" constituent. Also in 1954, Klosterman and Smith (69) reported the presence of HMG **39** (Fig. 2.10), following a similar treatment of an ethanol-dioxane extract of defatted flaxseed meal with sodium methoxide. However, no connection was made to SDG **15** being released under the same conditions. Accordingly, the existence of covalently linked SDG-HMG ester-linked oligomers (e.g., **16–20**) in flaxseed eluded discovery until two independent groups in 2001 (47,48) reported such moieties as the source of alkali-soluble SDG **15**.

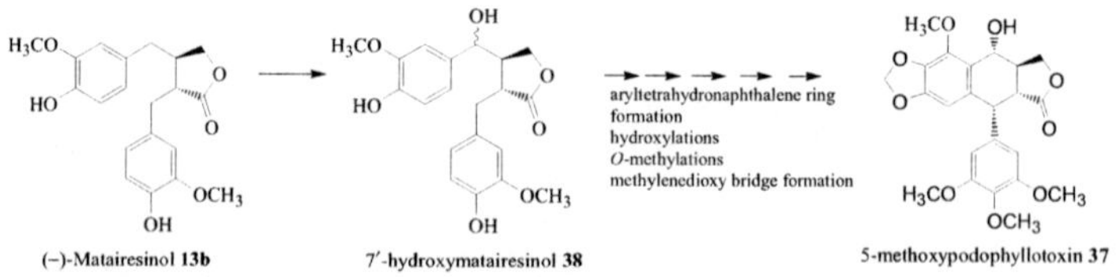

Fig. 2.9. Proposed biosynthetic pathway to 5-methoxypodophyllotoxin **37** in *Linum flavum*.

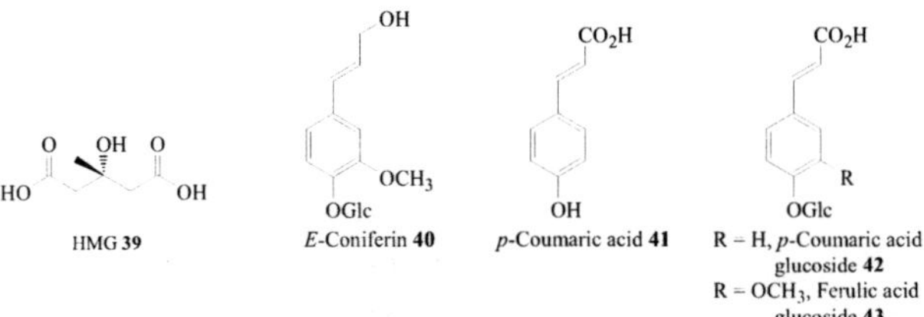

Fig. 2.10. HMG **39** and various phenylpropanoid derivatives **40–43**. *Abbreviation:* See Figure 2.2.

In this regard, the C_{18} reversed phase HPLC analysis of an aqueous-ethanol extract of mature flaxseed initially revealed a complex array of poorly resolvable UV-absorbing components (Fig. 2.11), which upon treatment with cold alkali was primarily converted into two SDG diastereomers (peaks 1a and 1b; 47). Repeated chromato-

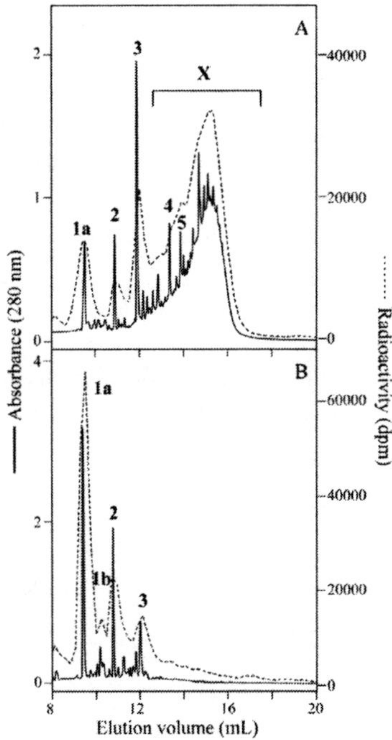

Fig. 2.11. C18 reversed-phase HPLC analyses of stage 4 developing flaxseed aqueous ethanol solubles following uptake and metabolism of L-[U-^{14}C]-phenylalanine **26**. **(A)** aqueous ethanol extract *before* base hydrolysis; **(B)** aqueous ethanol extract *after* base hydrolysis; 1a = SDG **15a**; 1b = SDG diastereomer **15b**; 2 = 6a-HMG SDG **16**; 3 = 6a, 6a'-di-HMG SDG **17**; 4 = dimer-1 **18**; 5 = dimer-2 **19**. *Abbreviations:* See Figure 2.2. *Source:* Adapted from Ford *et al.* (47).

graphic separation of the poorly resolved components ultimately afforded the 6a-HMG SDG derivative **16** (peak 2), the 6a,6a′ di-HMG SDG **17** (peak 3), the two dimeric moieties **18** and **19** (peaks 4 and 5), in addition to other detectable (but not further characterized) higher oligomers (**X**). Following aqueous ethanol extraction of the crude flaxseed, the SDG-HMG oligomeric constituents could be readily separated from other impurities in the extract by treatment with HCl to bring the pH ~3.0, which precipitated out the components of interest (Lewis laboratory, unpublished results). This precipitation methodology thus provides a simple but effective way of partially purifying the SDG-HMG oligomeric constituents.

As indicated above, one interesting observation, made upon alkali hydrolysis of the SDG-containing oligomeric components (e.g., **16–20**), was that two diastereomeric forms of SDG were released. Recognition of the presence of two SDG diastereomers had been made (70) prior to our studies, but the significance was unknown. Accordingly, subsequent enzymatic cleavage of the glucose moieties of each SDG diastereomer **15a** and **15b** (Fig. 2.12), gave two forms of secoisolariciresinol **14**, namely the (+)- and (–)-enantiomers in an ~99:1 ratio, thereby resolving the issue of diastereomer identification (47). This interesting finding potentially implied the presence of two distinct pathways to the SDG-linked oligomers, that is, one utilizing (+)-secoisolariciresinol **14a** and a minor pathway employing the corresponding (–)-antipode **14b**.

On the basis of these findings, together with our knowledge of secoisolariciresinol **14** and matairesinol **13** biosynthesis in *Forsythia*, *Thuja*, *Sesamum*, *Podophyllum*, and *Linum* species, it was of considerable interest to then begin to define how SDG-HMG oligomers were being biosynthesized in developing flaxseed tissue. This was not only for the purpose of defining the biosynthetic routes, but also for developing biotechnological approaches in order to increase secoisolariciresinol **14**, SDG **15**, and matairesinol **13** levels in various organisms, including foodstuffs.

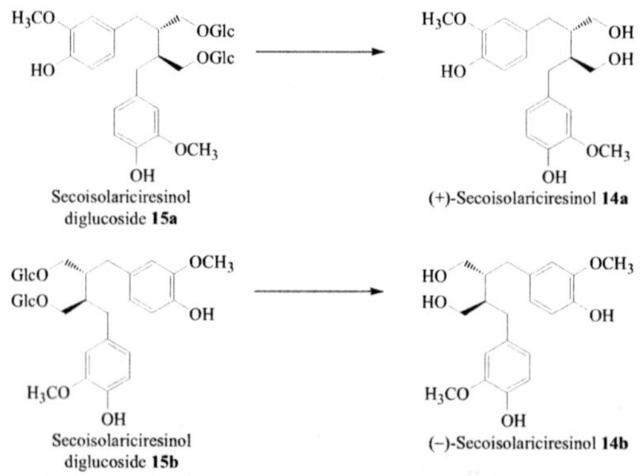

Fig. 2.12. Enzymatic cleavage of SDG diastereomers **15a** and **15b** to give (+)-secoisolariciresinol **14a** and (–)-secoisolariciresinol **14b**, respectively (47). *Abbreviation:* See Figure 2.2.

The approach undertaken was thus twofold: (i) to correlate SDG-HMG oligomer formation with flaxseed developmental stage, including analyzing the effects of the uptake and metabolism of [U-^{14}C]-phenylalanine **26**, the presumed precursor of sec-oisolariciresinol **14**; and (ii) to employ both an enzymatic and a molecular biological strategy to obtain the corresponding genes and enzymes involved in the pathway(s) to the SDG-HMG oligomers **16–20**.

SDG-HMG Oligomer Formation and Uptake/Metabolism of [U-^{14}C]-Phenylalanine in Developing Flaxseed

To begin our studies on flaxseed lignan biosynthesis, it was first important to establish whether SDG **15** and the SDG-HMG ester-linked oligomers **16–20** accumulating in flaxseed were formed *de novo* in this tissue or, alternatively, were transported from other parts of the plant. This question was addressed via *in vivo* administration of the presumed lignan precursor, L-[U-^{14}C]-phenylalanine **26**, to stems, leaves, and individual developing flaxseed capsules at distinct developmental stages (47). Following uptake and metabolism by the plant for periods up to 16 h, the aqueous ethanol extracts of each tissue, before and after base hydrolysis, were analyzed for free and alkali releasable [^{14}C]-SDG **15**. In this way, it was established that neither stem nor leaf tissue converted L-[U-^{14}C]-Phe **26** into [^{14}C]-SDG **15**, whereas the individual flaxseed capsules were able to do so depending upon the developmental stage examined (47).

From a developmental perspective, flaxseed maturation is of considerable interest. Flaxseed formation is not simultaneously synchronized, and thus in any given plant the seeds (and their developing capsules) are all at distinct developmental stages during the growth of each plant's corymbrose inflorescence. For this reason, we evaluated different possible methodologies to segregate the seed at different developmental stages, this being conveniently achieved by utilizing sizing screens. In this way, five developmental stages (stages 1–5, Fig. 2.13A) could be segregated, as well as a sixth representing the fully mature seed with a completely formed embryo. Flaxseed at the last (sixth) developmental stage has a brown seed coat, whereas the earlier stages including stage 5 are of a translucent vanilla color with visible cotyledons. Using this approach, analysis of the aqueous ethanolic solubles of each developmental stage established that SDG **15** itself was barely detectable at any point, but was releasable upon alkali treatment from stage 2 onwards (Fig. 2.13B).

The metabolism of [U-^{14}C]-phenylalanine **26** at stage 4 of developing flaxseed tissue also established that most of the radioactivity detected was present in the various oligomeric SDG-HMG (e.g., **16–20**) components, whose partial alkaline hydrolysis again mainly afforded the two SDG diastereomeric forms **15a-15b** (peaks 1a and 1b; Fig. 2.11), together with smaller amounts of the two SDG-HMG–containing components **17** and **16** (peaks 3 and 2). In addition to these components, and depending upon the developmental stage, we also found varying amounts of *E*-coniferyl alcohol **29**, *E*-coniferin **40** (with the glycosidic linkage at the phenolic group and not at C-9),

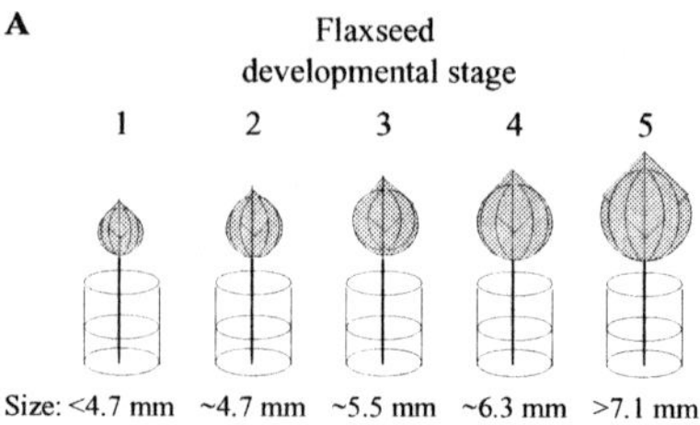

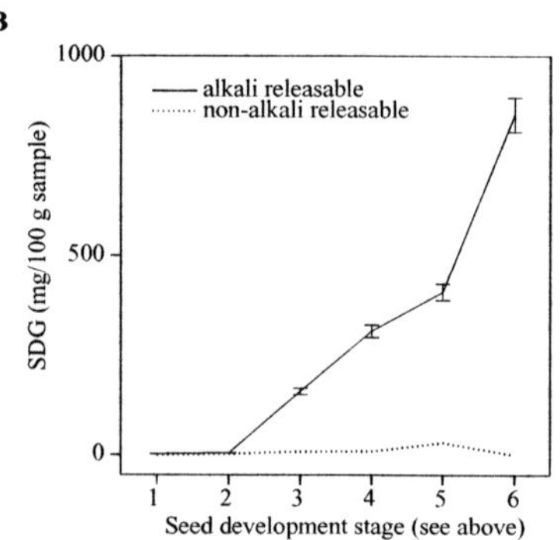

Fig. 2.13. Representative depiction of L-[U-^{14}C]-phenylalanine **26** administration experiment to developing flaxseed. (**A**) Capsules were placed separately in wells of a Costar® ELISA plate. Capsules are not drawn to scale, and are illustrated only to show the five different stages of seed development; (**B**) a plot of SDG **15** levels, both free and releasable upon alkali hydrolysis, as a function of flaxseed development stage. Relative SDG **15** levels for seed developmental stage 6 represent those in fully mature flaxseeds. *Abbreviation:* See Figure 2.2. *Source:* Adapted from Ford *et al.* (47).

as well as *p*-coumaric acid **41**, *p*-coumaric acid glucoside **42**, and ferulic acid glucoside **43**, respectively (47). On the other hand, neither pinoresinol **27**, lariciresinol **28**, nor secoisolariciresinol **14** were present in other than trace quantities, suggesting that they are rapidly metabolized when formed.

These studies thus established that flaxseed tissue was directly involved in the biosynthesis of the lignan component of the SDG-HMG oligomers (e.g., **16–20**), with [U-^{14}C]-Phe **26** being metabolized *de novo* to afford the secoisolariciresinol **14** moiety. These data, when taken together, thus permitted formulation of a possible biosynthetic scheme to the SDG-HMG oligomers (e.g., **16–20**) as follows: initial formation of secoisolariciresinol **14** via stereoselective coupling of two *E*-coniferyl alcohol **29** moieties, followed by subsequent NADPH-dependent reductive steps to afford lariciresinol **28** and secoisolariciresinol **14**, respectively. These could then undergo sequential glucosylation at C-9 and C-9′, to afford SDG **15**, with subsequent introduction of HMG **39** moieties via participation of acyltransferase(s), thereby generating the corresponding oligomers such as **16–20** (Fig. 2.2).

Enzymology and Molecular Biology of 8-8′ Linked Lignan Formation in Developing Flaxseed

Stereoselective Coupling and Dirigent Proteins. Based on the above, flax genomic DNA was obtained, in order to employ a PCR-guided strategy to obtain putative dirigent genes, this being employed because the dirigent genes lack intron. This used primers designed on consensus sequences for 18 previously identified (+)-pinoresinol forming dirigent proteins (7), and resulted in two putative dirigent genes being obtained (71). One of these, *Flax–1* was expressed in *Drosophila melanogaster* cells in functional form, and was demonstrated capable of converting *E*-coniferyl alcohol **29** into (+)-pinoresinol **27a**, in the presence of an oxidase/oxidant. Its occurrence in flaxseed tissue may thus explain the formation of the small amounts of (–)-secoisolariciresinol **14b** observed, following enzymatic cleavage of the glucosyl moieties attached to SDG **15** [i.e., through formation of (–)-secoisolariciresinol **14b** as in Fig. 2.5].

On the other hand, preliminary purification of proteins present in flaxseed tissue afforded a fraction engendering preferential formation of (–)-pinoresinol **27b** over that of the (+)-antipode **27a** (~2:1). Work is underway to obtain the corresponding encoding gene, and to compare it with the known (+)-pinoresinol-forming systems. In any event, these data reveal that two distinct proteins are involved in (+)- and (–)-pinoresinol **27a/b** formation in developing flaxseed tissue.

Enantiospecificity of Pinoresinol/Lariciresinol Reductases. Further evidence for distinct pathways to the (+)- and (–)-forms of secoisolariciresinol **14** in flaxseed was also obtained from studies of the enantiospecificity of putative flaxseed PLR. In this regard, incubation of (±)-[3,3′-O^{14}CH$_3$]-pinoresinols **27a/b** with partially purified flaxseed soluble protein extracts, in the presence of NADPH, gave interesting results; the (+)-form of [3,3′-O^{14}CH$_3$]-pinoresinol **27a** was converted to (+)-[3,3′-O^{14}CH$_3$]-lariciresinol **28a**, but not to (–)-secoisolariciresinol **14b**. On the other hand, the (–)-antipode of [3,3′-O^{14}CH$_3$]-pinoresinol **27b** was converted to (+)-[3,3′-O^{14}CH$_3$]-secoisolariciresinol **14a**, presumably via (–)-lariciresinol **28b** (which was not detected) (71). The results are summarized in Figure 2.14: the top panels (A,B) depict chiral separation of the (+)- and (–)-forms of lariciresinol **28** and secoisolariciresinol **14**, respectively, whereas the lower panels (C,D) indicate that radiolabel was only associated with the (+)-enantiomer in each case. These data thus provide good preliminary evidence for two distinct pathways to the (+)- and (–)-antipodes of secoisolariciresinol **14a/14b**, respectively.

In further support of this contention, a gene encoding a PLR homologue (*plr–Lu*) was obtained from a cDNA library constructed from 2-wk-old developing flaxseed using a PCR-guided strategy (71). The gene was obtained using degenerate primers, designed on the basis of amino acid sequence consensus regions of previously cloned PLR from *F. intermedia* (62) and *T. plicata* (63). The gene (*plr–Lu*) had an ORF with 936 base pairs, whose encoded protein displays 73.8% similarity and 60.5% identity at the amino acid level to that of *Forsythia, plr-Fi1*. Assays of the recombinant protein

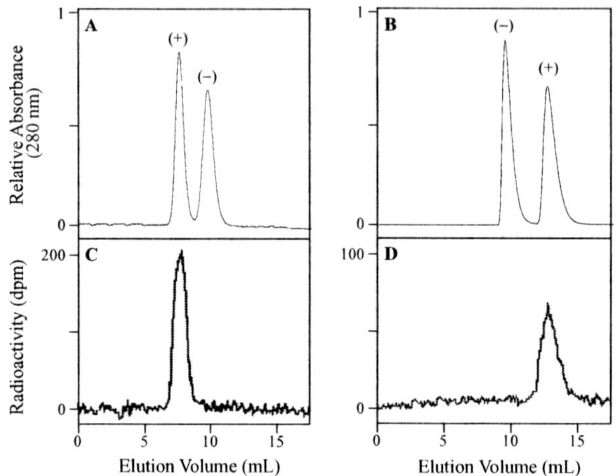

Fig. 2.14. Chiral column HPLC analysis of lariciresinols **28** and secoisolaricresinols **14**. Synthetic (**A**) (±)-lariciresinols **28a/b** and (**B**) (±)-secoisolariciresinols **14a/b**; (**C**) (+)-[3,3'-O^{14}CH$_3$]-lariciresinol **28a** and (**D**) (+)-[3,3'-O^{14}CH$_3$]-secoisolariciresinol **14a** formed after incubation of (±)-[3,3'-O^{14}CH$_3$]-pinoresinols **27a/b** with partially purified flaxseed soluble protein extracts, in the presence of NADPH.

plr–Lu when overexpressed in *E. coli*, revealed characteristics that were not observed with the previously cloned *Forsythia* and western red cedar reductases. When incubated with racemic (±)-pinoresinols **27a/b**, in the presence of NADPH, no product formation (lariciresinol **28** nor secoisolariciresinol **14**) occurred. Both (+)- and (–)-antipodes of pinoresinol **27a/b** were then separated by chiral HPLC, and reincubation of each separately revealed that (–)-pinoresinol **27b** was now converted into (–)-lariciresinol **28b**, whereas the (+)-antipode **27a** did not serve as substrate. In neither case was secoisolariciresinol **14** formation noted, nor was it formed when (–)-lariciresinol **28b** was used as an enantiomerically pure substrate. These data thus suggest a level of control over the reductive steps not previously observed in the *Forsythia* and/or western red cedar examples. It can be provisionally envisaged that distinct PLRs are required for each reductive step, rather than being engendered by bifunctional enzymes as before. Work is underway to complete defining the enzymology of the reductive processes utilizing pinoresinol **27** and lariciresinol **28**.

SDG and SDG-HMG Oligomer Formation. SDG **15** in flaxseed can be formed from secoisolariciresinol **14** by glycosylation of the hydroxyl group at the C-9, C-9' positions. Assaying partially purified enzyme fractions, in the presence of UDP[1-^{3}H] glucose and (±)-secoisolariciresinol **14**, resulted in formation of [^{3}H]-SDG **15**, the identity of which was confirmed by MALDI-TOF, ^{1}H and ^{13}C NMR spectroscopic analyses. Work is underway to complete the purification of the glucosyltransferase(s) involved.

In the future, it will also be of interest to establish how the different oligomeric forms of SDG-HMG are biosynthesized, including the types and numbers of acyltransferases involved. This is the subject of continuing work, as most effort has been dedicated thus far in establishing the biochemical pathways to the lignan component of the SDG-HMG oligomers.

Summary

Over the last decade or so, many of the basic transformations involved in secoisolariciresinol **14** and matairesinol **13** biosynthesis have been elucidated; this includes identifying proteins and genes involved in stereoselective coupling of monolignols, and subsequent enantiospecific reductive and dehydrogenative transformations to afford lariciresinol **28**, secoisolariciresinol **14**, and matairesinol **13**, respectively. With the proteins, enzymes, and genes on hand, much of our knowledge of lignan biosynthesis in various species (including flaxseed) has thus now reached a good level of understanding and clarification. Further studies in flaxseed lignan biosynthesis will be needed to complete the identification of the genes/proteins involved in generating (–)-pinoresinol **27b** formation and the subsequent transformations that occur to give mainly (+)-secoisolariciresinol **14a**. A further scientific goal will be to establish, at the most detailed mechanistic level, how the various SDG-HMG oligomers are formed including identification of the enzymes and genes involved. This is a particularly interesting question, given that there is no precedence in the literature for HMG-containing copolymers/oligomers with any other natural product.

Finally, it can be anticipated that our growing knowledge of the biosynthetic pathways to the SDG-HMG oligomers **16–20** and to podophyllotoxin **1** will be of importance in furthering their utility in offsetting the various human diseases that they are either effective or protective against. Certainly, the knowledge of these processes, and the ability to modulate or introduce such pathways into various organisms, will increase their availability for human use.

Acknowledgments

We gratefully acknowledge the National Science Foundation (MCB9976684), the United States Department of Agriculture (99-35103-8037), the United States Department of Energy (DE FG03-97ER20259), and McIntire Stennis for generous support of this study, as well as the Lewis B. and Dorothy Cullman and G. Thomas Hargrove Center for Land Plant Adaptation Studies.

References

1. Thompson, L.U., M.M. Seidl, S.E. Rickard, L.J. Orcheson, and H.H.S. Fong, Antitumorigenic Effect of a Mammalian Lignan Precursor from Flaxseed, *Nutr. Cancer* 26:159–165 (1996).
2. Ayres, D.C., and J.D. Loike, *Chemistry and Pharmacology of Natural Products. Lignans: Chemical, Biological and Clinical Properties*, Cambridge University Press, Cambridge, UK, 1990.

3. Lewis, N.G., and L.B. Davin, Lignans: Biosynthesis and Function, in *Comprehensive Natural Products Chemistry*, edited by D.H.R. Barton, K. Nakanishi, and O. Meth-Cohn, Elsevier Science, London, 1999, Vol. 1, pp. 639–712.

4. Davin, L.B., D.L. Bedgar, T. Katayama, and N.G. Lewis, On the Stereoselective Synthesis of (+)-Pinoresinol in *Forsythia suspensa* from Its Achiral Precursor, Coniferyl Alcohol, *Phytochemistry 31*:3869–3874 (1992).

5. Davin, L.B., and N.G. Lewis, Lignin and Lignan Biochemical Pathways in Plants: An Unprecedented Discovery in Phenolic Coupling, *An. Acad. Bras. Ci. 67(Suppl. 3)*:363–378 (1995).

6. Davin, L.B., H.B. Wang, A.L. Crowell, D.L. Bedgar, D.M. Martin, S. Sarkanen, and N.G. Lewis, Stereoselective Bimolecular Phenoxy Radical Coupling by an Auxiliary (Dirigent) Protein Without an Active Center, *Science 275*:362–366 (1997).

7. Gang, D.R., M.A. Costa, M. Fujita, A.T. Dinkova-Kostova, H.-B. Wang, V. Burlat, W. Martin, S. Sarkanen, L.B. Davin, and N.G. Lewis, Regiochemical Control of Monolignol Radical Coupling: A New Paradigm for Lignin and Lignan Biosynthesis, *Chem. Biol. 6*:143–151 (1999).

8. Halls, S.C., and N.G. Lewis, Secondary and Quaternary Structures of the (+)-Pinoresinol-Forming Dirigent Protein, *Biochemistry 41*:9455–9461 (2002).

9. Kim, M.K., J.-H. Jeon, M. Fujita, L.B. Davin, and N.G. Lewis, The Western Red Cedar (*Thuja plicata*) 8-8′ DIRIGENT Family Displays Diverse Expression Patterns and Conserved Monolignol Coupling Specificity, *Plant Mol. Biol. 49*:199–214 (2002).

10. Kim, M.K., J.-H. Jeon, L.B. Davin, and N.G. Lewis, Monolignol Radical-Radical Coupling Networks in Western Red Cedar and *Arabidopsis* and Their Evolutionary Implications, *Phytochemistry 61*:311–322 (2002).

11. Moinuddin, S.G.A., S. Hishiyama, M.-H. Cho, L.B. Davin, and N.G. Lewis, Synthesis and Chiral HPLC Analysis of the Antiviral Lignans, Larreatricins, 8′-*epi*-Larreatricins, 3,3′-Didemethoxyverrucosins and *meso*-3,3′-Didemethoxynectandrin B in the Creosote Bush (*Larrea tridentata*), *Org. Biomol. Chem. 1* (in press) (2003).

12. Wang, C.-Z., L.B. Davin, and N.G. Lewis, Stereoselective Phenolic Coupling in *Blechnum spicant*: Formation of 8-2′ Linked (–)-*cis*-Blechnic, (–)-*trans*-Blechnic and (–)-Brainic Acids, *J. Chem. Soc. Chem. Commun.* 113–114 (2001).

13. Davin, L.B., C.-Z. Wang, G.L. Helms, and N.G. Lewis, [^{13}C]-Specific Labeling of 8-2′ Linked (–)-*cis*-Blechnic, (–)-*trans*-Blechnic and (–)-Brainic Acids in the Fern *Blechnum spicant*, *Phytochemistry 62*:501–511 (2003).

14. Beutner, K.R., and G. von Krogh, Current Status of Podophyllotoxin for the Treatment of Genital Warts, *Semin. Dermatol. 9*:148–151 (1990).

15. Canel, C., R.M. Moraes, F.E. Dayan, and D. Ferreira, Podophyllotoxin, *Phytochemistry 54*:115–120 (2000).

16. O'Dwyer, P.J., B. Leyland-Jones, M.T. Alonso, S. Marsoni, and R.E. Wittes, Etoposide (VP-16-213). Current Status of an Active Anticancer Drug, *New Eng. J. Med. 312*:692–700 (1985).

17. Schacter, L., Etoposide Phosphate: What, Why, Where and How? *Semin. Oncol. 23(6 Suppl. 13)*:1–7 (1996).

18. Williams, S.D., R. Birch, L.H. Einhorn, L. Irwin, F.A. Greco, and P.J. Loehrer, Treatment of Disseminated Germ Cell Tumors with Cisplatin, Bleomycin and Either Vinblastine or Etoposide, *New Engl. J. Med. 316*:1435–1440 (1987).

19. Young, R.C., Etoposide in the Treatment of non-Hodgkin's Lymphomas, *Semin. Oncol. 19 (Suppl. 13)*:19–25 (1992).

20. Train, P., J.R. Henrick, and W.A. Archer, Medicinal Uses of Plants by Indian Tribes of Nevada, U.S. Dept. Agric., Washington, D.C., 1941.

21. Craigo, J., M. Callahan, R.C.C. Huang, and A.L. DeLucia, Inhibition of Human Papillomavirus Type 16 Gene Expression by Nordihydroguaiaretic Acid Plant Lignan Derivatives, *Antiv. Res. 47*:19–28 (2000).

22. Gnabre, J.N., J.N. Brady, D.J. Clanton, Y. Ito, J. Dittmer, R.B. Bates, and R.C.C. Huang, Inhibition of Human Immunodeficiency Virus Type 1 Transcription and Replication by DNA Sequence-Selective Plant Lignans, *Proc. Natl. Acad. Sci., USA 92*:11239–11243 (1995).

23. Hwu, J.R., W.N. Tseng, J. Gnabre, P. Giza, and R.C.C. Huang, Antiviral Activities of Methylated Nordihydroguaiaretic Acids. 1. Synthesis, Structure Identification, and Inhibition of Tat-Regulated HIV Transactivation, *J. Med. Chem. 41*:2994–3000 (1998).

24. McDonald, R.W., W. Bunjobpon, T. Liu, S. Fessler, O.E. Pardo, I.K. Freer, M. Glaser, M.J. Seckl, and D.J. Robins, Synthesis and Anticancer Activity of Nordihydroguaiaretic Acid (NDGA) and Analogues, *Anticancer Drug Des. 16*:261–270 (2001).

25. Oliveto, E.P., Nordihydroguaiaretic Acid: A Naturally Occurring Antioxidant, *Chem. Ind. 17*:677–679 (1972).

26. Hirata, F., K. Fujita, Y. Ishikura, K. Hosoda, T. Ishikawa, and H. Nakamura, Hypocholesterolemic Effect of Sesame Lignan in Humans, *Atherosclerosis 122*:135–136 (1996).

27. Valsaraj, R., P. Pushpangadan, U.W. Smitt, A. Adsersen, S.B. Christensen, A. Sittie, U. Nyman, C. Nielsen, and C.E. Olsen, New Anti-HIV1, Antimalarial, and Antifungal Compounds from *Terminalia bellerica*, *J. Nat. Prod. 60*:739–742 (1997).

28. Adlercreutz, H., T. Fotsis, R. Heikkinen, J.T. Dwyer, M. Woods, B.R. Goldin, and S.L. Gorbach, Excretion of the Lignans Enterolactone and Enterodiol and of Equol in Omnivorous and Vegetarian Postmenopausal Women and in Women with Breast Cancer, *Lancet 2(8311)*:1295–1299 (1982).

29. Setchell, K.D.R., A.M. Lawson, F.L. Mitchell, H. Adlercreutz, D.N. Kirk, and M. Axelson, Lignans in Man and in Animal Species, *Nature 287*:740–742 (1980).

30. Stitch, S.R., J.K. Toumba, M.B. Groen, C.W. Funke, J. Leemhuis, J. Vink, and G.F. Woods, Excretion, Isolation and Structure of a New Phenolic Constituent of Female Urine, *Nature 287*:738–740 (1980).

31. Axelson, M., J. Sjovall, B.E. Gustafsson, and K.D. Setchell, Origin of Lignans in Mammals and Identification of a Precursor from Plants, *Nature 298*:659–660 (1982).

32. Borriello, S.P., K.D.R. Setchell, M. Axelson, and A.M. Lawson, Production and Metabolism of Lignans by the Human Fecal Flora, *J. Appl. Bacteriol. 58*:37–43 (1985).

33. Thompson, L.U., P. Robb, M. Serraino, and F. Cheung, Mammalian Lignan Production from Various Foods, *Nutr. Cancer 16*:43–52 (1991).

34. Rickard, S.E., L.J. Orcheson, M.M. Seidl, L. Luyengi, H.H.S. Fong, and L.U. Thompson, Dose-Dependent Production of Mammalian Lignans in Rats and *in vitro* from the Purified Precursor Secoisolariciresinol Diglycoside in Flaxseed, *J. Nutrition 126*:2012–2019 (1996).

35. Setchell, K.D.R., A.M. Lawson, E. Conway, N.F. Taylor, D.N. Kirk, G. Cooley, R.D. Farrant, S. Wynn, and M. Axelson, The Definitive Identification of the Lignans *trans*-2,3-*bis*(3-hydroxybenzyl)-γ-butyrolactone and 2,3-*bis*(3-hydroxybenzyl)butane-1,4-diol in Human and Animal Urine, *Biochem. J. 197*:447–458 (1981).

36. Ford, J.D., L.B. Davin, and N.G. Lewis, Plant Lignans and Health: Cancer Chemoprevention and Biotechnological Opportunities, in *Plant Polyphenols 2: Chemistry*

and Biology, edited by G.G. Gross, R.W. Hemingway, and T. Yoshida, Kluwer Academic / Plenum Publishers, New York, 1999, pp. 675–694.

37. Zamir, L.O., R. Tiberio, M. Fiske, A. Berry, and R.A. Jensen, Enzymatic and Nonenzymatic Dehydration Reactions of L-Arogenate, *Biochemistry 24*:1607–1612 (1985).

38. Bakke, J.E., and H.J. Klosterman, A New Diglucoside from Flaxseed, *Proc. N. Dak. Acad. Sci. 10*:18–22 (1956).

39. Thompson, L.U., S.E. Rickard, L.J. Orcheson, and M.M. Seidl, Flaxseed and Its Lignan and Oil Components Reduce Mammary Tumor Growth at a Late Stage of Carcinogenesis, *Carcinogenesis 17*:1373–1376 (1996).

40. Jenab, M., and L.U. Thompson, The Influence of Flaxseed and Lignans on Colon Carcinogenesis and β-Glucuronidase Activity, *Carcinogenesis 17*:1343–1348 (1996).

41. Ingram, D., K. Sanders, M. Kolybaba, and D. Lopez, Case-Control Study of Phyto-oestrogens and Breast Cancer, *Lancet 350*:990–994 (1997).

42. Pietinen, P., K. Stumpf, S. Männistö, V. Kataja, M. Uusitupa, and H. Adlercreutz, Serum Enterolactone and Risk of Breast Cancer: A Case-Control Study in Eastern Finland, *Cancer Epidemiol. Biomark. Prev. 10*:339–344 (2001).

43. Prasad, K., Oxidative Stress as a Mechanism of Diabetes in Diabetic BB Prone Rats: Effect of Secoisolariciresinol Diglucoside (SDG), *Mol. Cell. Biochem. 209*:89–96 (2000).

44. Prasad, K., Suppression of Phosphoenolpyruvate Carboxykinase Gene Expression by Secoisolariciresinol Diglucoside (SDG), a New Antidiabetic Agent, *Int. J. Angiol. 11*:107–109 (2002).

45. Prasad, K., Reduction of Serum Cholesterol and Hypercholesterolemic Atherosclerosis in Rabbits by Secoisolariciresinol Diglucoside Isolated from Flaxseed, *Circulation 99*:1355–1362 (1999).

46. Clark, W.F., A. Parbtani, M.W. Huff, E. Spanner, H. de Salis, I. Chin-Yee, D.J. Philbrick, and B.J. Holub, Flaxseed: A Potential Treatment for Lupus nephritis, *Kidney Int. 48*:475–480 (1995).

47. Ford, J.D., K.-S. Huang, H.-B. Wang, L.B. Davin, and N.G. Lewis, Biosynthetic Pathway to the Cancer Chemopreventive Secoisolariciresinol Diglucoside-Hydroxymethyl Glutaryl Ester-Linked Lignan Oligomers in Flax (*Linum usitatissimum*) Seed, *J. Nat. Prod. 64*:1388–1397 (2001).

48. Kamal-Eldin, A., N. Peerlkamp, P. Johnsson, R. Andersson, R.E. Andersson, L.N. Lundgren, and P. Åman, An Oligomer from Flaxseed Composed of Secoisolariciresinol Diglucoside and 3-Hydroxy-3-Methyl Glutaric Acid Residues, *Phytochemistry 58*:587–590 (2001).

49. Xia, Z.Q., M.A. Costa, J. Proctor, L.B. Davin, and N.G. Lewis, Dirigent-Mediated Podophyllotoxin Biosynthesis in *Linum flavum* and *Podophyllum peltatum*, *Phytochemistry 55*:537–549 (2000).

50. Kitagawa, S., S. Nishibe, R. Benecke, and H. Thieme, Phenolic Compounds from *Forsythia* Leaves (II), *Chem. Pharm. Bull. 36*:3667–3670 (1988).

51. Rahman, M.M.A., P.M. Dewick, D.E. Jackson, and J.A. Lucas, Lignans of *Forsythia intermedia*, *Phytochemistry 29*:1971–1980 (1990).

52. Gardner, J.A.F., G.M. Barton, and H. MacLean, The Polyoxyphenols of Western Red Cedar (*Thuja plicata* Donn.). I. Isolation and Preliminary Characterization of Plicatic Acid, *Can. J. Chem. 37*:1703–1709 (1959).

53. Gardner, J.A.F., E.P. Swan, S.A. Sutherland, and H. MacLean, The Polyoxyphenols of Western Red Cedar (*Thuja plicata* Donn.). III. Structure of Plicatic Acid, *Can. J. Chem. 44*:52–58 (1966).

54. MacLean, H., and K. Murakami, Lignans of Western Red Cedar (*Thuja plicata* Donn.). IV. Thujaplicatin and Thujaplicatin Methyl Ether, *Can. J. Chem. 44*:1541–1545 (1966).

55. Takagi, K., M. Kimura, M. Harada, and Y. Otsuka, *Pharmacology of Medicinal Herbs in East Asia*, Nanzando, Tokyo, 1982, pp. 187–188.

56. Ozaki, Y., J. Rui, Y. Tang, and M. Satake, Antiinflammatory Effect of *Forsythia suspensa* Vahl and Its Active Fraction, *Biol. Pharm. Bull. 20*:861–864 (1997).

57. Chan-Yeung, M., P.C. Giclas, and P.M. Henson, Activation of Complement by Plicatic Acid, the Chemical Compound Responsible for Asthma Due to Western Red Cedar (*Thuja plicata*), *J. Allergy Clin. Immunol. 65*:333–337 (1980).

58. Giclas, P.C., Effect of Plicatic Acid of Human Serum Complement Includes Interference with C1 Inhibitor Function, *J. Immunol. 129*:168–172 (1982).

59. Paré, P.W., H.-B. Wang, L.B. Davin, and N.G. Lewis, (+)-Pinoresinol Synthase: A Stereoselective Oxidase Catalysing 8,8′-Lignan Formation in *Forsythia intermedia*, *Tetrahedron Lett. 35*:4731–4734 (1994).

60. Kwon, M., L.B. Davin, and N.G. Lewis, *In situ* Hybridization and Immunolocalization of Lignan Reductases in Woody Tissues: Implications for Heartwood and Other Forms of Vascular Preservation, *Phytochemistry 57*:899–914 (2001).

61. Jiao, Y., L.B. Davin, and N.G. Lewis, Furanofuran Lignan Metabolism as a Function of Seed Maturation in *Sesamum indicum*: Methylenedioxy Bridge Formation, *Phytochemistry 49*:387–394 (1998).

62. Dinkova-Kostova, A.T., D.R. Gang, L.B. Davin, D.L. Bedgar, A. Chu, and N.G. Lewis, (+)-Pinoresinol/(+)-Lariciresinol Reductase from *Forsythia intermedia*: Protein Purification, cDNA Cloning, Heterologous Expression and Comparison to Isoflavone Reductase, *J. Biol. Chem. 271*:29473–29482 (1996).

63. Fujita, M., D.R. Gang, L.B. Davin, and N.G. Lewis, Recombinant Pinoresinol/Lariciresinol Reductases from Western Red Cedar (*Thuja plicata*) Catalyze Opposite Enantiospecific Conversions, *J. Biol. Chem. 274*:618–627 (1999).

64. Chu, A., A. Dinkova, L.B. Davin, D.L. Bedgar, and N.G. Lewis, Stereospecificity of (+)-Pinoresinol and (+)-Lariciresinol Reductases from *Forsythia intermedia*, *J. Biol. Chem. 268*:27026–27033 (1993).

65. Umezawa, T., L.B. Davin, and N.G. Lewis, Formation of Lignans, (–)-Secoisolariciresinol and (–)-Matairesinol with *Forsythia intermedia* Cell-Free Extracts, *J. Biol. Chem. 266*:10210–10217 (1991).

66. Xia, Z.-Q., M.A. Costa, H.C. Pélissier, L.B. Davin, and N.G. Lewis, Secoisolariciresinol Dehydrogenase Purification, Cloning and Functional Expression: Implications for Human Health Protection, *J. Biol. Chem. 276*:12614–12628 (2001).

67. Molog, G.A., U. Empt, S. Kuhlmann, W. van Uden, N. Pras, A.W. Alfermann, and M. Petersen, Deoxypodophyllotoxin 6-Hydroxylase, a Cytochrome P450 Monooxygenase from Cell Cultures of *Linum flavum* Involved in the Biosynthesis of Cytotoxic Lignans, *Planta 214*:288–294 (2001).

68. Klosterman, H.J., and C.O. Clagett, A New Carbohydrate from Flaxseed, *Proc. N. Dak. Acad. Sci. 8*:20 (1954).

69. Klosterman, H.J., and F. Smith, The Isolation of β-Hydroxy-β-Methylglutaric Acid from the Seed of Flax (*Linum usitatissimum*), *J. Amer. Chem. Soc. 76*:1229–1230 (1954).

70. Bambagiotti-Alberti, M., S.A. Coran, C. Ghiara, G. Moneti, and A. Raffaelli, Investigation of Mammalian Lignan Precursors in Flax Seed: First Evidence of Secoisolariciresinol Diglucoside in Two Isomeric Forms by Liquid Chromatography/Mass Spectrometry, *Rapid Commun. Mass Spectrom.* 8:929–932 (1994).
71. Ford, J.D., Cancer Chemopreventive Flax Seed Lignans: Delineating the Metabolic Pathway(s) to the SDG-HMG Ester-Linked Polymer, Ph.D. Thesis, Washington State University, 2001.

Chapter 3

Dietary Sources and Metabolism of α-Linolenic Acid

Stephen C. Cunnane

Department of Nutritional Sciences, Faculty of Medicine, University of Toronto, Toronto, M5S 3E2, Canada

Introduction

Some major issues and recent advances in the field of the basic metabolic and nutritional importance of α-linolenic acid are reviewed in this chapter. Although consumer and scientific interest in the nutritional and health attributes of traditional high α-linolenic acid flaxseed has increased in the past decade, there are still relatively few papers reporting well-controlled experiments on this subject. Those that do exist are the primary material for other chapters in this book. Therefore, much of this chapter refers to studies with α-linolenic acid or flaxseed oil, and there will be infrequent reference to studies in which flaxseed *per se* was utilized. Health attributes will be noted in brief because they are covered in other chapters of this volume or in recent reviews (1,2). Published material on α-linolenic acid metabolism that was recently reviewed and published elsewhere (1) will not be rereviewed here in any detail.

α-Linolenic acid (18:3n-3) derives most of its importance in nutrition and health because it is the "parent" n-3 polyunsaturated fatty acid (PUFA). It is the main component of traditional high α-linolenic acid flaxseed, comprising 22–24% of the weight of the mature seed. This makes traditional flaxseed the richest commonly accessible food source of α-linolenic acid. Other oilseeds, like perilla, and edible plants such as purslane are also rich in α-linolenic acid (Table 3.1), but they are not as widely available in the food supply as flaxseed.

Historical Review

In the late 1920s researchers found that total absence of fat from the diet led to reproducible symptoms in rodents, including growth retardation, increased water loss across the skin, and scaling of the skin. At the time, methods for fatty acid analysis were relatively crude compared with modern day capillary gas chromatography. Nevertheless, careful dietary studies and analysis of tissue composition soon demonstrated that three fatty acids were involved in preventing or correcting fat deficiency symptoms: linoleic acid (18:2n-6), arachidonic acid (20:4n-6), and α-linolenic acid. The distinction between α-linolenic acid and linoleic acid was soon recognized as arising from their different number and location of double bonds. Likewise it was soon established that animals on purified diets could not make linoleic acid or α-

TABLE 3.1

α-Linolenic Acid as a Percentage of Total Fatty Acids in Foods

Food	Total fatty acids (%)	Food	Total fatty acids (%)
Fats and oils			
Traditional flaxseed	45–60	Corn	1–2
Perilla	50	High oleic sunflower	<1
Mustardseed	13	Safflower	<1
Canola	8–12	Olive	<1
Soybean	5–7	Cottonseed	<1
Margarines	0–7	Palm	<1
Low α-linolenic flaxseed	1–2	Sunflower	<1
Lard	1–2	Peanut	<1
Pulses, grains, nuts and cereals			
Beans	30–50	Rye	3–4
Lentils	16	Oats	2
Walnuts	11	Rice	1
Barley	6	Almonds	< 1
Wheat	4	Peanuts	< 1
Vegetables and fruits			
Purslane	55	Cucumbers	26
Turnip	58	Bananas	22
Mushrooms	55	Potatoes	17
Green beans	40	Green peppers	12
Spinach	45	Apples	5–11
Broccoli	40	Green peas	3–4
Eggs and dairy products			
Cow's milk	1–2	Traditional eggs	<1
		Eggs enriched with n-3	3–5
Meat			
Rabbit	10	Pork, ham, bacon	<2
Lamb	2–3	Chicken	<1
Beef	1–2		
Fish			
Halibut	3–4	Molluscs	1–2
Herring	1–2	Tuna	< 1
Mackerel	1–2	Cod	< 1
Sardines	1–2	Haddock	< 1
Shellfish	1–5	Salmon	< 1

Source: Modified with permission from Cunnane (1), Delion *et al.* (76).

linolenic acid. Initially it was not known that arachidonic acid was derived from linoleic acid so these three fatty acids became identified as "essential fatty acids" or vitamin F.

Once it was learned that arachidonic acid was synthesized from linoleic acid, linoleic became recognized as the "parent" n-6 PUFA and the single n-6 PUFA truly essential in the diet. Strictly speaking, arachidonic acid was then delisted as an essential fatty acid. In practice, arachidonic acid and all the other derived n-3 and n-6 PUFA are still frequently called essential fatty acids even though this is not consistent with the fact that they can be synthesized from dietary linoleic or α-linolenic acid. The problem is that linoleic acid and α-linolenic acid are not *always* able to meet the requirements for normal growth and development: that is, despite the existence of the biochemical pathway converting linoleic acid or α-linolenic acid to its respective long chain PUFA, there is not always adequate *capacity* of that pathway (see section on Synthesis of Long Chain n-3 Polyunsaturated Fatty Acids). Hence, arachidonic acid and other long chain n-6 and n-3 PUFA are conditionally essential or, as recently proposed, "conditionally indispensable" (3,4), especially during early development (see Chapter 8). However, at other times (i.e., in healthy young adults) there is little or no good evidence that long chain PUFA are dietarily essential. It is reasonable to speculate that as one moves into old age, a dietary source of long chain PUFA may once again become more important (see section on Perspective).

In fat-deficient rats, α-linolenic acid was shown not to restore growth quite as well as linoleic acid and was much less effective in preventing or correcting the dermal symptoms of fat deficiency (5). Despite the fact that rodents could not make α-linolenic acid, it was difficult to consistently demonstrate symptoms of its deficiency when linoleic acid or other n-6 PUFA were in the diet. Hence, the status of α-linolenic acid as an essential fatty acid was questioned. Long-term studies in rats eventually clearly demonstrated the now classical symptoms of n-3 PUFA deficiency (mainly visual and learning deficits; see section on Clinical n-3 PUFA Deficiency). Because of ethical considerations, including the length of these studies, they were much harder to do in humans. Hence, the dietary essentiality of n-3 PUFA, particularly α-linolenic acid, in humans became controversial and was doubted by many, a situation that persisted well into the 1980s. The human dietary requirement for α-linolenic acid is still really only extrapolated from rodent studies.

One important oversight in the dietary methodology used to investigate the dietary requirement for n-6 and n-3 PUFA occurred early in research on this subject and, for most studies employing a "fat deficiency" model, persists to this day. A cardinal rule of research methodology involving nutritional deficiencies is that a diet complete in all known nutrients except the one under study be provided. If this requires that the diet have purified ingredients, so be it. This is the way vitamin, mineral, and amino acid requirements have been established. Indeed this is the way the requirement for n-3 PUFA was established. The fact is that it was difficult to consistently demonstrate symptoms of n-3 PUFA deficiency in animals or humans when other fat sources including n-6 PUFA were present in the diet. The converse, however, was never done until five years ago (6); that is, the effects of n-6 PUFA deficiency in the presence of other fatty acids, *particularly* n-3 PUFA were evaluated. This is an important point; the dietary requirement for linoleic acid or for n-6 PUFA in general is based on prevention or correction of symptoms of *combined* n-3 and n-6 PUFA deficiency (i.e., classical essential fatty acid deficiency). However, there is good reason to believe that the requirement for linoleic acid has been set too high, essentially because the symptoms of n-6 PUFA deficiency are more severe in the concomitant absence of n-3 PUFA, i.e. during essential fatty acid deficiency than during n-6 PUFA deficiency.

The point is that it has long been known that n-3 PUFA, particularly α-linolenic acid, and linoleic acid interact synergistically at low intakes (7,8) but become competitive at higher intakes. Thus, low dietary levels of α-linolenic acid *reduce* rather than increase the requirement for n-6 PUFA (see section on Nutritional Factors Affecting α-Linolenic Acid Metabolism). This very important observation on synergism between low intakes of linoleic and α-linolenic acids dates back more than 50 yr but is largely ignored today. This oversight has indirectly led, I believe, to significantly overestimating dietary linoleic acid requirements.

Dietary Sources and Intake

α-Linolenic acid is found in relatively high amounts (<50% of fatty acids) in only a few oilseeds, notably traditional flaxseed and perilla (Table 3.1). Canola and soybean oils contain modest amounts of α-linolenic acid (5–10% of total fatty acids) but, in various processed and unprocessed forms, are by far the most widely consumed oilseed sources of α-linolenic acid. Food products made from unhydrogenated canola or soybean oils, including salad oils and margarines, are important sources of dietary α-linolenic acid in Western or affluent societies (Table 3.2). In other common edible oils, α-linolenic acid is present at levels below 2% of total fatty acids.

Green vegetables and a few pulses have proportionately high amounts of α-linolenic acid (30–60%) but low to extremely low fat content, making them relatively poor dietary sources of α-linolenic acid. α-Linolenic acid is stored more in animal fat than in lean muscle, and grazing animals get a high proportion of their fatty acid intake as α-linolenic acid so meat provides a fairly significant dietary source of α-linolenic acid (Table 3.2).

TABLE 3.2

Food Sources That Account for the Daily Intake of α-Linolenic Acid in North America

Source	Total daily intake of α-linolenic acid (%)
Fats, oils, salad dressings, mayonnaise, shortenings	19
Milk, cheese, butter, yogurt	19
Margarine	16
Meat (chicken, beef, pork, lamb)	12
Fruits and vegetables	11
Bread and baked goods	10
Potatoes	4
Sauces	3
Fish and seafood	3
Eggs	2
Other	1

Source: Modified from Cunnane (1), Tinoco (79).

Foods made with vegetable oils are relatively expensive and risk degradation if they contain high amounts of PUFA. Animal fats such as butter, lard, and beef tallow have gone out of favor for food processing because of the demand for products with no animal content and because of their high saturated fatty acid content which is perceived to raise risks of cardiovascular disease. As a result, consumption of edible vegetable oils has risen steadily over the past 25 yr. However, unless unsaturated vegetable oils are at least partially hydrogenated to "rigidify" them, unsaturated vegetable oils have two disadvantages: (i) they do not have desirable physical properties for baked products, and (ii) their content of PUFA increases the risk of peroxidation and impairs the organoleptic properties of the final product. This demand by the food industry for monounsaturated and/or partially hydrogenated vegetable oils has put a premium on economical oils with low amounts of PUFA and has led to the selection of oilseeds for low content of α-linolenic acid.

Canola oil is currently in high favor for this reason. If not partially hydrogenated, it has 8–12% α-linolenic acid and is a major dietary source of α-linolenic acid in North America and Europe, where it is known as "low erucic acid rapeseed" oil. Effectively, there is a competition between maintaining oils in the diet that are rich in α-linolenic acid so we can meet the dietary needs for α-linolenic acid, and reducing α-linolenic acid in oilseeds so as to maintain desirable properties for food processing and flavors.

Although α-linolenic acid cannot be synthesized *de novo* in animals (i.e., from acetate), it can be synthesized by chain elongation of shorter chain homologues with a double bond at the n-3 carbon. Three such homologues, dodecaenoic acid (12:1n-3), tetradecatrienoic acid (14:3n-3), and hexadecatrienoic acid (16:3n-3) have been studied. The latter two both are easily converted to α-linolenic acid and to longer chain n-3 PUFA (9,10). There are no known dietary sources of tetradecatrienoic acid so, although its conversion to α-linolenic acid is metabolically noteworthy, it is not nutritionally relevant. There are dietary sources of hexadecatrienoic acid that can account for as much as 14% of the fatty acids in edible green vegetables such as spinach and broccoli. Hence, α-linolenic acid is definitely being synthesized in humans consuming green vegetables. The green vegetables containing hexadecatrienoic acid also contain substantial amounts of α-linolenic acid so there appear to be no foods containing hexadecatrienoic acid as the exclusive n-3 PUFA. Nevertheless, depending on the consumption by humans of foods containing hexadecatrienoic acid, the n-3 PUFA intake, especially that of vegetarians, is 2–10% higher than commonly described.

Legrand and colleagues (11) have tantalizingly suggested another route of α-linolenic acid synthesis. Because lauric acid (12:0) can be desaturated at the n-3 carbon, the n-3 fatty acid dodecaenoic acid (12:1n-3) can be synthesized endogenously by the rat. Sprecher (9) did not observe conversion of 12:1n-3 to α-linolenic acid *in vitro* but Legrand *et al.* believe that this step may be possible. If so, complete endogenous synthesis of α-linolenic acid from acetate is possible, at least in the rat.

Epidemiological surveys employing validated food-intake questionnaires as well as fat disappearance data are the principal ways by which actual intakes of nutrients like α-linolenic acid are calculated (12,13). These studies show that mean α-linolenic acid intake in the United States is about 1.6 g/d, two thirds of which is from "man-made" produce and one third of which is unprocessed. The processed foods rich in α-linolenic acid are inevitably loosely linked to intake of *trans* fatty acids. Recent recommendations for α-linolenic acid intake suggest that this level represents an "adequate" intake (i.e., meets known requirements for n-3 PUFA). It is not really clear, however, that we have a good measure of the requirements for long chain n-3 PUFA or that we understand the factors controlling the efficacy of α-linolenic acid conversion to long chain n-3 PUFA. Furthermore, linoleic acid and energy intake are major variables affecting the requirement for n-3 PUFA.

The fatty acid profile of human adipose tissue is considered to be the most reliable indicator of PUFA intake, a view recently well supported specifically for α-linolenic acid (14,15).

Synthesis of Long Chain n-3 PUFA

Many of the clinically relevant or healthful features of n-3 PUFA are attributable more to n-3 long chain PUFA than to α–linolenic acid itself. In individuals not consuming dietary sources of n-3 long chain PUFA (e.g., no fish), it has therefore long been of interest to understand how effectively α-linolenic acid is converted to long chain n-3 PUFA.

The pathway converting α-linolenic acid to long chain n-3 PUFA involves an alternating series of desaturations (double bond insertion) followed by chain elongation (addition of two carbons; Figure 3.1). This pathway was elucidated mostly in rat liver microsomes but has been demonstrated in many species, in essentially all tissues, and in virtually all cell types except erythrocytes. The first step, a Δ^6 desaturation, is widely considered to be the rate limiting step. The desaturases contain iron and are the terminal part of the electron transport chain. A variety of nutritional, metabolic, and endocrine perturbations affect their activity and, hence, the conversion of α-linolenic acid to long chain n-3 PUFA (16). Anabolic hormones like insulin stimulate desaturation while catabolic hormones like glucocorticoids inhibit desaturation.

For many years, the final stage in this pathway during which n-3 docosapentaenoic acid (22:5n-3) is converted to docosahexaenoic acid (22:6n-3) was widely assumed to involve a simple desaturation by a putative Δ^4 desaturase. However, direct proof of Δ^4 desaturation has always been lacking. Elegant chemical synthesis of appropriate tracer fatty acids was employed by Sprecher and colleagues who demonstrated 10 yr ago that the equivalent of Δ^4 desaturation occurred *via* chain elongation of n-3 docosapentaenoic acid to n-3 tetracosapentaenoic acid (24:5n-3), subsequent Δ^6 desaturation to tetracosahexaenoic acid (24:6n-3), and finally chain shortening to docosahexaenoic acid. These steps apparently involve peroxisomes as obligatory organelles in the synthesis of docosahexaenoic acid (17). Although confirmed in principle by other groups (18) and by the impairment of docosahexaenoic acid synthesis in peroxisomal disorders such as Zellweger syndrome (19), this "peroxisomal retro-

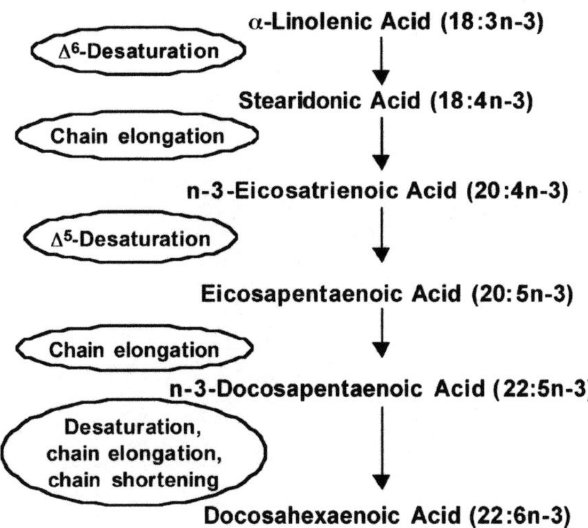

Fig 3.1. Pathway of desaturation and chain elongation involved in converting α-linolenic acid to long chain n-3 polyunsaturated fatty acids.

conversion-chain shortening" pathway remains controversial (20) and still cannot be considered as conclusively established.

The biochemical evidence is well established for conversion of shorter chain to longer chain n-3 PUFA in many species including humans of all ages. It is difficult, however, to extrapolate from the existence of this pathway to its actual *capacity*. There are several key questions: How much α-linolenic acid can be converted to eicosapentaenoic acid or to docosahexaenoic acid? In the absence of long chain n-3 PUFA intake, what is the "bioequivalence" of α-linolenic acid to long chain n-3 PUFA (2,21,22)? For instance, if one consumes 1 g/d of α-linolenic acid, what proportion is converted to long chain n-3 PUFA? If one doubles the α-linolenic acid intake, does one double conversion? What are the factors limiting this conversion?

Dietary supplementation and tracer studies provide useful answers to these questions but both have their limitations. Both types of study have been done abundantly in animal models, both *in vivo* and *in vitro*. Comparison of these results with studies done in humans makes it clear that marked differences exist among species and preclude easy extrapolation from one to the other (23). The focus here will be on human studies.

Many dietary supplementation studies have been done using adult humans in which an oil rich in α-linolenic acid, usually traditional flaxseed oil, is added to the diet over several weeks during which blood samples are taken for physiological and fatty acid analyses (Fig. 3.2). The studies pooled in Figure 3.2 were selected from the literature because they were relatively similar in design, having employed α-linolenic acid supplements of 9–21 g/d for 4–6 wk, and because they report analyses

                                      S.C. Cunnane

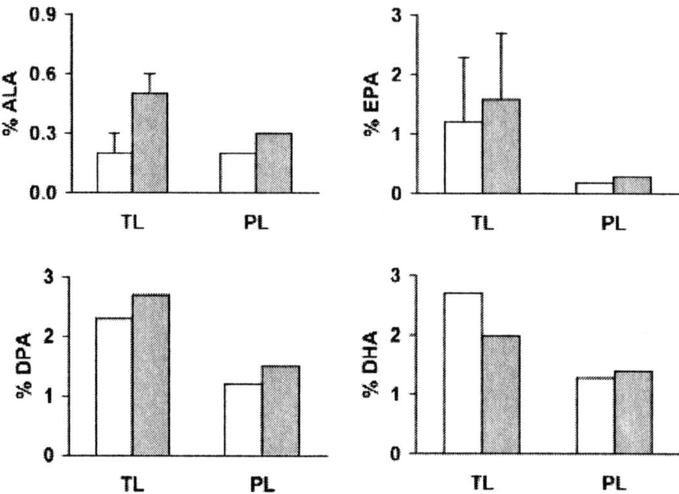

Fig 3.2. Change in n-3 polyunsaturated fatty acids in blood platelets after dietary supplementation with α-linolenic acid. The original papers reported effects after providing 9–21 g/d of α-linolenic acid over a period of 2–6 wk (96,100–103). Abbreviations: TL, total lipids; PL, phospholipids. *Source:* Reproduced with permission from Cunnane (1).

of the same blood lipid fatty acid profiles. In summary, they show that, during dietary supplementation with substantial amounts of α-linolenic acid, α-linolenic acid itself is raised more in blood triglycerides and cholesteryl esters (3–4-fold) than in phospholipids (1.5–2-fold). Hence, a statistically significant increase in α-linolenic acid in serum lipids can be anticipated under these conditions. However, averaging the results of these human studies, the long chain n-3 PUFA, eicosapentaenoic acid, n-3 docosapentaenoic acid, and docosahexaenoic acid are not significantly increased by dietary supplementation with α-linolenic acid. Of course, as indicated by the variability in the data, some of these studies did observe increases in long chain n-3 PUFA but, on average, no significant effect was seen (Fig. 3.2). Ostensibly, the same results have been reported for the n-3 PUFA composition of platelets (Fig. 3.3) and other blood cells after dietary supplementation with α-linolenic acid.

What does a change (or a lack of change) in the serum level of a particular fatty acid in blood samples taken before and after a supplementation study mean; that is, How should we interpret these dietary supplementation data? Clearly, we can say that α-linolenic acid rises after α-linolenic acid supplementation, so dietary α-linolenic acid directly influences blood α-linolenic acid levels. An increase of 2–4-fold in α-linolenic acid is the maximum that can realistically be expected, because supplements of >20 g/d are unlikely to be sustainable. To my knowledge, a dose response relationship has not been reported, so we do not know whether half as much α-linolenic acid raises blood α-linolenic acid half as much. The degree to which tis-

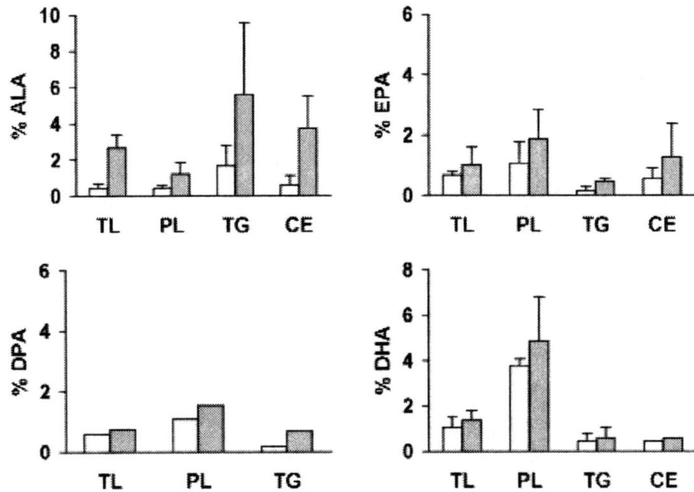

Fig. 3.3. Summary of the change in n-3 polyunsaturated fatty acids in blood after dietary supplementation with α-linolenic acid. The original papers reported effects after providing 9–21 g/d of α-linolenic acid over a period of 4–6 wk (93–99). Abbreviations: TL, total lipids; PL, phospholipids; TG, triglycerides; CE, cholesteryl esters. *Source:* Reproduced with permission from Cunnane (1).

sue α-linolenic acid changes under these conditions is unknown. Intakes of 35–40 g/d α-linolenic acid for 2–6 wk indicate plasma α-linolenic acid can rise >6-fold (24–26) but eicosapentaenoic acid may not rise at all (26). Changes in docosahexaenoic acid were not reported in these studies. A low dose (90 mg/d) of α-linolenic acid raised α-linolenic acid itself in elderly patients with putative n-3 PUFA deficiency but also did not significantly raise n-3 long chain PUFA (27). Hence, insufficient or excess doses of α-linolenic acid do not seem to be the reason for a lack of change in n-3 long chain PUFA during α-linolenic acid supplementation.

Although mean blood levels of long chain n-3 PUFA did not change significantly in these studies (Figs. 3.2, 3.3), whether some α-linolenic acid was converted to long chain n-3 PUFA which subsequently went elsewhere or was oxidized is unknown. The simple conclusion that no n-3 docosapentaenoic acid or docosahexaenoic acid was synthesized because their blood levels did not change is probably incorrect. There are metabolic controls on fatty acid levels in serum such that increased synthesis of docosahexaenoic acid could well lead to increased docosahexaenoic acid uptake elsewhere, leaving little or no change in serum levels despite its increased synthesis. That said, supplementation with fish oils containing docosahexaenoic acid certainly raises serum docosahexaenoic acid in humans, so the negligible change in serum docosahexaenoic acid after α-linolenic acid supplementation probably means negligible *net* synthesis (i.e., increased clearance to tissues or to oxidation). An added factor is that an α-linolenic acid supplement of 9–21 g/d is about 10 times the average daily intake of α-linolenic acid in North America and might well *suppress* desaturation-chain elongation through "substrate inhibition." Again, the human studies using dietary supplements are unclear on this point.

Radiolabeled or stable isotope tracers also provide useful information on metabolism of α-linolenic acid and synthesis of long chain n-3 PUFA in humans, but they also have their own limitations. By definition, "tracers" are physically distinct but chemically identical to the substance being traced (the tracee). Hence, they behave as the tracee would but very low amounts are needed, so mass effects (i.e., the potential for substrate inhibition as probably occurs with dietary supplementation using α-linolenic acid) do not occur. As with dietary supplementation, multiple sampling, preferably over an extended period, gives the most useful information. However, a tracer is usually (but not necessarily) given at a single time point. Metabolism of this single dose will ultimately follow a time course, rising in blood and tissues, getting converted to downstream products, each with a peak value, and each eventually disappearing. The point is that, unless it is done frequently, the timing of blood or tissue sampling will have a marked impact on the results. Hence, interpreting the "steady state" behavior of the tracee from the behavior of the tracer can be difficult. This preamble is relevant to understanding the limitations on interpreting tracer data on long chain n-3 PUFA synthesis in humans because most studies have few sampling points and rarely sample from sites other than blood. Nevertheless, though it is ill advised, most studies attempt to indicate what proportion of long chain n-3 PUFA can be made from α-linolenic acid.

The advantage of a tracer study is that there is no better way to answer a largely *qualitative* question such as Can term or premature infants make n-3 PUFA from α-linolenic acid? This was a major point of contention during the 1990s when the field struggled with rationalizing the addition of long chain n-3 PUFA to artificial infant milk formula. The hard clinical or biochemical evidence that long chain n-3 PUFA were essential for normal growth of *healthy, term* infants was equivocal. Many in the field resorted to empirically valid but, ultimately circumstantial reasons to justify the inclusion of long chain n-3 PUFA in formulas.

The application of stable isotope methodology using gas chromatography-combustion-isotope ratio mass spectrometry or quadrupole gas chromatography–mass spectrometry to PUFA metabolism had just become established in the 1990s. Reports using stable isotopes of different PUFA soon appeared showing that healthy term and preterm infants could make docosahexaenoic acid from α-linolenic acid (28–30); therefore, qualitatively, it was clear infants could make long chain n-3 PUFA from α-linolenic acid, a point that was established some years earlier in adults (31). The immediate problem was whether enough docosahexaenoic acid was being made so that an exogenous (dietary) source was unnecessary to meet the needs of normal infant development. That is a *quantitative* question that, despite heroic efforts to interpret limited kinetic data, could not and still cannot be answered for adults or infants. See Chapter 8 (α-Linolenic Acid in Brain Function and Infant Development) for a more detailed discussion.

Getting valid *quantitative* answers is an enduring limitation of tracer data obtained from one body compartment (blood), no matter how detailed the sampling and kinetic analysis (32–35). The bottom line is that human tracer studies or simultaneous tracer and dietary modification studies (36) have largely confirmed the main qualitative parameters of n-3 PUFA metabolism established over many decades in *in vitro* animal models; that is, long chain n-3 PUFA inhibit conversion of α-linolenic acid to long chain n-3 PUFA by end product inhibition. High intake of α-linolenic acid will inhibit its own conversion to long chain n-3 PUFA by substrate inhibition. Comparative tracer studies within the same model show that it takes 7–10 times as much dietary α-linolenic acid as docosahexaenoic acid to match accumulation of docosahexaenoic acid in the developing primate brain (37).

Thus, two totally opposing conclusions of studies using either dietary α-linolenic acid supplementation or labeled α-linolenic acid are still common: (i) either there is no synthesis or (ii) there is adequate synthesis of n-3 PUFA from dietary α-linolenic acid. Both conclusions are wrong. The tracer studies prove that there is some synthesis of long chain n-3 PUFA. The newly synthesized long chain n-3 PUFA may be shunted out of plasma so that there is no *net* change in blood levels, but their synthesis is occurring. Existence of a pathway in no way proves that the products of that pathway are being made in adequate amounts to meet all circumstances. Periods or physiological states such as early development clearly show that the capacity for long chain n-3 PUFA is not necessarily adequate; hence, they are found not only in the milk of all mammals but are now included in milk formulas as well.

Perhaps the most striking outcome of tracer studies with α-linolenic acid is that they show that it undergoes abundant metabolism outside the n-3 PUFA pathway, a process known as "carbon recycling" (see section on α-Linolenic Acid Metabolism—Other Aspects Including Carbon Recycling). Only tracer methodology could have led to these observations which, in time, may lead us to novel roles of α-linolenic acid not apparent from the currently known functions of n-3 PUFA.

Nutritional Factors Affecting α-Linolenic Acid Metabolism

Important nutritional influences on α-linolenic acid metabolism occur aside from the effects of α-linolenic acid itself or long chain n-3 PUFA on conversion of α-linolenic acid to long chain n-3 PUFA (substrate and end-product inhibition, respectively).

Energy Balance

By far the most important nutritional influence on α-linolenic acid metabolism is energy status. Several animal models (1) and occasional reports from human studies demonstrate that inadequate nutrition resulting in energy deficit will deplete body stores of α-linolenic acid and n-3 PUFA in general. The main reason is that among the common dietary fatty acids and in a variety of species, α-linolenic acid is relatively easily β-oxidized (i.e., consumed as a fuel; Table 3.3). The tracer oxidation data in

TABLE 3.3
β-Oxidation of Common Dietary Fatty Acids[a]

Reference	SA	PA	OA	EA	LA	ALA
Humans, *in vivo* (80)	70	87	106	127	100	146
Humans, *in vivo* (81)	30	—	160	—	100	—
Humans, *in vivo* (33)	—	—	122	143	100	169
Rat, *in vivo* (82)	—	50	—	—	100	—
Rat, *in vivo* (83)	49	63	116	—	100	135
Rat, *in vivo* (84)	21	53	—	—	100	—
Rat, liver mitochondria (85)	—	—	85	—	100	140
Rat, liver mitochondria (86)[b]	29	75	54	—	100	157
Rat, liver mitochondria (87)	4	7	15	—	100	—
Rat, perfused heart (88)	—	94	93	72	100	113
Rat, serum, liver, and brain (89)[c]	—	109	—	—	100	333
Catfish, *in vivo* (90)	—	—	—	—	100	161
Mean	34	67	94	114	100	191

[a]Each study used ^{14}C- or ^{13}C-labeled fatty acids and all reported data for linoleic acid and at least one of the other fatty acids are shown. Because different species, fatty acids, or methodologies were used, the data within each study were normalized to values obtained for linoleic acid (100%).
[b]Oxidation of eicosapentaenoic acid was equivalent to that of palmitic acid.
[c]Serum, liver, and brain values were pooled; data were reported as water soluble ^{14}C.
Abbreviations: SA, stearic acid; PA, palmitic acid; OA, oleic acid; EA, elaidic acid; LA, linoleic acid; ALA, α-linolenic acid; (—) not reported.

Table 3.3 are comparative, not absolute; they show that compared with other fatty acids, β-oxidation consumes a higher proportion of α-linolenic acid than of linoleic, oleic, palmitic, or stearic acids. These data do not tell us what proportion of our daily intake of α-linolenic acid is oxidized or when we may be in negative α-linolenic acid balance (i.e., when more α-linolenic acid is being oxidized than consumed), yet this occurs relatively easily and completely prevents conversion of α-linolenic acid to long chain n-3 PUFA.

In animal studies, it is actually fairly simple to measure the percentage of dietary α-linolenic acid that is oxidized to respiratory CO_2. The method is called whole body fatty acid balance and was first described in the mid-1990s (38–40). The principle is that if the intake of α-linolenic acid (or linoleic acid) over a specified period is known, its oxidation can be calculated by difference from the amount that accumulates during the study (balance) period. To measure accumulation, one needs to know whole body content at the start and at the end of the balance period, deriving accumulation by difference. In practice, this means that pairs of animals are used, one killed at the start and one at the end of the balance period. The accumulation of α-linolenic acid is the difference between the starting and ending values. Of course, α-linolenic acid is converted to long chain n-3 PUFA which also accumulate. If there are no long chain n-3 PUFA in the diet starting well before the actual balance period, then the long chain n-3 PUFA that accumulate during the balance period were derived from the α-linolenic acid consumed. Excretion of α-linolenic acid should also be measured but, in healthy animals, it rarely exceeds 2%, so excretion has little impact on the whole body utilization of α-linolenic acid relative to intake.

Two conditions have to be met in order to be able to do whole body analysis: (i) one cannot have a dietary source of the α-linolenic acid precursor, hexadecatrienoic acid, and (ii) one has to be able to accurately measure whole body fatty acid content. In practice, this limits this method to animal studies in which carcass analysis can be done and in which purified diets are used. We have shown that by estimating whole body linoleic acid (or α-linolenic acid) content indirectly, under certain conditions, whole body balance can be applied to living humans (41). The balance method can also be used with saturated or monounsaturated fatty acids, but the amount of their endogenous synthesis is unknown. This is a potentially big variable over which one has no control but does not apply to balance studies of α-linolenic acid or linoleic acid.

Several whole body α-linolenic acid balance studies have shown that when α-linolenic acid is provided to young free-feeding rats at just over the recommended intake, about 80% completely disappears (i.e., is β-oxidized as a fuel). An example is shown in Table 3.4. Thus, the overwhelming majority of dietary α-linolenic acid is normally oxidized. Pregnancy is a condition of heightened demand for many nutrients including n-3 PUFA, and whole body balance analysis in rats shows that n-3 PUFA are conserved during pregnancy. Conversely, undernutrition, whether by weight cycling, fasting-refeeding during pregnancy, or by zinc deficiency, all increase β-oxidation of α-linolenic acid such that its use as a fuel *exceeds* its retention in the body n-3 PUFA pool as a whole (i.e., there is net loss of n-3 PUFA;

TABLE 3.4

An Example of Whole Body Balance Analysis of n-3 PUFA in Rats Consuming α-Linolenic Acid as Their Only Dietary Source of n-3 PUFA

Variable	α-Linolenic acid	Long chain n-3 PUFA
Intake	6731 (mg)	0 (mg)
Excretion	149 (mg) 2.2 (% intake)	31 (mg) —
Body Content	592 (mg, start) 1330 (mg, finish)	96 (mg, start) 191 (mg, finish)
Accumulation	738 (mg) 10.9 (% intake)	95 (mg) 1.4 (% intake)
Oxidation	5700 (mg) 84.9 (% intake)	19[a] (mg) —

[a]Estimated as 20% of accumulation.
Abbreviation: PUFA, polyunsaturated fatty acids; (—), not reported.
Source: Modified with permission from Cunnane and Anderson (6).

1,16). The crucial point is that this loss of α-linolenic acid during weight cycling occurs despite sufficient, although intermittent, dietary α-linolenic acid intake. Thus, even a modest, transitory energy deficit markedly alters α-linolenic acid utilization such that all dietary α-linolenic acid can be oxidized and tissue n-3 PUFA become depleted.

In whole body α-linolenic acid balance studies, it is preferable, though not essential, to have no dietary long chain n-3 PUFA. If long chain n-3 PUFA are absent from the diet, comparison of their accumulation to the intake of α-linolenic acid during the balance period provides a good estimate of the capacity to synthesize long chain n-3 PUFA. Typically, the values we have observed range from 1–5%, depending on the experimental conditions, with more conversion occurring during pregnancy (42). Hence, in addition to indicating how energy balance affects α-linolenic acid homeostasis, whole body fatty acid balance analysis provides a good estimate of the *capacity* for long chain n-3 PUFA synthesis, a useful measure that is hard to obtain by any other method.

That undernutrition could lead to net depletion of n-3 PUFA primarily by increasing oxidation of α-linolenic acid seems counterintuitive. Why is α-linolenic acid not conserved so that deficiency of n-3 PUFA could be more easily prevented? There is no simple answer, but comparative studies of long chain fatty acid oxidation using diverse methodology in different species unequivocally show that α-linolenic acid is the most easily oxidized of the common long chain dietary fatty acids (Table 3.3). At least this gives us clear and consistent evidence that it would be difficult to avoid oxidizing α-linolenic when fatty acid oxidation is stimulated.

At present, α-linolenic acid is thought of as a vitamin, that is, as a nutrient serving an essential function but which cannot be synthesized in sufficient amounts. Throughout evolution, vitamins could have become nutritionally and metabolically important only if their dietary supply was assured (i.e., if there was little or no risk of a serious deficiency). Species cannot evolve to depend on nutrients if the supply of such nutrients is not dependable. Hence, like others (43,44), I would argue that until the twentieth century when food processing industries started affecting food selection, n-3 PUFA, like other vitamins, were unlikely to ever have been seriously limiting in the human food supply. Hence, humans evolved under conditions in which α-linolenic acid was a secondary n-3 PUFA in the sense that eicosapentaenoic and docosahexaenoic acids were in the human diet in sufficient amounts that assuring their synthesis from α-linolenic acid was never a problem. Before the twentieth century and certainly in our early evolution, α-linolenic acid may well have been dispensable because there already was an abundant supply of long chain n-3 PUFA in the diet (44–46). Mammals therefore developed cellular and molecular machinery that readily permitted β-oxidation of α-linolenic acid. This dispensability is, of course, no longer the situation, so we are seeing the consequences of chronically insufficient intake of long chain n-3 PUFA and problems of inadequate conversion of α-linolenic acid to long chain n-3 PUFA arising in part because α-linolenic acid is so readily β-oxidized.

Nutritional Cofactors Influencing Desaturation-Chain Elongation

Several experimental approaches, including whole body fatty acid balance analysis, have shown that conversion of α-linolenic acid to long chain n-3 PUFA is under the influence of a number of nutritional cofactors. This influence appears to be primarily exerted at the level of the desaturase enzymes, the activity of which depends on several essential nutrients. The desaturases contain Fe (47). Thus, Fe deficiency impairs synthesis of long chain PUFA (48). The desaturases depend on NADH as an electron donor which is a process highly sensitive to adequate Zn. Thus, Zn deficiency also markedly impairs synthesis of long chain PUFA, particularly those derived from linoleic acid (49). Mg and vitamin B_6 are also implicated for reasons that have not yet been elucidated. The influence of these cofactors on long chain PUFA synthesis has been studied more for n-6 than for n-3 PUFA, but because the desaturases are putatively the same for both PUFA series, the implications are similar.

Three series of unsaturated fatty acids have long been known to compete at least for the Δ^6 desaturase but probably for the Δ^5 desaturase as well. They are the n-9 fatty acids, primarily oleic acid (18:1n-9), linoleic acid (n-6 PUFA), and α-linolenic acid (n-3 PUFA). In liver microsomal preparations, the desaturases prefer n-3 PUFA over n-6 PUFA, and n-6 PUFA over n-9 fatty acids, so high intakes of linoleic acid largely prevent oleic acid desaturation and also compete with α-linolenic acid, thereby limiting α-linolenic acid conversion to long chain n-3 PUFA. Conversely, high intakes of α-linolenic acid inhibit linoleic acid desaturation. As long as there is balance in the intakes of these families of unsaturated fatty acids, appropriate amounts of long chain PUFA should, in principle, be formed.

In practice, affluent Western nations appear to have an excess intake of linoleic acid and a low intake of long chain n-3 PUFA. This creates a situation of double jeopardy for sustaining adequate long chain n-3 PUFA because not only are insufficient amounts of preformed long chain n-3 PUFA being consumed but high intakes of linoleic acid are effectively preventing adequate conversion of α-linolenic acid to long chain n-3 PUFA.

It is important to place the competition between n-6 PUFA and n-3 PUFA for desaturation–chain elongation in context. This competition is well established but does not occur at *low* intakes of these nutrients. At very low intakes, that at near or below their requirement, linoleic acid and α-linolenic acid are *synergistic*; the presence of each helps minimize the impact of a low intake of the other. Thus, small amounts of α-linolenic acid reduce the requirement for linoleic acid (7,8; see sections on Introduction and Historical Overview).

α-Linolenic Acid Metabolism—Other Aspects Including Carbon Recycling

The main routes of α-linolenic acid metabolism are primarily β-oxidation followed in distant second place by conversion to long chain n-3 PUFA (Fig. 3.4). The over-

                                    S.C. Cunnane

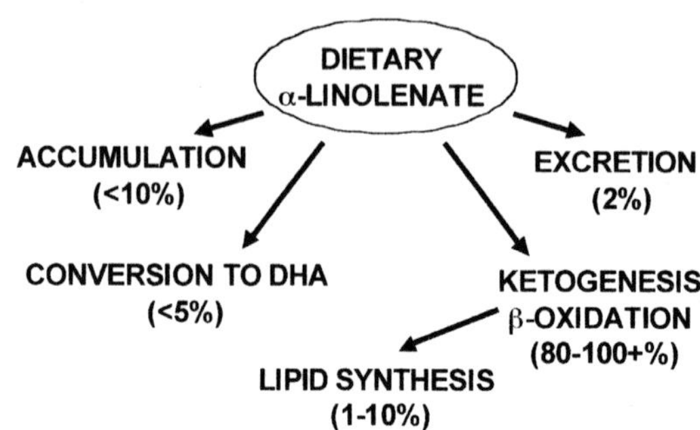

Fig. 3.4. Overview of the metabolic fate of α-linolenic acid emphasizing utilization during early postnatal development; appropriate values in adults shift towards β-oxidation. Abbreviation: DHA, docosahexaenoic acid.

whelming emphasis of research on α-linolenic acid over the past five decades has been on its conversion to long chain n-3 PUFA *via* desaturation and chain elongation. Evidence collected in the past decade demonstrates that oxidation is the predominant fate of α-linolenic acid and can consume all available dietary intake, preventing any net conversion to long chain n-3 PUFA.

Peroxidation is also a well-established route of α-linolenic acid degradation (50,51). This requires its exposure to air and heat or other conditions promoting free radical attack on the double bonds in PUFA. As such, purified α-linolenic acid or extracted flaxseed oil is at constant risk of rancidity. The same goes for milled flaxseed because the α-linolenic acid is then exposed to the air. Even small amounts of peroxidation produce sufficient amounts of volatile products to confer an unpleasant smell. Conversely, intact flaxseed has much greater resistance to lipid peroxidation. Many of the practical nutritional applications of flaxseed involve baking it into products such as bread or muffins. Despite the brief exposure of milled flaxseed to elevated temperatures during baking, there is little or no loss of α-linolenic acid after baking (50), suggesting the matrix of the seed provides enough antioxidant protection.

Recent research suggests that α-linolenic acid has still other routes of metabolism. The first example is one in which radiolabeled orally administered α-linolenic acid was reported to be relatively highly concentrated in skin and hair (52). Compared with other tissues, skin contains a relatively high proportion of α-linolenic acid (3.5%). After fat and muscle, skin is actually the next largest pool of α-linolenic acid (38), so relatively high concentrations of labeled α-linolenic acid in skin could be expected. There may be a link between these observations and the long-established use of various preparations of flaxseed or flaxseed oil to improve the coat quality of show animals including horses and dogs. If the α-linolenic acid reaching the skin does not all stay as an n-3 PUFA, perhaps its beneficial effect on coat and hair quality relates to a product of its peroxidation which causes a shiny, healthy appearance.

A further route of α-linolenic acid metabolism shown by mass balance tracer studies is the large amount of α-linolenic acid appearing in lipids synthesized *de novo* (i.e., fatty acids like palmitic acid (16:0), stearic acid (18:0), and oleic acid, and in cholesterol; 1,42,53,54). This is especially notable during early development and is an interesting, unexpected, and still unexplained route of α-linolenic acid metabolism that was first reported over 25 yr ago (55–57; Table 3.5). At that time, the proportion of the tracer α-linolenic acid that was "carbon recycled" into *de novo* lipid synthesis was not established. A recent report indicates that, within a week of orally administering labeled α-linolenic to suckling rats, 90% has been fully oxidized, about 9% is in newly synthesized lipids, and the remainder (<1%) is in α-linolenic acid and long chain n-3 PUFA (1). In this model, the whole body mass tracer balance assessment of α-linolenic metabolism was somewhat confounded by the presence of long chain n-3 PUFA in the maternal milk consumed by the suckling rats. However, similar results have been reported in primates consuming no long chain n-3 PUFA (53), and in isolated astrocytes cultured in media devoid of docosahexaenoic acid (58). Hence, there is clearly abundant metabolism of α-linolenic acid that cannot be explained within the commonly accepted framework of n-3 PUFA metabolism. Parallel results have been seen for recycling of linoleic acid carbon even in markedly n-6 PUFA-deficient adult rats (59), suggesting that carbon recycling of both these PUFA is an obligatory phenomenon in need of further exploration and explanation.

α-Linolenic acid may also be converted to products that improve pathogen resistance, including jasmonic acid and products of a 9-lipoxygenase (60–63).

TABLE 3.5
Pooled Analysis of Published Data for Carbon Recycling of α-Linolenic Acid[a]

Source	DNL/DHA[b]	Reference
Brain		
Rat	7.1	Menard *et al.* (54)
Rat	8.3	Sinclair (55)
Rat	5.0	Dhopeswarkar and Subramanian (56,57)
Rhesus	5.6	Sheaff-Greiner *et al.* (53)[c]
Liver		
Rat	1.1	Menard *et al.* (54)
Rhesus	1.4	Sheaff-Greiner *et al.* (53)[c]
Catfish	11.8	Bandyopadhyay *et al.* (90)
Lung		
Rat	20.0	Menard *et al.* (54)
Retina		
Rhesus	1.5	Sheaff-Greiner *et al.* (53)[c]

[a]*de novo* lipid synthesis (fatty acids and cholesterol) compared with docosahexaenoic acid synthesis; the data are for samples analyzed 48 h after dosing with the tracer except for the report by Sheaff-Greiner *et al.* (53) in which data are for 5 d after dosing.
[b]de novo lipogenesis (DNL) versus docosahexaenoic acid (DHA) synthesis.
[c]Sheaff-Greiner *et al.* (53) did not report data for carbon recycling into cholesterol, only for carbon recycling into saturated and monounsaturated fatty acids. They used a diet containing no long chain n-3 polyunsaturated fatty acids, which were not controlled for in the other studies reported here.

    S.C. Cunnane

Clinical Deficiency of n-3 PUFA

The absence of n-3 PUFA leads to reproducible pathology in rodents, but it usually takes multigenerational studies to obtain the symptoms which are still relatively mild. For instance, growth retardation and reproductive impairment is uncommon in rats even in the third generation despite clearly reduced levels of n-3 PUFA in tissues (64–66). The symptoms are now well established as relating mostly to sensory, learning, and behavioral impairment [i.e., impaired peripheral and central nervous system function (67)]. Vision as a function of n-3 PUFA adequacy has been widely studied because of the high levels of docosahexaenoic acid in the membranes of the rod outer segments in photoreceptors of the eye. Other sensory systems such as olfaction and hearing are now under study as well.

Development of the visual system in human infants is also vulnerable to adequacy of n-3 PUFA, which does not take three generations to demonstrate. Humans have very large brains relative to body size (12–14% of birth weight). As a result, they appear to be more vulnerable to n-3 PUFA deficiency than, for instance, rodents. Nevertheless, in humans, it has been challenging to do the necessary invasive studies to demonstrate diet-structure-function relationships for n-3 PUFA. As a result, knowledge of the symptoms of n-3 PUFA deficiency in humans has been difficult to obtain. In fact, the only two cases reported as such do not really agree on pathology (Table 3.6). It is interesting that both cases were in young children (6–7 years old), and both individuals required enteral nutrition for an extended period. Both received enteral nutrition that appears to have been inadequate in n-3 PUFA, and both had low serum n-3 PUFA before the symptoms of deficiency of n-3 PUFA were recognized. One had the anticipated neurological deficit, but the other did not. In one of these cases, excess intake of linoleic acid probably contributed to n-3 PUFA deficiency. The lack of consistency in these two cases and the absence of reports of other cases emphasize the extraordinary capacity of mammals in general, but humans in particular, to conserve function in the face of low n-3 PUFA intake.

Our body fatness is one important form of n-3 PUFA "insurance" that is commonly overlooked because the percentage of total fatty acids in fat as n-3 PUFA is usually very low. However, a child weighing 20 kg and having 20% body fat will have 4 kg of body fat containing just over 3 kg of fatty acids. If 1% of those fatty acids are n-3 PUFA, that is 30 g, most of which will be α-linolenic acid. The daily requirement for α-linolenic acid in adult humans is thought to be about 1.5 g so it should not be more than 1 g in a child. Assuming a requirement of 1 g/d and absolutely zero intake of n-3 PUFA (a situation almost impossible to achieve now that enteral and parenteral nutrition is made using soybean oil as the lipid source), a healthy child will still have a 30-d reserve of n-3 PUFA in fat alone. Enteral nutrition solutions were often inadequate in n-3 PUFA in the past, but this should rarely if ever be the case now. Therefore, in my view, the main variable affecting body reserves of n-3 PUFA is total body fatness. α-Linolenic acid is more easily mobilized from body fat than most common fatty acids (68), so this store should be accessible during n-3 PUFA deficien-

TABLE 3.6
Clinical Symptoms of Deficiency of n-3 PUFA

Effects of ALA supplementation	Holman's case (91)[a]	Bjerve's case (92)[b]
Before ALA supplementation		
Symptoms	Numbness, muscle pain, visual blurring	No neurological symptoms but low weight gain
Dose of ALA	0.66% of fatty acids; LA/ALA = 115:1	0.71% of fatty acid intake
n-3 PUFA level	Low ALA; normal DHA	all n-3 PUFA were 10–20% of normal.
After ALA supplementation		
Symptoms	Disappearance of neurological symptoms; 10–15% increase in nerve conduction velocity	Normalized weight gain
Dose of ALA	6.9% of fatty acids; LA/ALA = 6:1	0.71% of fatty acids
n-3 PUFA level	Raised 3–7-fold in serum phospholipids	Raised 2–10-fold in serum total lipids

[a]6-year-old girl on enteral nutrition for 12 mo.
[b]7-year-old girl with metachromatic leukodystrophy on enteral nutrition for 4 yr.
Abbreviations: ALA, α-linolenic acid; DHA, docosahexaenoic acid; LA, linoleic acid; PUFA, polyunsaturated fatty acid.
Source: Used with permission from Cunnane (1).

cy and appears to be a significant reason for needing long-term depletion studies before symptoms of n-3 PUFA deficiency will be observed.

A chronically sick child or a premature infant will probably have <20% body fat. Thus, it is in the sick or artificially nourished child that n-3 PUFA symptoms are most likely to occur (Table 3.6). Symptoms of n-3 PUFA deficiency have been reported in adults, especially the elderly undergoing long-term nutritional support. These symptoms were complicated by additional nutrient deficiencies (27), but again point to sickness as an important factor precipitating n-3 PUFA deficiency, probably because it disturbs energy balance and exacerbates energy demand.

A surprising cause of mild tissue n-3 PUFA depletion is specific n-6 PUFA deficiency (Bazinet, *et al.*, unpublished data). During the course of a study of the nutritional and metabolic impact of specific n-6 PUFA deficiency, we collected samples suitable for whole body fatty acid balance analysis and used them to look at n-3 PUFA balance. This analysis showed that despite slightly higher α-linolenic acid intake in the n-6 PUFA-deficient rats than in controls receiving adequate n-3 and n-6 PUFA, the n-6 PUFA-deficient rats accumulated less than half the amount of α-linolenic acid compared with the controls. Some of the α-linolenic acid deficit was due to more conversion to long chain n-3 PUFA, but about 35% of the deficit was due to increased oxidation of α-linolenic acid. This is perhaps an unexpected influence of n-6 PUFA deficiency on n-3 PUFA metabolism but is consistent with earlier observations showing the interdependence of linoleic and α-linolenic acid homeostasis (7,8). Extreme n-6 PUFA deficiency leads to marked n-6 PUFA depletion from body stores and seems to put an increased demand on n-3 PUFA of which at least α-linolenic acid is easily mobilized from fat stores and also relatively easily β-oxidized. Hence, during exclusive n-6 PUFA deficiency, body n-3 PUFA stores became depleted despite adequate n-3 PUFA intake.

Functions Attributable to α-Linolenic Acid Itself

The persistent question with α-linolenic acid is Does it do anything by itself or is it merely a dietary precursor to the long chain n-3 PUFA which perform all the important functions attributable to n-3 PUFA? As alluded to in the previous section, n-3 PUFA clearly have a role in the central nervous and sensory systems (see Chapter 8). Other than that, their main role seems to be to keep a brake on metabolism of n-6 PUFA and on signalling molecules derived from arachidonic acid, principally the "2 series" eicosanoids [prostaglandins, thromboxane A_2, and leukotrienes (44)]. This is the main reason regular intake of marine fish (the richest natural food source of eicosapentaenoic acid) is recommended to minimize risk of heart attack. Eicosapentaenoic acid reduces platelet clotting tendency and inhibits much of the inflammatory effect of eicosanoids derived from arachidonic acid, both of which contribute to heart attack risk. Compared with eicosapentaenoic acid, α-linolenic acid is less effective at inhibiting arachidonic acid release from membrane phospholipids or its metabolism through the arachidonic acid cascade, but is a precursor to isoprostanes and lipoxygenase products. α-Linolenic acid is not efficiently converted to eicosapentaenoic acid, and the currently high dietary tolerance for linoleic acid in the United States (17 g/d) is unlikely to reverse this situation. Is there a case that α-linolenic acid itself has useful functions? Yes, for the following four reasons.

(i) The synergism between low intakes of linoleic acid and α-linolenic acid (7,8) suggests that α-linolenic acid has an as yet poorly understood role in the homeostasis of linoleic acid and possibly other n-6 PUFA. This synergism may extend to higher intakes of α-linolenic acid as well (69). Given the widespread literature to the contrary, the idea that α-linolenic acid has a noncompetitive role in n-6 PUFA metabolism will, understandably, be controversial but deserves thorough investigation.

(ii) Epidemiological studies are uniformly supportive of a beneficial effect of α-linolenic acid in minimizing risk of heart disease (see Chapters 12,15). This is also true for cancer, except in the case of prostate cancer (see below and see Chapter 11). One report of the association between dietary factors and causes of death in the Framingham study indicates that α-linolenic acid (grams per day as well as percentage of energy intake) is the only dietary fatty acid negatively associated with all cause mortality (12; Table 3.7). Raised intake of long chain n-3 PUFA was not as strongly

TABLE 3.7
Relative Risk of Death from Four Major Causes Compared with Quintiles of Daily
α-Linolenic Acid Intake[a]

Quintile[b]	α-Linolenic acid intake	CHD	CVD	Cancer	All causes
1	0.87 g/d	1.00	1.00	1.00	1.00
	0.42% energy	1.00	1.00	1.00	1.00
2	1.27 g/d	0.96	0.93	1.34	0.96
	0.54% energy	0.72	0.86	1.12	0.86
3	1.58 g/d	0.56	0.66	0.85	0.69
	0.63% energy	0.80	0.97	0.72	0.85
4	1.93 g/d	0.96	0.88	1.14	0.89
	0.73% energy	0.61	0.66	0.90	0.75
5	2.80 g/d	0.66	0.61	0.87	0.69
	0.98% energy	0.58	0.66	0.78	0.68

[a]There was a significant trend toward reduced all cause mortality as a function of α-linolenic acid intake expressed as both grams/day and as a percentage of energy ($P < 0.05$). There was also a significant trend toward reduced mortality from CHD and CVD as α-linolenic acid intake increased but only as a percentage of energy intake ($P < 0.05$). No other statistically significant trends were reported.
[b]1251–1253 individuals/quintile.
Abbreviations: CHD, coronary heart disease; CVD, cardiovascular diseases other than CHD.
Source: Modified from Dolecek (12).

associated with reduced all cause mortality as was α-linolenic acid. The same concept arises from work on breast cancer in which higher levels of α-linolenic acid in breast tissue biopsies were more strongly associated with preventing metastasis of breast cancer than were long chain n-3 PUFA (70; see Chapter 11). These examples are cited as evidence that α-linolenic acid has a role in human nutrition and health related to reducing risk of the chronic killer diseases that is not directly or exclusively attributable to its conversion to long chain n-3 PUFA.

I believe that the association between high α-linolenic acid and increased prostate cancer risk arises because the fat in less-expensive cuts of red meat is a significant dietary source of α-linolenic acid. Hence, higher intake of red meat, which is also associated with greater risk of prostate cancer, is necessarily associated with a higher intake of α-linolenic acid. Though it has been portrayed that way (71,72), high intake of α-linolenic acid in those that eat red meat does not necessarily make α-linolenic acid a risk factor for prostate cancer but, rather, a good marker of animal fat intake. In fact, α-linolenic acid and flaxseed supplements are being used to *treat* prostate and breast cancer (see Chapter 1).

(iii) α-Linolenic acid is the main if not the only n-3 PUFA in the diet of at least a billion vegetarians worldwide. Despite not consuming much if any fish, vegetarians do not have a higher prevalence of chronic killer diseases than nonvegetarians. In simple terms, something is right about their lifestyles from which nonvegetarians could benefit. Hence, other factors being equal, the case cannot be made that vegetarians (nonconsumers of long chain n-3 PUFA) are at a long-term health disadvantage. Besides the intake of n-3 PUFA, many factors are involved in the long-term health risks of different lifestyles. To be sure, long chain n-3 PUFA provide valuable insurance of lower risk of heart disease and cancer but not more so than α-linolenic acid (Table 3.7). Long chain n-3 PUFA may be more cardioprotective than α-linolenic acid under some conditions (i.e., perhaps when lives are sedentary), but this is not necessarily the case as reported recently for platelet function (73).

(iv) By virtue of being the predominant or the only dietary n-3 PUFA for a large proportion of the world's population, α-linolenic acid is an important nutritional brake on the metabolism of n-6 PUFA. This braking action occurs at higher intakes by way of competition with linoleic acid and other n-6 PUFA and seems to be essential to reduce the risk of several killer diseases in affluent societies, especially cardiovascular disease (43,44). It does not contradict the synergistic effects of α-linolenic acid and linoleic acid at low intakes, though clearly there must be a point at which the relationship switches from being cooperative to competitive.

The bottom line is that we do not know all functions of n-3 PUFA, let alone α-linolenic acid itself. Little α-linolenic acid seems to be converted to long chain n-3 PUFA in humans, but the recent identification of carbon recycling as a major pathway of α-linolenic acid utilization suggests that its metabolism and function could yet hold some surprises. Furthermore, we have good epidemiological evidence of its healthfulness (1,12,13), some of which suggests a distinct role of α-linolenic acid. Hence, it would be shortsighted to disregard the benefits of increasing α-linolenic acid intake in favor of promoting fish intake when there is no realistic need or hope of converting the world's large vegetarian population to fish intake.

One recently described example of a "novel" function of α-linolenic acid is in the very green leaves where it concentrates and can be modified to form jasmonic acid which is toxic to predatory insects and other pathogens, thereby protecting the green leaves and the plant as a whole (60–63,74). This might go some way toward explaining why α-linolenic acid is present in relatively high concentrations in so many different green leafed plants. It might also lead us toward new mechanisms by which α-linolenic acid contributes to immune defense systems in mammals.

Perspective

Opinion is often polarized around the nutritional and health merits of α-linolenic acid: either that it is entirely sufficient or that it is completely inadequate to meet the body's needs for n-3 PUFA. I have tried to show in this chapter that one should not feel forced to choose between α-linolenic and long chain n-3 PUFA; a great deal depends on the circumstances and on the food preferences of the individual. If someone chooses not to eat fish, then barring supplements, α-linolenic acid is going to be that individual's primary source of n-3 PUFA. If that person suffers a heart attack, some form of additional n-3 PUFA intake, preferably long chain n-3 PUFA, will usually be advantageous. The risk of that occurring increases with age, so it is advantageous to remain open-minded about preventive health strategies as one gets older. If one already eats fish regularly, there is no evidence that α-linolenic acid supplements confer additional benefit but they might (12,75). Much, therefore, relates to individual taste and risk factors. These risk factors influence the metabolic fate of α-linolenic acid (Fig. 3.4), thereby influencing the merit of additional α-linolenic acid intake.

The influence of aging on metabolism of α-linolenic acid is unclear but will become an increasingly important research and preventive health theme in the future. On the one hand, aging influences the ease with which α-linolenic acid can correct low docosahexaenoic acid levels in the rat brain (76). α-Linolenic acid also influences the survival of stroke-prone rats (77), an effect paralleling its beneficial effect in reducing all cause mortality in humans (12). On the other hand, despite widespread concern that lipid peroxidation contributes to aging, old age seems to have no discernible influence on the plasma content of α-linolenic acid in humans (78). These and other aspects of aging will need to be further explored in the near future.

Acknowledgments

The support and collaboration of several colleagues, particularly Dennis McIntosh, Lilian Thompson, David Jenkins, Tom Brenna, and Michael Crawford, is greatly appreciated. Research on α-linolenic acid that was performed by the author's group was supported at various times by NSERC, CIHR, The Flax Council of Canada, Flax Growers—Western Canada, Unilever-Lipton, Martek Biosciences, and Milupa AG. Mary Ann Ryan provided valuable technical assistance in the author's research.

References

1. Cunnane, S.C., The Contribution of α-Linolenic Acid in Flaxseed to Human Health, in *Flax—The Genus Linum*, edited by A. Muir and N. Westcott, Taylor and Francis, London, 2003, pp. 150–180.
2. Gerster, H., Can Adults Adequately Convert α-Linolenic Acid (18:3n-3) to Eicosapentaenoic Acid (20:5n-3) and Docosahexaenoic Acid (22:6n-3)? *Internat. J. Vit. Nutr. Res.* 68:159–173 (1997).
3. Cunnane, S.C., Young Scientist Award Lecture, Some New Themes and Issues in Research on Polyunsaturates: Synthesis, β-Oxidation, Carbon Recycling and Pure Linoleate Deficiency, *Can. J. Physiol. Pharmacol.* 74:629–639 (1996).
4. Cunnane, S.C., The Conditional Nature of the Dietary Need for Polyunsaturates: A Proposal to Reclassify "Essential Fatty Acids" as "Conditionally Indispensable" or "Conditionally Dispensable" Fatty Acids, *Br. J. Nutrition* 84:803–812 (2000).
5. Holman, R.T. (ed.), *Progress in the Chemistry of Fats and Other Lipids*, Pergamon Press, Oxford, 1971, pp. 275–340.
6. Cunnane, S.C., and M.J. Anderson, Pure Linoleate Deficiency in the Rat: Influence on Growth, Accumulation of n-6 Polyunsaturates and ^{14}C-Linoleate Oxidation, *J. Lipid Res.* 38:805–812 (1997).
7. Greenberg, S.M., C.E. Calbert, E.E. Savage, and H.J. Deuel, Jr., The Effect of Fat Level of the Diet on General Nutrition, *J. Nutr.* 41:473–486 (1950).

8. Bourre, J-M., M. Piciotti, O. Dumont, and G. Durand, Dietary Linoleic Acid and Polyunsaturated Fatty Acids in Rat Brain and Other Organs: Minimal Requirements of Linoleic Acid, *Lipids 25*:465–472 (1990).

9. Sprecher, H., The Synthesis and Metabolism of Hexadeca-4,7,10-trienoate, Eicosa-8,11,14-trienoate, Docosa-10,13,16-trienoate, and Docosa-6,9,12,15-tetraenoate in the Rat, *Biochim. Biophys. Acta 1*:519–530 (1968).

10. Cunnane, S.C., M.A. Ryan, K.A. Craig, S. Brookes, B. Koletzko, H. Demmelmair, J. Singer, and D.J. Kyle, Synthesis of Linoleate and α-Linolenate by Chain Elongation in the Rat, *Lipids 30*:781–783 (1995).

11. Legrand, P., D. Catheline, V. Rioux, and G. Durand, Lauric Acid is Desaturated by Hepatocytes and Rat Liver Homogenates, *Lipids 37*:569–572 (2002).

12. Dolecek, T.A., Epidemiological Evidence of Relationships Between Dietary Polyunsaturated Fatty Acids and Mortality in the Multiple Risk Factor Intervention Trial, *Proc. Soc. Exp. Biol. Med. 200*:177–182 (1992).

13. Hu, F.B., M.J. Stampfer, J.E. Manson, E.B. Rimm, A. Wolk, G.A. Colditz, C.H. Hennekens, and W.C. Willett, Dietary Intake of Alpha-Linolenic Acid and Risk of Fatal Ischemic Heart Disease Among Women, *Am. J. Clin. Nutr. 69*:890–897 (1999).

14. Guallar, E., A. Aro, F.J. Jimenez, J.M. Martin-Moreno, I. Salminen, P. van't Veer, A.F. Kardinaal, J. Gomez-Aracena, B.C. Martin, L. Kohlmeier, J.D. Kark, V.P. Mazaev, J. Ringstad, J. Guillen, R.A. Riemersma, J.K. Huttunen, M. Thamm, and F.J. Kok, Omega-3 Fatty Acids in Adipose Tissue and Risk of Myocardial Infarction: the EURAMIC Study, *Arterioscler. Thromb. Vasc. Biol. 19*:1111–1118 (1999).

15. Baylin, A., E.K. Kabagambe, X. Siles, and H. Campos, Adipose Tissue Biomarkers of Fatty Acid Intake, *Am. J. Clin. Nutr. 76*:750–757 (2002).

16. Cunnane, S.C. Metabolism and Function of α-Linolenic Acid in Humans, in *Flaxseed in Human Nutrition*, edited by S.C. Cunnane and L.U. Thompson, AOCS, Champaign, Illinois, 1995, pp. 99–127.

17. Voss, A., M. Reinhart, S. Sankarappa, and H. Sprecher, The Metabolism of 7,10,13,16,19-Docosapentaenoic Acid to 4,7,10,13,16,19-Docosahexaenoic Acid Is Independent of a Δ^4 Desaturase, *J. Biol. Chem. 266*:19995–20000 (1991).

18. Moore, S.A., E. Hurt, E. Yodor, H. Sprecher, and A.A. Spector, Docosahexaenoic Acid Synthesis in Human Skin Fibroblasts Involves Peroxisomal Retroconversion and Tetracosahexaenoic Acid, *J. Lipid Res. 36*:2433–2443 (1995).

19. Martinez, M., Abnormal Profiles of Polyunsaturated Fatty Acids in the Brain, Liver, Kidney and Retina of Patients with Peroxisomal Disorders, *Brain. Res. 583*:171–182 (1992).

20. Infante, J., and V. Huszagh, Zellweger Syndrome Knock-out Mouse Models Challenge Putative Peroxisomal β-Oxidation Involvement in Docosahexaenoic Acid (22:6ω3) Biosynthesis, *Mol. Genet. Metab. 72*:1–7 (2001).

21. Brenna, J.T., Efficiency of Conversion of α-Linolenic Acid to Long Chain n-3 Fatty Acids in Man, *Curr. Opin. Clin. Nutr. Metabol. Care 5*:127–132 (2002).

22. Crawford, M.A., K. Costeloe, K. Ghebremeskel, A. Phylactos, L. Skirvin, and F. Stacey Are Deficits of Arachidonic and Docosahexaenoic Acids Responsible for the Neural and Vascular Complications of Preterm Babies? *Am. J. Clin. Nutr. 66*:1032S–1041S (1997).

23. Horrobin, D.F., Y-S. Huang, S.C. Cunnane, and M.S. Manku, Essential Fatty Acids in Plasma, Red Blood Cells and Liver Phospholipids in Common Laboratory Animals as Compared to Humans, *Lipids 19*:806–811 (1984).

24. Adam, O., G. Wolfram, and N. Zollner, Effect of α-Linolenic Acid in the Human Diet on Linoleic Acid Metabolism and Prostaglandin Biosynthesis, *J. Lipid Res. 27*:421–426 (1986).

25. Budowski, P., N. Trostler, M. Lupo, N. Vaisman, and A. Eldor, Effect of Linseed Oil Ingestion on Plasma Lipid Fatty Acid Composition and Platelet Aggregability in Healthy Volunteers, *Nutr. Res. 4*:343–346 (1984).

26. Singer, P., W. Jaeger, I. Berger, H. Barleben, M. Wirth, E. Richter-Heinrich, S. Voigt, and W. Godicke, Effects of Dietary Oleic, Linoleic, and Alpha-Linolenic Acids on Blood Pressure, Serum Lipids, Lipoproteins, and the Formation of Eicosanoid Precursors in Patients with Mild Essential Hypertension, *J. Hum. Hypertens. 4*:227–233 (1990).

27. Bjerve, K.S., I. Lovold Mostad, and L. Thorensen, Alpha-Linolenic Acid Deficiency in Patients on Long Term Gastric Tube Feeding: Estimation of Linolenic Acid and Long Chain Unsaturated n-3 Fatty Acid Requirement in Man, *Am. J. Clin. Nutr. 45*:66–77 (1987).

28. Carnielli, V.P., D.J.L. Wattimena, I.H.T. Luijendijk, A. Boerlage, H.J. Degenhart, and P.J.J. Sauer, The Very Low Birth Weight Premature Infant is Capable of Synthesizing Arachidonic and Docosahexaenoic Acids from Linoleic and α-Linolenic Acids, *Pediatr. Res. 40*:169–171 (1996).

29. Salem, N., B. Wegher, P. Mena, and R. Uauy, Arachidonic and Docosahexaenoic Acids Are Biosynthesized from Their 18 Carbon Precursors in Human Infants, *Proc. Natl. Acad. Sci. 93*: 49–54 (1996).

30. Sauerwald, T.V., D.L. Hachey, and C. Jensen, Intermediates in Endogenous Synthesis of 22:6n-3 by Term and Preterm Infants, *Pediatr. Res. 41*:183–186 (1997).

31. Emken, E.A., R.O. Adlof, H. Rakoff, W.K. Rohwedder, and R.M. Gulley, Metabolism *in vivo* of Deuterium-Labelled Linolenic and Linoleic Acids in Humans, *Biochem. Soc. Trans. 18*:766–769 (1990).

32. Palowsky, R.J., J.R. Hibbeln, J.A. Novotny, and N. Salem, Jr., Physiological Compartmental Analysis of α-Linolenic Acid Metabolism in Adult Humans, *J. Lipid Res. 42*:1257–1265 (2001).

33. McCloy, U., Metabolism of [13]C Unsaturated Fatty Acids in Healthy Women, Ph.D. Thesis, University of Toronto, Toronto, 2002.

34. Burdge, G.C., and S.A. Wooten, Conversion of α-Linolenic Acid to Eicosapentaenoic, Docosapentaenoic and Docosahexaenoic Acids in Young Women, *Br. J. Nutr. 88*:411–420 (2002).

35. Burdge, G.C., A.E. Jones, and S.A. Wooten, Eicosapentaenoic and Docosapentaenoic Acids Are the Principal Products of Alpha-Linolenic Acid Metabolism in Young Men, *Br. J. Nutr. 88*:355–363 (2002).

36. Vermunt, S.H., R.P. Mensink, M.M. Simonis, and G. Hornstra, Effects of Dietary α-Linolenic Acid on the Conversion and Oxidation of ^{13}C-α-Linolenic Acid, *Lipids 35*:137–142 (2000).

37. Su, H.M., L. Bernardo, M. Mirmiran, X.H. Ma, T.N. Corso, P.W. Nathanielsz, and J.T. Brenna, Bioequivalence of Dietary α-Linolenic and Docosahexaenoic Acids as Sources of Docosahexaenoate Accretion in Brain and Associated Organs of Neonatal Baboons, *Pediatr. Res. 45*:1–7 (1999).

38. Cunnane, S.C., R. Ross, J.L. Bannister, and D.J.A. Jenkins, β-Oxidation of Linoleate in Obese Men Undergoing Weight Loss, *Am. J. Clin. Nutr. 73*:713–716 (2001).

39. Cunnane, S.C., and M.J. Anderson, The Majority of Linoleate in the Rat is β-Oxidized or Stored in Visceral Fat, *J. Nutr. 127*:146–152 (1997).

40. Chen, Z-Y., and S.C. Cunnane, Weight Cycling Progressively Depletes Carcass and Adipose Tissue Linoleic Acid and α-Linolenic Acid in Young Rats, *Br. J. Nutr.* 75:583–591 (1996).

41. Poumes-Balliaut, C., B. Langelier, F. Houlier, J.M. Alessandri, G. Durand, C. Latge, and P. Guesnet, Comparative Bioavailability of Dietary α-Linolenic and Docosahexaenoic Acids in the Growing Rat, *Lipids 36*:793–800 (2001).

42. Cunnane, S.C., Carbon Recycling: An Important Pathway in α-Linolenate Metabolism in Fetuses and Neonates, in *Fatty Acids: Physiological and Behavioral Functions*, edited by D.I. Mostovsky, S. Yehuda, and N. Salem, Jr., Humana Press, Totowa, New Jersey, 2001, pp. 145–159.

43. Simopoulos, A.P., Evolutionary Aspects of Diet and Essential Fatty Acids, in *Fatty Acids and Lipids—New Findings,* edited by T. Hamazaki and H. Okuyama, Karger, Basel, 2001, pp. 18–27.

44. Lands, W.E.M., Impact of Daily Food Choices on Health Promotion and Disease Prevention, in *Fatty Acids and Lipids—New Findings,* edited by T. Hamazaki and H. Okuyama, Karger, Basel, 2001, pp. 1–5.

45. Cunnane, S.C., and M.A. Crawford, Survival of the Fattest: Fat Babies Were the Key to Evolution of the Large Human Brain, *Comp. Biochem. Physiol.,* in press.

46. Broadhurst, C.L., Y. Wang, M.A. Crawford, S.C. Cunnane, J.E. Parkington, and W.F. Schmidt, Brain-Selective Nutrition from Marine, Lacustrine or Terrestrial Food Resources: Potential Impact on Early African *Homo sapiens, Comp. Biochem. Physiol. 131*:653–673 (2001).

47. Okayasu, T., M. Nagao, I. Ishibashi, and Y. Imai, Purification and Partial Characterization of Linoleoyl-CoA Desaturase from Rat Liver Microsomes, *Arch. Biochem. Biophys. 206*:21–28 (1981).

48. Cunnane, S.C., K.R. McAdoo, and D.F. Horrobin, Iron Intake Influences Essential Fatty Acids and Lipid Composition of Rat Plasma and Erythrocytes, *J. Nutr. 117*:1514–1519 (1987).

49. Cunnane, S.C., and J. Yang, Zinc Deficiency Impairs Whole Body Accumulation of Polyunsaturates and Increases the Utilization of $[1\text{-}^{14}C]$-Linoleate for *de novo* Lipid Synthesis in Pregnant Rats, *Can. J. Physiol. Pharmacol. 73*:1246–1252 (1995).

50. Dillard, C.J., E.E. Dumelin, and A.L. Tappel, Effect of Dietary Vitamin E on Expiration of Pentane and Ethane by the Rat, *Lipids 12*:109–114 (1977).

51. Chen, Z-Y., N. Ratnayake, and S.C. Cunnane, Evaluation of the Stability of Flaxseed During Baking, *J. Am. Oil Chem. Soc. 71*:629–632 (1994).

52. Fu, Z., and A.J. Sinclair, Novel Pathway of Metabolism of α-Linolenic Acid in the Guinea Pig, *Pediatr. Res. 47*:414–417 (2000).

53. Sheaff-Greiner, R.C., Q. Zhang, K.J. Goodman, D.A. Guissani, P.W. Nathanielsz, and J.T. Brenna, Linoleate, α-Linolenate and Docosahexaenoate Recycling into Saturated and Monounsaturated Fatty Acids Is a Major Pathway in Pregnant or Lactating Adults and Fetal or Infant Rhesus Monkeys, *J. Lipid Res. 137*:243–254 (1996).

54. Menard, C.R., K. Goodman, T. Corso, J.T. Brenna, and S.C. Cunnane, Recycling of Carbon into Lipids Synthesized *de novo* Is a Quantitatively Important Pathway of $[U\text{-}^{13}C]$-α-Linolenate Utilization in the Developing Rat Brain, *J. Neurochem. 71*:2151–2158 (1998).

55. Sinclair, A.J., Incorporation of Radioactive Polyunsaturated Fatty Acids into Liver and Brain of the Developing Rat, *Lipids 10*:175–184 (1975).

56. Dhopeshwarkar, G.A., and C. Subramanian, Metabolism of Linolenic Acid in the Developing Brain: Incorporation of Radioactivity from [1-^{14}C] Linolenic Acid into Brain Fatty Acids, *Lipids 10*:230–241 (1975).

57. Dhopeswarkar, G.A., and C. Subramanian, Metabolism of α-Linolenic Acid in Developing Brain. II. Incorporation of Radioactivity from [1-^{14}C]-α-Linolenic Acid into Brain, *Lipids 10*:242–247 (1975).

58. Willard, D.E., S.D. Harman, T.L. Kaduce, M. Preuss, S.A. Moore, M.E.C. Robbins, and A.A. Spector, Docosahexaenoic Acid Synthesis from n-3 Polyunsaturated Fatty Acids in Differentiated Rat Brain Astrocytes, *J. Lipid Res. 42*:1368–1376 (2001).

59. Cunnane S.C., D. Trotti, and M.A. Ryan, Specific Linoleate Deficiency in the Rat Does Not Prevent Substantial Carbon Recycling from ^{14}C-Linoleate into Sterols, *J. Lipid Res. 41*:806–811 (2000).

60. Gobel, C., I. Feussner, M. Hamberg, and S. Rosahl, Oxylipin Profiling in Pathogen-Infected Potato Leaves, *Biochim. Biophys. Acta 1584*:55–64 (2002).

61. Lee, J.Y., Y.S. Kim, and D.H. Shin, Antimicrobial Synergistic Effect of Linolenic Acid and Monoglyceride Against *Bacillus cereus* and *Staphalococcus aureus, J. Agric. Food Chem. 50*:2193–2199 (2002).

62. Blee, E., Impact of Phyto-Oxylipins in Plant Defense, *Trends Plant Sci. 7*:315–322 (2002).

63. Weber, H., Fatty Acid-Derived Signals in Plants, *Trends Plant Sci. 7*:217–224 (2002).

64. Lamptey, M.S., and B.L. Walker, A Possible Essential Role for Dietary Linolenic Acid in the Development of the Young Rat, *J. Nutr. 106*:86–93 (1976).

65. Bourre, J.M., G. Pascal, G. Durand, M. Masson, O. Dumont, and M. Piciotti, Alterations in the Fatty Acid Composition of Rat Brain Cells (Neurons, Astrocytes, and Oligodendrocytes) and of Subcellular Fractions (Myelin, Synaptosomes) Induced by a Diet Devoid of n-3 Fatty Acids, *J. Neurochem. 43*:342–348 (1984).

66. Moriguchi, T., J. Loewke, M. Garrison, J.N. Catalan, and N. Salem, Jr., Reversal of Docosaheaenoic Acid Deficiency in the Rat Brain, Retina, Liver and Serum, *J. Lipid Res. 42*:419–427 (2001).

67. Neuringer, M., W.E. Connor, C. van Petten, and L. Barstad, Dietary Omega-3 Fatty Acid Deficiency and Visual Loss in Infant Rhesus Monkeys, *J. Clin. Invest. 73*:272–276 (1984).

68. Raclot, T., and R. Groscolas, Differential Mobilization of White Adipose Tissue Fatty Acids According to Chain Length, Unsaturation and Positional Isomerism, *J. Lipid Res. 34*:1515–1526 (1993).

69. Bazinet, R., E. McMillan, R. Seebaransingh, A.M. Hayes, and S.C. Cunnane, Whole Body Oxidation of Linoleate and α-Linolenate in the Pig Varies Markedly with Weaning Strategy and Dietary α-Linolenate, *J. Lipid Res. 44*:314–319 (2003).

70. Bougnoux, P., S. Koscielny, V. Chajes, P. Descamps, C. Couet, and G. Calais, Alpha-Linolenic Acid Content of Adipose Breast Tissue: A Host Determinant of the Risk of Early Metastasis in Breast Cancer, *Br. J. Cancer 70*:330–334 (1994).

71. Giovannucci, E., E.B. Rimm, G.A. Colditz, M.J. Stampfer, A. Ascherio, C.C. Chute, and W.C. Willett, A Prospective Study of Dietary Fat and Risk of Prostate Cancer, *J. Natl. Cancer Inst. 85*:1571–1579 (1993).

72. De Stefani, E., H. Deneo-Pellegrini, P. Boffetta, A. Ronco, and M. Mendilaharsu, α-Linolenic Acid and Risk of Prostate Cancer: A Case-Control Study in Uruguay, *Cancer Epidemiol. Biomarkers Prev. 9*:335–338 (2000).

73. Mutanen, M., and R. Freese, Fats, Lipids and Blood Coagulation, *Curr. Opin. Lipidol.* *12*:25–29 (2001).

74. Imbusch, R., and M.J. Mueller, Formation of Isoprostane F(2)-like Compounds (Phytoprostanes F(1)) from α-Linolenic Acid in Plants, *Free Radic. Biol. Med.* *28*:720–726 (2000).

75. Bemelmans, W.J., J. Broer, E.J. Feskens, A.J. Smit, F.A. Muskiet, J.D. Lefrandt, V.J. Bom, J.F. May, and B. Meyboom-de Jong, Effect of an Increased Intake of Alpha-Linolenic Acid and Group Nutritional Education on Cardiovascular Risk Factors: The Mediterranean Alpha-Linolenic Acid Enriched Grieningen Dietary Intervention (MARGARIN) Study, *Am. J. Clin. Nutr. 75*:221–227 (2002).

76. Delion, S., S. Chalon, D. Guilloteau, J.C. Basnard, and G. Durand, Alpha-Linolenic Acid Dietary Deficiency Alters Age-Related Changes of Dopaminergic and Serotoninergic Neurotransmission in the Rat Frontal Cortex, *J. Neurochem. 66*:1582–1591 (1996).

77. Shimokawa, T., A. Moriuchi, T. Hori, M. Saito, Y. Naito, H. Kabasawa, Y. Nagae, M. Matsubara, and H. Okuyama, Effect of Dietary Alpha-Linolenate/Linoleate Balance on Mean Survival Time, Incidence of Stroke and Blood Pressure of Spontaneously Hypertensive Rats, *Life Sci. 43*:2067–2075 (1988).

78. Bjerve, K.S., K.J. Fougner, K. Midthjell, and K. Bonaa, n-3 Fatty Acids in Old Age, *J. Intern. Med. (Suppl. 225)*:191–196 (1989).

79. Tinoco, J., Dietary Requirements and Functions of α-Linolenic Acid in Animals, *Prog. Lipid Res. 21*:1–38 (1982).

80. Delaney, J.P., M.M. Windhauser, C.M. Champagne, and G.A. Bray, Differential Oxidation of Individual Dietary Fatty Acids in Humans, *Am. J. Clin. Nutr. 72*:905–911 (2000).

81. Jones, P.J., P.B. Pencharz, and M.T. Clandinin, Whole Body Oxidation of Dietary Fatty Acids: Implications for Energy Utilization, *Am. J. Clin. Nutr. 42*:769–777 (1985).

82. Cenedella, R.J., and A. Allen, Differences Between the Metabolism of Linoleic and Palmitic Acids: Utilization for Cholesterol Synthesis and Oxidation to Respiratory CO_2, *Lipids 4*:155–158 (1969).

83. Leyton, J., P.J. Drury, and M.A. Crawford, Differential Oxidation of Saturated and Unsaturated Fatty Acids *in vivo* in the Rat, *Br. J. Nutr. 57*:383–393 (1987).

84. Dupont, J., Fatty Acid Oxidation in Relation to Cholesterol Biosynthesis in Rats, *Lipids 1*:415–421 (1966).

85. Clouet, P., I. Niot, and J. Bezard, Pathway of α-Linolenic Acid Through the Mitochondrial Outer Membrane in the Rat Liver and Influence on the Rate of Oxidation, *Biochem. J. 263*: 867–873 (1989).

86. Gavino, G.R., and V.C. Gavino, Rat Liver Outer Mitochondrial Carnitine Palmitoyl-transferase Activity Towards Long-Chain Polyunsaturated Fatty Acids and Their CoA Esters, *Lipids 26*:266–270 (1991).

87. Bjorntorp, P., Rates of Oxidation of Different Fatty Acids by Isolated Rat Liver Mitochondria, *J. Biol. Chem. 243*:2130–2133 (1968).

88. Vasdev, S.C., and K.J. Kako, Incorporation of Fatty Acids into Rat Heart Lipids: *in vivo* and *in vitro* Studies, *J. Molec. Cellul. Cardiol. 9*:617–631 (1977).

89. Anderson, G.J., and W.E. Connor, Uptake of Fatty Acids by the Developing Rat Brain, *Lipids 23*:286–290 (1988).

90. Bandyopadhyay, G.K., J. Dutta, and S. Ghosh, Preferential Oxidation of Linolenic Acid Compared to Linoleic Acid in the Liver of Catfish *(Heteropneustes fossilis* and *Clarias batrachus)*, *Lipids 17*:733–740 (1982).

91. Holman, R.T., S.B. Johnson, and T.F. Hatch, A Case of Human Linolenic Acid Deficiency Involving Neurological Abnormalities, *Am. J. Clin. Nutr. 35*:617–623 (1982).

92. Bjerve, K.S., L. Thoresen, and S. Borsting, Linseed and Cod Liver Oil Induce Rapid Growth in a 7-Year-Old Girl with n-3 Fatty Acid Deficiency, *J. Parent. Enter. Nutr. 12*:521–525 (1988).

93. Beitz, J., H-J. Mest, and W. Forster, Influence of Linseed Oil Diet on the Pattern of Serum Phospholipids in Man, *Acta Biol. Med. Germ. 40*:K31–K35 (1981).

94. Kestin, M., P. Clifton, G.B. Belling, and P.J. Nestel, n-3 Fatty Acids of Marine Origin Lower Systolic Blood Pressure and Triglycerides but Raise LDL Cholesterol Compared with n-3 and n-6 Fatty Acids from Plants, *Am. J. Clin. Nutr. 51*:1028–1034 (1990).

95. Cunnane, S.C., S. Ganguli, C.R. Menard, A.C. Liede, M.J. Hamadeh, Z-Y. Chen, T.M.S. Wolever, and D.J.A. Jenkins, High α-Linolenic Acid Flaxseed (*Linum usitatissimum*): Some Nutritional Properties in Humans, *Br. J. Nutr. 69*:443–453 (1993).

96. Kelley, D.S., G.J. Nelson, J.E. Love, L.B. Branch, P.C. Taylor, P.C. Schmidt, B.E. Mackey, and J.M. Iacono, Dietary α-Linolenic Acid Alters Tissue Fatty Acid Composition but Not Blood Lipids, Lipoproteins or Coagulation Status in Humans, *Lipids 28*:533–537 (1993).

97. Mantzioris, E., M.J. James, R.A. Gibson, and L.G. Cleland, Dietary Substitution with an α-Linolenic Acid-Rich Vegetable Oil Increases Eicosapentaenoic Acid Concentrations in Tissues, *Am. J. Clin. Nutr. 59*:1304–1309 (1994).

98. Cunnane, S.C., M.J. Hamadeh, A.C. Liede, L.U. Thompson, T.M.S. Wolever, and D.J.A. Jenkins, Nutritional Attributes of Traditional Flaxseed in Healthy Young Adults, *Am. J. Clin. Nutr. 61*:62–68 (1995).

99. Layne, K.S., Y.K. Goh, J.A. Jumpsen, E.A. Ryan, P. Chow, and M.T. Clandinin, Normal Subjects Consuming Physiological Levels of 18:3(n-3) and 20:5(n-3) from Flaxseed or Fish Oils Have Characteristic Differences in Plasma Lipid and Lipoprotein Fatty Acid Levels, *J. Nutr. 126*:2130–2140 (1996).

100. Sanders, T.A.B., and F. Roshanai, The Influence of Different Types of n-3 Polyunsaturated Fatty Acids on Blood Lipids and Platelet Function in Healthy Volunteers, *Clin. Sci. 64*:91–99 (1983).

101. Chan, J.K., B.E. McDonald, J.M. Gerrard, V.M. Bruce, B.J. Weaver, and B.J. Holub, Effect of Dietary α-Linolenic Acid and Its Ratio to Linoleic Acid on Platelet and Plasma Fatty Acids and Thrombogenesis, *Lipids 28*:811–817 (1993).

102. Ferrier, L.K., L.J. Caston, S. Leeson, J. Squires, B.J. Weaver, and B.J. Holub, α-Linolenic Acid- and Docosahexaenoic Acid-Enriched Eggs from Hens Fed Flaxseed: Influence on Blood Lipids and Platelet Phospholipid Fatty Acids in Humans, *Am. J. Clin. Nutr. 62*:81–86 (1995).

103. Freese, R., and M. Mutanen, α-Linolenic Acid and Marine Long-Chain n-3 Fatty Acids Differ Only Slightly in Their Effects on Hemostatic Factors in Healthy Subjects, *Am. J. Clin. Nutr. 66*:591–598 (1997).

Chapter 4

Analysis and Bioavailability of Lignans

Lilian U. Thompson

Department of Nutritional Sciences, Faculty of Medicine, University of Toronto, Toronto, Ontario M5S 3E2, Canada

Introduction

Plant lignans are phenolic compounds with a 2,3-dibenzylbutane skeleton. Some can be metabolized to the mammalian lignans enterodiol (ED) and enterolactone (EL) by the bacterial flora in the colon (1; Fig. 4.1). They are of interest because of suggestions that they are protective against chronic diseases including cancer, cardiovascular disease, diabetes, and kidney disease (2–6). Although lignans are found in many plant foods, flaxseed is the richest source (7), and its increased consumption and health benefits have been attributed in part to its lignans. However, the optimum level and frequency of flaxseed and lignan intake necessary to produce the health benefits remain to be established. This is in part due to incomplete information regarding lignan bioavailability, including lignan absorption, distribution, metabolism, and excretion, because of difficulties in the analysis of lignans in flaxseed to determine intake levels, and in body fluids and tissues to determine lignan disposition. This chapter will discuss some methods used for lignan analysis and the current knowledge on bioavailability of lignans, particularly those in flaxseed.

Analysis

Food Samples

Since the discovery of mammalian lignans about 20 yr ago (8,9), it has been assumed that the only precursors of mammalian lignans are secoisolariciresinol diglycoside (SDG) and matairesinol (1). Recently, however, other precursors of mammalian lignans were found, although they differ in degree of conversion (10). The precursors and their percentage conversion after 24 h incubation with human fecal flora are lariciresinol, 101%; secoisolariciresinol (SECO), 72%; matairesinol, 62%; pinoresinol diglucoside, 55%; 7-hydroxymatairesinol (HMR), 15%; arctigenin glucoside, 5.5%; and syringaresinol diglucoside, 4.0% (Fig. 4.1). It is established that SDG is the major mammalian lignan precursor in flaxseed, with matairesinol and pinoresinol also present in very small amounts (7,11,12). Isolariciresinol, although present, is not metabolized to the mammalian lignans. Hence, for flaxseed lignan analysis, most methods have concentrated on the determination of SDG.

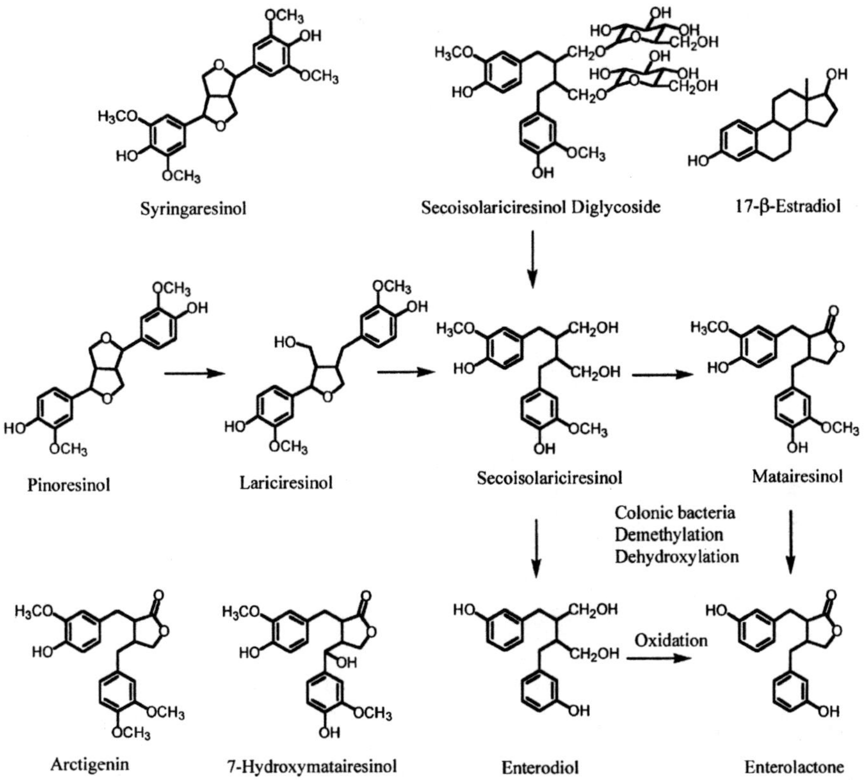

Fig. 4.1. Chemical structures of mammalian lignans and their precursors compared with 17-β estradiol.

The lignans must first be extracted from their biological matrix before separation and detection by chromatographic techniques [e.g., high performance liquid chromatography (LC) or gas chromatography (GC)]. This has been difficult in the case of flaxseed because of the complex polymeric structure of SDG (13–15). Unlike other major phytoestrogens such as the isoflavones, which are present as either free aglycones or as simple glycosides, the SDG in flaxseed exists as a "polymer" consisting of five SDG residues interconnected by four 3-hydroxyl-3-methyl glutaric acid residues (15). Incomplete extraction of SDG from this polymer may have been partly responsible for the large variation in SDG values reported for flaxseed in the literature (7,12,16–24; Table 4.1).

Table 4.1 summarizes some of the more recent methods used to analyze food lignans. In general, these methods utilize either enzyme (7,16–19), acid (12,18,20), or base (21–24) hydrolysis or their combination (18) before or after extraction with organic solvents such as methanol, dioxane/ethanol, ether, ethyl acetate, hexane, and acetonitrile. *In vitro* fermentation with human fecal inoculum has also been used not

TABLE 4.1
Some Methods of Food Lignan Analysis

Sample preparation	Stationary phase and detection	Mobile phase	Lignan concentration (mg/g flaxseed)[a]	Ref.
SPE extraction (C18 cartridge) Enzyme hydrolysis (β-glucuronidase) SPE extraction (C18 cartridge) TMS derivatization	HP-1 capillary (25 m × 0.2 mm, 0.11 μm) GC-FID GC-MS	Helium	0.29–0.95[b] 0.35–1.15[c]	7,17
Enzyme hydrolysis (β-glucuronidase) Solvent extraction (water and acetonitrile) SPE extraction (C18 cartridge)	Alltech C18 Econosil (250 × 4.6 mm, 5 μm) with guard cartridge (30 × 4.6 mm) LC-UV	Acetonitrile/0.1% glacial acetic acid aqueous 30:70	0.82	16
Enzyme hydrolysis (*Helix pomatia* juice) Solvent extraction (ether) Acid hydrolysis (2 M HCl) Solvent extraction (ether) Column chromatography (DEAE- Sephadex OH⁻ and QAE-Sephadex Ac⁻) TMS derivatization	BP-1 (0.2 mm × 12.5 m) ID-GC-MS-SIM	Helium	3.70	18
Solvent extraction (methanol/water 80:20) Enzyme hydrolysis (β-glucosidase) Solvent extraction (ether)	C8 reversed-phase column (100 × 4.6 mm, 300 Å) LC-MS	30% acetonitrile in 10 mM ammonium acetate	—	19

Acid hydrolysis (1.5 m HCl) Solvent extraction (ethyl acetate/ *tert*-butyl ether 1:1) TMS derivatization	Capillary column (100% polysiloxane, 0.25 mm × 15 m) GC-MS	Helium	7.4–12.6[d]	20
Solvent extraction (methanol/ water 80:20) Acid hydrolysis (1 M HCl) Solvent extraction (ethyl acetate/hexane 1:1)	Microsorb semipreparative C18 (250 × 10 mm, 5 µm) LC-DAD DB-1701 (J&W Scientific, 0.25 mm × 15 m, 0.25 µm) GC-MS	LC:A: water/glacial acetic acid 99.8:0.2; B: acetonitrile GC: Helium	—	12
Solvent extraction (1,4-dioxane/ 95% ethanol 50:50) Base hydrolysis (0.3 M NaOH) SPE (C18) extraction	Econosil RP C18 (250 × 4.6 mm, 5 µm) LC-UV/PDA	A: 5% acetonitrile in 0.01 M phosphate buffer, pH 2.8; B: acetonitrite	3.22–7.02	21
Solvent extraction (methanol/water 70:30) Base hydrolysis (0.1 M NaOH)	Symmetry C18 (250 × 4.6 mm, 5 µm) LC-PDA	A: 1% aqueous acetic acid; B: methanol	4.91 7.18–10.55	22 23
Solvent extraction (methanol/water 70:30) Base hydrolysis (1 M NaOH) Amberlite XAD-2 extraction	Hypersil RP18 (250 × 4.6 mm) LC-UV	A: 2% aqueous acetic acid; B: acetonitrile	—	24

[a]As secoisolariciresinol. [b]Enterodiol +, enterolactone + secoisolariciresinol. [c]Equivalent secoisolariciresinol. [d]Secoisolariciresinol + shonanin.

Abbreviations: BP-1, bonded phase column; FID, flame ionization detector; GC, gas chromatograph; ID-GC-MS-SIM, isotope dilution gas chromatograph mass spectrometer selected ion monitoring mode; LC, high performance liquid chromatograph; MS, mass spectrometer; PDA, diode array detector; SPE, solid phase extraction; TMS, trimethylsilyl; and UV, ultraviolet detector.

only to release the SDG from the polymer but also to metabolize it to the mammalian lignans, which are then analyzed (7,17).

Each of the methods has advantages and disadvantages. In the *in vitro* fermentation method (7,17), precursors other than SDG are metabolized to mammalian lignans at the same time as SDG. Therefore, analytical data using this technique represent the total amount of mammalian lignans produced from all precursors present in the food during the fermentation period. It is advantageous in that both known and unknown precursors of mammalian lignans can be accounted for. The data may also represent the amount of plant lignans that can be physiologically converted to mammalian lignans *in vivo*. It is good for screening foods for mammalian lignan precursors when the identity of the precursors is unknown. We used this method in 1991 to screen 68 plant foods for their potential to produce mammalian lignans (7). However, this indirect analytical method for lignan precursors has limitations for routine analysis. Until all the bacteria capable of metabolizing the different precursors to mammalian lignans are identified, the analysis will require the use of human fecal inoculum for *in vitro* fermentation. The analytical success will then be dependent on proper collection and handling of fecal inoculum so that the viability of bacteria is maintained. If fecal samples from different donors differ in bacterial profile and activity, then comparison of results from different batch analysis may not be possible. Therefore, samples to be compared should be analyzed at the same time using the same fecal inoculum.

In contrast, methods using enzymatic, acid, or base hydrolysis have the advantage of measuring the lignan precursors directly. The SDG content of flaxseed analyzed using acid or base hydrolysis methods is severalfold higher than that observed using *in vitro* fermentation or enzyme hydrolysis methods (Table 4.1). This indicates that enzyme hydrolysis alone does not completely release all the SDG and possibly other precursors from the food matrix. This finding also suggests that *in vitro* fermentation does not completely metabolize the precursors to mammalian lignans over a 24-h period. However, analysis of many flaxseed samples using enzyme (β-glucuronidase) hydrolysis produced results that are close to and related significantly to the total mammalian lignans produced using the *in vitro* fermentation method ($r = 0.572$, $P < 0.003$) (17). Thus, although the level of SDG from the enzyme hydrolysis method is low, it may represent the level that can be metabolized to mammalian lignans *in vivo*.

When the acid hydrolysis method is used, some SECO, the aglucone of SDG, is degraded to anhydrosecoisolariciresinol, which when analyzed can be considered as primarily derived from SECO (18,20). Because anhydrosecoisolariciresinol is identical to the naturally occurring lignan shonanin (3,4-divanyllyltetrahydrofuran) (20), acid hydrolysis methods measure not only SECO but also shonanin. This compound has not yet been demonstrated to be a precursor of mammalian lignan, although it has some biological activity (e.g., competing with 5-α dihydrotestosterrone for sex hormone binding globulin) and may be protective against prostate cancer (25). Furthermore, the optimum time for acid hydrolysis to obtain maximum yield of the aglucone lignan differs with foods (20). The yield of aglucone does not plateau with hydrolysis time but instead decreases as further hydrolysis breaks down the lignan. Hence, values obtained using acid hydrolysis represent only the amount of aglucone released at the specific hydrolysis time from the specific food and not necessarily the total amount of lignans in that food. The acid hydrolysis method may overestimate the lignans that can be metabolized to mammalian lignans *in vivo* or may underestimate the lignans due to breakdown that may have taken place during hydrolysis.

A comparison of available data bases for food lignans showed lower values for flaxseed and cereals and higher values for fruits and vegetables when analysis was done using the *in vitro* fermentation method than with the combined enzyme and acid hydrolysis method (26). This indicates that at high lignan levels, there is incomplete conversion of the precursors using the *in vitro* fermentation method, whereas precursors in foods other than SDG and matairesinol are not all accounted for by the current enzyme–acid hydrolysis method.

The base hydrolysis method has been used successfully in the analysis and isolation of flaxseed lignans (21–23). However, it has not yet been used extensively in the analysis of lignans in other foods. The enzyme hydrolysis method used in the analysis of foods other than flaxseed (19) found that the fruits and vegetables with a high content of SECO and matairesinol were similar to those foods that produced high levels of ED and EL using the *in vitro* fermentation method (7). This finding suggests agreement betweeen the two methods.

Previous rat studies on the anticancer effects of flaxseed compared the effect of a 5 or 10% flaxseed diet with that of SDG at levels present in the equivalent quantity of flaxseed (2). Because the flaxseed used contains 2 mg SDG/g based on analysis using the enzyme hydrolysis-HPLC-UV method, SDG was provided in a diet supplemented with 0.01 or 0.02% SDG or as a daily gavage of 1.5 or 3.0 mg, the equivalent daily intake of SDG in a 5 or 10% flaxseed diet. These levels of SDG produced effects similar to those of flaxseed, suggesting that the effect of flaxseed was primarily due to the mammalian lignans derived from SDG. Analysis using the base hydrolysis-GC-mass spectrometry (MS) method showed that the flaxseed used in these studies contained 13 mg SDG/g (Thompson, L.U., unpublished data). Hence, the actual intake of SDG from flaxseed per day would have been about 6.5 times more than the amount actually used in the studies. The significance of this observation is not clear, but it raises questions regarding the physiological implication of analytical data obtained by different methods of analysis.

Urine, Plasma, and Tissue Samples

Some recent methods for the determination of lignans in urine, plasma, and tissue samples are summarized in Table 4.2 (27–39). In urine and plasma, the lignans are generally conjugated with glucuronic acid or sulfate. However, lignan conjugates are often not directly analyzed, in part due to lack of standards for these conjugates. Instead, many available methods analyze the aglucones after overnight

TABLE 4.2
Some Methods of Analysis for Lignans in Blood, Urine, and Tissues

Sample preparation	Stationary phase and detection	Mobile phase	Ref.
Solvent extraction (ethanol) SPE extraction (Sep-Pak C18 cartridge) Column chromatography (DEAE-Sephadex Ac⁻ and SP-Sephadex H⁺) SPE extraction (Sep-Pak C18 cartridge) Column chromatography (QAE-Sephadex CO_3^-) TMS derivatization	BP-1 (0.2 mm × 12.5 m) ID-GC-MS	Helium	27, 28
Precipitation SPE extraction (C18 cartridge) Enzyme hydrolysis (β-glucuronidase) Column purification (DEAE-Sephadex) TMS derivatization	HP-1 capillary (25 m × 0.2 mm, 0.11 μm)	Helium	29
Enzyme hydrolysis (β-glucuronidase)	Hypersol C18 (150 × 3 mm, 3 μm) LC-CD	A: 50 mM sodium acetate, pH 4.8/methanol 80:20; B: 50 mM sodium acetate, pH 4.8/methanol/acetonitrile 40:40:20	30
Enzyme hydrolysis (β-glucuronidase and sulfatase) Solvent extraction diethyl ether)	Inertsil ODS-3 (150 × 3 mm, 3 μm) with Quick Release C18 (10 × 3 mm, 5 μm) LC-CD	A: 50 mM sodium acetate buffer pH 5.0/methanol 80:20; B: 50 mM sodium acetate buffer, pH 5.0/methanol/HCN 40:40:20	31
Enzyme hydrolysis with (*Helix pomatia* enzyme) SPE extraction (Sep-Pak tC18)	Symmetry C18 (100 × 4.6 mm, 3.5 μm) LC-MS-MS	A: methanol/0.1% HAc 90:10 with 0.1% isopropanol; B: 0.1% HAc with 1% isopropanol	32

(Continued)

hydrolysis with β-glucuronidase and sulfatase, a common source of which is *Helix pomatia*. The aglucones are purified from the buffered reaction medium by either solid phase extraction using C_{18} Sep-Pak cartridges or extraction with ethyl acetate, ether, or their combination. They are then analyzed by GC-MS or LC with various detectors such as UV, photodiode array (PDA), coulometric array (CoulArray), or MS.

TABLE 4.2
(*Cont.*)

Sample preparation	Stationary phase and detection	Mobile phase	Ref.
SPE extraction (Sep-Pak C18 cartridge) Enzyme hydrolysis (β-glucuronidase and sulfatase) SPE extraction (Sep-Pak C18 cartridge)	C8 reversed-phase column (150 × 4.6 mm, 300 Å) LC-MS	A: 10 mM ammonium acetate; B: Acetonitrile	33
Enzyme hydrolysis (β-glucuronidase/ sulfatase) SPE extraction (C18 or Oasis cartridge)	Prism reversed-phase column (50 × 3.0 mm, 5 μm) LC-MS/MS	A: 10.15 mM ammonium acetate; B: Acetonitrile/ methanol 1:1	34
Enzyme hydrolysis (β-glucuronidase and arylsulfatase) Solvent extraction (ethyl ether or diethyl ether)	HydroBond PS reversed-phase column (100 × 3.0 mm, 5 μm) with HydroBond PS C18 guard column (25 × 3.2 mm, 5 μm) LC-PDA-MS	1. A: Acetonitrile/methanol 1:1; B: 0.5% Aqueous acetic acid 2. A: Acetonitrile/methanol 1:1; B: Water	35
SPE extraction (RP-18 cartridge) Column chromatograph (DEAE-Sephadex A25) SPE extraction (RP-18 cartridge) Enzyme hydrolysis (β-glucuronidase and aryl sulfatase) SPE extraction (RP-18 cartridge)	RP-18 (250 × 4.6 mm, 5 μm) HPLC-UV	A: Water/methanol 84:16 with formic acid, pH2.8; B: Methanol	36
Enzyme hydrolysis (β-glucuronidase and sulfatase) Solvent extraction (diethyl ether)	TRF	TRF	37–39

Abbreviations: See Table 4.1; CD, coulometric array detector; ID, isotope dilution; TRF, time resolved fluoroimmunoassay.

GC-MS analysis is the main method, and has been used for about 20 y. When the isotope dilution (ID)-GC-MS-selected ion monitoring mode (ID-GC-MS-SIM) is used, it is quite specific and sensitive with detectable levels as low as 0.1 nM (27,28). Deuterated internal standards of the lignans are added to the extracted samples before silylation to form trimethylsilyl derivatives of the compounds and injection to the GC-MS. However, this method is very time consuming and labor intensive because of the extensive sample cleanup procedure, hydrolysis, and derivatization steps to volatilize the analytes. Hence, it is not suitable for the analysis of a large number of samples such as those in population studies. GC-MS and deuterated standards are also expensive and not available in many laboratories.

LC methods have the advantage of ease of use. They do not require an extensive sample cleanup procedure, and in many cases, samples can be injected directly into the LC after extraction. Both the aglucone and the conjugated lignans can be analyzed, because they need not be volatilized for analysis. LC coupled with UV or PDA detector (36) has been used because of its relatively lower equipment and operational costs. However, it lacks selectivity and sensitivity, particularly for measuring low baseline lignan levels in individuals consuming a typical Western diet that is traditionally low in lignans. LC with CoulArray detector is more sensitive and specific compared with LC-UV or LC-PDA systems (30,31). However, it is dependent on the redox potential, and reduced selectivity and sensitivity is a problem for some analytes in complex matrices such as serum. The highest sensitivity and selectivity with exceptional diagnostic ability can be achieved with the LC coupled with the MS detector and electrospray interface (ESI). Hence LC-ESI-MS has been suggested to be a better tool for measuring lignans considering its excellent sensitivity, specificity, ease of sample preparation, and higher sample throughput. Its disadvantages are the need for expensive equipment, which many laboratories cannot afford, as well as the extensive expertise required in the operation of the equipment. Its other limitation, which is also true for GC-MS, is the high cost or lack of commercially available lignan standards or internal standards (e.g., deuterated lignans), which are necessary for quantitation and accounting for analyte losses during the sample purification and chromatography.

Despite the advantages of LC- and GC-based methods, they remain laborious and expensive for screening a large number of samples, particularly in epidemiological studies. For this purpose, the time-resolved fluoroimmunoassay (TRF) method is particularly useful (37–39). This method depends on antigen-antibody interactions to form a soluble antigen-antibody complex. TRF has the advantage of being able to analyze many samples in a single analysis depending on whether 96- or 384-well plates are used. A batch of about 100 samples can be finished in a 4-h period. Commercial TRF kits are already available, which makes it suitable for routine analysis. The sensitivity of the assay is comparable to that of GC-MS, and the results relate significantly to values obtained by ID-GC-MS-SIM method. It has been used successfully in the analysis of urine and plasma in several epidemiological studies (40–43). However, a disadvantage is that the selectivity is sometimes decreased due to cross reactivity with other components in the sample. Also, single analytes are analyzed at a time, although this may be compensated by the large number of samples that can be analyzed per run. At present, an analytical kit for lignans is available only for EL. Hence, the TRF method is promising but needs further development for the analysis of other lignans.

Bioavailability

Absorption and Metabolism

After the discovery of the mammalian lignans ED and EL in human urine about 20 y ago (8,9), it was established that these lignans are dietary in origin. They cannot be formed in germ-free rats (44) or individuals who took antibiotics (45), indicating that intestinal bacteria are necessary for their synthesis. Bacterial action in the colon results in removal of glucose residues, demethylation, and dehydroxylation of the precursors to ED and EL (46; Fig. 4.1). Rats (47), monkeys (48), and humans (48) fed flaxseed experienced large increases in urinary lignan excretion of ED and EL, suggesting a high level of precursor in this food; these observations have been confirmed when flaxseed was compared with many plant foods (7,18). Subsequently, SDG, matairesinol, pinoresinol, and lariciresinol (10,47) were found to be precursors of ED and EL in flaxseed. Little is known regarding the types of bacteria responsible for metabolism of specific precursors, but *Lactobacilli*, *Bacteroides*, and *Bifidobacteria* are known to have β-glycosidase activity (49,50), and *Peptostreptococcus asp.* Strain SDG-1 and *Eubacterium sp.* Strain SDG-2 isolated from human fecal microbiota have been shown to convert SDG to ED (51). *In vitro* incubations of SDG with these bacteria also yielded five metabolites other than ED and EL, but they were not detected in the urine of rats gavaged with radiolabeled SDG (^{3}H-SDG) (52). The transformation of SDG or ED to EL occurred only under anaerobic and not under aerobic conditions or with sterilized fecal inoculum, indicating that it is due to the metabolic reaction of the bacteria under anaerobic conditions and not as a result of spontaneous chemical reaction (51).

After formation from precursors, mammalian lignans undergo enterohepatic circulation, that is, they are absorbed from the colon, transported to the liver by portal venous blood, and secreted in bile (1). The glycosides of lignans have not yet been found in biological fluids, suggesting that they are not absorbed intact and are hydrolyzed by the bacteria prior to absorption, as was observed for the isoflavone glycosides (53). The lignans are conjugated to glucuronic acid and sulfate in the liver with hepatic UDP-glucuronosyl transferase or sulfotransferase enzymes (1,54). This facilitates their clearance into the urine and bile. The presence of intestinal glucuronosyl transferase and sulfotransferase suggests that conjugation may also take place in the intestinal wall during uptake.

A portion of the circulating lignans reaches the kidney and is excreted in the urine. Hence, urinary lignan level has been used as an indicator of lignan intake, metabolism, and availability in humans and animals (2). As previously discussed (55; Chapter 5), oxidative metabolism of ED and EL also takes place in the liver. Studies *in vitro* have shown that rat liver microsomes can metabolize ED to three aromatic and four aliphatic monohydroxylated compounds and EL to six aromatic and six aliphatic monohydroxylated compounds (56). Some of these metabolites have been detected in the urine of rats administered ED or EL (10 mg/kg b.w.) intraduodenally and in the urine of rats gavaged with ED or EL or fed a diet containing flaxseed (36). However, only the aromatic monohydroxylated metabolites of ED and EL were detected in the urine of four human subjects fed 16 g flaxseed for 5 d (57). These metabolites represent <3% of the ED or EL fed to rats (36) or <5% of the total urinary lignans in the human subjects fed flaxseed (57). These results differ from those obtained when radiolabeled SDG was gavaged to rats. ED, EL, SECO, and four unidentified metabolites were detected, but none of the unidentified metabolites had mass spectra that matched those observed by others after incubation of ED or EL by liver microsomes (55,56) or of SDG by bacteria (51). When nonradioactive SDG or flaxseed was fed, only two of the unknown metabolites were observed in the urine of rats (52). The other metabolites may have been present but at too low concentrations to be detectable, because when a large volume of the urine was analyzed, one of the monohydroxylated ED metabolites was detected in the flaxseed-fed rats but not in the SDG-fed rats. The physiological significance of the low levels of oxidative metabolites of ED and EL remains to be determined. However, these studies further established that the main metabolites of SDG are ED and EL; therefore, the physiological effect of SDG is likely primarily due to these mammalian lignans.

Urinary lignan excretion of rats was determined 24 h after a single gavage of equivalent amounts (25 mg/kg b.w.) of SDG, SECO, HMR, matairesinol, and EL (58). All compounds were metabolized to ED and/or EL, but some HMR and SECO were also absorbed and excreted as such, suggesting that they may also exert some physiological effect without prior conversion to ED or EL. The urinary excretion of ED and EL was threefold higher after SECO administration than after administration of its glycoside SDG. EL was the major metabolite after matairesinol, HMR, and SDG administration, but ED was produced more after SECO administration. EL administered as such was excreted at the highest concentration compared with the other lignans. This suggests that compounds that require no or little microbial metabolism are more efficiently absorbed in the gastrointestinal tract. The absolute configuration of the lignans did not appear to change during intestinal microbial action.

Disposition of Lignans

The disposition of SDG at 12, 24, 36, and 48 h was determined in rats after gavaging them with ^{3}H-SDG acutely (immediately after overnight fasting) or chronically (10 d after they were gavaged with 1.5 mg unlabeled SDG/d) (59). The SDG metabolite distribution at 12 and 48 h after ^{3}H-SDG administration is summarized in Table 4.3.

Fecal and Urine Excretion. With acute exposure, maximum fecal radioactivity was observed at 12 h with no further significant increase at 48 h. In contrast, fecal radioactivity was negligible in the chronically fed group at 12 h but was similar to the acutely treated rats by 24 h. Hence, chronic treatment delays but does not reduce fecal excretion of SDG metabolites. It was suggested that at 12 h, higher bacterial

TABLE 4.3
Disposition of SDG Metabolites in Acute and Chronic Groups at 12 and 48 h
after ^{3}H-SDG

Samples	12 h		48 h	
	Acute	Chronic	Acute	Chronic
		(% of recovered ^{3}H-SDG dose)		
GI contents	32.9	83.7	3.1	4.1
Feces	52.3	0.2	61.1	65.0
Urine	7.3	2.6	32.1	26.9
Blood	0.4	0.7	0.2	0.3
GI organs	2.8	6.8	0.5	0.6
Other organs	4.3	5.9	3.0	3.2

Abbreviations: GI, gastrointestinal; SDG, secoisolariciresinol diglycoside.
Source: Summarized from Rickard and Thompson (59).

activity in the chronically fed group, as indicated by increased β-glucuronidase activity (60), may have increased the bacterial metabolism of the SDG and enhanced the absorption and enterohepatic circulation of lignans. However, between 12–24 h, the excess SDG substrate may have exceeded the metabolic capacity of the bacteria and hence decreased the metabolism and absorption and increased the excretion of SDG. At 24–48 h, most of the recovered radioactivity (50–65%) was excreted in the feces (Table 4.3), regardless of treatment.

Urinary excretion was 3–7% of the recovered dose at 12 h, which increased to 27–32% of the recovered dose at 48 h (59; Table 4.3). The increase over the 48 h was significant in the chronic group but not in the acute group. However, the urinary excretion at all four time points did not differ significantly between the two groups. The increase in urinary lignan excretion suggests removal of the lignans from the tissues over time. The proportion of radioactivity (12%) in the urine of the chronically treated group at 24 h is in agreement with the finding of 11.4% excretion of ED plus EL in rats given a daily gavage of 1.5 mg unlabeled SDG for 2 wk (61). This suggests that the main metabolites of ^{3}H-SDG are ED and EL, which was further confirmed in a later analysis of the radioactive lignan components of the urine (52).

Gastrointestinal Content and Tissues. At 12 h, a large proportion of the recovered activity was detected in the gastrointestinal contents, particularly in the cecum and colon of the chronic group (59; Table 4.3). Although the radioactivity decreased significantly at 48 h, a small amount still remained, indicating it takes more than 48 h to rid the body of lignans.

About 3–7% of the recovered dose was found in the gastrointestinal tissues at 12 h, which decreased to <1% at 48 h (Table 4.3). The highest level was found in the cecum throughout the 48 h period, which might be expected because bacterial metabolism of the lignans takes place in this rat organ. Higher cecal tissue radioactivity was observed in the chronic group than in the acute group (59).

Other Body Tissues. Nongastrointestinal tissues had about 4–6% of the recovered dose at 12 h, which declined to about 3% at 48 h (59; Table 4.3). Radioactivity was observed in all tissues analyzed (heart, liver, kidney, spleen, lung, adipose, mammary gland, ovary, uterus, skin, muscle, and brain), indicating that SDG metabolites are able to reach these tissues (Fig. 4.2). The highest radioactivity levels per gram of tissue were observed in the liver, kidney, and uterus compared with other nongastrointestinal tissues. This was expected because the liver is the site for lignan conjugation with glucuronide and metabolism, and a large portion of the metabolites are excreted by way of the kidney. Mammalian lignans have been shown to antagonize estrogen action in the uterus (62), so their high level in the uterus may also be expected. However, low activity was observed in the mammary gland, another hormone-sensitive tissue. This suggests that the effect of mammalian lignans on mammary cancer may be through their binding to the tumors rather than to the mammary gland. It is also possible that the lignan levels peak at times other than those chosen in this study, or the low levels may be a consequence of a constant flux of lignans into and out of the tissue, with no accumulation. Whereas at 12 h, the lowest radioactivity was found in the adipose tissue of acutely treated rats, this was three-fold higher in the chronically treated rats, perhaps due to transient storage of mammalian lignans. Although the adipose tissue is a primary site for peripheral estrogen production by the action of aromatase on andro-

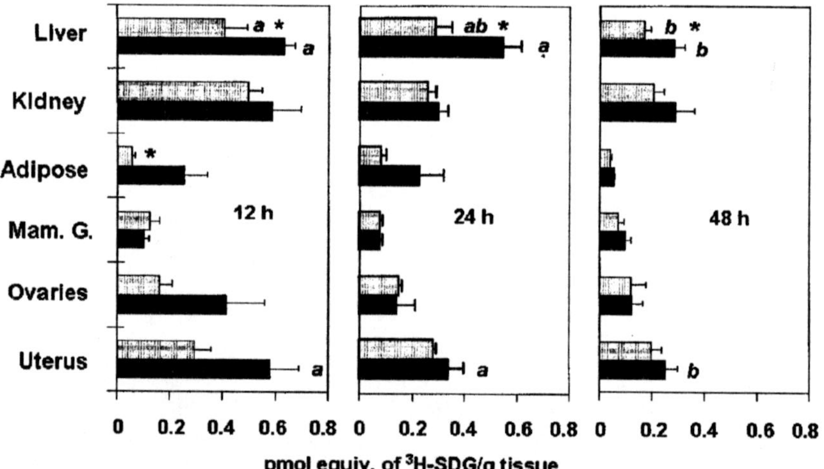

Fig. 4.2. Distribution of radioactivity over 48 h in some tissues of rats that were not exposed to secoisolariciresinol diglycoside (SDG) (acute, light bar) or that were gavaged daily with unlabeled SDG (1.5 mg/d) for 10 d (chronic, dark bar). Bars with an asterisk indicate a significant difference ($P < 0.05$) between acute and chronic groups for that tissue. Bars between panels with different letters represent significant differences ($P < 0.05$) betweeen time points for a tissue within a treatment group. *Abbreviation:* Mam. G, mammary gland. *Source:* Modified with permission from Rickard and Thompson (59).

gens and EL has antiaromatase activity (63,64), the physiological significance of the observed adipose levels remains to be elucidated. In general, chronic SDG intake increased the lignan levels in the tissues, particularly in the liver.

Blood Levels. Blood radioactivity represents <1% of recovered dose throughout the experimental period, and it is mostly in the plasma (Table 4.3). The highest level, which was observed at 12 h, decreased significantly at 48 h in both the acute and chronic treatment groups. Total clearance of lignans took more than 48 h because radioactivity was still detected at that time. A follow-up study showed that blood levels of lignans peaked at 9 h and were still higher than baseline at 24 h in both treatment groups (52; Fig. 4.3). However, at 24 h the blood levels remained high in the chronic treatment group whereas levels declined in the acute treatment group (52).

In agreement with the rat studies, plasma levels of healthy premenopausal women also peaked at 9 h on the first day of consuming 25 g flaxseed (acute intake) and did not change significantly at 12 or 24 h (29). On the eighth day of

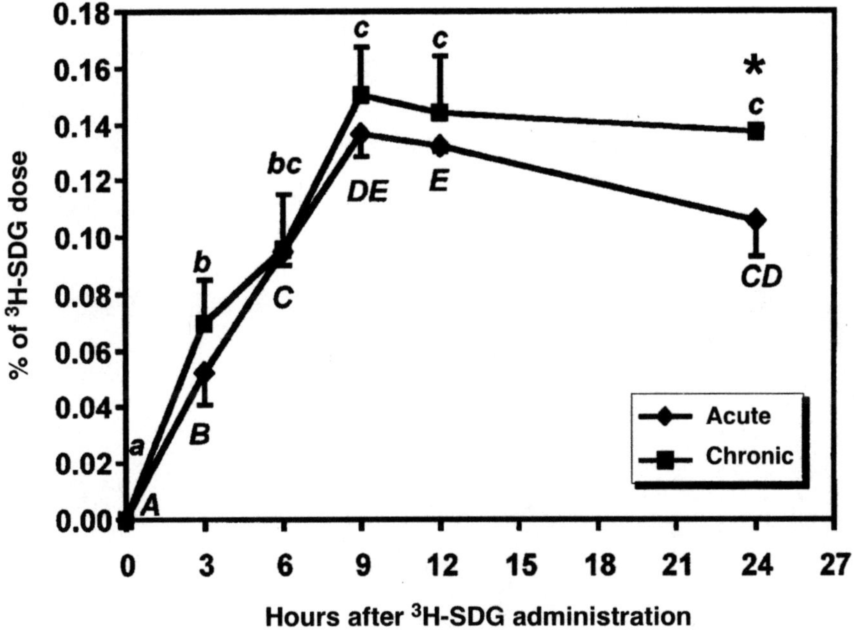

Fig. 4.3. Postprandial whole blood radioactivity levels (% of ³H-SDG dose) in rats that were not exposed to secoisolariciresinol diglycoside (SDG) (acute) or that were gavaged daily with unlabeled SDG (1.5 mg/d) for 10 d (chronic). Values were adjusted for baseline (0 h radioactivity). Symbols within a line not sharing a letter were significantly different (*P* < 0.05). *Source:* Modified with permission from Rickard and Thompson (52).

supplementation (chronic intake), plasma lignan levels did not differ significantly over the 24 h period, but, at zero time, the level was higher than the levels observed after 1 d of supplementation. These findings indicate that plasma lignan levels can be maintained with one daily dose of flaxseed after only 1 wk of supplementation. Also, plasma lignan levels were significantly related to urinary lignan excretion ($r = 0.54$, $P < 0.05$; $y = 2.20x + 663.37$), indicating that urinary lignan excretion may be used as an indicator of plasma lignan level. In another study, increases in blood levels of EL and ED were not observed until 8.5 h after ingestion of a cake containing soybean flour and cracked linseed (flaxseed) (65), in agreement with the above studies in rats (52) and humans (29). All of the above studies indicate that plasma lignan kinetics in response to consumption of flaxseed or SDG are comparable between rats and humans.

From the study with radiolabeled SDG, the plasma level was estimated to be 1 µM in rats fed 1.5 mg SDG/d, a value about 3000 times higher than peak estrogen levels in the rat (52), and sufficient to produce biological effects (62). However, the plasma levels in women fed 25 g flaxseed ranged from 17–519 nM with a mean of 97 nM (29). Whether this plasma level will produce the same effects as those observed in rat studies remains to be established. It is noted, however, that the consumption of 25 g flaxseed by postmenopausal breast cancer patients caused reductions in their tumor cell proliferation and c-erB labeling index (66); consumption of 10 g flaxseed by healthy premenopausal women influenced estrogen metabolism such that the ratio of urinary 2-hydroxy estrogen to 16α-hydroxy estrone is increased (67).

Maternal Transfer of Lignans. Studies in rats showed that suckling offspring of dams that were gavaged with radiolabeled SDG had significantly increased body radioactivity compared with offspring of dams gavaged with nonradioactive SDG (68). This suggests that mammalian lignans from dams can be transferred to the suckling offspring through the milk. At delivery, pregnant Japanese women who had consumed a soy-rich diet had isoflavones and lignans (ED and EL) in their plasma, umbilical cord, and amniotic fluid, indicating that phytoestrogens can cross from the maternal to the fetal compartment (69). Therefore, the presence of lignans in the maternal diet may have physiological implications in offspring.

Dose-Response to Flaxseed or Lignans. In rats fed 2.5, 5.0, or 10% flaxseed or gavaged with 1.1, 2.2, or 4.4 µmol SDG/d, levels equivalent to the respective flaxseed diets, the urinary ED + EL or ED + EL + SECO excretion increased linearly with consumption of from 0–5% flaxseed and from 0–2.2 µmol SDG/d, and then leveled off (61; Fig. 4.4). A similar trend in mammalian lignan production was observed when 1.1, 2.2., and 4.4. µmol SDG were subjected to *in vitro* fermentation using human fecal inoculum, thus resulting in a significant correlation ($r = 0.990$, $P < 0.02$) between *in vitro* production and *in vivo* excretion of mammalian lignans (62). This suggests that *in vivo* mammalian lignan production from foods may be predicted by

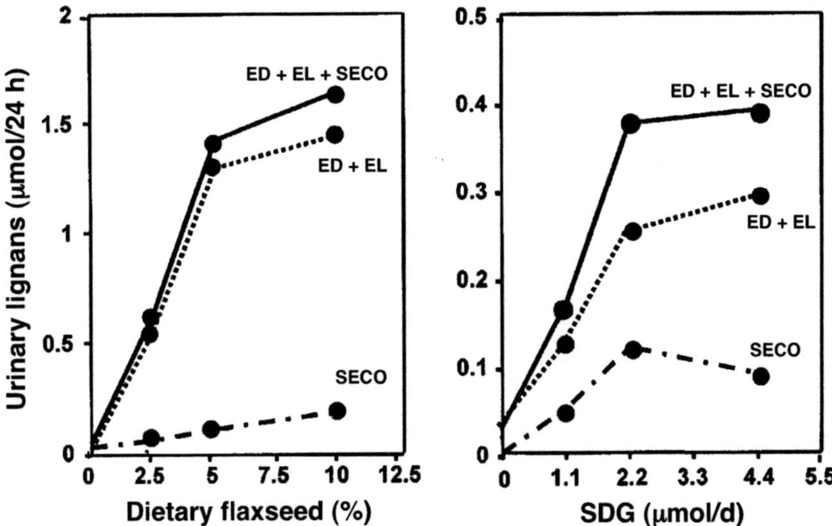

Fig. 4.4. Urinary lignan excretion of rats fed the basal high-fat diet supplemented with various levels of flaxseed or secoisolariciresinol diglycoside (SDG). *Abbreviations:* ED, enterodiol; EL, enterolactone; SECO, secoisolariciresinol. *Source:* Modified with permission from Rickard *et al.* (61).

results from *in vitro* fermentation and that *in vitro* methods can be used as a useful tool for screening foods for mammalian lignan-producing ability. However, the urinary lignan excretion in rats fed SDG was much lower than in rats fed flaxseed, accounting for only about 20% of the ED + EL excretion and 26–40% of the ED + EL + SECO excretion in flaxseed-fed rats. This could be due to a number of reasons including the presence of precursors other than SDG in flaxseed, underestimation of SDG in flaxseed, incomplete metabolism of SDG to mammalian lignans, and greater metabolism of SDG in flaxseed because of its high dietary fiber content.

In contrast to the above observation, the total urinary lignan excretion did not level off in a lupus nephritis mouse model given a daily gavage of 0.6, 1.2, and 4.8 mg SDG/mouse (70). The urinary lignan excreted was also mostly SECO instead of ED and EL. It is unclear if this effect is unique to this mouse model or if it is a consequence of the lupus disease. Nonetheless, it appears that, at high levels of SDG intake, the effect observed in this mouse model may be due more to SECO than to its ED and EL metabolites.

A study in rats gavaged with the lignan precursor HMR (3, 15, 25, and 50 mg/kg b.w.) showed a linear correlation between urinary EL level and the HMR dose (71). It did not plateau at the high HMR dose, indicating a greater capacity of the rats to metabolize the HMR than the SDG.

In agreement with the rat studies (61), a dose-dependent increase in urinary lignan was observed in premenopausal women who consumed 5, 15, and 25 g flaxseed (*r* =

0.72, $P < 0.001$) (29; Fig. 4.5), and in postmenopausal women who consumed 5 and 10 g flaxseed (72). A plateau was not observed, indicating that a higher level than 25 g/d may be required to reach the maximum limit of lignan production (29). The daily consumption of 5% flaxeed diet by rats is about equivalent to a daily consumption of 25 g flaxseed by humans. Hence the human data (29,72) are in agreement with data in rats (61), which showed linear response to flaxseed up to a 5% flaxseed diet.

Factors Affecting Lignan Bioavailability

Processing. Flaxseed has a hard seed coat so it has been assumed that the bioavailability of its components, including the lignans, will be low unless the seed is cracked or ground. However, no study has yet confirmed whether there are any differences in bioavailability from whole *vs.* ground seed. Despite this, studies on the effect of flaxseed have used ground flaxseed either in the raw form or incorporated in processed foods such as breads and muffins. Studies in healthy premenopausal women showed that the urinary lignan excretion after the consumption of 25 g ground flaxseeed either raw or in muffins or bread did not differ significantly (29; Fig. 4.5). This suggests high stability of the lignans under the baking conditions, or if there was any loss during baking, it was not high enough to cause significant differences in physiological response. Furthermore, when ground flaxseed was incorporated into pizza dough, pancakes, breads, and muffins, the production of mammalian lignans after *in vitro* fermentation with human fecal inoculum was directly related to the level of flaxseed (6.2–13.2%) despite differences in the cooking time (10-40 min) and temperature (190–205°C) of processing (73). In contrast, another study recovered only 73–75% of the SDG in bread loaves baked with 0, 4, 8, or 12% added flaxseed but recovered almost all of the

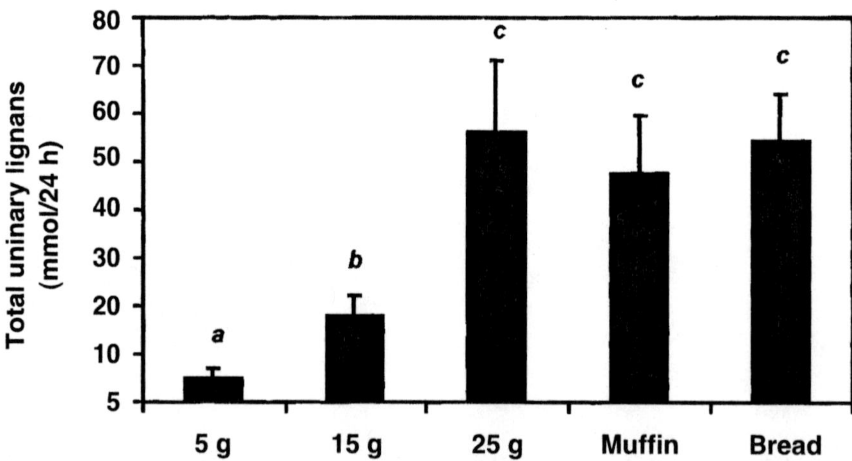

Fig. 4.5. Urinary lignan excretion in women consuming various levels of ground raw flaxseed or 25 g ground flaxseed in standard muffin or bread formulation. Bars with different letters are significantly different ($P \le 0.05$) *Source:* Drawn from data in Nesbitt *et al.* (29).

82% pure SDG incorporated into the bread formulation (22). No hydrolysis of the SDG complex was observed during baking. The reason for the discrepancies between the studies is not clear but may in part relate to the analytical method used to determine the loss of lignans. The latter study (22) analyzed the SDG directly using the base hydrolysis-LC-PDA method, whereas the former studies (29,73) measured mammalian lignans produced after *in vitro* fermentation or in the urine after human consumption *in vivo*. The method that is more appropriate to determine processing losses remains to be determined.

Methods of dehulling flaxseed have been developed to produce a hull fraction that has severalfold higher concentration of SDG than the intact seed (74). Because this fraction is also rich in dietary fiber that can enhance bacterial activity, increased bacterial metabolism of the SDG might be expected when it is consumed. However, the bioavailability of lignans in this hull fraction has not yet been tested.

Individual Variability. As reviewed by others (75; Chapter 6), many studies showed a large interindividual variation between subjects in plasma lignan levels and urinary lignan excretion, not only in response to flaxseed but also to other lignan-containing foods. Although ED and EL are the lignans produced in the highest concentration from SDG, the ratio of ED/EL in the plasma or urine varied with individuals (29,76), suggesting that the oxidation of ED to EL also differs with individuals, even within the same sex. Because lignan metabolism is dependent on gut microflora, the variation may be due to interindividual differences in their bacterial profile and their ability to metabolize lignans, which in turn may be influenced by enviromental and physiological factors including antibiotic use, gender, diet, and genetics.

The involvement of intestinal bacteria has been demonstrated by a reduction in serum or urinary lignans with the intake of antibiotics (46,77,78) and the inability of ileostomy subjects to convert the precursors to ED or EL (79). Gender differences were observed in a small study (n = 11) where women showed higher serum EL than men when fed a whole grain *vs.* refined diet (80). In a large Finnish study (2753 subjects) (81), baseline serum EL was lower (17.7 nmol/L) in men than in women (20.9 nmol/L). Lower serum EL in men (26 nmol/L) than in women (40 nmol/L) was also observed after consumption of whole rye bread for 4 w, although their mean urinary EL excretion did not differ (82). However, no gender difference in urinary lignan excretion was observed in a feeding study with 50 g flaxseed per day (83). Smokers had lower serum EL than did nonsmokers (81).

In a Finnish cross-sectional study, female subjects with normal weight had significantly higher EL than underweight or obese subjects (81). In obese subjects, the lower serum EL may be due to rapid transport of EL into preadipocyte cells. In men and women, a positive association was observed betweeen serum EL concentration and constipation. The slower intestinal motility in subjects with constipation is thought to result in a more complete metabolism and absoption of lignans and hence higher serum EL. In women, serum EL was positively and independently associated with subject age.

Diet. The metabolism of lignan-containing foods may also depend on the habitual or background diet of the individual. A large cross-sectional study showed that plasma EL has a positive relationship with daily vegetable servings, dietary fiber, alcohol, caffeine (as surrogate marker for coffee and tea consumption), and daily servings of foods in the Chenopodiaceae (e.g., spinach, beets), Juglandaceae (e.g., walnuts, pecans), Leguminosae (e.g., beans, peas), Pedaliaceae (e.g., sesame), and Vitaceae (e.g grapes) botanical groups (84). It was estimated that for each 10 g increase in fiber and each 50 mg increase in caffeine, the increases in plasma EL were 37 and 6.6%, respectively. Consumers of 0.5–1 alcoholic drink/d had a plasma EL level that was 131.4% higher than nondrinkers. Other studies showed a positive relationship of serum EL with the consumption of whole grain products, fruits, vegetables, and berries (81,85). In clinical studies, consumption of berries (86) and whole grains (80) has been shown to increase plasma levels and urinary excretion of EL.

Diets high in fruits, vegetables, and whole grains contain not only lignans but also high levels of dietary fiber. Dietary fiber may affect the absorption, reabsorption, and excretion of estrogen and phytoestrogens through its effect on the β-glycosidase and β-glucuronidase activity of the colonic microflora. Fiber can dilute the gut microflora, bind with phytoestrogens, and reduce the absorption and increase the excretion of lignans (87). Conversely, fermentable fiber may increase bacterial activity and enhance the metabolism and absorption of lignans, thereby increasing their bioavailabilty. Differences in diet may be responsible for the ethnic differences in metabolic and biological responses to lignans.

Overall, these observations suggest that certain foods in the individual's habitual diet may influence the blood or urinary levels of lignans, because they contain lignans or other components such as dietary fiber that may influence intestinal microflora. Therefore, the overall diet composition should be taken into consideration when conducting clinical studies to determine the bioavailability and biological effects of lignans.

Summary and Conclusions

Several methods have been developed for the analysis of lignans in flaxseed, other foods, urine, plasma, and tissues. Each of the methods has advantages and disadvantages, and the choice of method depends on the lignan concentration, the sample matrix, and available instrumentation. However, for a better comparison of analytical data from different laboratories, particularly in the case of SDG in flaxseed, there is a need to establish a standard method that is sensitive, selective, and physiologically relevant.

Many mammalian lignan precursors and metabolites have recently been identified, but SDG is still considered the major lignan in flaxseed, with ED and EL its primary metabolites. A linear dose response was observed in urinary lignan excretion with consumption of increasing levels of flaxseed up to 25 g in humans, or up to 5% flaxseed or 2.2 μmol purified SDG/d in rats, after which the excretion plateaued. In both rats and humans, plasma levels peaked at 9 h and remained high at 24 h. Once a day consumption of flaxseed or SDG for 1 wk stabilized the plasma levels. Lignans are detected in all body tissues, and it takes more than 48 h for the lignans to be cleared from the body. Chronic intake of flaxseed or SDG increased the bioavailabilty of lignans. Processing of flaxseed in baked products did not have a significant effect on urinary lignan excretion. Although many of the bioavailabity studies were done in the animal model, the results appear to be in agreement with those in humans.

There are large interindividual variations in the metabolism of lignans that may cause differences in physiological effects. The reasons are unclear but may include several factors such as variabilities in intestinal microflora, use of antibiotics, gender, constipation, and habitual diet. These factors should be taken into consideration when designing experiments to determine the physiological effects of flaxseed or lignans.

Acknowledgment

Zhen Liu helped to list the methods in Tables 4.1 and 4.2 and draw Figure 4.1. Work of the author was funded by Natural Sciences and Engineering Research Council of Canada.

References

1. Setchell, K.D.R., Discovery and Potential Clinical Importance of Mammalian Lignans, in *Flaxseed in Human Nutrition*, edited by S.C. Cunnane and L.U. Thompson, 1st edn., AOCS Press, Champaign, Illinois, 1995, pp. 82–98

2. Thompson, L.U., Flaxseed, Lignans, and Cancer, in *Flaxseed in Human Nutrition*, edited by L.U. Thompson and S.C. Cunnane, 2nd edn., AOCS Press, Champaign, Illinois, 2003, pp. 194–222.

3. Prasad, K., Flaxseed and Prevention of Experimental Hypercholesterolemic Atherosclerosis, in *Flaxseed in Human Nutrition*, edited by L.U. Thompson and S.C. Cunnane, 2nd edn., AOCS Press, Champaign, Illinois, 2003, pp. 259–272.

4. Stavro, P.M., and D.J., Jenkins, Flaxseed, Fiber and Coronary Heart Disease: Clinical Studies, in *Flaxseed in Human Nutrition*, edited by L.U. Thompson and S.C. Cunnane, 2nd edn., AOCS Press, Champaign, Illinois, 2003, pp. 288–300.

5. Prasad, K., Flaxseed Components in the Prevention of Experimental Diabetes, in *Flaxseed in Human Nutrition*, edited by L.U. Thompson and S.C. Cunnane, 2nd edn., AOCS Press, Champaign, Illinois, 2003, pp. 273–286.

6. Ogborn, M.R., Flaxseed and Flaxseed Products in Kidney Disease, in *Flaxseed in Human Nutrition*, edited by L.U. Thompson and S.C. Cunnane, 2nd edn., AOCS Press, Champaign, Illinois, 2003, pp. 301–318.

7. Thompson, L.U., P. Robb, M. Serraino, and F. Cheung, Mammalian Lignan Production from Various Foods, *Nutr. Cancer 16:*43–52 (1991).

8. Setchell, K.D.R., A.M. Lawson, F.L. Mitchell, H. Adlercreutz, D.N. Kirk, and M. Axelson, Lignans in Man and Animal Species, *Nature 287:*740–780 (1980).

9. Stitch, S.R., J.K. Toumba, M.B. Groen, C.W. Funke, J. Leemhuis, J. Vink, and G.F. Woods, Excretion, Isolation and Structure of a New Phenolic Constituent in Female Urine, *Nature 287:*738–740 (1980).

10. Heinonen, S., T. Nurmi, K. Liukkonen, K. Poutanen, K. Wahala, T. Deyam, S. Nishibe, and H. Adlercreutz, *In Vitro* Metabolism of Plant Lignans: New Precursor of Mammalian Lignans Enterolactone and Enterodiol, *J. Agric. Food Chem. 49*:3178–3186 (2001).

11. Mazur, W., Phytoestrogen Content in Foods, *Bailliere's Clin. Endocrinol. Metab. 12*:729–742 (1998).

12. Meagher, L.P., G.R. Beecher, V.P. Flanagan, and B.W. Li, Isolation and Characterization of the Lignans, Isolariciresinol and Pinoresinol, in Flaxseed Meal, *J. Agric. Food Chem. 47*:3173–3180 (1999).

13. Bakke, J.E., and H.J. Klosterman, A New Diglucoside from Flaxseed, *Proc. N. Dak. Acad. Sci. 10*:18–22 (1956).

14. Ford, J.D., K.S. Huang, H.B. Wang, L. Davin, and N.G. Lewis, Biosynthetic Pathway to the Cancer Chemopreventive Secoisolariciresinol Diglucoside-Hydroxymethyl Glutaryl Ester-Linked Lignan Oligomers in Flax (*Linum usitatissimum*) Seed, *J. Nat. Prod. 64*:1388–1397 (2001).

15. Kamal-Eldin, A., N. Peerlkamp, P. Johnson, R. Andersson, R.E. Andersson, L.N. Lundgren, and P. Aman, An Oligomer from Flaxseed Composed of Secoisolariciresinol Diglucoside and 3-Hydroxy-3-Methyl Glutaric Acid Residues, *Phytochemistry 58*:587–590 (2001).

16. Obermeyer, W.R., S.M. Musser, J.M. Betz, R.E. Casey, A.E. Pohland, and S.W. Page, Chemical Studies of Phytoestrogens and Related Compounds in Dietary Supplements: Flax and Chaparral, *Proc. Soc. Exp. Biol. Med. 208*:6–12 (1995).

17. Thompson, L.U., S.E. Rickard, F. Cheung, E.O. Kenaschuk, and W.R. Obermeyer, Variability in Anticancer Lignan Levels in Flaxseed, *Nutr. Cancer 27*:26–30 (1997).

18. Mazur, W.M., T. Fotsis, K. Wahala, S. Ojala, A. Salakka, and H. Adlercreutz, Isotope-dilution Gas-Chromatographic Mass-Spectrometric Method for the Determination of Isoflavonoids, Coumestrol and Lignans in Food Samples, *Anal. Biochem. 233*:169–180 (1996).

19. Horn Ross, P.L., S. Barnes, M. Lee, L. Coward, J.E. Mandel, J. Koo, E.M. John, and M. Smith, Assessing Phytoestrogen Exposure in Epidemiologic Studies: Development of a Data Base (United States), *Cancer Causes Control 11*:289-298 (2000).

20. Liggins, J., R. Grimwood, and S.A. Bingham, Extraction and Quantification of Lignan Phytoestrogens in Food and Human Samples, *Anal. Biochem. 287*:102–109 (2000).

21. Johnsson, P., A. Kamal-Eldin, L.N. Lundgren, and P. Aman, HPLC Method for Analysis of Secoisolariciresinol Diglucoside in Flaxseed, *J. Agric. Food Chem. 48:* 5216–5219 (2000).

22. Muir, A.D., and N.D. Westcott, Quantitation of the Lignan Secoisolariciresinol Diglucoside in Baked Goods Containing Flaxseed or Flax Meal, *J. Agric. Food Chem. 48*:4048–4052 (2000).

23. Westcott, N.D., and A.D. Muir, Variation in Flaxseed Lignan Concentration with Variety, Location and Year, in *Proceedings of the 56th Flax Institute of the United States*, Fargo, North Dakota, 1996, pp. 77–85.

24. Degenhardt, A., S. Habben, and P. Winterhalter, Isolation of the Lignan Secoisolariciresinol Diglucoside from Flaxseed (*Linum usitatissimum*) by High Speed Countercurrent Chromatography, *J. Chromatog. A 943*:299–302 (2002).

25. Schottner, M., D. Gansser, and G. Spiteller, Lignans from the Roots of *Urtica Dioca* and Their Metabolites Bind to Human Sex Hormone Binding Globulin (SHBG), *Planta Med. 63*:529–532 (1998).

26. Meagher, L.P., and G.R. Beecher, Assessment of Data on the Lignan Content of Foods, *J. Food Comp. Anal. 13*:935–947 (2000).

27. Adlercreutz, H., T. Fotsis, K. Wahala, T. Makela, G. Brunow, and T. Hase, Quantitative Determination of Lignans and Isoflavonoids in Plasma of Omnivorous and Vegetarian Women by Isotope Dilution Gas Chromatography-Mass Spectrometry, *Scand. J. Clin. Lab. Invest. 53* (Suppl. 215):5–18 (1993).

28. Adlercreutz, H., T. Fotsis, M.S. Kurzer, K.T. Wahala, T. Makela, and T. Hase, Isotope Dilution Gas Chromatographic-Mass Spectrometric Method for the Determination of Conjugated Lignans and Isoflavonoids in Human Feces, with Preliminary Results in Omnivorous and Vegetarian Women, *Anal. Biochem. 225*:101–108 (1995).

29. Nesbitt, P.D., Y. Lam, and L.U. Thompson, Human Metabolism of Mammalian Lignan Precursor in Raw and Processed Flaxseed, *Am. J. Clin. Nutr. 69*:549–555 (1999).

30. Gamache, P.H., and I.N. Acworth, Analysis of Phytoestrogens and Polyphenols in Plasma, Tissue, and Urine Using HPLC with Coulometric Array Detection, *Proc. Soc. Exp. Biol. Med. 217*:274–280 (1998).

31. Nurmi, T., and H. Adlercreutz, Sensitive High-Performance Liquid Chromatography Method for Profiling Phytoestrogen Using Coulometric Electrode Array Detection: Application to Plasma Analysis, *Anal. Biochem. 274*:110–117 (1999).

32. Saarinen, N.M., A. Smeds, S.I. Makela, J. Ammala, K. Hakala, J-M. Pihlava, E-L. Ryhanen, R. Sjoholm, and R. Santti, Structural Determinants of Plant Lignans for the Formation of Enterolactone *in vivo*, *J. Chromatog.* B *777*:311–319 (2002).

33. Horn Ross, P.L., S. Barnes, M. Kirk, L. Coward, J. Parsonnet, and R.A. Hiatt, Urinary Phytoestrogen Levels in Young Women from a Multiethnic Population, *Cancer Epidemiol. Biomark. Prev. 6*:339–345 (1997).

34. Valentin-Blasini, L., B.C. Blount, H.S. Rogers, and L. Needham, HPLC-MS/MS Method for the Measurement of Seven Phytoestrogens in Human Serum and Urine, *J. Exp. Anal. Environ. Epidemiol. 10*:799–807 (2000).

35. Franke, A.A., L.J. Custer, L.R. Wilkens, L.L. Marchand, A.M.Y. Nomura, M.T. Goodman, and L.N. Kolonel, Liquid chromatographic-Photodiode Array Mass Spectrometric Analysis of Dietary Phytoestrogens from Human Urine and Blood, *J. Chromatog.* B *777*:45–59 (2002).

36. Niemeyer, H.B., D. Honig, A. LangeBohmer, E. Jacobs, S.E. Kulling, and M. Metzler, Oxidative Metabolites of the Mammalian Lignans Enterodiol and Enterolactone in Rat Bile and Urine, *J. Agric. Food Chem. 48*:2910–2919 (2000).

37. Adlercreutz, H., G.J Wang, O. Lapcik, R. Hampl, K. Wahala, T. Makela, K. Lusa, M. Talme, and H. Mikola, Time-Resolved Fluoroimmunoassay for Plasma Enterolactone, *Anal. Biochem. 265*:208–215 (1998).

38. Stumpf, K., M. Uehara, T. Nurmi, and H. Adlercreutz, Changes in the Time-Resolved Fluoroimmunoassay of Plasma Enterolactone, *Anal. Biochem. 284*:153–157 (2000).

39. Uehara, M., O. Lapcik, R. Hampl, N. Al-Maharik, T. Makela, K. Wahala, H. Mikola, and H. Adlercreutz, Rapid Analysis of Phytoestrogens in Human Urine by Time Resolved Fluoroimmunoassay, *J. Steroid Biochem. Molec. Biol. 72*:273–282 (2000).

40. Pietinen, P., K. Stumpf, S. Mannisto, V. Kataja, M. Uusitupa, and H. Adlercreutz, Serum Enterolactone and Risk of Breast Cancer: A Case Control Study in Eastern Finland, *Cancer Epidemiol. Biomark. Prev. 10*:339–344 (2001).

41. Den Tonkelaar, I., L. Keinan-Boker, P. Van't Veer, C.J. Arts, H. Adlercreutz, J.H. Thijssen, and P.H. Peeters, Urinary Phytoestrogens and Postmenopausal Breast Cancer Risk, *Cancer Epidemiol. Biomark. Prev. 10*:223–228 (2001).

42. Vanharanta, M.S., T.A.Voutilainen, M. Lakka, H. Van der Lee, H. Adlercreutz, and J.T. Salonen, Risk of Acute Coronary Events According to Serum Concentration of

Enterolactone: A Prospective Population-Based Case-Control Study, *Lancet 354*:2112–2115 (1999).

43. Hulten, K.A., P. Winkvist, P. Lenner, R. Johansson, H. Adlercreutz, and G. Hallmans, An Incident Case-Referent Study on Plasma Enterolactone and Breast Cancer Risk, *Eur. J. Nutr. 41*:168–176 (2002).

44. Axelson, M., and K.D.R. Setchell, The Excretion of Lignans in Rats—Evidence for An Intestinal Bacterial Source for This New Group of Compounds, *FEBS Lett. 123*:337–342 (1981).

45. Setchell, K.D.R, A.M. Lawson, S.P. Borriello, R. Harkness, H. Gordon, D.M. Morgan, D.N. Kirk, H. Adlercreutz, L.C. Anderson, and M. Axelson, Lignan Formation in Man-Microbial Involvement and Possible Roles in Relation to Cancer, *Lancet 2*:4–7 (1981).

46. Borriello, S.P., K.D.R. Setchell, M. Axelson, and A.M. Lawson, Production and Metabolism of Lignans by the Human Fecal Flora, *J. Appl. Bacteriol. 58*:37–43 (1985).

47. Axelson, M., J. Sjovall, B.E. Gustafsson, and K.D.R. Setchell, Origin of Lignans in Mammals and Identification of a Precursor from Plants, *Nature 298*:659–660 (1982).

48. Setchell, K.D.R, A.M. Lawson, L.M. McLaughlin, S. Patel, D.N. Kirk, and M. Axelson, Measurement of Enterolactone and Enterodiol, the First Mammalian Lignans, Using Stable Isotope Dilution and Gas Chromatography-Mass Spectrometry, *Biomed. Mass Spectrom. 10*:227–235 (1983).

49. Barnes, S., J. Sfakianos, L. Coward, and M. Kirk, Soy Isoflavonoids and Cancer Prevention: Underlying Biochemical and Pharmacological Issues, *Adv. Exp. Med. Biol. 401*:87–100 (1996).

50. Xu, Y., K.S. Harris, H.J. Wang, P.A. Murphy, and S. Hendrich, Bioavailability of Soybean Isoflavones Depends on Gut Microflora in Women, *J. Nutr. 125*:2307–2315 (1995).

51. Wang, L.Q., M.R. Meselhy, Y. Li, G.-W. Qin, and M. Hattori, Human Intestinal Bacteria Capable of Transforming Secoisolariciresinol Diglucoside to Mammalian Lignan Enterodiol and Enterolactone, *Chem. Pharm. Bull. 48*:1606–1610 (2000).

52. Rickard, S.E., and L.U. Thompson, Urinary Composition and Postprandial Blood Changes in ^{3}H-Secoisolariciresinol Diglycoside (SDG) Metabolites in Rats Do Not Differ Between Acute and Chronic SDG Treatment, *J. Nutr. 130*:2299–2305 (2000).

53. Setchell, K.D.R., N.M. Brown, L. Zimmer-Nechemias, W.T. Brasher, B.E. Wolfe, A.S. Kirschner, and J.E. Heubi, Definitive Evidence for Lack of Absorption of Soy Isoflavone Glycosides in Humans, Supporting the Crucial Role of Intestinal Metabolism for Bioavailablity, *Am. J. Clin. Nutr. 76*:447–453 (2002).

54. Setchell, K.D., Phytoestrogens: The Biochemistry, Physiology, and Implications for Human Health of Soy Isoflavones, *Am. J. Clin. Nutr. 68*:1333–1346 (1998).

55. Metzler, M., Oxidative Metabolism of Lignans, in *Flaxseed in Human Nutrition*, edited by L.U. Thompson and S.C. Cunnane, 2nd edn., AOCS Press, Champaign, Illinois, 2003, pp. 117–125.

56. Jacobs, E., and M. Metzler, Oxidative Metabolism of the Mammalian Lignans Enterolactone and Enterodiol by Rat, Pig, and Human Liver Microsomes, *J. Agric. Food Chem. 47*:1071–1077 (1999).

57. Jacobs, E., S.E Kulling, and M. Metzler, Novel Metabolites of the Mammalian Lignan Enterolactone and Enterodiol in Human Urine, *J. Steroid Biochem. Mol. Biol. 68*:211–218 (1999).

58. Saarinen, N.M., A. Smeds, S.I. Makela, J. Ammala, K. Hakala, J.M. Puhlava, E.L. Ryhanen, R. Sjoholm, and R. Santti, Structural Determinants of Plant Lignans for the Formation of Enterolactone *in vivo*, *J. Chromatog. B 777*:311–319 (2002).

59. Rickard, S.E., and L.U. Thompson, Chronic Exposure to Secoisolariciresinol Diglycoside Alters Lignan Disposition in Rats, *J. Nutr. 128:*615–623 (1998).
60. Jenab, M., S.E. Rickard, L.J Orcheson, and L.U. Thompson, Flaxseed and Lignans Increase Cecal β-Glucuronidase Activity in Rats, *Nutr. Cancer 33:*154–158 (1999).
61. Rickard, S.E., L.J. Orcheson, M.M. Seidl, L. Luyengi, H.H. Fong, and L.U. Thompson, Dose Dependent Production of Mammalian Lignans in Rats and *in vitro* from the Purified Precursor Secoisolariciresinol Diglycoside from Flaxseed, *J. Nutr. 126:*2012–2019 (1996).
62. Adlercreutz, H., Y. Mousavi, J. Clark, K. Hockerstedt, E. Hamalainen, K. Wahala, T. Makela, and T. Hase, Dietary Phytoestrogens and Cancer: *in vitro* and *in vivo* Studies, *J. Steroid Biochem. Mol. Biochem. 41:*331–337 (1992).
63. Adlercreutz, H., C. Bannwart, K. Wahala, T. Makela, G. Brunow, T. Hase, P.J. Arosemena, J.T. Kellis, Jr., and L.E.. Vickery, Inhibition of Human Aromatase by Mammalian Lignans and Isoflavonoid Phytoestrogens, *J. Steroid Biochem. Mol. Biol. 44:*147–153 (1993).
64. Wang, C., T. Makela, T. Hase, H. Adlercreutz, and M.S. Kurzer, Lignans and Flavonoids Inhibit Aromatase Enzyme in Human Preadipocytes, *J. Steroid Biochem. Mol. Biol. 50:*205–212 (1994).
65. Morton, M., A. Matos-Ferreira, L. Abranches-Monteiro, R. Correia, N. Blacklock, P.S.F. Chan, C. Cheng, S. Lloyd, W. Chieh-Ping, and K. Griffiths, Measurement and Metabolism of Isoflavonoids and Lignans in the Human Male, *Cancer Lett. 114:*145–151 (1997).
66. Thompson, L.U., T. Li, J. Chen, and P.E. Goss, Biological Effects of Dietary Flaxseed in Patients with Breast Cancer, *Breast Cancer Res. Treat. 64:*50 (2000).
67. Haggans, C.J., E.J. Travelli, W. Thomas, M.C. Martini, and J.L. Slavin, The Effect of Flaxseed and Wheat Bran Consumption on Urinary Estrogen Metabolite in Premenopausal Women, *Cancer Epidemiol. Biomark. Prev. 9:*719–725 (2000).
68. Tou, J.C.L., J. Chen, and L.U. Thompson, Flaxseed and Its Lignan Precursor, Secoisolariciresinol Diglucoside, Affect Pregnancy Outcome and Reproductive Development in Rats, *J. Nutr. 128:*1861–1868 (1998).
69. Adlercreutz, H., T. Yamada, K. Wahala, and S. Watanabe, Maternal and Neonatal Phyto-estrogens in Japanese Women During Birth, *Am. J. Obstet. Gynecol. 180:*737–743 (1999).
70. Clark, W.F., A.D. Muir, N.D. Westcott, and A. Parbtani, A Novel Treatment for Lupus Nephritis: Lignan Precursor Derived from Flax, *Lupus 9:*429–436 (2000).
71. Saarinen, N.M., A. Warri, S.I. Makela, C. Eckerman, M. Reunanen, M. Ahotupa, S.M. Salmi, A.A. Franke, L. Kangas, and R. Santti, Hydroxymatairesinol, A Novel Enterolactone Precursor with Antitumor Properties from Coniferous Tree (*Picea abies*), *Nutr. Cancer 36:*207–216 (2000).
72. Hutchins, A.M., M. C. Martini, B.A. Olson, W. Thomas, and J.L. Slavin, Flaxseed Influences Urinary Lignan Excretion in a Dose-Dependent Manner in Postmenopausal Women, *Cancer Epidemiol. Biomark. Prev. 9:*1113–1118 (2000).
73. Nesbitt, P.D., and L.U. Thompson, Lignans in Homemade and Commercial Products Containing Flaxseed, *Nutr. Cancer 29:*222–227 (1997).
74. Madhusudhan, B., D. Wiesborn, J. Schwarz, K. Tostenson, and J. Gillespie, A Dry Mechanical Method for Concentrating the Lignan Secoisolariciresinol Diglucoside in Flaxseed, *Lebensm. Wiss. u.-Technol. 33:*268–275 (2000).
75. Hutchins, A.M., and J.L. Slavin, Effect of Flaxseed on Sex Hormone Metabolism, in *Flaxseed in Human Nutrition*, edited by L.U. Thompson and S.C. Cunnane, 2nd edn., AOCS Press, Champaign, Illinois, 2003, pp. 126–149.

76. Lampe, J.W., M.C. Martini, M.S. Kurzer, H. Adlercreutz, and J.L.Slavin, Urinary Lignan and Isoflavonoid Excretion in Premenopausal Women Consuming Flaxseed Powder, *Am. J. Clin. Nutr. 60:*122–128 (1994).

77. Adlercreutz, H., T. Fotsis, C. Bannwart, K. Wahala, T. Makela, G. Brunow., and T. Hase, Determination of Urinary Lignans and Phytoestrogen Metabolites, Potential Antiestrogens and Anticarcinogens, in Urine of Women on Various Habitual Diets, *J. Steroid Biochem. 25:*791–797 (1986).

78. Kilkkinen, A., P. Pietinen, T. Klaukka, J. Virtamo, P. Korhonen, and H. Adlercreutz, Use of Oral Antimicrobials Decreases Serum Enterolactone Concentration, *Am. Epidemiol. 155:*472–477 (2002).

79. Petterson, D., P. Aman, K.E. Knudsen, E. Lundin, J.X. Zhang, G. Hallmans, and H. Adlercreutz, Intake of Rye Bread by Ileostomists Increases Ileal Excretion of Fiber Polysaccharide Components and Organic Acids but Does Not Increase Plasma or Urine Lignans and Isoflavonoids, *J. Nutr. 126:*1594–1600 (1996).

80. Jacobs, Jr., D.R., M.A. Pereira, K. Stumpf, J.J. Pins, and H. Adlercreutz, Whole Grain Food Intake Elevates Serum Enterolactone, *Br. J. Nutr. 88:*111–116 (2002).

81. Kilkkinen, A., K. Stumpf, P. Pietinen, L.M. Valsta, H. Tapanainen, and H. Adlercreutz, Determinants of Serum Enterolactone, *Am. J. Clin. Nutr. 73:*1094–1100 (2001).

82. Juntunen, K.S., W.M. Mazur, K.H. Liukkonen, M. Uehara, K.S. Poutanen, H.C. Adlercreutz, and H.M Mykkanen, Consumption of Wholemeal Rye Bread Increases Serum Concentrations and Urinary Excretion of Enterolactone Compared with Consumption of White Bread in Healthy Finnish Men and Women, *Br. J. Nutr. 84:*839–846 (2000).

83. Cunnane, S.C., M.J. Hamadeh, A.C. Liede, L.U. Thompson, T.M. Wolever, and D.J. Jenkins, Nutritional Attributes of Traditional Flaxseed in Healthy Young Adults, *Am. J. Clin. Nutr. 61:*62–68 (1995).

84. Horner, N.K., A.R. Kristal, J.Prunty, H. Skor, J.D. Potter, and J.W. Lampe, Dietary Determinants of Plasma Enterolactone, *Cancer Epidemiol., Biomark. Prev. 11:*121–126 (2002).

85. Lampe, J.W, D.R. Gustafson, A.M. Hutchins, M.C. Martini, S. Li, K. Wahala, G.A. Grandits, J.D. Potter, and J.L. Slavin, Urinary Isoflavonoid and Lignan Excretion on a Western Diet: Relation to Soy, Vegetable and Fruit Intake, *Cancer Epidemiol. Biomark. Prev. 8:*699–707 (1999).

86. Mazur, W.M, M. Uehara, K. Wahala, and H. Adlercreutz, Phytoestrogen Content of Berries, and Plasma Concentrations and Urinary Excretion of Enterolactone After a Single Strawberry-Meal in Human Subjects, *Brit. J. Nutr. 83:*381–387 (2000).

87. Adlercreutz, H., K. Hockerstedt, C. Bannwart, S. Bloigu, E. Hamalainen, T. Fotsis, and A. Ollus, Effect of Dietary Components, Including Lignans and Phytoestrogens, on Enterohepatic Circulation and Liver Metabolism of Estrogen and on Sex Hormone Binding Globulin (SHBG), *J. Steroid Biochem. 27:*1135–1144 (1987).

Chapter 5

Oxidative Metabolism of Lignans

Manfred Metzler

Institute of Food Chemistry and Toxicology, University of Karlsruhe,
D-76128 Karlsruhe, Germany

Introduction

About 20 years ago, the mammalian lignans enterodiol (END, Fig. 5.1) and enterolactone (ENL) were identified in human urine and were shown to arise from the bacterial metabolism of the plant lignans secoisolariciresinol (SEC) and matairesinol (MAT) in the intestine (1). It was later disclosed that other plant lignans [e.g., pinoresinol (PIN) and lariciresinol (LAR)] are also converted to END and ENL by intestinal bacteria (2). The highest levels of mammalian lignans are achieved by the ingestion of flaxseed (1). Much interest has recently been focused on lignans, because epidemiological and experimental studies suggest that a diet rich in flaxseed may protect against certain diseases common in Western industrialized societies, such as breast and prostate cancer as well as coronary heart disease (3–5).

As summarized by Setchell (1), lignans occur mostly as glycosides in the plant. After ingestion of lignans, hydrolysis of the glycosides by gastric acid and bacterial glycosidases liberates the aglycones which are then converted to the mammalian lignans END and ENL by the intestinal bacteria. END and ENL are absorbed from the intestine and transported to the liver by the portal venous blood. The presence of significant amounts of glucuronides of the mammalian lignans observed in the portal venous blood of rats fed flaxseed (6) suggests that conjugation occurs in the enterocytes during uptake. Conjugation with glucuronic acid and sulfate takes place in the liver and facilitates the clearance of the lignans into bile and urine (6). However, detailed information about the metabolic fate of these important food-related lignans is scarce. In particular, the question of oxidative metabolism has only recently been addressed. Hydroxylated metabolites, if formed, may be of endocrinological or toxicological importance. Jacobs and Metzler (7) first demonstrated that END and ENL are good substrates for cytochrome P450, because hepatic microsomes from various species including humans gave rise to a variety of aromatic and aliphatic hydroxylation products of END and ENL *in vitro*. Likewise, oxidative metabolites were identified when SEC or MAT were incubated with microsomes (8). In the present chapter, the oxidative *in vitro* metabolism of these four lignans is reviewed, together with data on the formation of oxidative metabolites in rats and humans *in vivo*. Moreover, studies on the potential of the four lignans to cause genetic damage in *in vitro* systems are reviewed.

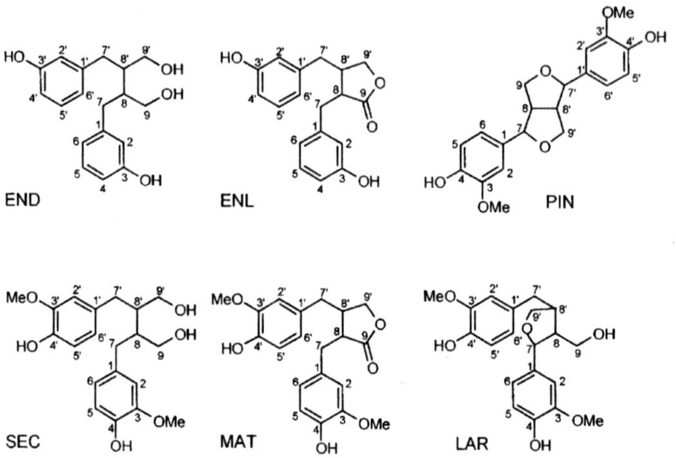

Fig. 5.1. Chemical structures of mammalian lignans (END, ENL) and plant lignans (SEC, MAT, LAR, PIN). END, enterodiol; ENL, enterolactone; SEC, secoisolariciresinol; MAT, matairesinol; LAR, lariciresinol; PIN, pinoresinol.

Oxidative Metabolism *in vitro*

Enterodiol

When END was incubated with hepatic microsomes from aroclor 1254-induced male Wistar rats and the incubation mixture analyzed by GC/MS and HPLC/MS, seven metabolites were identified as monohydroxylation products of END (7). Based on their mass spectra, three minor microsomal metabolites of END were tentatively identified as products of aromatic hydroxylation; comparison by GC/MS with synthetic reference compounds led to the unambiguous identification of the structures depicted in Figure 5.2. For two major and two minor microsomal END metabolites, the structures of aliphatic hydroxylation products were proposed; the exact position of the hydroxyl group could not yet be determined due to the lack of authentic reference compounds. When END was incubated with hepatic microsomes from untreated rats, pigs, and humans, none of the minor metabolites was detected; the only metabolites observed were the two products of aliphatic hydroxylation formed as major metabolites with induced rat liver microsomes (7).

Enterolactone

Incubation of ENL with aroclor 1254-induced rat liver microsomes gave rise to six major and six minor metabolites, which were separated by GC/MS and HPLC/MS and identified as products of aromatic and aliphatic monohydroxylation (7). By means

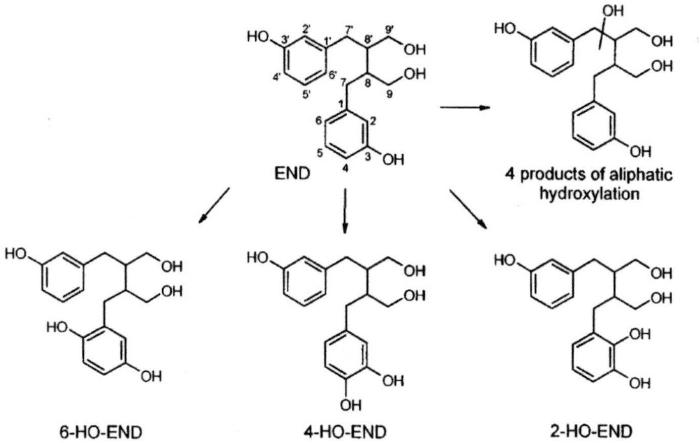

Fig. 5.2. Oxidative pathways in the metabolism of enterodiol (END). See Figure 5.1 for abbreviation.

of synthetic reference compounds, three of the six major metabolites were unequivocally identified as the aromatic hydroxylation products 6-hydroxy-ENL, 6'-hydroxy-ENL, and 4'-hydroxy-ENL (Fig. 5.3), whereas the other three major metabolites were products of aliphatic hydroxylation and could not be definitively identified. The minor metabolites comprised the remaining three aromatic hydroxylation products depicted in Figure 5.3 and three unidentified aliphatic hydroxylation products. With noninduced rat, pig, and human liver microsomes, aliphatic hydroxylation of ENL was much more pronounced than was aromatic hydroxylation (7).

Secoisolariciresinol

Incubation of SEC with hepatic microsomes from aroclor 1254-induced and noninduced male Wistar rats gave rise to one major, two minor, and traces of another four metabolites upon analysis by GC/MS (8). The major metabolite was lariciresinol (LAR, Fig. 5.4), unequivocally identified by comparison with an authentic standard in GC/MS, whereas one of the minor metabolites was tentatively identified as the aliphatic hydroxylation product 7-hydroxy-SEC based on its mass spectrum. A small amount of isolariciresinol (ISL, Fig. 5.4) was detected as the second minor metabolite, but this may be an artifact formed from LAR during sample workup. The four trace metabolites of SEC consisted of another aliphatic and three aromatic hydroxylation products according to their mass spectra. Human microsomes were able to metabolize SEC to LAR and traces of ISL. It is of interest that no metabolites arising from oxidative demethylation of SEC or from combined hydroxylation and demethylation were detected when the respective molecular ions and characteristic fragments were searched for in GC/MS.

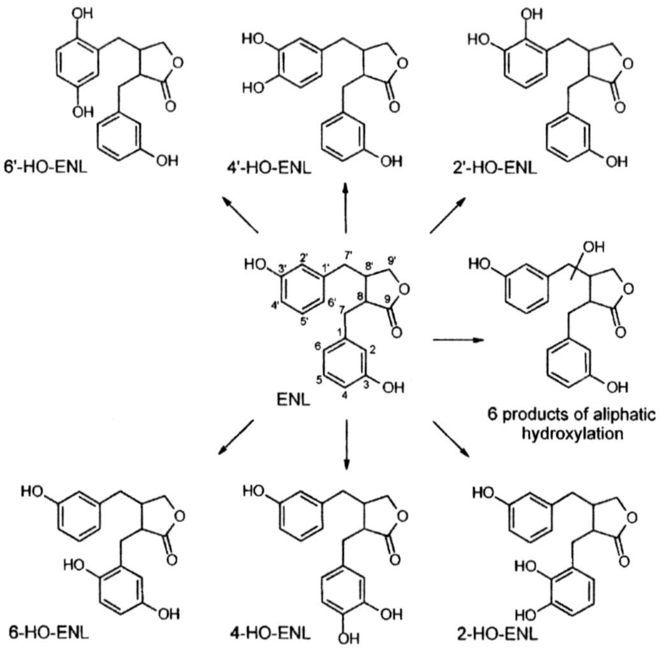

6'-HO-ENL   4'-HO-ENL   2'-HO-ENL

ENL   6 products of aliphatic hydroxylation

6-HO-ENL   4-HO-ENL   2-HO-ENL

Fig. 5.3. Oxidative pathways in the metabolism of enterolactone (ENL). See Figure 5.1 for abbreviation.

Matairesinol

When MAT was incubated with liver microsomes from aroclor 1254-induced and noninduced male Wistar rats, about 43 and 37%, respectively, of MAT was converted into biotransformation products as estimated by HPLC analysis. GC/MS analysis revealed the formation of at least eight metabolites (8). Major products were 7'-hydroxy-MAT, a metabolite with opened lactone ring, and two demethylation products of MAT (Fig. 5.5). Trace amounts of four aromatic hydroxylation products were also detected. Surprisingly, incubation of MAT with human hepatic microsomes did not yield oxidative metabolites.

Oxidative Metabolism *in vivo*

Enterodiol and Enterolactone

When END was injected at a dose of 10 mg/kg body weight into the duodenum of bile duct-catheterized female Wistar rats and the 6-h bile analyzed by GC/MS after cleanup and enzymatic hydrolysis of conjugated metabolites, two products of aliphatic hydroxylation of END and three products of aromatic hydroxylation type were detect-

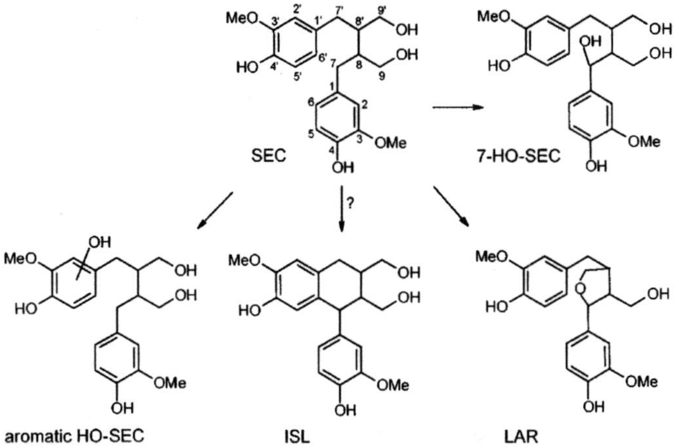

Fig. 5.4. Oxidative pathways in the metabolism of secoisolariciresinol (SEC). ISL, isolariciresinol; see Figure 5.1 for other abbreviations.

ed (9). The two aliphatic hydroxylation products had identical mass spectra; moreover, they had the same mass spectra and GC retention times as the two major aliphatic hydroxylation products of the microsomal incubations of END (see above). Based on the mass spectra, they were proposed to represent stereoisomers of 7-hydroxy-END (9). The three products of aromatic END hydroxylation from rat bile also

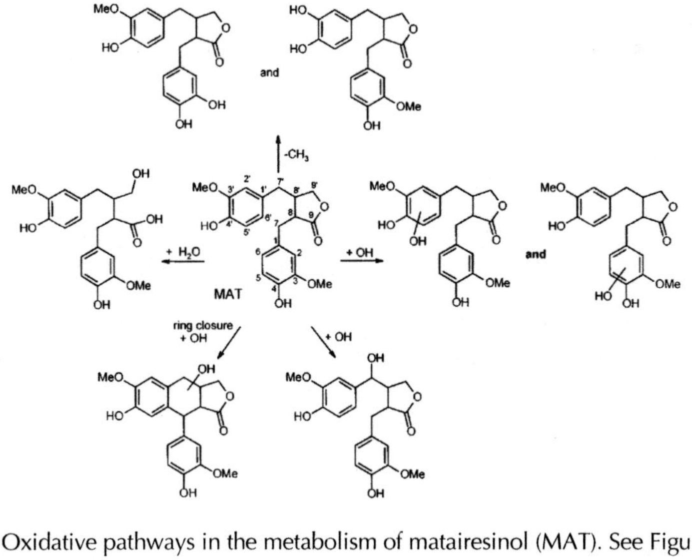

Fig. 5.5. Oxidative pathways in the metabolism of matairesinol (MAT). See Figure 5.1 for abbreviation.

cochromatographed with their microsomal counterparts and were 6-hydroxy-END, 4-hydroxy-END, and 2-hydroxy-END.

When the bile from rats dosed with ENL was analyzed by GC/MS, six products of aromatic and five of aliphatic monohydroxylation were detected (9). The aromatic hydroxylation products were identified as the para- and ortho-hydroxylated compounds and were thus identical with the microsomal metabolites (Fig. 5.3). The five ENL metabolites carrying the additional hydroxyl group at the aliphatic moiety cochromatographed with the corresponding microsomal metabolites in GC/MS. Two of them were tentatively identified as stereoisomers of 7'-hydroxy-ENL (9).

When END or ENL were orally administered to intact female Wistar rats at 10 mg/kg body weight, small amounts of both aromatic and aliphatic hydroxylation products of END and ENL were detected in the 24-h urine (9). The urinary metabolites had the same GC retention times and mass spectra as the biliary metabolites. Virtually the same set of oxidative metabolites of END and ENL was observed in the 24-h urine of female and male Wistar rats fed a diet containing 5% flaxseed. A detailed account is given by Niemeyer *et al.* (9). However, when the urine of female or male volunteers ingesting 16 g flaxseed/day for five days was analyzed by GC/MS, very small amounts of the three and six aromatic hydroxylation products of END and ENL, respectively, but none of the aliphatic hydroxylation products observed as microsomal metabolites or as biliary and urinary metabolites in rats was detectable (10). The reason for not finding aliphatic hydroxylation products of the mammalian lignans in the urine of humans is presently unclear. Moreover, as recent studies (Niemeyer, H.B., unpublished data) have shown, both mammalian and plant lignans form traces of aromatic, but not aliphatic, hydroxylation products spontaneously during enzymatic cleavage of conjugates, especially in the presence of ascorbic acid. It is presently unknown which proportion of the oxidative products of END and ENL in human urine represents true metabolites and which is due to artifacts.

Secoisolariciresinol and Matairesinol

When a single dose of 10 mg/kg body weight of SEC or MAT was orally administered to female Wistar rats and the 24-h urine analyzed by GC/MS after enzymatic hydrolysis of conjugates, the parent plant lignans were identified together with large amounts of the mammalian lignans END and ENL (Niemeyer, H.B., unpublished data). Notably, none of the oxidative metabolites of SEC and MAT, which were identified as microsomal metabolites (see above), could be detected in rat urine *in vivo*. A possible explanation for this finding is that SEC and MAT are rapidly metabolized, either to their glucuronides/sulfates in the intestinal wall or liver, or to END and ENL by the gut microflora. Consistent with this notion is the observation that no oxidative metabolites of SEC and MAT were detectable in the urine of human subjects after ingestion of a flaxseed diet for five days (Niemeyer, H.B., unpublished data).

Genotoxic Potential *in vitro*

Other phytoestrogens [e.g., genistein and coumestrol (11,12)] and structurally related synthetic diphenolic compounds [e.g., diethylstilbestrol (13) and bisphenol A (14)] have been reported to display genotoxic activity *in vitro*. Therefore, we have studied the effects of END, ENL, SEC, and MAT at certain endpoints for genotoxicity in cultured Chinese hamster V79 cells (15). The results are summarized in Table 5.1. At concentrations up to 100 µM, which was the limit of their solubility in these assays, neither the mammalian nor the plant lignans induced micronuclei or hypoxanthine phosphoribosyltransferase (hprt) mutations, or affected the microtubule system in cultured cells. In this respect, the lignans resembled the isoflavone daidzein and differed from genistein and coumestrol, which exhibited clastogenic and gene mutagenic activity in V79 cells (11).

Conclusion

The metabolic studies cited in this review demonstrate that END, ENL, SEC, and MAT are good substrates for cytochrome P450-mediated hydroxylation reactions when incubated with microsomes *in vitro*. Although not all the metabolites have been completely identified, it is clear that hydroxylation takes place at both aliphatic and aromatic positions of the parent lignan molecules. For END and ENL, aliphatic hydroxylation was favored over aromatic hydroxylation by microsomes from noninduced rat, pig, and human liver, whereas an increased formation of aromatic hydroxylation products was observed after induction of rat liver microsomes with aroclor 1254 (7). SEC and MAT were preferentially hydroxylated at the aliphatic moiety by both induced and noninduced rat liver microsomes.

TABLE 5.1

Summary of the Genotoxic Effects of Various Phytoestrogens and Synthetic Estrogens in Cultured V79 Cells at Different Endpoints *in vitro*

Endpoint	GEN	DAI	COM	DES	BPA	END[a]	ENL[a]	SEC[a]	MAT[a]
Induction of micronuclei									
CREST-positive	–	–	–	+++	++	–	–	–	–
CREST-negative	+++	–	++	–	–	–	–	–	–
Interference with MT									
Disruption of CMTC	–	–	–	+++	++	–	–	–	–
Mitotic spindle	–	–	–	+++	++	–	–	–	–
Cell-free MT assembly	–	–	–	+++	++	–	–	–	–
Induction of mitotic arrest	–	–	–	+++	++	–	–	–	–
Mutations at the hprt locus	(+)	–	++	n.d.	n.d.	–	–	–	–

Abbreviations: GEN, genistein; DAI, daidzein; COM, coumestrol; DES, diethylstilbestrol; BPA, bisphenol A; END, enterodiol; ENL, enterolactone; SEC, secoisolariciresinol; MAT, matairesinol; MT, microtubule; CMTC, cytoplasmic microtubule complex; CREST, calcinosis, Raynaud's phenomenon, oesophageal motility abnormalities, sclerodactyly and telangiectasia; hprt, hypoxanthine phosphoribosyltransferase; n.d., not determined. [a]END, ENL, SEC, and MAT were tested up to their limit of solubility (100 µM).

Crucial questions remain to be asked concerning the biological significance of oxidative lignan metabolism. First, it is important to know whether hydroxylation products of lignans are formed under *in vivo* conditions. Lignans already contain two hydroxyl groups and can therefore become conjugated without further hydroxylation. Small amounts of both types of oxidative metabolites (i.e., aliphatic and aromatic hydroxylation products) were detected in the bile and urine of rats dosed with END and ENL (9). It was estimated that the total amount of oxidative metabolites was less than 3% of the dosed lignans. Searches for hydroxylated END and ENL in the urine of humans fed a flaxseed diet have revealed the excretion of small amounts of aromatic but not aliphatic hydroxylation products (10), whereas 7-hydroxy-ENL and 7-hydroxy-MAT were tentatively identified in the urine of humans after ingestion of flaxseed (5). Further studies are needed to clarify this discrepancy and to better assess the role of oxidative lignan metabolites *in vivo*.

Another interesting aspect of the oxidative metabolism of lignans is the possible formation of biologically active metabolites. END and ENL are weakly estrogenic (1) and do not exhibit genotoxicity *in vitro* (15). However, the addition of a hydroxyl group to the diphenolic structure of the parent lignan may decrease or increase the estrogenic activity or confer genotoxic potential. Nothing is known to date about the genotoxic potential of hydroxylated lignans. In particular, the products of aromatic hydroxylation, which are hydroquinones and catechols, may be genotoxic, but hydroxylation at the benzylic position may also provide a metabolite (e.g., 7-hydroxy-ENL) for further activation to a genotoxin [e.g., by conjugation with sulfate in analogy to the situation with tamoxifen (16)]. In addition to exploring possible pathways for the metabolic activation of lignans and their *in vivo* significance, it should prove interesting to study the enzymology of the oxidative and conjugative metabolism (i.e., which isoforms of cytochrome P450 and UDP glucuronosyltransferases are involved in the formation of the various metabolites). If lignans share metabolic enzymes with other substrates (e.g., steroid estrogens), interaction with these compounds may occur.

Acknowledgment

Studies conducted in our laboratory were supported by the *Deutsche Forschungsgemeinschaft* (Grant Me 574).

References

1. Setchell, K.D.R., Discovery and Potential Clinical Importance of Mammalian Lignans, in *Flaxseed in Human Nutrition*, edited by S.C. Cunnane and L.U. Thompson, AOCS Press, Champaign, Illinois, 1995, pp. 82–98.
2. Heinonen, S., T. Nurmi, K. Liukkonen, K. Poutanen, K. Wähälä, T. Deyama, S. Nishibe, and H. Adlercreutz, *In vitro* Metabolism of Plant Lignans: New Precursors of Mammalian Lignans Enterolactone and Enterodiol, *J. Agric. Food Chem. 49*:3178–3186 (2001).
3. Adlercreutz, H., Phyto-Oestrogens and Cancer, *Lancet Oncol. 3*:364–373 (2002).
4. Ward, W.E., and L.U. Thompson, Dietary Estrogens of Plant and Fungal Origin: Occurrence and Exposure, in *Endocrine Disruptors, Part 1. The Handbook of*

Environmental Chemistry, edited by M. Metzler, Springer-Verlag, Berlin/Heidelberg, Germany, 2001, Vol. 3, pp. 101–128.

5. Adlercreutz, H., Phytoestrogens: Epidemiology and a Possible Role in Cancer Protection, *Environ. Health Perspect. 103(Suppl. 7)*:103–112 (1995).
6. Axelson, M., and K.D. Stechell, The Excretion of Lignans in Rats—Evidence for an Intestinal Bacterial Source for this New Group of Compounds, *FEBS Lett. 123*:337–342 (1981).
7. Jacobs, E., and M. Metzler, Oxidative Metabolism of the Mammalian Lignans Enterolactone and Enterodiol by Rat, Pig, and Human Liver Microsomes, *J. Agric. Food Chem. 47*:1071–1077 (1999).
8. Niemeyer, H., and M. Metzler, Oxidative Metabolism and Genotoxic Potential of Mammalian and Plant Lignans *in vitro, J. Chromatogr. B 777*:297–303 (2002).
9. Niemeyer, H.B., D. Honig, A. Lange-Böhmer, E. Jacobs, S.E. Kulling, and M. Metzler, Oxidative Metabolites of the Mammalian Lignans Enterodiol and Enterolactone in Rat Bile and Urine, *J. Agric. Food Chem. 48*:2910–2919 (2000).
10. Jacobs, E., S.E. Kulling, and M. Metzler, Novel Metabolites of the Mammalian Lignans Enterolactone and Enterodiol in Human Urine, *J. Steroid Biochem. Mol. Biol. 68*:211–218 (1999).
11. Kulling, S.E., and M. Metzler, Induction of Micronuclei, DNA Strand Breaks and HPRT Mutations in Cultured Chinese Hamster V79 Cells by the Phytoestrogen Coumoestrol, *Food Chem. Toxicol. 35*:605–613 (1997).
12. Kulling, S.E., B. Rosenberg, E. Jacobs, and M. Metzler, The Phytoestrogens Coumoestrol and Genistein Induce Structural Chromosomal Aberrations in Cultured Human Peripheral Blood Lymphocytes, *Arch. Toxicol. 73*:50–54 (1999).
13. Metzler, M., Genotoxic Potential of Natural and Synthetic Endocrine Active Compounds, in *Endocrine Disruptors, Part II. The Handbook of Environmental Chemistry*, edited by M. Metzler, Springer-Verlag, Berlin/Heidelberg, Germany, 2002, Vol. 3/M, pp. 187–207.
14. Pfeiffer, E., B. Rosenberg, S. Deuschel, and M. Metzler, Interference with Microtubules and Induction of Micronuclei *in vitro* by Various Bisphenols, *Mutat. Res. 390*:21–31 (1997).
15. Kulling, S.E., E. Jacobs, E. Pfeiffer, and M. Metzler, Studies on the Genotoxicity of the Mammalian Lignans Enterolactone and Enterodiol and Their Metabolic Precursors at Various Endpoints *in vivo, Mutat. Res. 416*:115–124 (1998).
16. Glatt, H., W. Davis, W. Meinl, H. Hermersdorfer, S. Venitt, and D.H. Phillips, Rat, but Not Human, Sulfotransferase Activates a Tamoxifen Metabolite to Produce DNA Adducts and Gene Mutations in Bacteria and Mammalian Cells in Culture, *Carcinogenesis 19*:1709–1713 (1998).

Chapter 6

Effects of Flaxseed on Sex Hormone Metabolism

Andrea M. Hutchins[a] and Joanne L. Slavin[b]

[a]Department of Nutrition, Arizona State University East, 7001 East Williams Field Road, Mesa, Arizona 85212, USA;
[b]Department of Food Science and Nutrition, University of Minnesota, 1334 Eckles Avenue, St. Paul, Minnesota 55108, USA

Introduction

The mammalian lignans, diphenolic compounds whose structure and molecular weight are similar to those of steroid estrogens, were first identified in the urine of women and vervet monkeys by two independent groups in 1980 (1–3). Both groups observed that the mammalian lignans were excreted in a cyclical pattern related to the menstrual cycle (1,2). Setchell *et al.* (3) also reported that excretion of the mammalian lignans seemed to be influenced by the reproductive cycle in vervet monkeys, although there was no apparent correlation with gonadal steroid excretion. This cyclical excretion pattern, and the observation that lower amounts of the mammalian lignans were excreted by postmenopausal, ovariectomized, or anovulatory women, led to the hypothesis that the mammalian lignans either had an ovarian origin or were somehow influenced by ovarian function. However, three reports (2–4) noted that mammalian lignans were also excreted by men in concentrations equal to those of the postmenopausal, ovariectomized, or anovulatory women, whereas children and infants had little or no lignan excretion. This led to the hypothesis that the mammalian lignans may not be ovarian in origin as previously thought, but that excretion may in some way be related to gonad development. Subsequent studies demonstrated that the mammalian lignans did not have an ovarian origin but are formed in the large intestine from plant precursors (5–8) and undergo enterohepatic circulation, similar to that of the endogenous steroid hormones. Secoisolariciresinol, a plant lignan found in the highest concentrations in flaxseed, is converted by the colonic microflora to the mammalian lignan enterodiol (6). Enterodiol can then be oxidized by the colonic microflora to form the mammalian lignan enterolactone (6,9). Flaxseed also contains small quantities of the plant lignan matairesinol, which is also converted by the colonic microflora to enterolactone (6,9). Similarities between the structure of the mammalian lignans and sex steroid hormones, along with the cyclical excretion pattern, led researchers to theorize that the mammalian lignans may interact with the endogenous sex steroid hormones (1–3).

Potential Effects on Estrogen Synthesis, Metabolism, and Bioavailability

After conjugation in the liver, estrogen conjugates are excreted in the bile along with the mammalian lignan conjugates. Approximately 80% of biliary estrogen conjugates are deconjugated by intestinal microflora and reabsorbed to undergo enterohepatic circulation (10,11). The primary estrogen conjugate form found in bile is sulfoglucuronide, followed by glucuronide and minor amounts of sulfates (11). In contrast, 98% of the estrogens measured in feces are in the unconjugated form (11). Because a majority of the biliary estrogen conjugates are sulfoglucuronides and glucuronides, β-glucuronidase plays an important role in the deconjugation and subsequent reabsorption of biliary estrogens (12). Therefore, the intestinal microflora play a critical role in the formation of estrogens as well as of the mammalian lignans. In women, more estradiol seems to be excreted in feces than in bile, suggesting that, along with the mammalian lignans, biologically active estrogens may be formed in the intestinal tract. In the case of estradiol, this production could be the result of the bacterial reduction of estrone because, in women, the amount of estrone excreted in the bile is higher than that excreted in the feces (11). Administration of antibiotics, which decreases or eliminates the intestinal microflora, causes a decrease in urinary estrogen excretion, as well as mammalian lignan excretion, and an increase in fecal estrogen excretion as a result of impaired bacterial steroid conjugate hydrolysis, necessary for the efficient reabsorption of estrogens from the intestinal tract (11).

The competition between lignans and estrogens for deconjugation by β-glucuronidase in the colon may offer a partial explanation for the protective effect of lignans. Jenab and Thompson (10) reported that the specific activity of β-glucuronidase in male Sprague-Dawley rats treated with azoxymethane was significantly greater in a group fed defatted flaxseed, and the total activity of β-glucuronidase was significantly higher in both the group fed defatted flaxseed and that fed regular-fat flaxseed compared with controls. Total urinary enterodiol and enterolactone excretion had a significant positive correlation with the total activity of β-glucuronidase ($R = 0.280$, $R^2 = 0.078$, $P < 0.036$) and the intake level of flaxseed ($R = 0.780$, $R^2 = 0.608$, $P < 0.0001$). A second study by Jenab *et al.* (13) reported that the specific and total activities of β-glucuronidase in the cecum of Sprague-Dawley rats were significantly correlated with the levels of flaxseed ($R = 0.539$, $P < 0.0008$ and $R = 0.599$, $P < 0.002$, respectively) and secoisolariciresinol-diglycoside ($R = 0.567$, $P < 0.007$ and $R = 0.435$, $P < 0.04$, respectively) in the diets. Even though flaxseed or its isolated lignan secoisolariciresinol increases β-glucuronidase activity, due to competition with plant and mammalian lignans, fewer estrogens may be deconjugated and more excreted in the feces, thus reducing circulating concentrations. The decrease in the enterohepatic circulation of estrogens could partially explain the reduction in breast cancer risk associated with lignan consumption.

Numerous studies have reported the plasma, urinary, or fecal concentrations of enterodiol and enterolactone in men and women, and most also noted the high inter-subject variability that exists (Tables 6.1, 6.2). Although wide ranges in the urinary

TABLE 6.1
Plasma Concentrations of Lignans in Various Populations and Dietary Groups[a]

Population	Number of subjects	Enterolactone (nmol/L)	Enterodiol (nmol/L)	Total lignans (nmol/L)
French men (69)	6	73 (37–127)	N/A	N/A
American man (70)				
Habitual diet	1	97.3 ± 6.0	46.6 ± 4.4	N/A
Diet supplemented with				
whole-wheat flaxseed bread	1	493.3 ± 24.2	154.0 ± 47.0	N/A
Pre- and postmenopausal Finnish women (17)				
Omnivores	14	28.5 (20.5–39.6)	1.4 (0.6–3.6)	N/A
Vegetarians	14	89.1 (38.5–209.0)	5.4 (2.4–11.8)	N/A
Postmenopausal Australian women (18)				
Flaxseed supplementation (25 g/d)	23	394 (140–819)	352 (6–1291)	N/A
Adults with ileostomies (71)				
Low-fiber diet	8	1.01 ± 0.23	0.12 ± 0.26	N/A
High-fiber diet	8	0.98 ± 0.32	0.17 ± 0.25	N/A
Adult men (36)				
Portuguese	50	13.1 (ND–55.7)	1.2 (ND–6.6)	N/A
Chinese	53	20.8 (0.37–365.8)	5.6 (ND–88.4)	N/A
British	36	20.8 (0.2–41.3)	N/A	N/A
Premenopausal Canadian women (14)				
Flaxseed supplementation (25 g/d)				
First day of intake				
0 h	9	6.9 ± 5.7	22.5 ± 2.7	N/A
24 h	9	19.7 ± 6.6	46.3 ± 9.3	N/A
Eighth day of intake				
0 h	9	24.5 ± 8.2	72.7 ± 19.3	N/A
24 h	9	25.8 ± 10.9	57.8 ± 14.9	N/A
Middle-aged Finnish men and women (72)				
Habitual diet	85	12.2 (10.4–19.3)	N/A	N/A
Low-fat, high vegetable and fruit diet (12 wk)	85	19.5 (16.1–31.5)	N/A	N/A
Premenopausal and postmenopausal women (73)				
Japanese	111	13.3 ± 15.6	N/A	N/A
Finnish	87	25.0 ± 16.6	N/A	N/A
Finnish men and women (74)				
Basal (low fruit, vegetable, grain) diet	7	10.3 ± 2.5	N/A	N/A
Strawberry meal—500 g (24 h following meal)	7	20.6 ± 6.2	N/A	N/A
Finnish adults (75)				
Men	1168	13.8	N/A	N/A
Women	1212	16.6	N/A	N/A

[a]Values are geometric means ± SEM, with SEM range in parentheses. Abbreviations: N/A, values not available or not measured; ND, values not detectable.

TABLE 6.2

Excretion of Lignans in Urine and Feces in Various Populations and Dietary Groups[a]

Population	Number of subjects	Enterolactone (nmol/24 h)	Enterodiol (nmol/24 h)	Total lignans (nmol/24 h)
Premenopausal Finnish women (76)[b]				
Omnivores	12	2470 (2100–2900)	214 (166–276)	2890 (2530–3300)
Vegetarians	11	3660 (2890–4620)	368 (290–468)	4160 (3340–5180)
Breast cancer	10	2130 (1730–2640)	142 (103–195)	2340 (1880–2900)
Postmenopausal Finnish women (76)[b]				
Omnivores	10	1710 (1380–2120)	161 (117–212)	1990 (1640–2410)
Vegetarians	10	7400 (4720–11,600)	436 (259–733)	8090 (5190–12,600)
Breast cancer	10	1750 (1210–2540)	128 (92–177)	2080 (1540–2820)
Premenopausal American women (76)[b]				
Omnivores	10	1920 (1660–2230)	244 (194–307)	2220 (1920–2570)
Lacto-ovovegetarians	10	4970 (3550–6950)	1030 (649–1637)	6780 (4920–9330)
Macrobiotics	13	19,900 (15,300–25,900)	7096 (5445–9247)	28,800 (23,000–36,100)
Postmenopausal American women (76)[b]				
Omnivores	11	2030 (1690–2440)	169 (130–219)	2070 (1690–2540)
Lacto-ovovegetarians	12	1560 (910–2670)	158 (94.60–265)	1900 (1150–3130)
Macrobiotics	7	19,400 (13,200–28,300)	3784 (2028–7063)	24,500 (16,300–37,000)

Oriental immigrant women in Hawaii (38, 77)[b]				
Premenopausal	13	370	77,100	480
		(300–500)	(57,400–1040)	(390–590)
Postmenopausal	3	200	56,500	270
		(130–310)	(43,100–74,100)	(190–370)
Postmenopausal English women (78)[c]	36	524 ± 628	206 ± 350	N/A
		(13–3126)	(1–1763)	
Japanese women (79)[b]	10	890	4170	1380
		(650–1230)	(3090–5620)	(1020–1860)
Japanese men (79)[b]	9	890	2200	1130
		(690–1140)	(1500–3300)	(850–1470)
American men (54)[d]				
Habitual diet	6	5400 ± 700	400 ± 100	N/A
Flaxseed bread supplemented diet (13.5 g/d)	6	53,600 ± 4400	9600 ± 1800	N/A
Australian men and women[b]				
Soy flour challenge (40 g/d)	12	11,200	N/A	N/A
		(2500–20,300)		
Habitual diet	12	11,030)	N/A	N/A
		(900–35,300		
Premenopausal American women (81)[b]				
Habitual diet	18	2550	620	2510
		(610–8360)	(190–3270)	(640–8140)
Flaxseed powder supplemented diet (10 g/d)	18	20,000	14,660	44,650
		(1300–56,300)	(3640–79,950)	(22,350–113,640)
Young Canadian adults (5 men, 5 women) (82)[c]				
Habitual diet	10	2540 ± 910	1370 ± 1000	3910 ± 1500
Control muffin supplemented diet	10	3800 ± 1190	4550 ± 2330	8350 ± 3170
Flaxseed muffin supplemented diet (50 g/d)	10	12,390 ± 3710	18,700 ± 6600	31,090 ± 7440

(continued)

TABLE 6.2 (continued)

Young American adults (11 men, 9 women) (83)[e]				
Control diet	20	550 ± 130 [530]	120 ± 130 [130]	670 ± 200 [630]
Carotenoid vegetable supplemented diet	20	1460 ± 150 [1310]	530 ± 150 [420]	1990 ± 220 [1890]
Cruciferous vegetable supplemented diet	20	2490 ± 140 [2240]	1370 ± 140 [840]	3860 ± 210 [3590]
Soy supplemented diet	20	710 ± 140 [700]	130 ± 130 [190]	840 ± 210 [850]
Young American men (84)[e]				
Habitual diets	17	552 ± 45 (201–4022)	147 ± 10 (37–850)	N/A
Tempeh	17	317 ± 45 (193–2957)	96 ± 10 (53–748)	N/A
Soybean pieces	17	328 ± 45 (224–3256)	114 ± 10 (89–1128)	N/A
Young American adults (7 men, 3 women) (85)[e]				
Control diet	10	778 ± 141 (356–1,867)	107 ± 40 (45–589)	886 ± 177
Legume/allium supplemented diet	10	713 ± 141 (274–1878)	98 ± 40 (41–327)	811 ± 177
Low fruit/vegetable supplemented diet	10	859 ± 141 (259–1634)	154 ± 40 (52–586)	1013 ± 177
High fruit/vegetable supplemented diet	10	1092 ± 141 (345–3,444)	258 ± 40 (65–1,155)	1350 ± 177
Young men (38)[b]				
Urban Finnish men	14	3600 (3180–5980)	149 (61–405)	N/A
Rural Japanese men	14	1158 (704–2,878)	188 (134–420)	N/A

Premenopausal American women (86)[b]				
Caucasian	14	4310	690	N/A
African American	15	1840	210	N/A
Latina	15	1090	80	N/A
Japanese	5	1260	190	N/A
Premenopausal Canadian women (14)[d]				
Habitual diets[e]	9	260 ± 110	50 ± 500	N/A
Flaxseed supplemented diet (5 g/d)	9	1810 ± 1090	3040 ± 1040	N/A
Flaxseed supplemented diet (15 g/d)	9	7280 ± 1930	10,080 ± 3340	N/A
Flaxseed supplemented diet (25 g/d)	9	10,860 ± 3950	43,610 ± 13,410	N/A
Flaxseed muffin supplemented diet (25 g/d)	9	13,450 ± 6420	31,850 ± 11,240	N/A
Flaxseed bread supplemented diet (25 g/d)	9	9770 ± 3210	42,230 ± 7270	N/A
Premenopausal and postmenopausal women (73)[d]				
Japanese	106	ND	N/A	N/A
Finnish	106	4900 ± 3100	N/A	N/A
Finnish men and women (74)[d]				
Basal (low fruit, vegetable, grain) diet	7	1494.3 ± 429.2	N/A	N/A
Strawberry meal—500 g (24 h following meal)	7	1421.7 ± 212.4	N/A	N/A
Postmenopausal American women (87)[b]				
Habitual diet	31	3421 (2866–4083)	365 (291–458)	3983 (3412–4649)
Habitual diet + 5 g flaxseed/day	31	24,663 (20,662–29,439)	1374 (1095–1723)	28,316 (24,258–33,053)
Habitual diet + 10 g flaxseed/day	31	56,247 (47,122–67,140)	3232 (2577–4054)	64,623 (55,361–75,433)

(continued)

TABLE 6.2 *(continued)*

Postmenopausal women, Netherlands (88)[f]				
Breast cancer cases	100	576.2 ± 360.4	N/A	N/A
Controls	300	565.6 ± 353.6	N/A	N/A
Fecal excretion				
American premenopausal women (89)[c]				
Control diet	13	0.64 (0.16–1.12) [0.45]	80 (0–160) [60]	0.73 (0.22–1.24) [0.55]
Flaxseed supplemented diet	13	10.30 (2.72–17.88) [6.82]	2560 (0–3660) [980]	12.87 (4.44–21.30) [10.34]
Finnish pre- and postmenopausal women (90)[b]				
Vegetarians	9	3.28 (2.70–3.86)	479.20 (347.00–611.40)	N/A
Omnivores	9	1.51 (1.09–1.94)	147.70 (79.50–227.20)	N/A

[a]Total lignans = enterodiol + enterolactone + matairesinol.
[b]Values are geometric means, (SEM-range).
[c]Values are arithmetic means (SD-range); [geometric mean].
[d]Values are arithmetic means ± SEM.
[e]Values are LSMeans ± SEM, (range) or [geometric mean].
[f]Values are mean ± SD, expressed as µmol/mol creatinine
Abbreviations: See Table 6.1.

excretion of total mammalian lignans and the ratios of individual mammalian lignans have been observed on various habitual diets and in relation to a flaxseed challenge (Table 6.2), Nesbitt *et al.* (14) reported that concentrations of plasma lignans correlated well with urinary excretion of lignans after eight days of supplementation with flaxseed ($R = 0.54$, $R^2 = 0.29$, $P \leq 0.05$; $y = 2.20x + 663.37$).

Unconjugated enterolactone and enterolactone-sulfate account for 21–25% of the total enterolactone in the circulation, and unconjugated enterodiol plus enterodiol-sulfate account for similar quantities of total enterodiol in circulation (15). Estrone-sulfate can be hydrolyzed at the cell membrane and has biological activity due to the presence of intracellular sulfatases; therefore, it is likely that enterolactone-sulfate and enterodiol-sulfate may have similar biological activity (15,16). Also, because the total mean amounts of diphenols (enterolactone, enterodiol, plus the isoflavonoids genistein, daidzein, *O*-desmethylangolensin, and equol) are 10–100 times the concentrations of total estrogens in plasma, it is logical that these compounds may have many biological effects in humans and animals (17). Morton *et al.* (18) also noted that the combined plasma concentrations of enterolactone and enterodiol can approach 500 ng/mL after flaxseed supplementation, a value that is approximately 10,000 times the concentration of free estradiol in postmenopausal women and similar to the concentration of tamoxifen observed in patients with breast cancer treated with 20 mg/d (54 µM/d) of the drug. Therefore, from the data currently available, it appears that the plasma concentrations of the mammalian lignans are high enough, especially after flaxseed supplementation, to produce biologic effects in humans.

in vitro Studies

Sex Hormone Binding Globulin (SHBG)

In an *in vitro* model, enterolactone stimulated SHBG synthesis in HepG2 human liver cancer cells (19). The concentrations of enterolactone used, 0.5–10 µM (0.15–2.98 µg/mL), were 10 times the amount of 17β-estradiol required for the same degree of stimulation. The maximal effect of enterolactone occurred at 5 µM (1.49 µg/mL), and a toxic effect was noted at concentrations greater than 10 µM (2.98 µg/mL). Plasma enterolactone concentrations of 15–70 nM (0.004–0.02 µg/mL) and 20–>1000 nM (0.005–0.3 µg/mL) for omnivores and vegetarians, respectively, have been reported (19); therefore, it is likely that these compounds occur in humans at concentrations high enough to produce physiologic responses. Because increases in SHBG concentrations reduce the biological activity of steroid hormones by decreasing the relative amount of the free fraction of the hormones (primarily estradiol and testosterone), lignan-induced increases in SHBG may result in the decreased biological activity of the steroid hormones.

Martin *et al.* (20) also examined the interactions between enterolactone, enterodiol, and SHBG in an *in vitro* model. Enterolactone was more efficient than enterodiol at displacing estradiol-binding by SHBG at concentrations of 2–50 μM (0.06–14.9 μg/mL) of enterolactone with IC_{50} values in the 10–50 μM (2.98–14.9 μg/mL) range. Their study suggested that increased plasma concentrations of the mammalian lignans may actually increase the free, and biologically more active, fraction of the hormones.

Schöttner *et al.* (21,22) reported that in addition to enterolactone, secoisolariciresinol and matairesinol also have moderate dose-dependent binding affinities for SHBG in an *in vitro* model. At concentrations of 250 mg/L, secoisolariciresinol (691 μM) displaced 60 ± 7% of ^3H-dihydrotestosterone (DHT) from its binding sites on the SHBG molecule, and enterolactone (838 μM) was slightly less effective, displacing 55 ± 3% of ^3H-DHT. Enterodiol (828 μM) demonstrated a weak binding affinity, displacing only 16 ± 6% of the ^3H-DHT molecules (21). The IC_{50} values were 620 μM (185 mg/L) for enterolactone, 230 μM (83 mg/L) for secoisolariciresinol, and 52 μM (19 mg/L) for matairesinol (22). Schöttner *et al.* (23) also synthesized a number of lignans to test their binding affinity to SHBG. They discovered that (±)-diastereoisomers are more active than the comparable meso compounds and that the most effective substitution pattern is the 4-hydroxy-3-methoxy substitution in the aromatic part of the structure. They also reported that the activity of the lignans increases with the decline in the polarity of the aliphatic part of the molecule. Therefore, although epidemiologic and *in vitro* studies suggest that mammalian lignans may stimulate SHBG production *in vivo*, thereby producing an antiestrogenic effect, the *in vitro* studies by Martin *et al.* (20) and Schöttner *et al.* (21,22) suggest that the mammalian lignans increase the free, and biologically more active, fraction of the endogenous steroid hormones by displacing the hormones from their binding sites on SHBG. Consequently, more research is needed to determine if, in relation to SHBG production and binding affinities, the mammalian lignans influence endogenous hormone activity in an antiestrogenic or estrogenic manner.

Growth of Estrogen Dependent Cells

The estrogenic and antiestrogenic effects of enterolactone and enterodiol in two primary cultures of normal cells and in the MCF-7 and T47D breast cancer cell lines were examined by Welshons *et al.* (24). Enterolactone and enterodiol demonstrated weak estrogenic effects (e.g., induction of progesterone receptor or cell growth) in each of the four cell lines with the half-maximal effect of enterolactone occurring at 10 μM (2.98 μg/mL) compared with 10 pM (2.72 pg/mL) for estradiol. Enterodiol was only 1/10 as active as enterolactone. Enterolactone could stimulate cell growth of both MCF-7 and T47D breast cancer cells at concentrations of 1 and 10 μM (0.3–2.98 μg/mL) in a concentration-dependent manner, and the antiestrogenic drug tamoxifen inhibited the enterolactone-stimulated growth. However, the inhibition by tamoxifen was overcome by administration of an excess of estradiol (100 nM; 27.23 ng/mL), and enterolactone inhibited the growth of both cell lines at concentrations of 100 μM (29.8 μg/mL), an inhibition that could not be overcome by an excess of estradiol.

Mousavi and Adlercreutz (25) reported that estradiol and enterolactone both stimulated cell growth in MCF-7 human breast cancer cells when added separately at concentrations of 1–2 µM (0.3–0.6 µg/mL) enterolactone and 1 nM (0.0003 µg/mL) estradiol (1000 times more enterolactone). However, when applied in combination, estradiol and enterolactone inhibited each other's growth-stimulating effect. When applied alone at concentrations above 10 µM (2.98 µg/mL), enterolactone inhibited the growth of the MCF-7 cells. Mousavi and Adlercreutz (25) suggested that the inhibition could be due to competition at the type II nuclear estrogen receptor or the competition of enterolactone-sulfate with estrone-sulfate for the sulfatases which would inhibit estradiol formation in the cells.

Enterolactone also stimulated the growth of MCF-7 cells as measured by cellular proliferation and stimulation of pS2 mRNA expression in the cells (26). Cellular proliferation in the presence of 1 µM (0.3 µg/mL) of enterolactone was 92% of that produced by 100 pM (27.2 pg/mL) of estradiol, and pS2 mRNA expression was 82% of that produced by estradiol. Enterodiol produced no growth stimulation in either assay.

In contrast to the previous studies, estradiol-induced DNA synthesis in MCF-7 cells was enhanced by 10–50 µM (2.98–14.9 µg/mL) enterolactone, but not by lower concentrations, in studies by Wang and Kurzer (27,28). Enterolactone inhibited DNA synthesis at high concentrations with an IC_{50} of 82 µM (24.4 µg/mL) (28). The stimulatory effects of enterolactone on DNA synthesis were significantly inhibited by 0.5–1.0 µM (0.2–0.4 µg/mL) tamoxifen, again suggesting that a known antiestrogen can counter the estrogenic effects of enterolactone.

Aromatase

Aromatase, an enzyme complex located in the endoplasmic reticulum of estrogen-producing cells, is a principal regulator of estrogen biosynthesis (i.e., the aromatization of androgens to estrogens) in humans. A member of the P450 mixed-function oxidase family, aromatase consists of the aromatase cytochrome P450 and the flavoprotein NADPH-P450 reductase. In humans, ovarian aromatase is the primary regulator of estrogen concentrations in premenopausal women and adipose aromatase is responsible for the regulation of estrogen concentrations in men and postmenopausal women. Aromatase is also found in the testis, placenta, and brain (29,30).

Enterolactone, and to a lesser extent enterodiol, inhibited human placental aromatase activity in a study conducted by Adlercreutz *et al.* (31). The effects of enterolactone were first noted at 1 µM (0.3 µg/mL) with a 6% inhibition of aromatase activity. The effects of enterodiol were also first noted at 1 µM (0.3 µg/mL) with a 4% inhibition of aromatase activity. The half-maximal effect of enterolactone occurred at 14 µM (4.2 µg/mL), and enterodiol had a half-maximal effect at 30 µM (9.1 µg/mL). Adlercreutz *et al.* (31) also performed a kinetic analysis of the mechanism of inhibition and demonstrated that enterolactone binds with 1/75–1/300 of the affinity of androstenedione or testosterone. Spectroscopic analysis (31) demonstrated that enterolactone displaces androstenedione on the P450 enzyme by binding to or near the active site of the enzyme.

Adlercreutz *et al.* (31) also tested the effect of enterolactone on estrone production in the human choriocarcinoma cell line JEG-3. With androstenedione as substrate and a 1–2 h incubation time, enterolactone inhibited estrone production by 22–27, 37–39, and 73–76% of that of the control level when the enterolactone concentration was 1, 10, and 100 µM, (0.3, 2.98, and 29.8 µg/mL), respectively. These results suggest that enterolactone can pass freely into the cell and either inhibit aromatase activity directly or down-regulate its gene expression, thereby reducing the amount of aromatase produced by the cell.

Wang *et al.* (32) confirmed the results of Adlercreutz *et al.* (31) using human preadipocytes. However, their results indicated that enterolactone was a less potent inhibitor of enzyme activity. In this study, enterolactone was shown to be a competitive inhibitor of androstenedione with 50% inhibition occurring at 74 µM (22.1 µg/mL), compared with 14 µM (4.2 µg/mL) from the study by Adlercreutz *et al.* (31), and a K_i of 14.4 µM (4.3 µg/mL), compared with the K_i of 6 µM (1.8 µg/mL) found by those investigators (31). Enterodiol was a weak inhibitor of aromatase enzyme activity with an I_{50} value greater than 100 µM (30.2 µg/mL; the K_i value was not determined). With a K_i value approximately 200 times the K_m of androstenedione, enterolactone binds preadipocyte aromatase less than 1% as tightly as does androstenedione.

Gansser and Spiteller (33) reported that at a concentration of 409 µM (148 µg/mL), secoisolariciresinol, isolated from *urtica dioica* roots, also weakly inhibited aromatase activity *in vitro*. A subsequent study by Mäkelä *et al.* (34), using synthesized theoretical precursors of enterolactone and enterodiol, reported that these theoretical precursors were weak inhibitors of human aromatase activity as well. Consequently, inhibition of aromatase activity by plant and mammalian lignans and the resultant down-regulation of estrogen biosynthesis may explain, in part, their reported effects on endogenous sex steroid hormones.

17β-*Hydroxysteroid Dehydrogenase*

17β-Hydroxysteroid dehydrogenase, a family of enzymes that catalyzes the reversible reactions between 17-hydroxy and 17-keto steroids, is necessary for the biosynthesis of androgens and estrogens (35). In particular, this family of enzymes is responsible for the conversion of dehydroepiandrosterone (DHEA) to androstenediol and the reversible conversion of androstenedione to testosterone and of estrone to estradiol.

In a study by Evans *et al.* (35), 100 µM (29.8 µg/mL) enterolactone reduced 17β-hydroxysteroid dehydrogenase activity in genital skin fibroblast monolayers by 98%, and 100 µM (30.2 µg/mL) enterodiol reduced activity by 79%. In the same study, a cocktail that contained 10 µM each of enterolactone (2.98 µg/mL), enterodiol (3.02 µg/mL), and six isoflavonoids (genistein, formononetin, biochanin A, daidzein, coumestrol, and equol) reduced 17β-hydroxysteroid dehydrogenase activity by 90%, and the IC_{50} for a cocktail of seven of the compounds (equol was not included) was approximately 3.5 µM (1.04 µg/mL enterolactone and 1.06 µg/mL enterodiol) of each of the compounds.

5α-*Reductase*

5α-Reductase is the enzyme responsible for the conversion of testosterone to 5α-DHT. Without 5α-reductase, the prostate cannot develop, grow, or function (35), and increased serum concentrations of DHT have been implicated as a risk factor for prostate cancer, another hormone-dependent cancer.

5α-Reductase activity in genital skin fibroblast monolayers was inhibited by 74% by 100 μM (29.8 μg/mL) of enterolactone, but enterodiol inhibited 5α-reductase activity by only 22% (35). In the same study, a cocktail that contained 10 μM each of enterolactone (2.98 μg/mL) and enterodiol (3.02 μg/mL) plus 10 μM each of six isoflavonoids (genistein, formononetin, biochanin A, daidzein, coumestrol, and equol) reduced 5α-reductase activity by 76%, and the IC_{50} for a cocktail of seven of the compounds (equol was not included) was approximately 2.9 μM (0.86 μg/mL enterolactone and 0.88 μg/mL enterodiol) of each of the compounds.

Using benign prostatic hyperplasia homogenates, Evans *et al.* (35) found that 100 μM of enterolactone (29.8 μg/mL) or enterodiol (30.2 μg/mL) inhibited 5α-reductase activity by 18% and 10%, respectively, at pH 5.5. However, the inhibition increased to 80% and 33%, respectively, at pH 7.5, and enterolactone had an IC_{50} of 14 μM (4.2 μg/mL). A cocktail that contained 10 μM of enterolactone (2.98 μg/mL), enterodiol (3.02 μg/mL), genistein, formononetin, biochanin A, daidzein, coumestrol, and equol produced 56% inhibition of 5α-reductase activity at pH 5.5 and 81% inhibition at pH 7.5. However, it is still questionable whether plasma concentrations of these compounds reach these levels. Evans *et al.* (35) and Morton *et al.* (36,37) reported that their studies revealed that concentrations of enterolactone in prostatic fluid reached only 2 μM (0.6 μg/mL) and concentrations of plasma lignans in women rarely exceeded 2 μM even after dietary supplementation with flax (18).

Effects on Nuclear Type II Estrogen-Binding Sites

The nuclear type II estrogen-binding sites have not been regarded as true estrogen receptors due to their moderate affinity and relatively high capacity (38). However, these receptors are involved in the regulation of estrogen-stimulated uterine growth (39,40). Adlercreutz *et al.* (19) reported that enterolactone, enterodiol, and matairesinol all compete effectively with estradiol for binding at the rat uterine estrogen nuclear type II binding site. At a concentration of 1 μM (0.36 μg/mL), matairesinol demonstrated the greatest competition, followed by enterolactone and enterodiol. Additionally, enterolactone and enterodiol bind to rat and human α-fetoprotein, competing with estradiol and estrone for their high affinity-binding site with inhibitory effects at concentrations ranging from 0.5–50 μM (41).

Animal Studies

Soon after the initial discovery and identification of enterolactone, Setchell *et al.* (7) assessed its estrogenic activity by measuring its effect on mouse uterine weight. The researchers reported that enterolactone, administered in doses ranging from 0.05–67.00 µM/kg (14.9–19,966 µg/kg) body weight, did not produce a significant increase in uterine weight.

Waters and Knowler (42) also reported that when administered alone (0.3–30 µg/rat; 89.4–8940 µM/rat), enterolactone was unable to produce an estrogenic effect, measured by the stimulation of uterine RNA synthesis. Additionally, when administered at the same time as, or up to 12 h before estradiol, enterolactone did not inhibit the stimulation of uterine RNA synthesis by estradiol. However, when administered 22 h before estradiol, enterolactone suppressed estradiol-stimulated RNA synthesis by approximately 50%, suggesting that enterolactone may have antiestrogenic effects.

Orcheson *et al.* (43) also found that flax and its component lignans or mammalian lignan products may have antiestrogenic effects. They reported that 1.5 and 3.0 mg (2.2–4.4 µM) secoisolariciresinol-diglycoside (SD) or 10% flaxseed were able to induce significant changes in estrous cycles in rats. Irregular or acyclic periods were experienced by 50, 66, and 66% of the animals in the 1.5 mg (2.2 µM) SD, 3.0 mg (4.4 µM) SD, and 10% flaxseed groups, respectively. Of the animals with regular cycles in these groups, the cycles were lengthened by 20–30%. These results support the role of lignans as antiestrogens at high concentrations.

Gestational and lactational exposure to 5% flaxseed or to SD produced an antiestrogenic effect in female Sprague-Dawley rats as evidenced by delayed puberty onset. It also lengthened the diestrus phase, whereas the same exposure to 10% flaxseed resulted in early puberty onset and lengthened the estrus phase of the estrous cycles, suggesting an estrogenic effect (44). These results were supported by two additional studies that also reported that lifetime exposure to a 5% flaxseed diet delayed puberty onset in female Sprague-Dawley rats, resulting in puberty onset occurring at a heavier body weight (45,46). In contrast, lifetime exposure to 10% flaxseed resulted in early puberty onset and puberty onset at a lighter weight. Lifetime exposure to 10% flaxseed also lengthened all estrous cycles, including the first estrous cycle, due to a prolonged estrus. The animals fed 10% flaxseed had higher serum estradiol concentrations and ovarian weight at puberty (postnatal day 50) and adulthood (postnatal day 132) compared with the 5% lifetime flaxseed and basal diet groups. In contrast, at weaning (postnatal day 21) serum estradiol concentrations were significantly lower in the group with lifetime exposure to 5% flaxseed. In male rats, lifetime exposure to 5% flaxseed resulted in a reduced adult prostate weight, whereas lifetime exposure to 10% flaxseed produced increased sex accessory gland, seminal vesicle, prostate, and testes weight compared with the control group. Therefore, the estrogenic or antiestrogenic effects of flaxseed appear to be dose-dependent.

Epidemiological Studies

Adlercreutz *et al.* (47) found a positive correlation between urinary enterolactone excretion and SHBG concentrations in premenopausal Finnish women and a negative correlation with percentage of free estradiol, again suggesting an antiestrogenic role for these compounds in human hormone metabolism. A negative correlation was also found between urinary enterolactone excretion and plasma free testosterone. These results led Adlercreutz *et al.* (47) to suggest that mammalian lignans may have an estrogenic effect in the liver, stimulating SHBG synthesis. The increase in SHBG would then reduce the concentrations of free estradiol and testosterone in plasma, producing an overall antiestrogenic effect. Also, high concentrations of diphenols in peripheral tissues may act as antiestrogens and inhibit aromatase enzyme activity, thus reducing the conversion of androgens to estrogens.

Positive correlations between lignan excretion in women and SHBG concentrations were reported in a second study by Adlercreutz *et al.* (19). The urinary excretion of lignans in pre- and postmenopausal Finnish women who were omnivores, vegetarians, or breast cancer survivors (women having had stage I and stage II cancers treated with surgical removal of the breast and now apparently healthy) were compared. Vegetarians had significantly higher urinary excretion of enterolactone and total lignans compared with the other two groups. Vegetarians also had significantly higher plasma SHBG values (70.3 nM/L) compared with the omnivores (31.1 nM/L) and breast cancer survivors (34.8 nM/L). After the confounding factor of body mass index was eliminated, a significant positive association between urinary excretion of enterolactone ($R = 0.391$; $R^2 = 0.130$), total lignans ($R = 0.382$; $R^2 = 0.146$), and plasma SHBG remained. Adlercreutz *et al.* (19) also reported significant negative correlations between SHBG concentrations and urinary 16α-hydroxyestrone excretion ($R = -0.390$; $R^2 = 0.152$) after eliminating the confounding factor of body mass index.

Flaxseed and Hormones in Humans

Although the *in vitro*, animal and epidemiological studies have suggested that supplementation with lignans, or flaxseed, may influence endogenous hormone concentrations, the few interventions studies that have been conducted in humans have not yielded convincing results. The extent of estrogenic or antiestrogenic activity exhibited by the lignans in humans may depend on the concentration of endogenous, and biologically more active, hormones (48,49). The average daily production of estrogens and androgens in premenopausal women at the midcycle peak ranges from 25–100 μg (50–52). These levels fall to 5–10 μg/d after menopause, and male androgen concentrations range from 2–25 μg/day (50,52,53). In premenopausal women, due to the high concentrations of endogenous hormones, lignans have to compete with the endogenous hormones for binding sites, and because their estrogenic activity is much lower than that of the endogenous hormones, the lignans exert an antiestrogenic effect. Conversely, the lignans may produce estrogenic effects in postmenopausal women and in men due to the lower concentrations of endogenous hormones found in these groups; therefore, there is less competition for binding sites.

Flaxseed and Hormones in Men

In a study by Shultz *et al.* (54), six healthy young men consumed a whole wheat/flaxseed bread (13.5 g flaxseed/day) daily for six weeks. Although the flaxseed supplementation resulted in a significant increase in urinary enterolactone and enterodiol concentrations (Table 6.2), no significant changes were noted for plasma total testosterone, free testosterone, or SHBG.

Flaxseed and Hormones in Women

Premenopausal Women

Phipps *et al.* (55) examined the effects of 0 or 10 g flaxseed/day on the menstrual cycle and serum hormone concentrations of 18 premenopausal women over the course of three menstrual cycles. The mean length of the luteal phase of the menstrual cycle was increased with flaxseed feeding but there was no change in the length of the follicular phase or overall cycle length. The only significant changes in serum hormone concentrations were an increase in the midfollicular phase testosterone concentration ($P = 0.004$) and the luteal phase progesterone/estradiol ratio ($P = 0.018$) during flaxseed consumption. There were no significant changes in early follicular, midfollicular, or luteal phase plasma estradiol or estrone concentrations, early follicular dehydroepiandrosterone sulfate (DHEAS), SHBG, or testosterone, luteal phase testosterone or progesterone, overall menstrual cycle length, or follicular phase length.

A more recent study by Haggans *et al.* (56) found that supplementation with 10 g flaxseed/day significantly increased the urinary 2:16α-hydroxyestrone ratio in 16 premenopausal women. Because estrogen is metabolized to 2-hydroxylated or 16α-hydroxylated metabolites along competing pathways, the researchers suggested that flaxseed supplementation can influence endogenous estrogen metabolism, particularly the 2-hydroxylation pathway, in premenopausal women.

Postmenopausal Women

Wilcox *et al.* (57) reported that 25 g flaxseed/day significantly increased the maturation of vaginal cells (determined by lateral wall vaginal smears) in 25 postmenopausal women but had no effect on luteinizing hormone or follicle stimulating hormone concentrations. These results suggest that in postmenopausal women the mammalian lignans may have tissue-specific estrogenic effects but no effect on hormones derived from the central nervous system.

Brzezinski *et al.* (58,59) conducted a 12-wk feeding study in postmenopausal women to determine if a phytoestrogen-rich diet could reduce or alleviate their climacteric complaints. A total of 165 women were randomly assigned to either the phytoestrogen-rich diet (containing tofu, soy drink, miso, and flaxseed) or a control diet. At the end of the study, the subjects on the phytoestrogen-rich diet experienced a significant decrease in hot flushes (1.82 ± 0.14 *vs.* 0.84 ± 0.10, $P < 0.001$), vaginal dryness ($P < 0.005$), serum estradiol concentrations (152 ± 14 *vs.* 139.8 ± 11, $P < 0.05$), and serum SHBG concentrations ($P < 0.005$). However, the control group also experienced a decrease in hot flushes ($1.71 \pm .21$ *vs.* 0.96 ± 0.14, $P < 0.005$) and serum estradiol concentrations (152 ± 14.4 *vs.* 140 ± 12.3, $P < 0.05$), suggesting that, at least in the case of the hot flushes, the decrease might be an artifact of participation in the study rather than a result of the intake of phytoestrogens. In contrast to the previous studies, Arjmandi *et al.* (60) reported that flaxseed supplementation at 38g/d for 6 wk did not influence serum follicular stimulating hormone or 17β-estradiol concentrations in 38 postmenopausal women.

A more recent study reported that consumption of 5 or 10 g flaxseed/day can influence serum hormone concentrations in postmenopausal women (61). A group of 28 postmenopausal women consumed either 0, 5, or 10 g flaxseed/day for 7 wk each in addition to their habitual diets in a randomized, crossover study. The 5 and 10 g flaxseed diets reduced serum concentrations of estradiol by 2.90 pg/mL ($P = 0.0386$) and 3.62 pg/mL ($P = 0.0096$), respectively. The 10-g flaxseed diet reduced serum concentrations of estrone-sulfate by 0.10 ng/mL ($P = 0.0185$) and increased serum prolactin concentrations by 2.20 μg/L ($P = 0.0123$). Serum concentrations of estrone, SHBG, progesterone, androstenedione, testosterone, free testosterone, DHEA, and DHEAS were not significantly altered with flaxseed feeding. Additionally, serum estradiol concentrations were negatively correlated with urinary enterodiol ($R = -0.2226$, $R^2 = 0.0496$, $P = 0.0458$), enterolactone ($R = -0.3163$, $R^2 = 0.1000$, $P = 0.004$), and total lignan (enterodiol + enterolactone + matairesinol) excretion ($R = -0.3396$, $R^2 = 0.1154$, $P = 0.0019$). There were no correlations between serum estrone-sulfate or serum prolactin concentrations and excretion of urinary enterodiol, enterolactone, or total lignans.

Haggans *et al.* (62) reported that mean urinary 2-hydroxyestrogen excretion in this same group of postmenopausal women was significantly higher with the 10-g flaxseed diet (14.39 μg/24 h) than with the 5-g flaxseed diet (12.09 μg/24 h, $P < 0.05$) or the control period (10.72 μg/24 h, $P < 0.0005$). In addition, the 2-hydroxyestrogen excretion increased in a linear fashion from the control, through the 5- and 10-g flaxseed diets ($P < 0.0005$, $R^2 = 0.815$). Although there were no significant changes in the urinary excretion of 16α-hydroxyestrone, the mean urinary 2/16α-hydroxyestrogen ratio was significantly higher for the 10-g flaxseed diet (4.85) than for the control period (4.02, $P < 0.05$). This ratio also showed a linear, dose-response relationship between the control and the 5- and 10-g flaxseed diets ($P < 0.05$, $R^2 = 0.769$). Both papers suggested that long-term supplementation with flaxseed in postmenopausal women can influence estrogen metabolism.

Relationship to Hormone Replacement Therapy

The previous studies explored the effects of high lignan intake on hormone concentrations and menopausal symptoms in postmenopausal women; however, these women were not receiving hormone replacement therapy. The average age of women at natural menopause is 49–51 yr, but the average life expectancy of women in developed countries is approximately 80 yr (51,53,63–65). Consequently, most women live close to one-third of their lives after their last menstrual period (66). A meta-analysis of epidemiological studies has reported that up to 50% of post-menopausal women are currently using hormone replacement therapy or have used it at sometime in the past (65). Although the typical American diet does not contain large quantities of lignans, certain dietary patterns (e.g., vegetarianism) do result in significant lignan intake. It is estimated that by the year 2020 there will be close to 46 million postmenopausal women in the United States alone (66), and as the percentage of the American female population over the age of 50 increases, the effects of lignans on synthetic hormone metabolism in women will become an area requiring study.

Currently, the effects of consumption of a diet high in lignans on the beneficial results of hormone replacement therapy in humans are unexplored. Use of hormone replacement therapy has been shown to relieve the vasomotor symptoms (e.g., hot flushes) that some women experience (51,52,66). Typically, the risks associated with menopause and the symptoms that indicate menopause have been assumed to occur universally. However, some researchers (53,63,67) suggest that some menopausal experiences (e.g., hot flushes) vary from culture to culture. Lock (67) reports that fewer postmenopausal Japanese women report having experienced hot flushes during menopause compared with postmenopausal Canadian women. Postmenopausal Japanese women also have lower rates of heart disease and hip fractures compared with American and Canadian women (63). However, only approximately 4% of postmenopausal Japanese women are using hormone replacement therapy (67,68). Although these differences may be due to differences in lifestyle or other factors, Adlercreutz *et al.* (68) have suggested that a diet high in phytoestrogens may explain why menopausal symptoms are infrequent in Japanese women. High intake of phytoestrogens, which may act in an estrogenic manner in postmenopausal women, may reduce the need for hormone replacement therapy.

However, in women who are currently using hormone replacement therapy, the serum concentrations of estrogens are similar to those of premenopausal women. These higher concentrations of serum hormones could result in the lignans acting in an antiestrogenic manner and interfering with the action of the synthetic hormones, resulting in the need to increase the hormone replacement dosage administered.

Summary

Since the identification of mammalian lignans in human urine in 1981, evidence supporting their role as modulators of endogenous sex steroid hormones has increased. However, the most convincing results have come from *in vitro*, animal, and epidemiological studies. Results of the few intervention studies that have been conducted have been mixed; therefore, further research, in particular long-term intervention trials, is needed to provide clarification for this relationship.

References

1. Stitch, S.R., J.K. Toumba, M.B. Groen, C.W. Funke, J. Leemhuis, J. Vink, and G.F. Woods, Excretion, Isolation and Structure of a New Phenolic Constituent of Female Urine, *Nature 287*:738–740 (1980).
2. Setchell, K.D.R., A.M. Lawson, F.L. Mitchell, H. Adlercreutz, D.N. Kirk, and M. Axelson, Lignans in Man and in Animal Species, *Nature 287*:740–742 (1980).
3. Setchell, K.D.R., R. Bull, and H. Adlercreutz, Steroid Excretion During the Reproductive Cycle and in Pregnancy of the Vervet Monkey (*Ceropithecus aethiopus pygerethus*), *J. Steroid Biochem. 12*:375–384 (1980).
4. Axelson, M., and K.D.R. Setchell, Conjugation of Lignans in Human Urine, *FEBS Lett. 122*:49–53 (1980).
5. Axelson, M., and K.D.R. Setchell, The Excretion of Lignans in Rats—Evidence for an Intestinal Bacterial Source for This New Group of Compounds, *FEBS Lett. 123*:337–342 (1981).
6. Axelson, M., J. Sjövall, B.E. Gustafsson, and K.D.R. Setchell, Origin of Lignans in Mammals and Identification of a Precursor from Plants, *Nature 298*:659–660 (1982).
7. Setchell, K.D.R., A.M. Lawson, S.P. Borriello, R. Harkness, H. Gordon, D.M.L. Morgan, D.N. Kirk, H. Adlercreutz, L.C. Anderson, and M. Axelson, Lignan Formation in Man—Microbial Involvement and Possible Roles in Relation to Cancer, *Lancet 2*:4–7 (1981).
8. Setchell, K.D.R., A.M. Lawson, S.P. Borriello, H. Adlercreutz, and M. Axelson, Formation of Lignans by Intestinal Microflora, in *Colonic Carcinogenesis: Falk Symposium 31*, edited by R.A. Malt and R.C.N. Williamson, MTP Press, Lancaster, 1982, pp. 93–97.
9. Bannwart, C., H. Adlercreutz, T. Fotsis, K. Wähälä, T. Hase, and G. Brunow, Identification of *O*-Desmethylangolensin, a Metabolite of Daidzein, and of Matairesinol, One Likely Plant Precursor of the Animal Lignan Enterolactone, in Human Urine, *Finn. Chem. Lett.* 120–125 (1984).
10. Jenab, M., and L.U. Thompson, The Influence of Flaxseed and Lignans on Colon Carcinogenesis and Beta-Glucuronidase Activity, *Carcinogenesis 17*:1343–1348 (1996).
11. Adlercreutz. H., F. Martin, P. Jävenpää, and T. Fotsis, Steroid Absorption and Enterohepatic Recycling, *Contraception 20*:201–224 (1979).
12. Hughes, R.E., Dietary Fibre and Female Reproductive Physiology, in *Dietary Fibre Perspectives Reviews & Bibliography*, edited by V.J. Burley and John Libbey, London, 1990, Vol. 2, pp. 76–86.
13. Jenab, M., S.E. Rickard, L.J. Orcheson, and L.U. Thompson, Flaxseed and Lignans Increase Cecal β-Glucuronidase Activity in Rats, *Nutr. Cancer 33*:154–158 (1999).
14. Nesbitt, P.D., Y. Lam, and L.U. Thompson, Human Metabolism of Mammalian Lignan Precursors in Raw and Processed Flaxseed, *Am. J. Clin. Nutr. 69*:549–555 (1999).
15. Adlercreutz, H., T. Fotsis, J. Lampe, K. Wähälä, T. Mäkelä, G. Brunow, and T. Hase, Quantitative Determination of Lignans and Isoflavonoids in Plasma of Omnivorous and Vegetarian Women by Isotope Dilution Gas-Chromatography Mass-Spectrometry, *Scand. J. Clin. Lab. Invest. 53*:5–18 (1993).
16. Adlercreutz, H., Y. Mousavi, J. Clark, K. Hockerstedt, E. Hamalainen, K. Wähälä, T. Mäkelä, and T. Hase, Dietary Phytoestrogens and Cancer: *in vitro* and *in vivo* Studies, *J. Steroid Biochem. Molec. Biol. 41*:331–337 (1992).

17. Adlercreutz, H., H. Markkanen, S. Watanabe, J. Lampe, K. Wähälä, T. Mäkelä, and T. Hase, Determination of Lignans and Isoflavonoids in Plasma by Isotope Dilution Gas Chromatography-Mass Spectrometry, *Cancer Detect. Prev. 18*:259–271 (1994).

18. Morton, M.S., G. Wilcox, M.L. Wahlqvist, and K. Griffiths, Determination of Lignans and Isoflavonoids in Human Female Plasma Following Dietary Supplementation, *J. Endocrinol. 142*:251–259 (1994).

19. Adlercreutz, H., Y. Mousavi, J. Clark, K. Höckerstedt, E. Hämäläinen, K. Wähälä, T. Mäkelä, and T. Hase, Dietary Phytoestrogens and Cancer: *in vitro* and *in vivo* Studies, *J. Steroid Biochem. Molec. Biol. 41*:331–337 (1992).

20. Martin, M.E., M. Haourigui, C. Pelissero, C. Benassayag, and E.A. Nunez, Interactions Between Phytoestrogens and Human Sex Steroid Binding Protein, *Life Sci. 58*:429–436 (1996).

21. Schöttner, M., D. Gansser, and G. Spiteller, Lignans from the Roots of *Urtica dioica* and Their Metabolites Bind to Human Sex Hormone Binding Globulin (SHBG), *Planta Med. 63*:529–532 (1997).

22. Schöttner, M., and G. Spiteller, Lignans Interfering with 5α-Dihydrotestosterone Binding to Human Sex Hormone-Binding Globulin, *J. Nat. Prod. 61*:119–121 (1998).

23. Schöttner, M., D. Gansser, and G. Spiteller, Interaction of Lignans with Human Sex Hormone Binding Globulin (SHBG), *Z. Naturforsch. [C] 52*:834–843 (1997).

24. Welshons, W.V., C.S. Murphy, R. Koch, G. Calaf, and V.C. Jordan, Stimulation of Breast Cancer Cells *in vitro* by the Environmental Estrogen Enterolactone and the Phytoestrogen Equol, *Breast Cancer Res. Treat. 10*:169–175 (1987).

25. Mousavi, Y., and H. Adlercreutz, Enterolactone and Estradiol Inhibit Each Other's Proliferative Effect on MCF-7 Breast Cancer Cells in Culture, *J. Steroid Biochem. Mol. Biol. 41*:615–619 (1992).

26. Sathyamoorthy, N., T.T.Y. Wang, and J.M. Phang, Stimulation of pS2 Expression by Diet-Derived Compounds, *Cancer Res. 54*:957–961 (1994).

27. Wang, C., and M.S. Kurzer, Effects of Phytoestrogens on DNA Synthesis in MCF-7 Cells in the Presence of Estradiol or Growth Factors, *Nutr. Cancer 31*:90–100 (1998).

28. Wang, C., and M.S. Kurzer, Phytoestrogen Concentration Determines Effects on DNA Synthesis in Human Breast Cancer Cells, *Nutr. Cancer 28*:236–247 (1997).

29. Yen, S.S.C., Chronic Anovulation Caused by Peripheral Endocrine Disorders, in *Reproductive Endocrinology: Physiology, Pathophysiology, and Clinical Management*, edited by S.S.C. Yen, R.B. Jaffe, and R.L. Barbieri, W.B. Saunders, Philadelphia, PA, 1999, pp. 479–509.

30. Brodie, A.M.H., Aromatase Inhibitors in the Treatment of Breast Cancer, *J. Steroid Biochem. Molec. Biol. 49*:281–287 (1994).

31. Adlercreutz, H., C. Bannwart, K. Wähälä, T. Mäkelä, G. Brunow, T. Hase, P.J. Arosemena, J.T. Kellis, and L.E. Vickery, Inhibition of Human Aromatase by Mammalian Lignans and Isoflavonoid Phytoestrogens, *J. Steroid Biochem. Mol. Biol. 44*:147–153 (1993).

32. Wang, C., T. Mäkelä, T. Hase, H. Adlercreutz, and M.S. Kurzer, Lignans and Flavonoids Inhibit Aromatase Enzyme in Human Preadipocytes, *J. Steroid Biochem. Mol. Biol. 50*:205–212 (1994).

33. Gansser, D., and G. Spiteller, Aromatase Inhibitors from *urtica dioica* Roots, *Planta Med. 61*:138–140 (1995).

34. Mäkelä, T.H., K.T. Wähälä, and T.A. Hase, Synthesis of Enterolactone and Enterodiol Precursors as Potential Inhibitors of Human Estrogen Synthetase (Aromatase), *Steroids 65*:437–441 (2000).

35. Evans, B.A., K. Griffiths, and M.S. Morton, Inhibition of 5 Alpha-Reductase in Genital Skin Fibroblasts and Prostate Tissue by Dietary Lignans and Isoflavonoids, *J. Endocrinol. 147*:295–302 (1995).

36. Morton, M.S., P.S. Chan, C. Cheng, N. Blacklock, A. Matos-Ferreira, L. Abranches-Monteiro, R. Correia, S. Lloyd, and K. Griffiths, Lignans and Isoflavonoids in Plasma and Prostatic Fluid in Men: Samples from Portugal, Hong Kong, and the United Kingdom, *Prostate 32*:122–128 (1997).

37. Morton, M.S., A. Matos-Ferreira, L. Abranches-Monteiro, R. Correia, N. Blacklock, P.S. Chan, C. Cheng, S. Lloyd, W. Chieh-ping, and K. Griffiths, Measurement and Metabolism of Isoflavonoids and Lignans in the Human Male, *Cancer Lett. 114*:145–151 (1997).

38. Herman, C., H. Adlercreutz, B.R. Goldin, S.L. Gorbach, K.A.V. Höckerstedt, S. Watanabe, E.K. Hämäläinen, M.H. Markkanen, T.H. Mäkelä, K. Wähälä, T.A. Hase, and T. Fotsis, Soybean Phytoestrogen Intake and Cancer Risk, *J. Nutr. 125*:757S–770S (1995).

39. Markaverich, B.M., and J.H. Clark, Two Binding Sites for Estradiol in Rat Uterine Nuclei: Relationship to Uterotropic Response, *Endocrinology 105*:1458–1462 (1979).

40. Markaverich, B.M., S. Upchurch, and J.H. Clark, Progesterone and Dexamethasone Antagonism of Uterine Growth: A Role for a Second Nuclear Binding Site for Estradiol in Estrogen Action, *J. Steroid Biochem. 14*:125–132 (1981).

41. Gareau, B., G. Vallett, H. Adlercreutz, K. Wähälä, T. Mäkelä, C. Benassayag, and E.A. Nunez, Phytoestrogens: New Ligands for Rat and Alpha-Fetoprotein, *Biochem. Biophys. Acta* 1094:339–345 (1991).

42. Waters, A.P., and J.T. Knowler, Effect of a Lignan (HPMF) on RNA Synthesis in the Rat Uterus, *J. Reprod. Fertil. 66*:379–381 (1982).

43. Orcheson, L., S. Richard, M. Seidl, S.F. Cheung, L. Luyengi, H. Fong, and L.U. Thompson, Estrus Cycle and Organ Changes in Rats Fed Flaxseed, Its Mammalian Lignan Precursor or Tamoxifen, *FASEB J. 7*:A291 (Abstr.) (1993).

44. Tou, J.C.L., J. Chen, and L.U. Thompson, Flaxseed and Its Lignan Precursor Secoisolariciresinol Diglycoside Affects Pregnancy Outcome and Reproductive Development in Rats, *J. Nutr. 128*:1861–1868 (1998).

45. Tou, J.C.L., J. Chen, and L.U. Thompson, Dose, Timing, and Duration of Flaxseed Exposure Affect Reproductive Indices and Sex Hormone Levels in Rats, *J. Toxicol. Environ. Health 56* (Part 1):555–570 (1999).

46. Tou, J.C.L., and L.U. Thompson, Exposure to Flaxseed or Its Lignan Component During Different Developmental Stages Influences Rat Mammary Gland Structures, *Carcinogenesis 20*:1831–1835 (1999).

47. Adlercreutz, H., K. Höckerstedt, C. Bannwart, S. Bloigu, E. Hämäläinen, T. Fotsis, and A. Ollus, Effect of Dietary Components, Including Lignans and Phytoestrogens, on Enterohepatic Circulation and Liver Metabolism of Estrogens, and on Sex Hormone Binding Globulin (SHBG), *J. Steroid Biochem. 27*:1135–1144 (1987).

48. Adlercreutz, H., Western Diet and Western Diseases: Some Hormonal and Biochemical Mechanisms and Associations, *Scand. J. Clin. Lab. Invest. 50* (Suppl. 201):3–23 (1990).

49. Whitten, P.L., and F. Naftolin, Dietary Estrogens—A Biologically Active Background for Estrogen Action, in *New Biology of Steroid Hormones*, edited by R.B. Hochberg and F. Haftolin, Raven Press, New York, 1991, pp. 155–167.

50. F. Anderson, Kinetics and Pharmacology of Estrogens in Pre- and Postmenopausal Women, *Int. J. Fertil. 38* (Suppl. 1):55–64 (1993).

51. Leach, R.E., and S.L. Hendrix, Hormone Replacement Therapy, in *Clinical Reproductive Gynecology*, edited by S.N. Hajj and W.J. Evans, Appleton and Lange, Norwalk, CT, 1993, pp. 59–73.

52. Goldfien, A., and S.E. Monroe, Ovaries, in *Basic and Clinical Endocrinology*, edited by F.S. Greenspan and G.J. Strewler, Appleton & Lange, Stamford, CT, 1997, pp. 434–486.

53. Barnes, R., and S.N. Hajj, Menopause: Pathophysiology of Menopause, in *Clinical Reproductive Gynecology*, edited by S.N. Hajj and W.J. Evans, Appleton & Lange, Norwalk, CT, 1993, pp. 31–41.

54. Shultz, T.D., W.R. Bonorden, and W.R. Seaman, Effect of Short-Term Flaxseed Consumption on Lignan and Sex Hormone Metabolism in Men, *Nutr. Res. 11*:1089–1100 (1991).

55. Phipps, W.R., M.C. Martini, J.W. Lampe, J.L. Slavin, and M.S. Kurzer, Effect of Flaxseed Ingestion on the Menstrual Cycle, *J. Clin. Endocrinol. Metab. 77*:1215–1219 (1993).

56. Haggans, C.J., E.J. Travelli, W. Thomas, M.C. Martini, and J.L. Slavin, The Effect of Flaxseed and Wheat Bran Consumption on Urinary Estrogen Metabolites in Premenopausal Women, *Cancer Epidemiol. Biomarkers Prev. 9*:719–725 (2000).

57. Wilcox, G., M.L. Wahlqvist, H.G. Burger, and G. Medley, Oestrogenic Effects of Plant Foods in Postmenopausal Women, *Br. Med. J. 301*:905–906 (1990).

58. Brzezinski, A., H. Adlercreutz, R. Shaoul, A. Shmueli, A. Reosler, and J.G. Schenker, Phytoestrogen-rich Diet: A Possible Alternative for Hormone Replacement Therapy. *Menopause 3*(4):251 (Abstr.) (1996).

59. Brzezinski, A., H. Adlercreutz, R. Shaoul, A. Rösler, A. Shmueli, V. Tanos, and J.G. Schenker, Short-term Effects of Phytoestrogen-rich Diet on Postmenopausal Women, *Menopause 4*:89–94 (1997).

60. Arjmandi, B.H., S. Juma, E.A. Lucas, L. Wei, S. Venkatesh, and D.A. Khan, Flaxseed Supplementation Positively Influences Bone Metabolism in Postmenopausal Women, *JANA 1*:27–32 (1998).

61. Hutchins, A.M., M.C. Martini, B.A. Olson, W. Thomas, and J.L. Slavin, Flaxseed Consumption Influences Endogenous Hormone Concentrations in Postmenopausal Women, *Nutr. Cancer 39*:58–65 (2001).

62. Haggans, C.J., A.M. Hutchins, B.A. Olson, W. Thomas, M.C. Martini, and J.L. Slavin, Effect of Flaxseed Consumption on Urinary Estrogen Metabolites in Postmenopausal Women, *Nutr. Cancer 33*:188–195 (1999).

63. Barrett-Connor, E., Epidemiology and the Menopause: A Global Overview, *Int. J. Fertil. 38* (Suppl. 1):6–14 (1993).

64. Brenner, P.F., The Menopausal Syndrome, *Obstet. Gynecol. 72*:6S–11S (1988).

65. Collaborative Group on Hormonal Factors in Breast Cancer, Breast Cancer and Hormone Replacement Therapy: Collaborative Reanalysis of Data from 51 Epidemiological Studies of 52,705 Women with Breast Cancer and 108,411 Women Without Breast Cancer, *Lancet 350*:1047–1059 (1997).

66. Jaffe, R.B., Menopause and Aging, in *Reproductive Endocrinology: Physiology, Pathophysiology, and Clinical Management*, 4th edn., edited by S.S.C. Yen, R.B. Jaffe, and R.L. Barbieri, W.B. Saunders, Philadelphia, PA, 1999, pp. 301–320.

67. Lock, M., Contested Meanings of the Menopause, *Lancet 337*:1270–1272 (1991).

68. Adlercreutz, H., E. Hämäläinen, S. Gorbach, and B. Goldin, Dietary Phyto-Oestrogens and the Menopause in Japan, *Lancet 339*:1233 (1992).

69. Dehennin, L., A. Reiffsteck, M. Jondet, and M. Thibier, Identification and Quantitative Estimation of a Lignan in Human and Bovine Semen, *J. Reprod. Fert. 66*:305–309 (1982).

70. Atkinson, D.A., H.H. Hill, and T.D. Shultz, Quantification of Mammalian Lignans in Biological Fluids Using Gas Chromatography with Ion Mobility Detection, *J. Chromatogr. 617*:173–179 (1993).

71. Pettersson, D., P. Aman, K.E. Knudsen, E. Lundin, J.X. Zhang, G. Hallmans, H. Harkonen, and H. Adlercreutz, Intake of Rye Bread by Ileostomists Increases Ileal Excretion of Fiber Polysaccharide Components and Organic Acids but Does Not Increase Plasma or Urine Lignans and Isoflavonoids, *J. Nutr. 126*:1594–1600 (1996).

72. Stumpf. K., P. Pietinen, P. Puska, and H. Adlercreutz, Changes in Serum Enterolactone, Genistein, and Daidzein in a Dietary Intervention Study in Finland, *Cancer Epidemiol. Biomarkers Prev. 9*:1369–1372 (2000).

73. Uehara, M., Y. Arai, S. Watanabe, and H. Adlercreutz, Comparison of Plasma and Urinary Phytoestrogens in Japanese and Finnish Women by Time-Resolved Fluoroimmunoassay, *BioFactors 12*:217–225 (2000).

74. Mazur, W.M., M. Uehara, K.T. Wahala, and H. Adlercreutz, Phyto-Oestrogen Content of Berries, and Plasma Concentrations and Urinary Excretion of Enterolactone After a Single Strawberry-Meal in Human Subjects, *Br. J. Nutr. 83*:381–387 (2000).

75. Kilkkinen, A., K. Stumpf, P. Pietinen, L.M. Valsta, H. Tapanainen, and H. Adlercreutz, Determinants of Serum Enterolactone Concentration, *Am. J. Clin. Nutr. 73*:1094–1100 (2001).

76. Adlercreutz, H., T. Fotsis, C. Bannwart, K. Wähälä, T. Mäkelä, G. Brunow, and T. Hase, Determination of Urinary Lignans and Phytoestrogen Metabolites, Potential Antiestrogens and Anticarcinogens, in Urine of Women on Various Habitual Diets, *J. Steroid Biochem. 25*:791–797 (1986).

77. Goldin, B.R., H. Adlercreutz, S.L. Gorbach, M.N. Woods, J.T. Dwyer, T. Conlon, E. Bohn, and S.N. Gershoff, The Relationship Between Estrogen Levels and Diets of Caucasian American and Oriental Immigrant Women, *Am. J. Clin. Nutr.* 44:945–953 (1986).

78. Cassidy, A., S.A. Bingham, K. Setchell, and D. Watson, Urinary Plant Oestrogen Excretion in Post-menopausal Women, *Proc. Nutr. Soc. 50*:105A (1991).

79. Adlercreutz, H., H. Honjo, A. Higashi, T. Fotsis, E. Hämäläinen, T. Haseawa, and H. Okada, Urinary Excretion of Lignans and Isoflavonoid Phytoestrogens in Japanese Men and Women Consuming Traditional Japanese Diet, *Am. J. Clin. Nutr. 54*:1093–1100 (1991).

80. Kelly, G.E,. C. Nelson, M.A. Waring, G.E. Joannou, and A.Y. Reeder, Metabolites of Dietary (Soya) Isoflavones in Human Urine, *Clin. Chim. Acta 223*:9–22 (1993).

81. Lampe, J.W., M.C. Martini, M.S. Kurzer, H. Adlercreutz, and J.L. Slavin, Urinary Lignan and Isoflavonoid Excretion in Premenopausal Women Consuming Flaxseed Powder, *Am. J. Clin. Nutr. 60*:122–128 (1994).

82. Cunnane, S.C., M.J. Hamadeh, A.C. Liede, L.U. Thompson, T.M.S. Wolever, and D.J.A. Jenkins, Nutritional Attributes of Traditional Flaxseed in Healthy Young Adults, *Am. J. Clin. Nutr. 61*:62–68 (1994).

83. Kirkman, L.M., J.W. Lampe, D.R. Campbell, M.C. Martini, and J.L. Slavin, Urinary Lignan and Isoflavonoid Excretion in Men and Women Consuming Vegetable and Soy Diets, *Nutr. Cancer 24*:1–12 (1995).

84. Hutchins, A.M., J.L. Slavin, and J.W. Lampe, Urinary Isoflavonoid Phytoestrogen and Lignan Excretion after Consumption of Fermented and Unfermented Soy Products, *J. Am. Diet. Assoc. 95*:545–551 (1995).

85. Hutchins, A.M., J.W. Lampe, M.C. Martini, D.R. Campbell, and J.L. Slavin, Vegetables, Fruits, and Legumes: Effect on Urinary Isoflavonoid Phytoestrogen and Lignan Excretion, *J. Am. Diet. Assoc. 95*:769–774 (1995).
86. Horn-Ross, P.L., S. Barnes, M. Kirk, L. Coward, J. Parsonnet, and R.A. Hiatt, Urinary Phytoestrogen Levels in Young Women from a Multiethnic Population, *Cancer Epidemiol. Biomark. Prev. 6*:339–345 (1997).
87. Hutchins, A.M., M.C., Martini, B.A. Olson, W. Thomas, and J.L. Slavin, Flaxseed Influences Urinary Lignan Excretion in a Dose-dependent Manner in Postmenopausal Women, *Cancer Epidemiol. Biomark. Prev. 9*:1113–1118 (2000).
88. den Tonkelaar, I., L. Keinan-Boker, P. Van't Veer, C.J.M. Arts, H. Adlercreutz, J.H.H. Thijssen, and P.H.M. Peeters, Urinary Phytoestrogens and Postmenopausal Breast Cancer Risk, *Cancer Epidemiol. Biomark. Prev. 10*:223–228 (2001).
89. Kurzer, M.S., J.W. Lampe, M.C. Martini, and H. Adlercreutz, Fecal Lignan and Isoflavonoid Excretion in Premenopausal Women Consuming Flaxseed Powder, *Cancer Epidemiol. Biomark. Prev. 4*:353–358 (1995).
90. Adlercreutz, H., T. Fotsis, M.S. Kurzer, K. Wähälä, T. Mäkelä, and T. Hase, Isotope Dilution Gas Chromatographic Mass Spectrometric Method for the Determination of Unconjugated Lignans and Isoflavonoids in Human Feces, with Preliminary Results in Omnivorous and Vegetarian Women, *Anal. Biochem. 225*:101–108 (1995).

Chapter 7

Nutritional and Hematological Effects of Flaxseed

Uma S. Babu and Paddy W. Wiesenfeld

U.S. Food and Drug Administration, Center for Food Safety and Applied Nutrition, 8301 Muirkirk Road, Laurel, Maryland 20708, USA

Introduction

Food uses of flaxseed (FS) in the United States, Europe, and Asia include it as a component in some brands of cereals, in specialty breads and cookies, as a seed dressing, and in other bakery products (1,2). Flaxseed is also sold in health food stores as FS oil. Flaxseed comprises about 35% lipids, 20–30% protein, 35–50% fiber, and 6% ash or minerals, by weight. Alpha-linolenic acid (ALA) contributes to more than 50% of the total fatty acids of the FS lipids. In addition to these nutrient components, there are non-nutrient and antinutrient components in FS. The major non-nutrient component is the lignan precursor, namely secoisolariciresinol; the antinutrient components are the phytic acids, linatine and linustatin (vitamin B6 antagonists), and the cyanogenic glycosides (1,3). The growing popularity of FS is due to the possible health benefits (3,4), which include decreased risk of cardiovascular disease (1,5–12) and cancer (10,13–17), antiviral and bactericidal activity (18), anti-inflammatory activity (19–23), laxative effect (1,7,24), hypoglycemic effect (7,25), and prevention of menopausal symptoms and osteoporosis (26).

The controversy about the nutritional benefit and safety of FS is due in part to its complex nature. Nutrients and non-nutrients of FS can have both beneficial and/or adverse effects depending upon dose, timing, and length of exposure. Among the bioactive components of FS is the n-3 fatty acid, α-linolenic acid (ALA, C18:3). ALA from FS is bioavailable, and can be metabolized to other very long chain n-3 fatty acids, which may alter tissue fatty acid composition and/or function (27–34). Some of the non-nutrient components such as fiber can bind divalent metals and thus potentially affect the mineral status. Similarly, phytate from the fiber can inhibit iron absorption and may compromise the iron indices. Other examples of antinutrient components in FS that could have an adverse health effect are linatine and linustatin that can bind vitamin B6, thus causing B6 deficiency (3).

In this chapter, we will discuss some of the nutritional and hematological effects of FS, such as iron status, serum and tissue fatty acids including n-3/n-6 ratio, serum cholesterol and triglyceride levels, liver and kidney function, mineral and vitamin status, platelet structure and function, and laxative effects and blood glucose levels. Some of these topics are discussed in greater depth in other chapters in this book.

Effects of FS on Hemoglobin, Hematocrit, and Total Red Blood Cell (RBC) Counts

Although FS is rich in fiber, which could inhibit iron absorption (35–38), none of the indicators of iron status was affected in rats that were fed moderate or high FS for 8 wk postweaning or during gestation, lactation, and until 90 d of age, respectively (28,39). In fact, 10% dietary FS increased hematocrit and RBC counts among young rats fed the diet for 8 wk (39; Table 7.1). Other fibers such as sugar-beet fiber were also shown to not affect iron status among men fed either 26 or 40 g dietary fiber/day, for 4–5 wk (40,41). The ratio of dietary phytate to ascorbic acid is very critical to maintain the iron status (42). It is possible that the studies that showed no detrimental effect of dietary FS or other fibers on iron status had the desired ratio of ascorbic acid to phytate to overcome the negative effects of phytate on iron absorption (35). Furthermore, the ratio of soluble to insoluble fibers in FS may be such that it does not inhibit iron absorption (43). On the other hand, flaxseed oil or linseed oil without the fiber either increased or did not interfere with iron status of 5- or 18-day-old piglets or healthy volunteers (31,44,45). Similarly, rats fed high ALA diet from weanling to 4 mon of age showed no deleterious effects on hematological indices, including mean corpuscular volume, hemoglobin concentrations, hematocrit, mean corpuscular hemoglobin concentration, and RBC counts (46). Overall, studies thus far indicate that fiber-rich FS consumption may not compromise iron status when consumed for a short period of time and at a low to moderate dose, especially when included in a balanced diet.

Fatty Acid Metabolism of FS

The effects of dietary FS and other oils containing ALA on fatty acid metabolism in humans, and in various species are presented in Table 7.2. Rats fed FS or FS oil diets have been shown to have a dose-dependent decrease in serum and tissue linoleic acid (LA, C18:2) and an increase in ALA compared with control rats (5,27,28,47,48; Table 7.2). In addition, rats fed FS and defatted flaxseed meal (FLM) diets had a significant reduction in serum arachidonic acid (AA, C20:4) (5,27,28,47,48; Table 7.2). There is

TABLE 7.1

Red Blood Cells (RBC), Hemoglobin, and Hematocrit of Young Female Rats Fed Either Flaxseed (FS) or Flaxseed Meal (FLM)[a]

Parameter	0% FS	5% FS	10% FS	6.2% FLM
RBC	$6.88 \pm 0.70^b(5)$	$7.39 \pm 0.33^a(9)$	$7.59 \pm 0.30^a(10)$	$7.37 \pm 0.31^{a,b}(10)$
Hemoglobin	$13.54 \pm 1.48(5)$	$14.40 \pm 0.57(9)$	$14.82 \pm 0.92(10)$	$14.27 \pm 0.27(10)$
Hematocrit	$37.34 \pm 3.56^b(5)$	$40.06 \pm 1.44^{a,b}(9)$	$41.79 \pm 2.24^a(10)$	$40.45 \pm 1.49^{a,b}(10)$

[a]Results are expressed as mean ± SD. Means with different superscripts within the same row are significantly different at $P < 0.05$.
Source: Babu *et al.* 2000 (39).

some disagreement on the effects of FS, FS oil, and FLM diets on eicosapentanoic acid (EPA, C20:5) or docosahexanoic acid (DHA, C22:6) production in both animal and human studies. In most of the cases, FS diets produced a significant increase in EPA compared with the controls (5,25,27,28,34,44,49–52; Table 7.2). However, FS or FS oil diets do not always produce an increase in serum or tissue DHA (28,44,50,51; Table 7.2). Rats fed 10 or 40% FS, and 20% FS oil diets showed a significant increase in serum EPA and DHA (5,27,47). However, in other rodent studies, 7% FS oil or 40% FS diets increased EPA, but not DHA (28,48). In other animal species (e.g., pigs, steers, and chickens) FS-enriched diets resulted in an increase in tissue concentration of ALA and EPA (44,52,53). In another chicken study, FS feeding resulted in a significant increase in tissue ALA, EPA, and docosapentaenoic acid (DPA, C22:5) (54). Similarly, dietary FS oil resulted in a significant increase in ALA and EPA in muscle and adipose tissue, but only a small change in DHA in cattle (29). Healthy human males fed FS had a three- and twofold increase, respectively, in plasma EPA and DHA (34,49). However, in another study, healthy subjects consuming 50 g/d FS showed an increase in plasma ALA, EPA, and DPA, but no increase in DHA (7). Other studies using FS have reported an increase in serum ALA, a decrease in eicosatrienoic acid (C20:3), but no change in AA, EPA, or DHA (50,55). Several reasons for a lack of increase in DHA have been suggested including a higher oxidation rate of ALA, a lower rate of conversion to *n*-3 long chain metabolites, or a significant retroconversion of DHA to EPA (56).

Flaxseed-Derived ALA Influences on *n*-3/*n*-6 Fatty Acid Ratio

Gerster (56) reported that in humans, a diet rich in *n*-6 polyunsaturated (PUFA) reduced synthesis of EPA and DHA by as much as 40–50% and that the ideal dietary ratio for *n*-6/*n*-3 fatty acids should not exceed 4–6:1. The *n*-6/*n*-3 ratio of current Western diets is much higher (57). The dietary LA to ALA fatty acid ratio significantly influences fatty acid elongation and desaturation, specifically to EPA and DHA (8,28,52,58). A ratio of LA to ALA of 4:1 or less has been shown to be optimal for elongation of ALA to EPA (57). Changes in *n*-3/*n*-6 ratios in tissues caused by altering dietary lipids, including the addition of FS or FLM, have been shown to impact developmental, physiological, and immunological functions (59–61). AA, EPA, and DHA are known to be essential for early development of the brain, retina, and spermatozoa (56,58,62–65). DHA derived from ALA has been demonstrated to be essential for brain development and visual function, whereas AA derived from LA is essential for neonatal growth and ovulation (66). Reproductive effects of ALA were demonstrated by increased duration and number of ovulations, increased gestational length, decreased number of ova released, and reduced number and size of ovarian follicles compared with the effects of dietary *n*-6 fatty acids (66–68). In regard to the male reproductive system, decreased testicular size and a loss of fertility in male rats fed ALA-supplemented diets have been reported (69). However, the number of cauda epididymal sperm has been shown to increase with FS in male rats (70).

TABLE 7.2
Effects of Dietary Flaxseed (ALA) on LA, ALA, AA, EPA, and DHA[a]

Study	ALA dose	LA	ALA	AA	EPA	DHA
Human						
Cunnane *et al.* 1993 (25)	50 g FS/d; 4 wk	ND	TG 1.2 → 3.5 P 0.4 → 0.75	ND	TG 0. → 0.4 P 0.7 → 1.6	TG 0.8 → 1.2 P 3.5 → 4.0
Kelley *et al.* 1993 (55)	21 g/d ALA (6.3% energy)	S 26 → 25 MNC 5.7 → 6.8	S 0.2 → 3.2 MNC 0.2 → 0.6	S 6.0 → 5.5 MNC 4.2 → 23.8	S 0.6 → 0.6 MNC 0.4 → 0.4	S 0.8 → 1.1 MNC 1.4 → 0.9
Cunnane *et al.* 1994 (7)	50 g FS/d; 4 wk	ND	TG 3.0 → 5.4	ND	TG 0.1 → 0.4	TG 0.3 → 0.2
Mantzioris *et al.* 1994 (49)	13.7 g/d ALA from FSO; 4 wk	P 22 → 21 TG 18.4 → 16.8	P 0. → 1.1 TG 0.8 → 6.3	P 11 → 11 TG 1.5 → 1.4	P 0.8 → 1.9 TG 0.2 → 0.6	P 2.9 → 3.3 TG 0.5 → 0.6
Allman *et al.* 1995 (50)	40 g FSO/d; 23 d	PLT 4.3 → 5.2	PLT 0.2 → 0.4	PLT 23 → 22	PLT 0.5 → 1.2	PLT 1.9 → 1.7
Ferrier *et al.* 1995 (33)	20% FS chicken; 4 eggs/day; each egg contained 10.7% ALA	PLT 5 → 4.9	PLT 0.2 → 0.3	PLT 24.5 → 24.2	PLT 0.17 → 0.23	PLT 1.5 → 2.0
Mantzioris *et al.* 2000 (34)	9.0 g/d ALA (3% energy); 4 wk	P 21 → 18 MNC 6.4 → 5.9 PLT 7.4 → 6.8	P 0.2 → 0.55 MNC 0.02 → 1.0 PLT 0.08 → 0.2	P 9.5 → 8.8 MNC 22 → 20.8 PLT 22.2 → 21.3	P 0.98 → 2.98 MNC 0.4 → 1.11 PLT 1.9 → 2.3	P 3.2 → 5.4 MNC 2.4 → 3.33 PLT 1.9 → 3.1
Thies *et al.* 2001 (51)	FS 37% of total fat, 12 wk	P 22 → 22.8	ND	P 9 → 9.26	P 0.83 → 1.24	P 3.28 → 3.28
Tarpila *et al.* 2002 (106)	1.3 g/100 g FS + 5 g/100 g FSO; 4 wk 20% energy from FS	S 28.3 → 29.7	S 1.28 → 1.43	S 6.0 → 6.1	S 1.21 → 1.34	S 2.83 → 2.93

Rat						
L'Abbe *et al.* 1991 (47)	20% LSO; 16 wk	H 14 → 12.2 L 16.8 → 11.6	H 0.3 → 2.6 L 0.3 → 4.2	H 16.6 → 15.2 L 18.0 → 12.0	H 0 → 0.2 L 0.2 → 0	H 6.5 → 10.1 L 3.4 → 7.7
Ratnayake *et al.* 1992 (5)	40% FS; 90 d (0, 10, 20, 40% FS)	L 22.9 → 14.1 H 18.4 → 17.9 K 29.5 → 14.9	L 0.3 → 7.9 H 0.8 → 6.8 K 0.4 → 16.0	L 10.8 → 7.4 H 16.2 → 10.8 K 6.9 → 5.8	L 0.1 → 3.4 H 0.3 → 0.6 K 0 → 1.0	L 1.1 → 3.0 H 3.1 → 5.4 K 0.3 → 0.5
Babu *et al.* 1997 (27)	10% FS; 8 wk	S 27.4 → 25.3	S 0.0 → 3.24	S 30.6 → 26.2	S 0.0 → 2.43	S 0.65 → 3.1
Dell *et al.* 2001 (48)	7% FSO; 40 d	A 28.6 → 21.7 B 1.5 → 2.9	A 2.9 → 38.9 B 0.1 → 0.3	A 0.3 → 0.1 B 13.4 → 11.8	ND	A 0.05 → 0.1 B 16.4 → 16.0
Wiesenfeld *et al.* 2003 (in press) (28)	40% FS; 8 wk (0, 20, 40% FS; 13, 26% FLM)	S 18.73 → 16.05 L 25.98 → 15.67	S 2.97 → 10.57 L 2.05 → 12.21	S 32.56 → 13.48 L 17.37 → 11.61	S 0.36 → 8.84 L 0.79 → 3.07	S 5.49 → 3.48 L 8.57 → 8.15
Steer						
Scollan *et al.* 2001 (29)	21.3% LSO	M 11.4 → 10.1 A 0.94 → 0.86	M 2.1 → 4.3 A 0.37 → 0.65	M 5.1 → 4.5 ND	M 2.3 → 3.6 ND	M 0.55 → 0.63 ND
Neonatal piglets						
Innis and Dyer 1999 (44)	10% canola oil; 18 d	P 28.7 → 19.7 TG 40 → 18	P 0.8 → 1.5 TG 4.7 → 5.9	P 9.7 → 7.2 TG 1.8 → 0.9	P 0.2 → 0.8 TG 0.4 → 0.9	P 2.3 → 2.3 TG 0.3 → 0.5
Pig						
Matthews *et al.* 2000 (52)	100 g/kg FS; 65 d	P 34.1 → 31.4 L 18.1 → 18.3	P 1.7 → 7.9 L 0.8 → 4.4	P 6.8 → 4.5 L 17.2 → 12.5	P 0.7 → 2.5 L 1.2 → 5.9	P 0.8 → 1.7 L2.2 → 2.2

[a]Fatty acids expressed as g% of total fatty acids. Abbreviations: ALA, alpha-linolenic acid; LA, linoleic acid; AA, Arachidonic Acid; EPA, eicosapentaenoic acid; DHA, docosahexaenoic acid; FS, flaxseed; FSO, flaxseed oil; FLM, flaxseed meal; TG, plasma triglyceride fatty Acids; P, plasma fatty acids; S, serum fatty acids; PLT, fatty acids, platelets; MNC, mononuclear cells phospholipid fatty acids; H, heart fatty acids; L, liver fatty acids; B, brain fatty acids; K, kidney fatty acids; A, adipose fatty acids; M, muscle fatty acids; ND, not determined; NR, not reported.

Effects of FS on Serum Cholesterol

The effects of dietary ALA, FS, and FS oil on serum lipids and lipoprotein fractions of humans and various animal species are shown in Table 7.3. Among the reported health effects of FS and FS oil consumption in humans and animals are the reduction in atherosclerosis and serum cholesterol (5–11,28,32, 47,71,72; Table 7.3). In two different studies, hyperlipidemic subjects fed FS or partially defatted FLM had a significant reduction in serum total cholesterol and LDL cholesterol without a reduction in HDL-C (6,11). In another study, healthy adults had decreased plasma total cholesterol and LDL, but no change in triglycerides, upon consumption of 50 g of FS/d (7). The authors suggested that mucilage in FS could be hypocholesterolemic in a manner analogous to other soluble fibers. Most of the rat studies that reported a cholesterol-lowering effect used very large amounts of FS or FS oil, or included additional dietary cholesterol. Moderate levels of FS, FS oil, or FLM did not seem to alter serum total cholesterol or HDL levels in healthy human subjects or rats (8,30,31,33,39,48,55,73–75; Table 7.3). Thus, both human and animal data suggest that FS is more effective in lowering cholesterol in hypercholesterolemic situations. The variable effects of FS or FS oil on serum lipids in both humans and animals may be due to the complex nature of FS, the ratio of C18:2/C18:3 (or *n*-6/*n*-3 ratio) in the test diets, other dietary variables such as fat and cholesterol content, as well as the appropriateness of the animal model and/or the human subject population (hyperlipidemic *vs.* healthy). It is possible that lignans, fiber, and vegetable proteins present in the FS and FLM could have major roles in reducing serum cholesterol in animal models and/or in humans.

Effects of FS on Kidney and Liver Function Enzymes

Alpha-linolenic acid from FS is a precursor for group 3 prostanoids whereas *n*-6 fatty acids are precursors for group 2 prostanoids (76). Group 3 prostanoids are known inhibitors of inflammation (66). In addition, the lignans might be an important contributing factor in the anti-inflammatory activity of FS, through the monocyte chemoattractant protein pathway (77). Recent reviews have summarized the role of dietary soy and flaxseed and their phytoestrogens in renal disease protection in animals and humans (23,78,79). Both FS and FS oil have been shown to successfully retard the progression of renal injury among rats that were subjected to 5/6 nephrectomy, as measured by reduced proteinuria and increased glomerular filtration rate. Both FS and FS oil were shown to preserve glomerular architecture over the 20-wk study period, FS being more protective than FS oil. Hypertension and dyslipidemia were also prevented by FS and FS oil, further suggesting the role for FS in alleviating renal injury (19). Similarly, efficacy of FS in MRL/lpr mice has been established by Hall *et al.* (80). In MLR/lpr mice, with features similar to glomerulonephritis of human systemic lupus erythematosus, 15% FS was shown to increase glomerular filtration rate, decrease proteinuria and reduce mortality in mice fed FS diet for 14 wk. Likewise, both short-term (4 wk) and long-term (2 yr) studies conducted with lupus nephritis patients showed that dietary FS resulted in nonsignificant reduction in proteinuria and

TABLE 7.3
Changes in Serum Lipid Fractions by Flaxseed or Flaxseed Oil

			Lipid class			
Study	Dose	Time	TC	HDL-C	LDL-C	TG
Human						
Singer *et al.* 1990 (32)	38 g/d ALA	2 wk	↓[a]	NS[a]	↓[a]	↓[a]
Kelley *et al.* 1993 (55)	21 g/d ALA (6.3% of energy)	56 d	NS	NS	NS	NS
Bierenbaum *et al.* 1993 (6)	15 g/d FS and 3 slices of 10% FS-containing bread	3 mon	↓[a]	NS[a]	↓[a]	↓[a]
Cunnane *et al.* 1994 (7)	50 g/d	4 wk	↓[a]	NS	↓[a]	NS
Ferrier *et al.* 1995 (33)	10.7% ALA/d from 4 eggs	2 wk	NS	NS	NS	ND
Mantzioris *et al.* 1994 (49)	13.7 g/d ALA from FSO (5% of energy)	4 wk	NS	↑	NS	NS
Layne *et al.* 1996 (30)	4.3 g/d FS oil	3 mon	NS	NS	NS	NS
Freese and Mutanen 1997 (31)	5.9 g/d ALA	4 wk	NS	NS		NS
Nestel *et al.* 1997 (9) (overweight subjects)	20 g/d FSO	4 wk	NS	↓	NS	NS
Jenkins *et al.* 1999 (11) (hyperlipidemic)	50 g/d partially defatted FS	3 wk	↓	NS	↓	↑
Lucas *et al.* 2002 (74)	40 g/d FS	3 mon	↓	NS	NS	NS
Rat						
Dell *et al.* 2001 (49)	7% FSO	40 d	NS	ND	ND	NS
Babu *et al.* 2000 (39)	10% FS	56 d	NS	NS	ND	↑
Kritchevsky *et al.* 1991 (75)	20% FSO	3 wk	NS	NS	ND	NS
L'Abbe *et al.* 1991 (47)	20% LSO	16 wk	↓	↓	ND	ND
Ratnayake *et al.* 1992 (5)	40% FS	90 d	↓	↓	↓	↓
Wiesenfeld *et al.* 2003 (in press) (28)	40% FS	90 d	↓	↓	ND	NS
Rabbit						
Prasad *et al.* 1998 (72)	7.5 g/d FS with additional 1% cholesterol in the diet	8 wk	↓	NS	↓	↑
Prasad *et al.* 1999 (102)	15 mg secoisolariciresinol/kg body weight/d	8 wk	↓	↑	↓	NS

[a]Within group comparison. Abbreviations: FS, flaxseed; FSO, flaxseed oil; LSO, linseed oil, TC, total cholesterol; HDL-C, high density cholesterol; LDL-C, low density cholesterol; TG, triglycerides; ND, not determined; NS, not significant; ↓, decrease; ↑, increase.

significant reduction of serum creatinine, with a concomitant reduction in platelet activating factor, thus suggesting the usefulness of FS in human renal diseases (21,81).

There are a few studies describing the impact of dietary ALA on liver function in certain animal models. Short-term feeding of Sprague-Dawley rats with ALA has been shown to increase β-oxidation of fatty acids in the liver and decrease activity of enzymes related to fatty acid synthesis, such as glucose-6-phosphate dehydrogenase and acetyl-CoA carboxylase, leading to lowering of serum and liver lipids in those rats (82–84). In addition to these enzymes, dietary linseed oil (20% by weight) was also shown to increase phenobarbital and acetone-induced cytochrome P-450 enzyme in the rat liver (85). In contrast to these potential beneficial effects of ALA on hepatic enzymes, dietary perilla oil containing ALA increased serum alanine aminotransferase (ALT) in mice treated with bacterial lipopolysaccharide and D-galactosamine (86), which resulted in higher mortality in these mice. However, in a recent multigeneration study, 40% dietary FS was shown to increase ALT but not aspartate aminotransferase or alkaline phosphatase in 90-day-old rats that were fed a high FS diet during gestation, lactation, and postweaning periods (28). The increase in ALT was statistically significant compared only with the control group, but the biological significance of this change was questionable because there were no other signs of illness in that group of rats. Overall, FS, FS oil, or other oils containing ALA may exert positive effects on liver function, with the exceptions of certain treatments in which the *n*-6 fatty acids and prostanoids derived from these fatty acids may be protective (86).

Effects of FS on Mineral and Vitamin Metabolism

Flaxseed is very rich in fiber containing phytic acid, which is known to bind positively charged minerals such as calcium and zinc (3,87,88). However, FS is also the richest source of lignans, which are known to have estrogenic and antiestrogenic activities (89,90). These phytoestrogens have been shown to be useful in preventing bone resorption and in improving bone density (91,92). Furthermore, ALA in the FS might exert positive effect on bone formation *via* reduced prostaglandin E_2 (PGE_2) (93,94). Thus, one could postulate that the negative effects of fiber and/or phytic acid on mineral balance might be nullified by the lignans and the fatty acid portion of FS. Postmenopausal women consuming 40 g FS/d for 3 mon did not show any negative effects on biomarkers of bone metabolism (74). Similarly, rat studies have indicated that moderate or high dietary FS did not affect serum calcium levels or bone density, thus suggesting that the phytate of FS did not limit the bioavailability of minerals for bone mineralization (28,39,88,95). Similar to this lack of a negative effect of FS on calcium status, a 90-d study in rats has shown no impact of 40% dietary FS on tibia or liver zinc concentrations (5). Lower dietary calcium and slightly higher zinc content of FS, which resulted in a ratio of [phytate] [calcium]: [zinc], was speculated to be noninhibitory to the zinc status of rats in that study.

Long chain fatty acids present in FS and those produced by further elongation and desaturation are highly susceptible to oxidation by free radicals, which results in increased antioxidant requirement to compensate for higher utilization by these fatty acids. This oxidative stress by ALA has been demonstrated by increased urinary thiobarbituric acid reacting substances (TBARS), and decreased tissue α-tocopherol in rats that were fed 40% FS for 90 d (5). Similarly, another study has shown decreased liver vitamin E among rats that were fed 40% FS during gestation, lactation, and up to 90 d of age (28; Table 7.4). However, FS meal containing all the other components of FS without the lipids had no impact on vitamins A or E in serum or liver. A study in which rats were fed 20% of total fat as PUFA, from either linseed or menhaden oil, showed decreased heart and plasma vitamin E, the effect being more pronounced with the menhaden oil (96). A human trial with hyperlipidemic subjects showed a slight but statistically significant reduction in serum protein thiol content upon consumption of 50 g partially defatted flaxseed/day for 3 wk (11). In contrast to the studies that showed a negative impact of high dietary FS or ALA on antioxidant status, moderate consumption of FS or linseed oil did not impact vitamin E or TBARS in plasma or urine in human volunteers (7), rats (39; Table 7.5), or monkeys (97). The lignans from FS are known to have antioxidant properties (3,98–101), which also might increase binding activity of the tocopherol-binding protein (102). Thus, the lignan portion of FS might counteract the lipid portion in preventing the deleterious effect on vitamin E status, especially when FS is added to the diet in moderate amounts.

Although high dietary FS or FS oil has been shown to reduce the tissue vitamin E, and to increase the urinary TBARS, dietary linseed oil did not appear to affect the vitamin A or β-carotene status. This was demonstrated in one of the rat studies that tested the impact of different types of fats of animal and plant origin on plasma and tissue concentrations of vitamin A and β-carotene (103). In that study, all the fats tested enhanced the absorption of β-carotene compared to the control group. However, there was no difference in the amounts of β-carotene in the liver or lungs from the linseed oil and soy oil groups. Vitamin A concentrations were similar in plasma, kidneys, lungs, spleens, and livers from the two groups fed linseed oil and soy oil (103). These results and those from Tomassi and Olson (104) as well as Wiesenfeld *et al.* (28; Table 7.4) suggest that linseed oil or FS do not pose any stress on vitamin A status compared with the *n*-6 fatty acid containing diets. However, in general, saturated fatty acids appeared to significantly facilitate the absorption of β-carotene, conversion to vitamin A, and accumulation in spleen compared with unsaturated fatty acids (103).

The cyanogenic glucosides present in FS, namely, linustatin, neolinustatin, and linamarin, can be toxic (105). The toxicity comes from the release of hydrogen cyanide (a potent respiratory inhibitor), by the action of the enzyme β-glycosidase, which is present in the seed itself. Although a twofold increase in urinary thiocyanate excretion was observed by consumption of 50 g FS/d for 4 wk among young healthy female volunteers, there was no increase in serum thiocyanate in volunteers upon consumption of 1.3 g ground FS and 5 g FS oil/100 g food, for 4 wk (25,106). Different cultivars of flaxseed were shown to produce varying amounts of cyanide, which was almost totally abolished by boiling the flaxseed homogenate, due to inactivation of β-glycosidase (107). Furthermore, cyanide was not detectable in bread, due to the baking process (107), thus indicating that consumption of baked products containing FS may not present cyanide toxicity.

TABLE 7.4

Serum and Tissue Vitamins A and E in 90-Day-Old (F1) Rats Fed Flaxseed (FS) or Flaxseed Meal (FLM) During Gestation, Lactation, and Post-Weaning Periods[a,b]

Dietary group	Serum vitamin A (µg/mL)	Serum vitamin E (µg/mL)	Liver vitamin A (µg/g)	Liver vitamin E (µg/g)	Heart vitamin E (µg/g)
Control	0.31 ± 0.08 (11)	9.18 ± 2.12 (11)	93.52 ± 34.66 (11)[ab]	56.64 ± 14.83(11)[a]	49.34 ± 6.86(11)[ab]
20% flaxseed	0.33 ± 0.06 (8)	9.86 ± 3.1 (8)	85.04 ± 34.94 (10)[ab]	54.04 ± 13.08 (10)[a]	53.26 ± 6.18 (10)[a]
13% flaxseed meal (11)[ab]	0.26 ± 0.07 (6)	7.76 ± 2.61 (6)	106.25 ± 20.76 (11)[a]	59.03 ± 10.12 (11)[a]	51.68 ± 5.19
40% flaxseed	0.30 ± 0.11 (6)	6.59 ± 2.10 (6)	80.62 ± 31.90 (9)[ab]	44.92 ± 22.37 (9)[b]	45.6 ± 5.85 (9)[b]
26% flaxseed meal	0.36 ± 0.13 (8)	9.36 ± 3.26 (8)	71.75 ± 6.15 (10)[b]	73.37 ± 20.49 (10)[a]	53.13 ± 6.29 (11)[a]

[a]Results are expressed as means ± SD. Number of animals/group indicated in parentheses. Heart vitamin A content was below detectable limits.

[b]Different superscripts within each column indicate that the mean values were significantly different at $P < 0.05$. Statistical comparisons: FS comparisons (0, 20, and 40%); low-dose comparisons (20% FS and 13% FLM); FLM comparisons (0, 13, and 26%); high-dose comparisons (40% FS and 26% FLM). *Source*: Wiesenfeld *et al.* 2002 (28).

TABLE 7.5

Dietary, Plasma, and Liver Vitamin E Levels of Rats Fed Either Flaxseed (FS) or Flaxseed Meal (FLM)[a]

Vitamin E	0% FS	5% FS	10% FS	6.2% FLM
Diet (mg/100 g)	8.25	8.32	8.39	8.27
Plasma	31.49 ± 6.34 (10)	36.90 ± 13.58 (7)	38.80 ± 6.20(6)	34.37 ± 9.97(8)
µg/mg cholesterol	12.32 ± 2.98 (10)	14.07 ± 4.53 (7)	15.18 ± 2.49 (6)	13.06 ± 3.17 (8)
Liver (µg/g)	57.59 ± 8.25 (10)[c]	60.76 ± 9.53 (10)[b,c]	73.83 ± 15.23 (10)[a]	64.35 ± 4.84 (10)[b]

[a]Results are expressed as mean ± SD. Means with different superscripts within the same row are different at $P < 0.05$. Plasma vitamin E levels are expressed as both SI units and µg/mg cholesterol. *Source*: Babu *et al.* 2000 (39).

Another antinutrient in FS, linatine, is known to be a vitamin B6 antagonist (108). Marginal vitamin B6 deficiency was observed among pigs that were fed diets containing 30% by weight linseed, whereas the same diet caused no change in B6 status in rats (109). However, neither of the two species showed any clinical signs of vitamin B6 deficiency. The differences in the B6 status of the two species were attributed to the lower consumption of linseed by rats as well as the lower requirement for B6 by rats than pigs (109). Similarly, rats that were fed 40% FS showed no signs of vitamin B6 deficiency in another 90-d dietary study (5).

Laxative Effects of FS

Among other health benefits, FS has been used as an effective laxative agent (1). Cunnane *et al.* (7) have reported a significantly higher number of bowel movements among young healthy adults consuming 25 g FS/d for a period of 3 wk. Similarly, linseed, as a part of daily fiber mix, was shown to be effective in routine laxative therapy for elderly long-care patients (110,111). Additionally, raw linseed oil has been commonly used in livestock as a laxative for treatment of intestinal impactions. Treatment of horses with 2.5 mL raw linseed oil/kg, twice/day, for a 2-wk period was shown to have greater laxative effect than mineral oil at the same concentration, suggesting its use in the treatment of intestinal obstructions (24). Thus, FS may be used either alone or in conjunction with other laxative therapies to alleviate constipation and intestinal problems in people or animals.

Effects of FS and ALA on Platelet Structure, Function, and Aggregation

Impact of dietary ALA on eicosanoid production and platelet aggregation is presented in Table 7.6. Fatty acids are used not only for energy and development, but are also precursors for hormones and intracellular messengers. Linoleic acid is metabolized to AA, a precursor of the proinflammatory eicosanoids, such as PGE_2, thromboxane A_2 (TXA_2), and leukotriene B_4 (112), whereas ALA is metabolized to EPA and DHA, which are precursors for anti-inflammatory eicosanoids PGE_3 and PGF_3 (66). Dietary FS or ALA has been reported to reduce the synthesis of TXA_2, TXB_2, PGE_2 and increase the concentration of noninflammatory prostanoid such as PGE_3 (34,47,113–128; Table 7.6). Flaxseed might alter eicosanoid production and function by several mechanisms. Eicosapentanoic acid produced from ALA may replace AA as a substrate for TXB_2 production (50) or competitively inhibit AA conversion to PGE_2, TXB_2, and LTB_4 (34,49,50,113). Some of these proinflammatory eicosanoids are known to be potent platelet aggregatory factors (129,130). Platelets are known to play a very important role in the human vascular system, by participating in thrombosis and haemostasis. A change in platelet activity, which can be accomplished by changing the fatty acid composition, has been proposed to affect the risk of coronary thrombosis (131,132). Suppression of platelet activation using aspirin has been

TABLE 7.6
Changes in Eicosanoid Production and Platelet Aggregation by Dietary ALA[a]

Study	TXA$_2$	TXB$_2$	PGE$_2$	PGE$_3$/PGI$_3$	Aggregation
Human					
Renaud and Nordoy 1983 (134)	ND	ND	ND	ND	↓
Freese *et al.* 1994 (135)	ND	ND	ND	ND	↓
Freese and Mutanen 1997 (31)	ND	■	ND	ND	■
Allman *et al.* 1995 (50)	ND	ND	ND	ND	↓
Chan *et al.* 1993 (138)	ND	■	ND	ND	ND
Budowski *et al.* 1984 (113)	ND	↓	ND	ND	↓
Ferretti and Flanagan 1996 (115)	ND	↓	ND	ND	ND
Li *et al.* 1999 (45)	ND	■	ND	ND	■
Mantzioris *et al.* 2000 (34)	ND	↓	↓	ND	ND
Bierenbaum *et al.* 1993 (6)	ND	ND	ND	ND	↓
Bjerve *et al.* 1987 (114)	■	■	ND	↑	ND
Clark *et al.* 1995 (81)	ND	ND	ND	ND	↓
Equine					
Hansen *et al.* 2002 (139)	ND	ND	ND	ND	■
Henry *et al.* 1990 (116)	ND	↓	ND	ND	↓
Morris *et al.* 1989 (117)	ND	↓	ND	ND	ND
Rabbit					
Vas Dias *et al.* 1982 (136)	ND	ND	ND	ND	↓
Rat					
L'Abbe *et al.* 1991 (47)	ND	↓	ND	ND	ND
Moore *et al.* 1991 (119)	ND	↓	ND	ND	ND
Magrum and Johnston 1983 (120)	ND	ND	↓	ND	ND
Opmeer *et al.* 1984 (121)	ND	ND	↓	ND	ND
Ikeda *et al.* 1995 (118)	↓	■	ND	ND	ND
Streptozotocin-induced diabetic rats					
Ikeda *et al.* 1996 (122)	↓	ND	ND	ND	ND
Leaver *et al.* 1991 (123)	ND	ND	ND	↑	ND
Mouse					
Fritsche and Johnston 1989 (124)	ND	ND	↓	ND	ND
Watanabe *et al.* 1993 (125)	ND	ND	↓	ND	ND
Hall *et al.* 1993 (80)	ND	ND	ND	ND	↓
Pig					
Nolan *et al.* 1995 (126)	ND	↓	ND	ND	ND
Salmon					
Bell *et al.* 1996 (127)	ND	ND	↓	↑	ND
Henderson *et al.* 1996 (128)	ND	ND	↓	↑	ND

[a]Abbreviations: ALA, alpha-linolenic acid; TXA$_2$, thromboxane; TXB$_2$, thromboxane B$_2$; PGE$_2$, prostaglandin E$_2$; PGE$_3$/PGI$_3$, prostaglandin E$_3$/prostaglandin I$_3$; ND, not determined; ↓, decrease; ↑, increase; ■, no change.

demonstrated to offer protection against myocardial infarction and death (133). Dietary ALA has been shown to increase platelet EPA and reduce platelet aggregation to thrombin, ADP, and collagen among farmers and other healthy male volunteers (134,135; Table 7.6). Female MRL/lpr mice fed 15% FS diet for 16 wk and lupus patients consuming either 15, 30, or 45 g FS/d for 4 wk showed reduced platelet aggregation to platelet activating factor; mortality among these mice was also decreased compared with that in the control diet group (80,81; Table 7.6). Other studies involving healthy young male subjects or New Zealand white rabbits consuming FS/linseed oil showed increased ALA, EPA, and EPA to AA ratio in the platelets, and decreased platelet aggregation in response to collagen, but not to ADP (50,136; Table 7.6). Additionally, polymorphonuclear leukocytes (PMN) from male weanling rats fed ALA containing perilla oil for 6 wk were shown to produce significantly lower platelet activating factor upon stimulation with calcium ionophore than PMN from rats fed safflower oil, which was attributed to lower levels of arachidonic acid and possibly to lower transacylase activity of PMN (137). However, in contrast to the studies that showed an inverse relationship between dietary ALA and platelet aggregation, studies involving smaller concentrations of dietary ALA did not inhibit ADP- or collagen-induced platelet aggregation or bleeding time in healthy human volunteers or horses (31,55,138,139). Data from studies thus far suggest that the addition of FS to the diet might be beneficial in decreasing cardiovascular disease by many mechanisms including inhibition of platelet aggregation as well as lowering of total and LDL cholesterol (6,12,140).

Effects of FS on Glucose and Insulin Levels

Oxidative stress is known to be one of the factors involved in the etiology of type II diabetes, which is characterized by hyperglycemia, impaired glucose tolerance, insulin resistance, and hyperlipidemia (141–145). The indicators of oxidative stress in diabetic patients include, but are not limited to, increased superoxide dismutase activity, increased TBARS, decreased glutathione and α-tocopherol concentrations. Oxidative stress was shown to induce the expression of cyclin-dependent kinase inhibitor p21 mRNA in isolated rat pancreatic islet cells as well as in Zucker diabetic fatty (ZDF) male rats. Overexpression of p21 in pancreatic islet cells was shown to impair the function of beta cells by suppressing the insulin gene transcription and insulin biosynthesis in pancreatic islets and in ZDF rats (144). Among the FS components, lignan precursor, secoisolariciresinol diglucoside (SDG) has been shown to have antioxidant activity as demonstrated by reduced pancreatic chemiluminescence that was induced by tertbutyl hydroperoxide, and a reduction in serum and pancreatic malondialdehyde. This reduction in oxidative stress by SDG treatment was well correlated with lowered incidence of diabetes among the diabetic-prone BioBreeding (BBdp) rats by about 71% (99). Prasad (143) has further reported the ability of 40 mg/kg body weight SDG (derived from FS) to reduce the incidence of diabetes to 2 of 10 compared with 100% of the untreated ZDF rats, until 72 d of age. The average ages for the onset of diabetes in untreated and SDG-treated ZDF rats were 69 and 84 d, respectively, thus demonstrating that SDG delayed the development of diabetes among the ZDF rats (143). Hypoglycemic effect of FS has been demonstrated in

TABLE 7.7
Impact of Dietary Flaxseed and Its Components on Glucose and Insulin Levels[a]

Study	Species	Daily dose	Blood glucose	Insulin
Cunnane *et al.* 1994 (7)	Human	50 g/FS, 4 wk	■	ND
Cunnane *et al.* 1993 (25)	Human	50 g/FS, 4 wk	↓	ND
Ezaki *et al.* 1999 (147)	Elderly human	3 g/d ALA, 10 mon	■	■
McManus *et al.* 1996 (148)	Type II diabetic	35 mg ALA/kg body weight, 3 mon	■	■
Lemay *et al.* 2002 (146)	Menopausal	40 g FS/d, 4 mon	↓	■
Nestel *et al.* 1997 (9)	Obese	26% FS, 4 wk	■	↑
Ikeda *et al.* 1995 (118)	Streptozotocin (STZ)-induced diabetic rats	Intravenous infusion of perilla oil	■	■
Prasad *et al.* 2000 (99)	Diabetic prone rats	22 mg/kg body weight SDG, up to 19 wk	↓	ND
Ratnayake *et al.* 1992 (5)	Rats	40% FS	■	ND
Prasad 2001 (143)	Zucker diabetic rats	40 mg SDG/kg body weight	■	ND
Schumacher *et al.* 1997 (24)	Horses	2.5 mL linseed oil/kg body weight, twice, 2 wk	■	ND

[a]ND, Not determined; ↓, decrease; ↑, increase; ■, no change; SDG, secoisolariciresinol diglucoside. For other abbreviations see Table 7.6.

healthy female undergraduate students consuming 50 g/d ground FS for 4 wk or 25 g FS mucilage in water, and among menopausal women consuming 40 g/d FS for 2 mon (25,146). Another study involving 10 healthy young adult volunteers showed that intake of 50 g/d of FS for 4 wk was helpful in lowering the blood glucose measured 30 min after breakfast. However, this regimen did not cause significant changes in fasting blood glucose or total area under the glucose tolerance curve performed for 4 h (7). Fatty acid part of FS is unlikely to contribute to the hypoglycemic effects. This speculation is based on the studies that showed no change in insulin, and no change or an increase in blood glucose by linseed oil or perilla oil in elderly or diabetic subjects, mice, STZ-induced diabetic rats, and horses (24,118,147–149; Table 7.7). In a recent 12-yr cohort study involving male health professionals, saturated fat intake was associated with a higher risk of type II diabetes. However, dietary oleic acid, *trans* fatty acids and long chain *n*-3 fatty acids including ALA were not associated with diabetes risk (150).

In addition to showing the nutritional and hematological effects described in this chapter, FS may be helpful in lowering the risk of hormone-related cancers, such as breast, colon, and prostate cancers, as demonstrated in animal models (13,151,152). Dietary FS is also known to have anti-inflammatory properties and therefore might play a role in reducing autoimmune diseases (19,21–23,79–81). These effects of FS are discussed in other chapters in this book. Overall, from the studies conducted thus far, moderate quantities of dietary FS or ALA may be safe or beneficial from a nutritional and hematological perspective. There may be some concerns if consumed at very high levels and over extended time, without compensating for possible loss of antioxidant vitamins. Furthermore, very large amounts of FS consumption may not be desirable due to the estrogenic effects caused by lignans.

Summary

This chapter has dealt with the nutritional and hematological effects of dietary FS and ALA. Moderate amounts of FS may not pose any health risks in terms of iron status, although FS is very rich in fiber, which has the potential to inhibit iron absorption. Dietary ALA from FS may be metabolized to other very long chain fatty acids, thus possibly offering the benefits achieved by fish oil fatty acids, namely EPA and DHA. Dietary FS and ALA may lower the incidence of coronary heart disease by several mechanisms, including cholesterol lowering and reduction in platelet aggregation. Flaxseed may also positively influence kidney and liver function, *via* production of noninflammatory factors. Fiber-rich FS may not cause negative impact on bone mineralization, possibly due to the estrogenic effects conferred by lignans. Higher intake of FS may cause oxidative stress due to long chain fatty acids, especially if there are insufficient antioxidant vitamins in the diet. Flaxseed might be useful as a laxative for those suffering from constipation. From the data obtained in animal studies, FS and its lignans appear to be useful in delaying the onset of diabetes. Despite the potential beneficial effects, high FS consumption may not be prudent because of the possible oxidative stress due to ALA and the estrogenic activity of lignans.

References

1. Carter, J.F., Potential of Flaxseed and Flaxseed Oil in Baked Goods and Other Products in Human Nutrition, *Cereal Foods World 38*:753–759 (1993).
2. Nesbitt, P.D., and L.U. Thompson, Lignans in Homemade and Commercial Products Containing Flaxseed, *Nutr. Cancer 29*:222–227 (1997).
3. Thompson, L.U., Potential Health Benefits and Problems Associated with Antinutrients in Foods, *Food Res. Internat. 26*:131–149 (1993).
4. Brezinski, A., and A. Debi, Phytoestrogens: The "Natural" Selective Estrogen Receptor Modulators? *Eur. J. Obstet. Gynecol. Reprod. Biol. 85*:47–51 (1999).
5. Ratnayake, W.M.N., W.A. Behrens, P.W.F. Fischer, M.R.L L'Abbe, P. Mongeau, and J.L. Beare-Rogers, Chemical and Nutritional Studies of Flaxseed (variety Linott) in Rats, *J. Nutr. Biochem. 3*:232–240 (1992).
6. Bierenbaum, M.L., R. Reichstein, and T.R. Watkins, Reducing Atherogenic Risk in Hyperlipidemic Humans with Flax Seed Supplementation: A Preliminary Report, *J. Am. Coll. Nutr. 12*:501–504 (1993).
7. Cunnane, S.C., M.J. Hamadeh, A.C. Liede, L.U. Thompson, T.M.S. Wolever, and D.J.A. Jenkins, Nutritional Attributes of Traditional Flaxseed in Healthy Young Adults, *Am. J. Clin. Nutr. 61*:62–68 (1994).
8. Harris, W.S., *n*-3 Fatty Acids and Serum Lipoproteins: Human Studies, *Am. J. Clin. Nutr. 65*(Suppl.):1645S–1654S (1997).

9. Nestel, P.J., S.E. Pomeroy, T. Sasahara, T. Yamashita, T.Y. Liang., A.M. Dart, G.L. Jenning, M. Abbery, and J.D. Cameron, Arterial Compliance in Obese Subjects is Improved with Dietary Plant n-3 Fatty Acid from Flaxseed Oil Despite Increased LDL Oxidizability, *Arterioscler. Thromb. Vasc. Biol. 17*:1163–1170 (1997).

10. Craig, W.J., Health-promoting Properties of Common Herbs, *Am. J. Clin. Nutr. 70* (Suppl.):491S–499S (1999).

11. Jenkins, D.J., C.W. Kendall, E. Vidgen, S. Agarwal, A.V. Rao, R.S. Rosenberg, E.P. Diamandis, R. Novokmet, C.C. Mehling, T. Perera, L.C. Griffin, and S.C. Cunnane, Health Aspects of Partially Defatted Flaxseed, Including Effects on Serum Lipids, Oxidative Measures, and *ex vivo* Androgen and Progestin Activity: A Controlled Crossover Trial, *Am. J. Clin. Nutr. 69*:395–402 (1999).

12. Hasler, C.M., S. Kundrat, and D. Wool, Functional Foods and Cardiovascular Disease, *Curr. Atheroscler. Rep. 2*:467–475 (2000).

13. Serraino, M., and L.U. Thompson, The Effect of Flaxseed Supplementation on the Initiation and Promotional Stages of Mammary Tumorigenesis, *Nutr. Cancer 17*:153–159 (1992).

14. Serraino, M., and L.U. Thompson, Flaxseed Supplementation and Early Markers of Colon Carcinogenesis, *Cancer Lett. 63*:159–165 (1992).

15. Jenab, M., and L.U. Thompson, The Influence of Flaxseed and Lignans on Colon Carcinogenesis and β-Glucoronidase Activity, *Carcinogenesis 17*:1343–1348 (1996).

16. Thompson, L.U., S.E. Rickard, L.J. Orcheson, and M.M. Seidl, Flaxseed and Its Lignan and Oil Components Reduce Mammary Tumor Growth at a Late Stage of Carcinogenesis, *Carcinogenesis 17*:1373–1376 (1996).

17. Thompson, L.U., Experimental Studies on Lignans and Cancer, *Baillieres Clin. Endocrinol. Metab. 12*:691–705 (1998).

18. Adlercreutz, H., T. Fotsis, C. Bannwart, K. Wahala, T. Makela, G. Brunow, and T. Hase, Determination of Urinary Lignans and Phytoestrogen Metabolites, Potential Antiestrogens and Anticarcinogens, in Urine of Women on Various Habitual Diets, *J. Steroid Biochem. 25*:791–797 (1986).

19. Ingram, A.J., A. Parbtani, W.F. Clark, E. Spanner, M.W. Huff, D.J. Philbrick, and B.J. Holub, Effects of Flaxseed and Flax Oil Diets in a Rat 5/6 Renal Ablation Model, *Am. J. Kidney Dis. 25*:320–329 (1995).

20. Clark, W.F., A.D. Muir, N.D. Westcott, and A.A. Parbtani, Novel Treatment for Lupus Nephritis: Lignan Derived from Flax, *Lupus 9*:429–436 (2000).

21. Clark, W.F., C. Kortas, A.P. Heidenheim, E. Spanner, and A. Parbtani, Flaxseed in Lupus Nephritis: A Two-Year Nonplacebo-Controlled Crossover Study, *J. Am. Coll. Nutr. 20*:143–148 (2001).

22. Ogborn, M.R., E. Nitschmann, H. Weiler, D. Leswick, and N. Bankovic-Calic, Flaxseed Ameliorates Interstitial Nephritis in Rat Polycystic Kidney Disease, *Kidney Int. 55*:417–423 (1999).

23. Ranich, T., S.J. Bhathena, and M.T. Velasques, Protective Effects of Dietary Phytoestrogens in Chronic Renal Disease, *J. Ren. Nutr. 11*:183–193 (2001).

24. Schumacher, J., F.J. DeGraves, and J.S. Spano, Clinical and Clinicopathologic Effects of Large Doses of Raw Linseed Oil as Compared to Mineral Oil in Healthy Horses, *J. Vet. Intern. Med. 11*:296–299 (1997).

25. Cunnane, S.C., S. Ganguli, C. Menard, A.C. Liede, M.J. Hamadeh, Z.Y.Chen, T.M. Wolever, and D.J. Jenkins, High Alpha-Linolenic Acid Flaxseed (*Linum usitatissimum*): Some Nutritional Properties in Humans, *Br. J. Nutr. 69*:443–453 (1993).

26. Kurzer, M.S., and X. Xu, Dietary Phytoestrogens, *Annu. Rev. Nutr. 17*:353–381 (1997).

27. Babu, U.S., V.K. Bunning, P.W. Wiesenfeld, R.B. Raybourne, and M. O'Donnell, Effect of Dietary Flaxseed on Fatty Acid Composition, Superoxide, Nitric Oxide Generation and Antilisterial Activity of Peritoneal Macrophages from Female Sprague-Dawley Rats, *Life Sci. 60*:545–554 (1997).

28. Wiesenfeld, P.W., U.S. Babu, T.F.X. Collins, R. Sprando, M.W. O'Donnell, T.J. Flynn, T. Black, and N. Olejnik, Flaxseed Increased α-Linolenic Acid and Eicosapentanoic Acid and Decreased Arachidonic Acid in Serum and Tissues of Rat Dams and Offspring, *Food Chem. Tox.*, in press (2003).

29. Scollan, N.D., N-J. Choi, E. Kurt, A.V. Fisher, M. Enser, and J.D. Wood, Manipulating the Fatty Acid Composition of Muscle and Adipose Tissue in Beef Cattle, *Br. J. Nutr. 85*:115–124 (2001).

30. Layne, K.S., Y.K. Goh, J.A. Jumpsen, E.A. Ryan, P. Chow, and M.T. Clandinin, Normal Subjects Consuming Physiological Levels of 18:3(n-3) and 20:5(n-3) from Flaxseed or Fish Oils Have Characteristic Differences in Plasma Lipid and Lipoprotein Fatty Acid Levels, *J. Nutr. 26*:2130–2140 (1996).

31. Freese, R., and M. Mutanen, Small Effects of Linseed Oil or Fish Oil Supplementation on Postprandial Changes in Hemostatic Factors, *Thromb. Res. 85*:147–152 (1997).

32. Singer, P., W. Jaeger, I. Berger, H. Barleben, M. Wirth, E. Richter-Heinrich, S. Voigt, and W. Godicke, Effects of Dietary Oleic, Linoleic and Alpha-Linolenic Acids on Blood Pressure, Serum Lipids, Lipoproteins and the Formation of Eicosanoid Precursors in Patients with Mild Essential Hypertension, *J. Hum. Hypertens. 4*:227–233 (1990).

33. Ferrier, L.K., L.J. Caston, S. Leeson, J. Squires, B.J. Weaver, and B.J. Holub, α-Linolenic Acid and Docosahexaenoic Acid-enriched Eggs from Hens Fed Flaxseed: Influence on Blood Lipids and Platelet Phospholipid Fatty Acids in Humans, *Am. J. Clin. Nutr. 62*:81–86 (1995).

34. Mantzioris, E., L.G. Cleland, R.A. Gibson, M.A. Neumann, M. Demasi, and M.J. James, Biochemical Effects of a Diet Containing Foods Enriched with n-3 Fatty Acids, *Am. J. Clin. Nutr. 72*:42–48 (2000).

35. Haddad, E.H., L.S. Berk, J.D. Kettering, R.W. Hubbard, and W.R. Peters, Dietary Intake and Biochemical, Hematologic, and Immune Status of Vegans Compared with Nonvegetarians, *Am. J. Clin. Nutr. 70*(Suppl. 3):586S–593S (1999).

36. Sandstead, H.H., Causes of Iron and Zinc Deficiencies and Their Effects on Brain, *J. Nutr. 130*(Suppl. 2S):347S–349S (2000).

37. Hallberg, L., L. Rossander, and A-B. Skanberg, Phytates and Inhibitory Effect of Bran on Iron Absorption in Man, *Am. J. Clin. Nutr. 45*:988–996 (1987).

38. Reddy, M.B., and J.D. Cook, Assessment of Dietary Determinants of Nonheme-Iron Absorption in Humans and Rats, *Am. J. Clin. Nutr. 54*:723–728 (1991).

39. Babu, U.S., G.V. Mitchell, P. Wiesenfeld, M.Y. Jenkins, and H. Gowda, Nutritional and Hematological Impact of Dietary and Defatted Flaxseed Meal in Rats, *Int. J. Food. Sci. Nutr. 51*:109–117 (2000).

40. Cossack, Z.T., A. Rojhani, and A.O. Musaiger, The Effects of Sugar-Beet Fibre Supplementation for Five Weeks on Zinc, Iron and Copper Status in Human Subjects, *Eur. J. Clin. Nutr. 46*:221–225 (1992).

41. Coudray, C., J. Bellanger, C. Castiglia-Delavaud, C. Remesy, M. Vermorel, and Y. Rayssignuier, Effect of Soluble or Partly Soluble Dietary Fibres Supplementation on Absorption and Balance of Calcium, Magnesium, Iron and Zinc in Healthy Young Men, *Eur. J. Clin. Nutr. 51*:375–380 (1997).

42. Heath, A.L., C.M. Skeaff, S.M. O'Brien, S.M. Williams, and R.S. Gibson, Can Dietary Treatment of Non-Anemic Iron Deficiency Improve Iron Status? *J. Am. Coll. Nutr.* *20*:477–484 (2001).

43. Gralak, M.A., M. Leontowicz, M. Morawiec, E. Bartnikowska, and G.W. Kulasek, Comparison of the Influence of Dietary Fibre Sources with Different Proportions of Soluble and Insoluble Fibre on Ca, Mg, Fe, Zn, Mn and Cu Apparent Absorption in Rats, *Arch. Tierernahr.* *49*:293–299 (1996).

44. Innis, S.M., and R.A. Dyer, Dietary Canola Oil Alters Hematological Indices and Blood Lipids in Neonatal Piglets Fed Formula, *J. Nutr.* *29*:1261–1268 (1999).

45. Li, D., A. Sinclair, A. Wilson, S. Nakkote, F. Kelly, L. Abedin, N. Mann, and A. Turner, Effect of Dietary Alpha-Linolenic Acid on Thrombotic Risk Factors in Vegetarian Men, *Am. J. Clin. Nutr.* *69*:872–882 (1999).

46. Sakai, K., H. Okuyama, K. Kon, N. Maeda, M. Sekiya, T. Shiga, and R.C. Reitz, Effects of High Alpha-Linolenate and Linoleate Diets on Erythrocyte Deformability and Hematological Indices in Rats, *Lipids 25*:793–797 (1990).

47. L'Abbe, M.R., K.D. Trick, and J.L. Beare-Rogers, Dietary (n-3) Fatty Acids Affect Rat Heart, Liver and Aorta Protective Enzyme Activities and Lipid Peroxidation, *J. Nutr.* *121*:1331–1340 (1991).

48. Dell, C.A., S.S. Likhodii, K. Musa, M.A. Ryan, W.M. Burnham, and S.C. Cunnane, Lipid and Fatty Acid Profiles in Rats Consuming Different High-Fat Ketogenic Diets, *Lipids 36*:373–378 (2001).

49. Mantzioris, E., M.J. James, R.A. Gibson, and L.G. Cleland, Dietary Substitution with a Linolenic Acid-Rich Vegetable Oil Increases Eicosapentaenoic Acid Concentrations in Tissues, *Am. J. Clin. Nutr.* *59*:1304–1309 (1994).

50. Allman, M.A., M.M. Pena, and D. Pang, Supplementation with Flaxseed Oil Versus Sunflower Seed Oil in Healthy Young Men Consuming a Low Fat Diet: Effects on Platelet Composition and Function, *Eur. J. Clin. Nutr.* *19*:169–178 (1995).

51. Thies, F., G. Nebe-von-Caron G, J.R. Powell, P. Yaqoob, E.A. Newsholme, and P.C. Calder, Dietary Supplementation with Eicosapentaenoic Acid, but Not with Other Long-Chain n-3 or n-6 Polyunsaturated Fatty Acids, Decreases Natural Killer Cell Activity in Healthy Subjects Aged >55 y, *Am. J. Clin. Nutr.* *73*:539–548 (2001).

52. Matthews, K.R., D.B. Homer, F. Thies, and P.C. Calder, Effect of Whole Linseed (*Linum usitatissimum*) of Finishing Pigs on Growth Performance and on Quality and Fatty Acid Composition of Various Tissues, *Br. J. Nutr.* *83*:637–643 (2000).

53. Ajuyah, A.O., R.T. Hardin, and J.S. Sim, Effect of Dietary Full-Fat Flaxseed With and Without Antioxidant on the Fatty Acid Composition of Major Lipid Classes of Chicken Meats, *Poult. Sci.* *72*:125–136 (1993).

54. Gonzalez-Esquerra, R., and S. Lesson, Effect of Menhaden Oil and Flaxseed in Broiler Sensory Quality and Lipid Composition of Poultry, *Br. Poult. Sci.* *41*:481–488 (2000).

55. Kelley, D.S., G.J. Nelson, J.E. Love, L.B. Branch, P.C. Taylor, P.C. Schmidt, B.E. Mackey, and J.M. Iacono, Dietary α-Linolenic Acid Alters Tissue Fatty Acid Composition, but Not Blood Lipids, Lipoproteins or Coagulation Status in Humans, *Lipids 28*:533–537 (1993).

56. Gerster, H., Can Adults Adequately Convert α-Linolenic Acid (18:3, n-3) to Eicosapentaenoic Acid (20:5, n-3) and Docosahexaenoic Acid (22:6, n-3)? *Int. J. Vitam. Nutr. Res.* *68*:159–173 (1998).

57. Simopoulos, A.P., Symposium: Role of Poultry Products in Enriching the Human Diet with *n*-3 PUFA: Human Requirement for *n*-3 Polyunsaturated Fatty Acids, *Poult. Sci.* *79*:961–970 (2000).

58. Cunnane, S.C., C.R. Menard, S.S. Likhodii, J.T. Brenna, and M.A. Crawford, Carbon Recycling into *de novo* Lipogenesis is a Major Pathway in Neonatal Metabolism of Linoleate and Alpha-Linolenate, *Prostaglandins Leukot. Essent. Fatty Acids 60*:387–392 (1999).

59. Yaqoob, P., and P.C. Calder, The Effects of Fatty Acids on Lymphocyte Functions, *Int. J. Biochem. 25*:1705–1714 (1993).

60. Jeffery, N.M., P. Sanderson, E.J. Sherrington, E.A. Newsholme, and P.C. Calder, The Ratio of n-6 to n-3 Polyunsaturated Fatty Acids in the Rat Diet Alters Serum Lipid Levels and Lymphocyte Functions, *Lipids 31*:737–745 (1996).

61. Wander, R.C., J.A. Hall, J.L. Gradin, S. Du, and D.E. Jewell, The Ratio of Dietary n-6 to (n-3) Fatty Acids Influences Immune System Function, Eicosanoid Metabolism, Lipid Peroxidation and Vitamin E Status in Aged Dogs, *J. Nutr. 127*:1198–1205 (1997).

62. Greiner, R.C., J. Winter, P.W. Nathanielsz, and J.T. Brenna, Brain Docosahexaenoate Accretion in Fetal Baboons: Bioequivalence of Dietary Alpha-Linolenic and Docosahexaenoic Acids, *Pediat. Res. 42*:826–834 (1997).

63. Abedin, L., E.L. Lien, A.J. Vingrys, and A.J. Sinclair, The Effect of Dietary Alpha-Linolenic Acid Compared with Docosahexaenoic Acid on Brain, Retina, Liver and Heart in the Guinea Pig, *Lipids 34*:475–482 (1999).

64. Cheon, S-H., M-H. Huh, Y-B. Lee, J-S. Park, H-S. Sohn, and C-W. Chung, Effect of Dietary Linoleate/Alpha-Linolenate Balance on the Brain Lipid Composition, Reproductive Outcome and Behavior of Rats During Their Prenatal and Postnatal Development, *Biosci. Biotech. Biochem. 64*:2290–2297 (2000).

65. Innis, S.M., The Role of Dietary n-6 and n-3 Fatty Acids in the Developing Brain, *Dev. Neurosci. 22*:474–480 (2000).

66. Abayasekara, D.R., and D.C. Wathes, Effects of Altering Dietary Fatty Acid Composition on Prostaglandin Synthesis and Fertility, *Prostaglandins Leukot. Essent. Fatty Acids 61*:275–287 (1999).

67. Trujillo, E.P., and K.S. Broughton, Ingestion of n-3 Polyunsaturated Fatty Acids and Ovulation in Rats, *J. Reprod. Fertil. 105*:197–203 (1995).

68. Allen, K.G., and M.A. Harris, The Role of *n*-3 Fatty Acids in Gestation and Parturition, *Exp. Biol. Med. 226*:498–506 (2001).

69. Leat, W.M.F., C.A. Northrop, F.A. Harrison, and R.W. Cox, Effect of Dietary Linoleic and Linolenic Acids on Testicular Development in the Rat, *Q.J. Exp. Physiol. 68*:221–231 (1983).

70. Sprando, R.L., T.F. Collins, T.N. Black, N. Olejnik, J.I. Rorie, P. Wiesenfeld, U.S. Babu, and M. O'Donnell, The Effect of Maternal Exposure to Flaxseed on Spermatogenesis in F(1) Generation Rats, *Food Chem. Tox. 38*:325–334 (2000).

71. Prasad, K., Dietary Flax Seed in Prevention of Hypercholesterolemic Atherosclerosis, *Atherosclerosis 132*: 69–76 (1997).

72. Prasad, K., S.V. Mantha, A.D. Muir, and N.D. Westcott, Reduction of Hypercholesterolemic Atherosclerosis by CDC-Flaxseed with Very Low Alpha-Linolenic Acid, *Atherosclerosis 136*:367–375 (1998).

73. Clandinin, M.T., A. Foxwell, Y.K. Goh, K. Layne, and J.A. Jumpsen, Omega-3 Fatty Acid Intake Results in a Relationship Between the Fatty Acid Composition of LDL Cholesterol Ester and LDL Cholesterol Content in Humans, *Biochem. Biophys. Acta 1346*:247–252 (1997).

74. Lucas, E.A., R.D. Wild, L.J. Hammond, D.A. Khalil, S. Juma, B.P. Daggy, B.J. Stoecker, and B.H. Arjmandi, Flaxseed Improves Lipid Profile Without Altering

Biomarkers of Bone Metabolism in Postmenopausal Women, *J. Clin. Endocrinol. Metab.* 87:1527–1532 (2002).

75. Kritchevsky, D., S.A. Tepper, and D.M. Klurfeld, Influence of Flaxseed on Serum and Liver Lipids in Rats, *J. Nutr. Biochem.* 2:133–134 (1991).

76. Suryaprabha, P., U.N. Das, G. Ramesh, K.V. Kumar, and G.S. Kumar, Reactive Oxygen Species, Lipid Peroxides and Essential Fatty Acids in Patients with Rheumatoid Arthritis and Systemic Lupus Erythematosus, *Prostaglandins Leukot. Essent. Fatty Acids* 43:251–255 (1991).

77. Frazier-Jessen, M.R., and E.J. Kovacs, Estrogen Modulation of JE/Monocyte Chemoattractant Protein-1 mRNA Expression in Murine Macrophages, *J. Immunol. 154*: 1838–1845 (1995).

78. Velasquez, M.T., and S.J. Bhathena, Dietary Phytoestrogens: A Possible Role in Renal Disease Protection, *Am. J. Kidney Dis.* 37:1056–1068 (2001).

79. Pollen, S.M., Renal Disease in Small Animals: A Review of Conditions and Potential Nutrient and Botanical Interventions, *Altern. Med. Rev.* 6(Suppl.):S46–S61 (2001).

80. Hall, A.V., A. Parbtani, W.F. Clark, E. Spanner, M. Keeney, I. Chin-Yee, D.J. Philbrick, and B.J. Holub, Abrogation of MRL/lpr Lupus Nephritis by Dietary Flaxseed, *Am. J. Kidney Dis.* 22:326–332 (1993).

81. Clark, W.F., A. Parbtani, M.W. Huff, E. Spanner, H. de Salis, I. Chin-Yee, D.J. Philbrick, and B.J. Holub, Flaxseed: A Potential Treatment for Lupus Nephritis, *Kidney Int.* 48:475–480 (1995).

82. Kabir, Y., and T. Ide, Activity of Hepatic Fatty Acid Oxidation Enzymes in Rats Fed Alpha-Linolenic Acid, *Biochim. Biophys. Acta 1304*:105–119 (1996).

83. Ide, T., M. Murata, and M. Sugano, Stimulation of the Activities of Hepatic Fatty Acid Oxidation Enzymes by Dietary Fat Rich in Alpha-Linolenic Acid in Rats, *J. Lipid Res.* 37:448–463 (1996).

84. Takeuchi, H., T. Nakamoto, Y. Mori, M. Kawakami, H. Mabuchi, Y. Ohishi, N. Ichikawa, A. Koike, and K. Masuda, Comparative Effects of Dietary Fat Types on Hepatic Enzyme Activities Related to the Synthesis and Oxidation of Fatty Acid and to Lipogenesis in Rats, *Biosci. Biotechnol. Biochem.* 65:1748–1754 (2001).

85. Chen, H.W., C.K. Lii, M.H. Wu, C.C. Ou, and L.Y. Sheen, Amount and Type of Dietary Lipid Modulate Rat Hepatic Cytochrome P-450 Activity, *Nutr. Cancer* 29:174–180 (1997).

86. Watanabe, S., and H. Okuyama, Effect of Dietary Alpha-Linolenate/Linoleate Balance on Endotoxin-Induced Hepatitis in Mice, *Lipids* 26:467–471 (1991).

87. Urbano, G., M. Lopez-Jurado, P. Aranda, C. Vidal-Valverde, E. Tenorio, and J. Porres, The Role of Phytic Acid in Legumes: Antinutrient or Beneficial Function, *J. Physiol. Biochem.* 56:283–294 (2000).

88. Ward, W.E., Y.V. Yuan, A.M. Cheung, and L.U. Thompson, Exposure to Purified Lignan from Flaxseed (*Linum usitatissimum*) Alters Bone Development in Female Rats, *Br. J. Nutr.* 86:499–505 (2001).

89. Chiechi, L.M., Dietary Phytoestrogens in the Prevention of Long-Term Postmenopausal Diseases, *Int. J. Gynaecol Obstet.* 67:39–40 (1999).

90. Adlercreutz, H., Phytoestrogens: Epidemiology and a Possible Role in Cancer Protection, *Environ. Health Perspect. 103*(Suppl. 7):103–112 (1995).

91. Arjmandi, B.H., The Role of Phytoestrogens in the Prevention and Treatment of Osteoporosis in Ovarian Hormone Deficiency, *J. Am. Coll. Nutr. 20*(Suppl. 5):398S–402S (2001).

92. Humfrey, C.D., Phytoestrogens and Human Health Effects: Weighing Up the Current Evidence, *Nat. Toxins 6*:51–59 (1998).

93. Watkins, B.A., Y. Li, K.G. Allen, W.E. Hoffmann, and M.F. Seifert, Dietary Ratio of (n-6)/(n-3) Polyunsaturated Fatty Acids Alters the Fatty Acid Composition of Bone Compartments and Biomarkers of Bone Formation in Rats, *J. Nutr. 130*:2274–2284 (2000).

94. Albertazzi, P., and K. Coupland, Polyunsaturated Fatty Acids: Is There a Role in Postmenopausal Osteoporosis Prevention, *Maturitas 42*:13–22 (2002).

95. Ward, W.E., Y.V. Yuan, A.M. Cheung, and L.U. Thompson, Exposure to Flaxseed and Its Purified Lignan Reduces Bone Strength in Young but Not Older Male Rats, *J. Toxicol. Environ. Health A. 63*:53–65 (2001).

96. Javouhey-Donzel, A., L. Guenot, V. Maupoll, L. Rochette, and G. Rocquelin, Rat Vitamin E Status and Heart Lipid Peroxidation: Effect of Dietary Alpha-Linolenic Acid and Marine *n*-3 Fatty Acids, *Lipids 28*:651–655 (1993).

97. Kaasgaard, S.G., G. Holmer, C.E. Hoy, W.A. Behrens, and J.L. Beare-Rogers, Effects of Dietary Linseed Oil and Marine Oil on Lipid Peroxidation in Monkey Liver *in vivo* and *in vitro*, *Lipids 27*:740–745 (1992).

98. Kitts, D.D., Y.V. Yuan, A.N. Wijewickreme, and L.U. Thompson, Antioxidant Activity of the Flaxseed Lignan Secoisolariciresinol Diglycoside and Its Mammalian Lignan Metabolites Enterodiol and Enterolactone, *Mol. Cell Biochem. 202*:91–100 (1999).

99. Prasad, K., Oxidative Stress as a Mechanism of Diabetes in Diabetic BB Prone Rats: Effect of Secoisolariciresinol Diglucoside (SDG), *Mol. Cell Biochem. 209*:89–96 (2000).

100. Wang, L., Mammalian Phytoestrogens: Enterodiol and Enterolactone, *J. Chromatogr. B Analyt. Technol. Biomed. Life Sci. 777*:289 (2002).

101. Prasad, K., Reduction of Serum Cholesterol and Hypercholesterolemic Atherosclerosis in Rabbits by Secoisolariciresinol Diglucoside Isolated from Flaxseed, *Circulation 99*:1355–1362 (1999).

102. Clement, M., and J.M. Bourne, Graded Dietary Levels of RRR-Gamma-Tocopherol Induce a Marked Increase in the Concentrations of Alpha- and Gamma-Tocopherol in Nervous Tissues, Heart, Liver and Muscle of Vitamin-E-Deficient Rats, *Biochim. Biophys. Acta 1334*:173–181 (1997).

103. Schweigert, F.J., A. Baumane, I. Buchholz, and H.A. Schoon, Plasma and Tissue Concentrations of Beta-Carotene and Vitamin A in Rats Fed Beta-Carotene in Various Fats of Plant and Animal Origin, *J. Environ. Pathol. Toxicol. Oncol. 19*:87–93 (2000).

104. Tomassi, G., and J.A. Olson, Effect of Dietary Essential Fatty Acids on Vitamin A Utilization in the Rat, *J. Nutr. 113*:697–703 (1983).

105. Oomah, B.D., G. Mazza, and E.O. Kenaschuk, Cyanogenic Compounds in Flaxseed, *J. Agric. Food Chem. 40*:1346–1348 (1992).

106. Tarpila, S., A. Aro, I. Salminen, A. Tarpila, P. Kleemola, J. Akkila, and H. Adlercreutz, The Effect of Flaxseed Supplementation in Processed Foods on Serum Fatty Acids and Enterolactone, *Eur. J. Clin. Nutr. 56*:157–165 (2002).

107. Chadha, R.K., J.F. Lawrence, and W.M. Ratnayake, Ion Chromatographic Determination of Cyanide Released from Flaxseed Under Autohydrolysis Conditions, *Food Addit. Contam. 12*:527–533 (1995).

108. Klosterman, H.J., G.L. Lamoureux, and J.L. Parsons, Isolation, Characterization, and Synthesis of Linatine: A Vitamin B6 Antagonist from Flaxseed (*Linum usitatissimum*), *Biochem. 6*:170–177 (1967).

109. Bishara, H.N., and H.F. Walker, The Vitamin B6 Status of Pigs Given a Diet Containing Linseed Meal, *Br. J. Nutr. 37*:321–331 (1977).

110. Meier, P., W.O. Seiler, and H.B. Stahelin, Bulk-Forming Agents as Laxatives in Geriatric Patients, *Schweiz. Med. Wochenschr. 120*:314–317 (1990).

111. Wirths, W., T. Berglar, A. Dieckhues, and G. Bauer, Fiber-Rich Snacks with Reference to Their Effect on the Digestive Activity and Blood Lipids of the Elderly, *Z. Gerontol. 18*:107–110 (1985).

112. James, M.J., R.A. Gibson, and L.G. Cleland, Dietary Polyunsaturated Fatty Acids and Inflammatory Mediator Production, *Am. J. Clin. Nutr. 71*(Suppl.):343S–348S (2000).

113. Budowski, P., N. Trostler, M. Lupo, N. Vaisman, and A. Eldor, Effect of Linseed Oil Ingestion on Plasma Lipid Fatty Acid Composition and Platelet Aggregability in Healthy Volunteers, *Nutr. Res. 4*:343–346 (1984).

114. Bjerve, K.S., S. Fischer, and K. Alme, Alpha-Linolenic Acid Deficiency in Man: Effect of Ethyl Linolenate on Plasma and Erythrocyte Fatty Acid Composition and Biosynthesis of Prostanoids, *Am. J. Clin. Nutr. 46*:570–576 (1987).

115. Ferretti, A., and V.P. Flanagan, Antithromboxane Activity of Dietary Alpha-Linolenic Acid: A Pilot Study, *Prostaglandins Leukot. Essent. Fatty Acids 54*:451–455 (1996).

116. Henry, M.M., J.N. Moore, E.B. Feldman, J.K. Fischer, and B. Russell, Effect of Dietary Alpha-Linolenic Acid on Equine Monocyte Procoagulant Activity and Eicosanoid Synthesis, *Circ. Shock 32*:173–188 (1990).

117. Morris, D.D., M.M..Henry, J.N. Moore, and K. Fischer, Effect of Dietary Linolenic Acid on Endotoxin-Induced Thromboxane and Prostacyclin Production by Equine Peritoneal Macrophages, *Circ. Shock 29*:311–318 (1989).

118. Ikeda, A., K. Inui, Y. Fukuta, Y. Kokuba, and M. Sugano, Effects of Intravenous Perilla Oil Emulsion on Nutritional Status, Polyunsaturated Fatty Acid Composition of Tissue Phospholipids, and Thromboxane A2 Production in Streptozotocin-Induced Diabetic Rats, *Nutrition 11*:450–455 (1995).

119. Moore, J.N., J.A. Cook, M.M. Henry, H.T, Jonsson, K.M. Spicer, W.C. Wise, and P.V. Halushka, Effects of a Linseed Oil Enriched Diet on Endotoxin-Induced Sequelae: Differential *in vitro* and *in vivo* Effects, *Eicosanoids 4*:47–55 (1991).

120. Magrum, L.J., and P.V. Johnston, Modulation of Prostaglandin Synthesis in Rat Peritoneal Macrophages with Omega-3 Fatty Acids, *Lipids 18*:514–521 (1983).

121. Opmeer, F.A., M.J. Adolfs, and I.L. Bonta, Regulation of Prostaglandin E2 Receptors *in vivo* by Dietary Fatty Acids in Peritoneal Macrophages from Rats, *J. Lipid Res. 25*:262–268 (1984).

122. Ikeda, I., K. Mitsui, and K. Imaizumi, Effect of Dietary Linoleic, Alpha-Linolenic and Arachidonic Acids on Lipid Metabolism, Tissue Fatty Acid Composition and Eicosanoid Production in Rats, *J. Nutr. Sci. Vitaminol.* (Tokyo), *42*:541–551 (1996).

123. Leaver, H.A., A. Howie, and N.H. Wilson, The Biosynthesis of the 3-Series Prostaglandins in Rat Uterus after Alpha-Linolenic Acid Feeding: Mass Spectroscopy of Prostaglandins E and F Produced by Rat Uteri in Tissue Culture, *Prostaglandins Leukot. Essent. Fatty Acids 42*:217–224 (1991).

124. Fritsche, K.L., and P.V. Johnston, Modulation of Eicosanoid Production and Cell-Mediated Cytotoxicity by Dietary Alpha-Linolenic Acid in BALB/c Mice, *Lipids 24*:305–311 (1989).

125. Watanabe, S., K. Onozaki, S. Yamamoto, and H. Okuyama, Regulation by Dietary Essential Fatty Acid Balance of Tumor Necrosis Factor Production in Mouse Macrophages, *J. Leukoc. Biol. 53*:151–156 (1993).

126. Nolan, M.R., S. Kennedy, W.J. Blanchflower, and D.G. Kennedy, Lipid Peroxidation, Prostacyclin and Thromboxane A2 in Pigs Depleted of Vitamin E and Selenium and Supplemented with Linseed Oil, *Br. J. Nutr.* 74:369–380 (1995).

127. Bell, J.G., B.M. Farndale, J.R. Dick, and J.R. Sargent, Modification of Membrane Fatty Acid Composition, Eicosanoid Production, and Phospholipase A Activity in Atlantic Salmon (*Salmo salar*) Gill and Kidney by Dietary Lipid, *Lipids 31*:1163–1171 (1996).

128. Henderson, R.J., J.G. Bell, and M.T Park, Polyunsaturated Fatty Acid Composition of the Salmon (*Salmo salar* L.) Pineal Organ: Modification by Diet and Effect on Prostaglandin Production, *Biochim. Biophys. Acta 1299*:289–298 (1996).

129. Sheu, J.R., G. Hsiao, M.Y. Shen, W.Y. Lin, and C.R. Tzeng, The Hyperaggregability of Platelets from Normal Pregnancy is Mediated Through Thromboxane A2 and Cyclic AMP Pathways, *Clin. Lab. Haematol.* 24:121–129 (2002).

130. Brothers, T.E., J.G. Robison, B.M. Elliott, and J.M. Boggs, Preoperative Thromboxane A2/Prostaglandin H2 Receptor Activity Predicts Early Graft Thrombosis, *J. Vasc. Surg.* 27:317–325 (1998).

131. Kerins, D.M., L. Roy, G.A. FitzGerald, and D.J. Fitzgerald, Platelet and Vascular Function During Coronary Thrombolysis with Tissue-Type Plasminogen Activator, *Circulation 80*:1718–1725 (1989).

132. Holub, B.J., Platelet Composition and Function, in *Flaxseed in Human Nutrition*, edited by S.C. Cunnane and L.U. Thompson, AOCS Press, Champaign, Illinois, 1995, pp. 128–135.

133. Anonymous, Collaborative Overview of Randomised Trials of Antiplatelet Therapy. III: Reduction in Venous Thrombosis and Pulmonary Embolism by Antiplatelet Prophylaxis Among Surgical and Medical Patients, Antiplatelet Trialists' Collaboration, *B.M.J.* 308:235–246 (1994).

134. Renaud, S., and A. Nordoy, "Small is Beautiful": Alpha-Linolenic Acid and Eicosapentaenoic Acid in Man, *Lancet 1*(8334):1169 (1983).

135. Freese, R., M. Mutanen, L.M. Valsta, and I. Salminen, Comparison of the Effects of Two Diets Rich in Monounsaturated Fatty Acids Differing in Their Linoleic/Alpha-Linolenic Acid Ratio on Platelet Aggregation, *Thromb. Haemost.* 71:73–77 (1994).

136. Vas Dias, F.W., M.J. Gibney, and T.G. Taylor, The Effect of Polyunsaturated Fatty Acids on the n-3 and n-6 Series on Platelet Aggregation and Platelet and Aortic Fatty Acid Composition in Rabbits, *Atherosclerosis 43*:245–257 (1982).

137. Oh-hashi, K., T. Takahashi, S. Watanabe, T. Kobayashi, and H. Okuyama, Possible Mechanisms for the Differential Effects of High Linoleate Safflower Oil and High Alpha-Linolenate Perilla Oil Diets on Platelet-Activating Factor Production by Rat Polymorphonuclear Leukocytes, *J. Lipid Mediat. Cell Signal 17*:207–220 (1997).

138. Chan, J.K., B.E. McDonald, J.M. Gerrard, V.M. Bruce, B.J. Weaver, and B.J. Holub, Effect of Dietary Alpha-Linolenic Acid and Its Ratio to Linoleic Acid on Platelet and Plasma Fatty Acids and Thrombogenesis, *Lipids 28*:811–817 (1993).

139. Hasler, C.M., S. Kundrat, and D. Wool, Functional Foods and Cardiovascular Disease, *Curr. Atheroscler. Rep.* 2:467–475 (2000).

140. Simopoulos, A.P., Essential Fatty Acids in Health and Chronic Disease, *Am. J. Clin. Nutr.* 70(Suppl. 3):560S–569S (1999).

141. Seghrouchni, I., J. Drai, E. Bannier, J. Riviere, P. Calmard, I. Garcia, J. Orgiazzi, and A. Revol, Oxidative Stress Parameters in Type I, Type II and Insulin-Treated Type 2 Diabetes Mellitus:Insulin Treatment Efficiency, *Clin. Chim. Acta 321*:89–96 (2002).

142. Shinomiya, K., M. Fukunaga, H. Kiyomoto, K. Mizushige, T. Tsuji, T. Noma, K. Ohmori, M. Kohno, and S. Senda, A Role of Oxidative Stress-Generated Eicosanoid in

the Progression of Arteriosclerosis in Type 2 Diabetes Mellitus Model Rats, *Hypertens. Res.* *25*:91–98 (2002).

143. Prasad, K., Secoisolariciresinol Diglucoside from Flaxseed Delays the Development of Type 2 Diabetes in Zucker Rat, *J. Lab. Clin. Med.* *138*:32–39 (2001).

144. Kaneto, H., Y. Kajimoto, Y. Fujitani, T. Matsuoka, K. Sakamoto, M. Matsuhisa, Y. Yamasaki, and M. Hori, Oxidative Stress Induces p21 Expression in Pancreatic Islet Cells: Possible Implication in Beta-Cell Dysfunction, *Diabetologia* *42*:1093–1097 (1999).

145. Kakkar, R., S.V. Mantha, J. Radhi, K. Prasad, and J. Kalra, Increased Oxidative Stress in Rat Liver and Pancreas During Progression of Streptozotocin-Induced Diabetes, *Clin. Sci.* (London), *94*:623–632 (1998).

146. Lemay, A., S. Dodin, N. Kadri, H. Jacques, and J.C. Forest, Flaxseed Dietary Supplement Versus Hormone Replacement Therapy in Hypercholesterolemic Menopausal Women, *Obstet. Gynecol.* *100*:495–504 (2002).

147. Ezaki, O., M. Takahashi, T. Shigematsu, K. Shimamura, J. Kimura, H. Ezaki, and T. Gotoh, Long-Term Effects of Dietary Alpha-Linolenic Acid from Perilla Oil on Serum Fatty Acids Composition and on the Risk Factors of Coronary Heart Disease in Japanese Elderly Subjects, *J. Nutr. Sci. Vitaminol.* (Tokyo), *45*:759–772 (1999).

148. McManus, R.M., J. Jumpson, D.T. Finegood, M.T. Clandinin, and E.A. Ryan, A Comparison of the Effects of n-3 Fatty Acids from Linseed Oil and Fish Oil in Well-Controlled Type II Diabetes, *Diabetes Care* *19*:463–467 (1996).

149. Ikemoto, S., M. Takahashi, N. Tsunoda, K. Maruyama, H. Itakura, and O. Ezaki, High-Fat Diet-Induced Hyperglycemia and Obesity in Mice: Differential Effects of Dietary Oils, *Metabolism* *45*:1539–1546 (1996).

150. van Dam, R.M., W.C. Willett, E.B. Rimm, M.J. Stampfer, and F.B. Hu, Dietary Fat and Meat Intake in Relation to Risk of Type 2 Diabetes in Men, *Diabetes Care* *25*:417–424 (2002).

151. Demark-Wahnefried, W., D.T. Price, T.J. Polascik, C.N. Robertson, E.E. Anderson, D.F. Paulson, P.J. Walther, M. Gannon, and R.T. Vollmer, Pilot Study of Dietary Fat Restriction and Flaxseed Supplementation in Men with Prostate Cancer Before Surgery: Exploring the Effects on Hormonal Levels, Prostate-Specific Antigen, and Histopathologic Features, *Urology* *58*:47–52 (2001).

152. Rickard, S.E., Y.V. Yuan, J. Chen, and L.U. Thompson, Dose Effects of Flaxseed and Its Lignan on N-Methyl-N-Nitrosourea-Induced Mammary Tumorigenesis in Rats, *Nutr. Cancer* *35*:50–57 (1999).

α-Linolenic Acid in Brain Function and Infant Development

Stephen C. Cunnane

Department of Nutritional Sciences, Faculty of Medicine, University of Toronto, Toronto, Canada M5S 3E2

Introduction

Traditional flaxseed is a rich natural source of α-linolenic acid (18:3n-3). An overview of the basic metabolism and nutritional importance of α-linolenic acid is given in Chapter 3. The present chapter will focus on how α-linolenic acid is utilized during infant development, especially by the developing brain. Several excellent reviews of the role of n-3 polyunsaturated fatty acids (PUFA) in infant development have been published in the last few years (1–6). They describe in detail the way that n-3 PUFA are utilized in the infant and explore the emerging evidence of exactly how the principal long chain n-3 PUFA, docosahexaenoic acid (22:6n-3) functions in the developing brain and eye. Liberal reference is made in these reviews to animal models for more detailed metabolic and biochemical information. My goal here, therefore, is to focus on themes that remain controversial and to discuss new topics that have received somewhat less attention in other reviews. Brief mention will also be made concerning how n-3 PUFA are involved in the function of the adult and aging human brain.

As described elsewhere (7,8), my premise here is twofold: First, the function of PUFA in general and α-linolenic acid in particular is set in the context of a dual and sometimes competing role as both membrane constituents and energy substrates. Overall energy balance of the individual as well as the high energy requirements of some organs have inescapably significant effects on the metabolism of PUFA. Nowhere is this more apparent than for the brain, nor at any time more so than during early development. Second, the dietary requirement for long chain PUFA is conditional on developmental age and the presence of pathology (9); infants have a well-established need for long chain PUFA that is harder to demonstrate in healthy adults. Even among infants it is clear that differences exist, with premature infants probably having a higher requirement than term infants. Hence, the term—essential fatty acid—will not be used here because it does not distinguish adequately between parent and long chain PUFA nor does it incorporate the conditional nature of the requirement for PUFA (9).

In general, n-3 PUFA seem to be most important during periods of rapid growth and in excitable cells, notably in the photoreceptor and brain. Thus, much of what has been learned about their biological role in mammals has been learned during early postnatal development. In such cells or organs, n-3 PUFA exist in predominantly one form, docosahexaenoic acid, and in predominantly one membrane-based matrix — phospholipids. All tissues, including excitable tissues, almost always have detectable levels of n-3 PUFA shorter and less unsaturated than docosahexaenoic acid, but abundant evidence indicates that the latter is the key functional n-3 PUFA in these cells. Therefore, the main issue in discussing the relevance of α-linolenic acid to early infant and brain development is the extent to which it is an adequate dietary precursor to docosahexaenoic acid, particularly when α-linolenic acid is the sole n-3 PUFA in the diet. The efficacy of the conversion of α-linolenic acid to docosahexaenoic acid has been a central point of controversy in human infant nutrition over the past three decades (10), indeed, ever since a role of n-3 PUFA in behavior and learning first became apparent.

The actual mechanism(s) by which docosahexaenoic acid functions in membrane phospholipids has barely been touched, let alone understood, but focusing exclusively on this role in understanding how n-3 PUFA contribute to brain function has several pitfalls. First, eicosapentaenoic acid (20:5n-3) is a key regulatory PUFA but does not have a major presence in membranes. It may well have beneficial functions in the brain that are unrelated to being a precursor to docosahexaenoic acid (11). Second, although there is little doubt that docosahexaenoic acid has a key role in the function of excitable tissues, it is still relatively easily β-oxidized (12,13) and other data suggest it has antioxidant and perhaps even prooxidant functions (14). Thus, we should be careful to maintain a broad perspective in exploring how n-3 PUFA contribute to normal development, especially in the brain and sensory systems.

Historical Review

Prior to the invention of milk formulas to replace mother's milk, suckling infants always consumed the full range of n-3 PUFA albeit at levels that vary widely across cultures (15). The content of long chain n-3 PUFA certainly varies depending on their intake from the maternal diet, but no one has reported that they are absent from milk even in those who have a low to negligible intake of long chain n-3 PUFA (i.e., vegetarians). Ruminant milk has long been used as a replacer for human milk, but even that contains some long chain n-3 PUFA as well as α-linolenic acid. In the 1970s, formula manufacturers began to "humanize" milk formulas and reduce their allergenicity (16). Cow's milk was replaced by whey- and soy-based constituents, and oils such as corn and sunflower were included as a source of fat energy and n-6 PUFA. These milk formulas contained only trace amounts of n-3 PUFA because, at the time, there was little or no consensus that n-3 PUFA had a role in nutrition and early development. That situation gradually changed with more focused attention on the role of n-3 PUFA in brain structure, function, and behavior (17–21), but there remained considerable resistance to accepting n-3 PUFA as legitimate nutrients needed for normal development.

Eventually, soybean oil became the dominant oil of choice for infant milk formulas but, among n-3 PUFA, it contains only α-linolenic acid, albeit at a reasonable ratio with linoleic acid of about 8:1. Milk formulas have been developed not only for the healthy term infant but for the preterm and small for gestational age infant as well, and the latter contain other oils providing medium as well as long chain fatty acids. In the early 1990s, European milk formula manufacturers responded to the demand of many scientists, some government regulatory agencies, and consumers that, in addition to linoleic acid and α-linolenic acid, milk formulas contain long chain n-3 PUFA. Permission to produce milk formulas containing long chain n-3 PUFA was granted in 2001 in the United States and in 2002 in Canada.

Since the 1980s, milk formulas have contained n-3 PUFA but, before that, at least a generation of infants in westernized countries who were not breast-fed were nourished through the suckling period without adequate dietary provision of n-3 PUFA. As many as three generations of non–breast-fed infants received no long chain n-3 PUFA. Indeed, this will continue because not all milk formulas presently contain long chain n-3 PUFA and those that do are more expensive.

Availability of α-Linolenic Acid to the Fetus and Neonate

Like all other long chain fatty acids, α-linolenic acid is transferred to the developing fetus via the placenta. As a proportion of the PUFA in blood on the fetal side of the placenta, α-linolenic acid is relatively low compared not only with other n-3 PUFA but also compared with α-linolenic acid on the maternal side of the placenta. The net result is that the fetus accumulates very little α-linolenic acid. This is most evident in the fatty acid profile of body fat at birth in which α-linolenic acid is not only lower than in adults but also lower than the principal long chain n-3 PUFA, docosahexaenoic acid (Table 8.1). Although perhaps surprising, this is not a peculiarity of α-linolenic acid or even of n-3 PUFA because fat at birth also has a relative dearth

TABLE 8.1

Uniquely Low α-Linolenic Acid and Linoleic Acid in Body Fat and Brain Lipids Compared with Levels in Liver in Newborns or Body Fat in Adults[a]

Fatty acid	Adult body fat	Newborn body fat	Newborn brain	Newborn liver
Palmitic acid (16:0)	20.1	43.0	28.5	11.4
Palmitoleic acid (16:1n-7)	4.2	14.3	3.2	0.9
Stearic acid (18:0)	4.7	4.1	18.2	4.4
Oleic acid (18:1n-9)	41.7	27.6	20.0	12.4
Linoleic acid (18:2n-6)	15.4	2.0	0.9	6.0
Arachidonic acid (20:4n-6)	0.3	0.7	11.2	3.1
α-Linolenic acid (18:3n-3)	1.5	<0.1	<0.1	0.2
Docosahexaenoic acid (22:6n-3)	0.1	0.4	8.4	2.0

[a]% Composition data; Pooled from (27,75–78).

of linoleic acid and a higher content of the principle long chain n-6 PUFA (arachidonic acid, 20:4n-6).

The main interpretation for the emphasis on long chain PUFA in the fetus is that the placenta selectively transfers these fatty acids compared with the parent PUFA (22,23). The placenta and preterm infant have the capacity to desaturate and chain elongate α-linolenic acid (24), but this would seem to be a secondary cause of the "biomagnification" of both series of long chain PUFA toward the fetus (25) that would come into play only when maternal production or intake of long chain n-3 PUFA was severely limited. Conceivably, there may also be selective oxidation of α-linolenic acid by the fetus but, at present, there is no evidence of the latter. The point is that whether they originate from maternal synthesis, maternal diet, or placental synthesis, net influx and accumulation of PUFA to the near-term human fetus favors long chain PUFA. Teleologically, these observations fit with the growing demand for long chain PUFA by the developing fetus, especially its brain.

Milk from women in Western countries contains about 1% α-linolenic acid (range 0.2–2.0%) whereas non-Western women have milk containing about twice as much α-linolenic acid (15). Little is known about the composition of fatty acids in infant tissues or the rate at which infants accumulate different fatty acids, but it is clear that α-linolenic acid starts to rise rapidly in infant body fat soon after birth. How the rate of α-linolenic acid accumulation in the infant varies with milk (maternal or formula) α-linolenic acid content is unknown.

α-Linolenic Acid and Long Chain n-3 Polyunsaturates in the Brain

Of all the tissues measured, the brain seems to have the lowest content of α-linolenic acid (Table 8.1). This is true irrespective of developmental stage (fetal, newborn, adult). Rates of n-3 PUFA accumulation in the human brain during fetal and early postnatal development have been tracked in detail (19,26), and the influence on brain fatty acid profiles of breast versus formula feeding has been reported (27,28).

With the exception of the retinal photoreceptor, no organ favors long chain n-3 PUFA more than the brain. In part, this is attributable to the brain's almost exclusive incorporation of PUFA into membrane phospholipids. The generally low content of α-linolenic acid in membranes of all tissues strongly suggests that α-linolenic acid itself is of minimal importance to membrane function. Nevertheless, α-linolenic acid is transported into the brain (29), and modest elevations of brain α-linolenic acid can be achieved by raising α-linolenic acid intake. After oral dosing of rats with α-linolenic acid labeled with ^{2}H, ^{14}C, or ^{13}C, its uptake into the brain has been reported by some researchers (29–31) but not by others doing essentially identical studies (32,33). The point is not so much whether or not there is brain uptake of α-linolenic acid (which there must be if it is found in the brain) but why there are such low levels in the brain. The obvious but perhaps overly simplistic explanation is that it does little or nothing of importance in the brain so it does not accumulate there.

Furthermore, because α-linolenic acid can be converted either to ketone bodies (34) or converted to long chain n-3 PUFA (35,36), its content in the brain is well regulated by one or both of these routes.

Infant milk formulas containing no docosahexaenoic acid result in a lower percentage of docosahexaenoic acid in the infant brain during the first 6 mon of life (27,28) This reduction relative to the percentage of docosahexaenoic acid in the brain at birth represents, in quantitative terms, about a 50% lower accumulation of docosahexaenoic acid (37). Although the level of α-linolenic acid, especially in relation to the level of linoleic acid, may not have been in an "ideal" range in these formulas, these results show that the main deficit that accounts for the lower brain docosahexaenoic acid when an infant is not breast-fed is the absence of dietary docosahexaenoic acid. Ethically and logistically it is nearly impossible to imagine how to get access to brain tissue in such a way as to be able to determine the lowest brain docosahexaenoic acid level at which optimal (or even adequate) brain development occurs. There do not yet appear to be any reports of brain docosahexaenoic acid levels in infants consuming a formula containing docosahexaenoic acid, but they will probably demonstrate that brain docosahexaenoic acid will be within normal limits seen in breast-fed infants. Scientifically, the issue therefore becomes one of attempting to determine the brain docosahexaenoic acid threshold below which optimal brain function is threatened but, ethically, there is no longer a valid excuse for excluding a dietary source of docosahexaenoic acid during early infant development.

Why Is a Dietary Requirement for Long Chain n-3 PUFA Controversial in the Infant?

Despite a clear indication since the early 1990s that docosahexaenoic acid has an important role in human brain and visual development (see section on Disorders of Brain Development Involving n-3 PUFA) and recent global agreement that it can (but is not required to be) put into milk formulas, there are five main reasons why this subject remains controversial.

(i) The biochemical pathway for synthesis of docosahexaenoic acid exists in humans of all ages from the premature infant to the aging adult. This is most clearly demonstrated by providing labeled α-linolenic acid and showing that labeled long chain n-3 PUFA including docosahexaenoic acid are present in blood at time points varying from hours to days later (24,38–40). Until this was done, it was not known from dietary studies (see Chapter 3) whether humans, especially infants, actually had sufficient desaturation-chain elongation capacity to make even minimally detectable amounts of docosahexaenoic acid. We now know that this capacity does exist; at least, we know that it exists in principle, that is, under some, perhaps ideal, conditions. The problem with tracer methodology is that it is not easy to determine the *quantitative capacity* of the pathway relative to the demand for the product of that pathway (i.e., docosahexaenoic acid in this example). In fact, we have reason to doubt that the capacity to convert α-linolenic acid to docosahexaenoic acid is sufficient to meet the demand in human infants (37; Table 8.2).

(ii) The best techniques available to study brain development and function in infants are still relatively crude relative to the subtlety of the behavioral or developmental differences that occur as intake of n-3 PUFA, let alone long chain n-3 PUFA, decreases from optimal through adequate to suboptimal and frankly deficient. Because of this difficulty in matching functional deficits to the clear structural deficit (low brain docosahexaenoic acid) that occurs during n-3 PUFA deficiency, we in fact often do not know when a functional deficit exists. Part of the problem is the ethical constraint on experiments in infants, but part is the variability within existing behavioral and developmental testing procedures.

(iii) The body has significant stores of docosahexaenoic acid at birth. This point is rarely discussed and, to be fair, there are few reports on which to base an estimate of docosahexaenoic acid stores in infants. Nevertheless, such reports do exist and with the better knowledge of infant body composition, especially percentage fat at birth, it has been possible to calculate that healthy term infants have a store of about 1 g of docosahexaenoic acid at birth (37; Table 8.3). We have estimated that the human brain accumulates about 5 mg/d of docosahexaenoic acid during the first 6 mon of life and that this represents about half the daily accumulation of docosahexaenoic acid in the whole body; insufficient data exist to provide an estimate beyond that age. Although docosahexaenoic acid absorption is very efficient in

TABLE 8.2

Calculated Daily Requirement for Docosahexaenoic Acid in the Human Infant Compared with the Absence of Evidence of Sufficient Capacity to Meet This Requirement *via* Synthesis from α-Linolenic Acid

Calculation of docosahexaenoic acid requirement	
Accumulation of docosahexaenoic acid (mg/d)[a]	10
Estimated requirement (mg/d)[b]	20
α-Linolenic acid intake from milk (mg/d)	400
Minimum required rate of conversion (%)[c]	5
Measured rate of conversion of α-linolenic acid to docosahexaenoic acid in different models (%)	
Rat[d]	1.4
Near term baboon fetus[e]	< 0.1
4 week old baboon neonate[f]	0.2
Adult humans[g]	< 1

[a]See Table 8.3.

[b]Based on an estimated 50% loss of docosahexaenoic acid due to carbon recycling (12) and/or complete β-oxidation (13).

[c]To make 20 mg/d of docosahexaenoic acid, assuming no dietary source of docosahexaenoic acid or n-3 PUFA other than α-linolenic acid.

[d](79).

[e](12).

[f](34).

[g](80,81).

TABLE 8.3
Change in Body Content of Docosahexaenoic Acid (mg) Between Birth and 6-mo old

Compartment	Newborn	6-mo old	Accumulation (mg/d)
Brain	720	1625	5.0
Liver	336	360	0.1
Body fat	1053	1200	0.8
Remaining lean tissue	1694	2500	4.4
Whole body	3803	5685	10.3

Source: Modified with permission from Cunnane *et al.* (37).

infants (<95%), a surprisingly high proportion of docosahexaenoic acid is catabolized (12,13). Assuming a conservative availability of docosahexaenoic acid of only 50%, an intake or production rate of double the whole body accumulation rate would therefore be needed (20 mg/d). Irrespective of synthesis or intake in milk, estimated body stores at birth therefore contain enough docosahexaenoic acid to meet this demand for at least 50 d (Table 8.3). The point is that, at least theoretically, the absence of docosahexaenoic acid synthesis or in milk could therefore easily go unnoticed in the healthy term infant for well over a month. Hence, lack of a demonstrable functional deficit in healthy infants not receiving a dietary supply of docosahexaenoic acid is not all that surprising and indicates only that, *under these ideal circumstances,* sufficient alternative sources of docosahexaenoic acid may exist to avert a problem. The crucial variable that makes this possible is the size of body fat stores at term (Fig. 8.1); reduce these by 90% as is common in a premature infant born 10 wk early, and one wipes out the insurance while simultaneously greatly increasing the risk of impaired neurodevelopment (25,41).

(iv) Adverse effects have been observed with long chain n-3 PUFA supplementation under certain and now unacceptable circumstances. In one study, supplementation with fish oils containing eicosapentaenoic acid without a concurrent source of arachidonic acid led to a somewhat smaller head circumference in infants (42). At the time, this study was a logical step to have taken when formula supplementation with long chain PUFA was first being contemplated in the early 1990s. However, it is now well recognized that it is safer (43) and more appropriate to provide docosahexaenoic acid directly and always in combination with arachidonic acid with both not exceeding average levels in human milk (both present at about 0.2–0.4%; 15). Nevertheless, this early report indicates why nutritional modification of milk formulas should be done cautiously. Unfortunately, it also remains as an unreasonable source of bias against supplementation of infant formulas with long chain n-3 PUFA because eicosapentaenoic acid has not even been tested experimentally for this purpose in the past 5 yr.

(v) Despite evidence that neurodevelopment in healthy term infants consuming a formula containing α-linolenic acid as the only n-3 PUFA cannot easily be distinguished from that in breast-fed infants, there is occasionally evidence that α-linolenic acid alone does not allow for normal neurological and physical development (44). Ironically, a similar observation with eicosapentaenoic acid (42) was used to caution

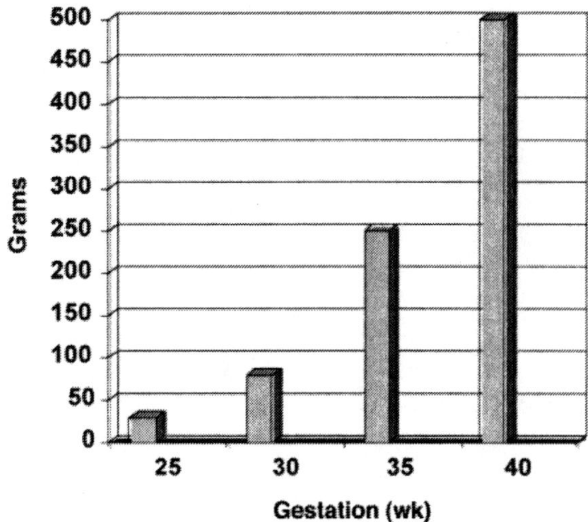

Fig. 8.1. Rapid accumulation of body fat in the third-term human fetus which will account for 14–18% of body weight at full-term birth (40 wk). The rate of gain in body fat during the third trimester represents a 20-fold increase during just 15 wk. Note that body fat is rare at birth in land-based species. *Source:* Modified with permission from Widdowson (83).

against inclusion of long chain n-3 PUFA in milk formulas (45). This problem associated with α-linolenic acid as the sole n-3 PUFA in milk formulas has not been reported often (and may not occur often) but remains a constant risk given that a large proportion of milk formulas that contain no long chain n-3 PUFA will continue to be produced.

Energy Requirements of the Human Brain

A big factor, perhaps the biggest, affecting the need for n-3 PUFA by the developing brain is its size and consequent energy demand. As a proportion of body weight, the brain is bigger at birth (12–14% of weight) and consumes a higher proportion of the body's total energy demand (74%) than at any time later in life (Table 8.4). Thus, the need for docosahexaenoic acid by excitable cells is magnified by the large size of the brain and, equally important, must be complemented by a suitable and reliable source of energy to fuel those cells. This directly affects the need for n-3 PUFA not only by the brain but by the body as a whole because α-linolenic acid is relatively easily β-oxidized (i.e., converted to respiratory CO_2). The apparent preference to oxidize α-linolenic relative to other common long chain fatty acids is evident from a variety of models (see Chapter 3). This is the main reason that tissue levels of α-linolenic acid are low — it is readily mobilized from stores and oxidized. That being the case, conditions that increase fat oxidation such as energy deficit and fasting, increase oxidation

                                  S.C. Cunnane

TABLE 8.4

The Brain's Energy Requirement Relative to the Body as a Whole

Weight			Metabolic rate		Brain's energy requirement
Body (kg)	Brain (g)	Brain (%)	Body (kcal/d)	Brain (kcal/d)	(% of body)
3.5	400	11.0	161	118	74
5.5	650	11.0	300	192	64
11	1045	9.5	590	311	53
19	1235	6.5	830	367	44
31	1350	4.4	1160	400	34
50	1360	2.7	1480	403	27
70	1400	2.0	1800	414	23

Source: Modified with permission from Holliday (82).

of α-linolenic acid (46; see Chapter 3).

This is directly relevant to getting sufficient n-3 PUFA into the brain during early development because, if there are no long chain n-3 PUFA in the formula, synthesis from α-linolenic acid becomes the sole means of obtaining them once they are used up from body stores. If those stores are small, such as in premature infants, the store of docosahexaenoic acid will be exhausted earlier, all the more so if the infant undergoes periods of energy deficit when some docosahexaenoic acid will, like all other fatty acids, be used for energy. Then there will be a competition for the available α-linolenic acid: Is it used to make energy or is it conserved to make docosahexaenoic acid? Available evidence in animal models suggests that more than 100% of dietary α-linolenic acid can be burned as a fuel, that is, that energy deficit can deplete body n-3 PUFA stores and completely prevent synthesis of docosahexaenoic acid (47,48). Body loss exceeding intake has not been demonstrated for α-linolenic acid in humans undergoing energy deficit (weight loss) but has been shown for linoleic acid (49), so it is reasonable to deduce that though an infant has the biochemical machinery to make docosahexaenoic acid, in the sick or premature infant, energy deficit may well obliterate conversion of α-linolenic acid to long chain n-3 PUFA because it is being used for energy. This problem is accentuated by the high energy demand by the brain, especially during early postnatal development, and illustrates the folly of depending on α-linolenic acid as the sole dietary n-3 PUFA during early development.

Infant Development During Total Dietary Deficiency of n-3 PUFA

Total Dietary n-3 PUFA Deficiency

Total dietary deficiency of n-3 PUFA is a common animal model to study the function and metabolism of these fatty acids. Although introduced much earlier (18), it was not really until the mid-1980s that this model began to produce reliable evidence that normal behavior and brain development depend on a dietary source of n-3 PUFA. Part of the challenge is that n-3 PUFA are tenaciously retained in membrane lipids, so dietary deficiency has to be extreme and prolonged before reproducible detrimental effects are seen. Because rodents and primates generally have low fat stores during early development, this is the time when they are most susceptible to n-3 PUFA deficiency. In contrast to these animal models, healthy human infants have significant body fat stores, usually accounting for about 20% of body weight. Because these infant body fat stores contain n-3 PUFA (Tables 8.1, 8.3), humans are moderately well protected against dietary deficit of n-3 PUFA and it is quite uncommon to see the classical physiological symptoms of n-3 PUFA depletion (impaired visual development, impaired learning, and attenuated response to noxious stimuli; Table 8.5). The now widespread use of enteral and parenteral nutrition containing at least α-linolenic acid, coupled with the relative insensitivity of most of our tools used to determine behavioral, memory, or learning deficits, means that few cases of total dietary n-3 PUFA have been reported and, to my knowledge, none in the past 15 yr. This is distinct from genetic disorders that may cause profound n-3 PUFA depletion and catastrophic mental retardation (see section on Inherited Disorders of Brain Development Involving n-3 PUFA).

The two best-known cases of dietary total n-3 PUFA deficiency are described in Chapter 3 (50,51). The symptoms of one of these cases are summarized here also because of their direct relevance to the question of n-3 PUFA and neurological function (Table 8.5). Although these two cases differ somewhat in symptoms, the important point is that they are rare now because enteral and parenteral nutrition contains at least α-linolenic acid.

Dietary Deficiency of Long Chain n-3 PUFA

This subject is discussed in some detail in the earlier section, Why Is a Dietary Requirement for Long Chain n-3 PUFA Controversial in the Infant?, but additional

TABLE 8.5

Symptoms Before and After α-Linolenic Acid Supplementation in a Clinical Case of Deficiency of n-3 PUFA in a 6-yr-old Girl on Enteral Nutrition for 12 mo

Parameters	Time point	Outcome
Symptoms	Before	Numbness, muscle pain, visual blurring
	After	Disappearance of neurological symptoms
		10–15% increase in nerve conduction velocity
α-Linolenic acid intake	Before	0.66% of fatty acids
		Linoleic acid/α-linolenic acid = 115:1
	After	6.9% of fatty acids
		Linoleic acid/α-linolenic acid = 6:1
n-3 PUFA in blood:	Before	Low α-linolenic acid; normal docosahexaenoic acid
	After	Raised 3–7-fold in serum phospholipids

Abbreviation: PUFA, polyunsaturated fatty acid.
Source: Modified from Holman *et al.* (50).

points are made here. In several cases of learning and developmental deficit attributed in the original work to a relative deficit of long chain n-3 PUFA, the deficit cannot fairly be attributed solely to a dietary absence of long chain n-3 PUFA. Confounders like high ratio of linoleic to α-linolenic acid (52) can suppress synthesis of long chain n-3 PUFA. The reference or control group has an important influence as well. Recent studies reporting minimal apparent benefit of supplementing formula-fed term infants with arachidonic acid and docosahexaenoic acid used breast-fed infants that, themselves, had a very low intake of docosahexaenoic acid (53,54). Hence, if the reference group, even though breast fed, has low intake of long chain n-3 PUFA, this biases the interpretation of a lack of difference in the outcome measures. Regardless of the seemingly normal development of term infants not receiving docosahexaenoic or arachidonic acid, the low brain accumulation of docosahexaenoic acid when it is not in the formula (27,28) warns that an increased risk of abnormal development exists in these infants whether or not existing techniques presently allow detection of such a risk.

As discussed earlier, perhaps the most important variable influencing the effect of a supplemental source of long chain PUFA is body fat stores because, in the healthy term infant, they not only provide the energy store that protects brain development during infancy but they also have a significant store of docosahexaenoic acid (and arachidonic acid; Table 8.3). Although it makes good ethical sense and minimizes other confounders, studying how supplemental long chain n-3 PUFA influence learning is least likely to succeed in healthy term infants because they are already well protected from a deficit of these fatty acids.

Inherited Disorders of Brain Development Involving n-3 PUFA

Zellweger syndrome is perhaps the best known example of the rare group of peroxisomal biogenesis disorders that have devastating consequences for brain development in part because of abnormal neuronal migration and dysmyelination. In Zellweger syndrome, functional peroxisomes are not produced. Blindness, deafness, and early death are common. The overlap in the profound neurological symptoms of Zellweger syndrome with those of the consequences of n-3 PUFA deficiency led to investigation of a possible link with brain n-3 PUFA deficiency (55). Docosahexaenoic acid supplementation has been attempted and, when started early enough, is highly beneficial in some cases (reviewed in 56). Because a step occurring in peroxisomes is now widely thought to be necessary for the synthesis of docosahexaenoic acid, clearly the absence of functional peroxisomes seems to be the reason not only for the neurodevelopmental deficit and markedly low tissue docosahexaenoic acid levels but also for the therapeutic benefit of supplemental dietary docosahexaenoic acid.

Questions still exist about the variability in the clinical therapeutic effectiveness of docosahexaenoic acid in Zellweger syndrome (57), about incomplete agreement between the clinical disorder and emerging gene knock-out models (58) and, indeed, whether peroxisomes are even obligatory organelles in the synthesis of docosahexaenoic acid (59). Resolution of these issues in careful clinical and experimental research should be rewarded by progress in understanding how n-3 PUFA are utilized by the developing neonate.

Phenylketonuria is a rare inherited disease in which phenylalanine accumulates because it is not adequately degraded to tyrosine due to the absence of phenylalanine hydroxylase. Because phenylalanine is present in many foods, to minimize the adverse effects of its excess in patients with phenylketonuria, a diet is provided in which phenylalanine is severely limited. Effectively, a low phenylalanine milk formula is needed during the first year of life and sometimes for most of the first 5 yr of life (i.e., during the critical period of maximal brain growth). These phenylalanine-restricted formulas are intended to be complete in other known essential nutrients but, at present, none contain long chain n-3 PUFA. Fatty acid analyses that we have done indicate a typical ratio of n-6 to n-3 PUFA (linoleic to α-linolenic acid) of about 20:1. This raises the issue of whether these formulas are complete in n-3 PUFA and whether residual mental deficits in children with phenylketonuria are attributable only to excess phenylalanine or could also arise either from an inappropriately high ratio of n-6 to n-3 PUFA or from the absence of long chain n-3 PUFA (60). The example of phenylketonuria raises a caution about all enteral or parenteral formulas used to correct diseases that are expressed as an imbalance, excess, or absence of a particular nutrient; it would be a great pity to be correcting one nutrient deficiency while simultaneously and inadvertently inducing another.

Disorders of the Adult and Aging Brain

The role of n-3 PUFA in disorders of brain function and behavior in adults and in degenerative diseases of aging are also under investigation. These disorders include schizophrenia, bipolar disorder, depression, Alzheimer's disease, memory loss, and high suicide and homicide risk (61,62). The wide spectrum of neurological disorders and symptoms of these disorders obviously implicates many different disease processes. Nevertheless, epidemiological studies suggest that problems of low intake or inadequate synthesis or metabolism of n-3 PUFA may be common features or may exacerbate underlying pathology (63–65). Matching animal models to the human situation and coping effectively with major confounders like placebo effects will be important challenges to overcome in solidifying a role of n-3 PUFA in these disorders. The overwhelming emphasis of research into the therapeutic effects of n-3 PUFA is on the role of docosahexaenoic acid, but it is becoming clear that other n-3 PUFA, particularly eicosapentaenoic acid (66), may have therapeutic benefits that are not directly attributable to its conversion to docosahexaenoic acid.

As the human brain ages, it is at increased risk of memory loss and reduced cognitive function. Because learning and cognition are partly dependent on adequate brain n-3 PUFA, particularly docosahexaenoic acid, considerable research interest has developed with the aim of determining loss and possible corrective benefits of supplemental dietary n-3 PUFA in degenerative neurological diseases such as Alzheimer's disease. Although plasma levels of n-3 PUFA may be normal in the elderly (67), there are promising results suggesting that docosahexaenoic acid supplements may delay the onset of memory loss with advanced age (68). Interest in this field will undoubtedly increase over the next decade as the proportion represented by the elderly increases, particularly in Western countries.

Several Possible Mechanisms

Studies modeling the structure of docosahexaenoic acid increasingly suggest that its functional uniqueness in excitable cells derives from its helical conformation (69,70). Its six double bonds equally distributed over the length of the molecule make this conformation unique relative to its nearest structural neighbors, the two docosapentaenoic acids (22:5n-3 or 22:5n-6). The high concentration of docosahexaneoic acid in the outer pigment epithelium of rod photoreceptors has led to concerted efforts to determine how docosahexaenoic acid participates in transducing light to chemical signals. This work increasingly implicates docosahexaenoic acid in a tight relationship with rhodopsin's own changes in conformation during photoreception. Rhodopsin is not present elsewhere in the central nervous system, so the mechanism of the effects of docosahexaenoic acid must differ in the brain from those in the eye. Nevertheless, the experimental approaches, if not the specific molecular targets to analyzing the molecular relationships and function of docosahexaenoic acid in membrane phospholipids, are similar.

Docosahexaenoic acid may contribute to optimal brain function via mechanisms affecting the brain's energy supply, namely glucose uptake (71). The mammalian brain has a high energy demand, especially during early postnatal development (Table 8.4), and it is likely that optimal brain function depends not only on optimal membrane organization but also on optimal energy supply to fuel the respiration of neurons. This recent paper (71) opens up an unexplored direction that has considerable promise to reveal the dual nature of the demand for PUFA, that is, improving both signal transduction and energy utilization.

There is a strong tendency to focus attention on docosahexaenoic acid as the sole PUFA mediator of normal brain function. Although resolving how docosahexaenoic acid functions as a constituent of membrane phospholipids in the brain will be heralded as a breakthrough, there are several reasons to continue to look broadly at how PUFA function in the brain. For instance, reports of therapeutic benefits of supplemental eicosapentaenoic acid are becoming more frequent in psychiatric illness (72) yet there is little or no evidence that these effects occur by raising brain docosahexaenoic acid. Superficially, this is puzzling because there are only low amounts of eicosapentaenoic acid in the brain and because it is usually thought to be moderately easily converted to docosahexaenoic acid. Hence, supplemental eicosapentaenoic acid should lead to raised brain docosahexaenoic acid but, as yet, apparently does not (63).

One possibility is that eicosapentaenoic acid is effective because it inhibits release of arachidonic acid from phospholipids (72). This was established many years ago and is seen as a principal mechanism by which eicosapentaenoic acid inhibits the production of proinflammatory and proaggregatory eicosanoids derived from arachidonic acid. The implication is that the same mechanism could operate in the brain and could modulate the release of arachidonic acid from phospholipids and/or its conversion to destructive eicosanoids.

Arachidonic acid is present at as high a concentration in the brain as docosahexaenoic acid, but its effects there remain far more mysterious than those of docosahexaenoic acid. This is mainly because it is difficult to change brain arachidonic acid by the classical approaches such as n-6 PUFA deficiency. Intriguingly, although the absence of long chain n-3 PUFA from milk formulas results in lower docosahexaenoic acid in the infant brain, the absence of arachidonic acid from these same formulas does not lower brain arachidonic acid (27,28). The reason for this distinction is completely unknown, but I predict that when it is understood, it will reveal much about the functional importance of arachidonic acid in the brain.

One tantalizing clue to the brain's ability to maintain normal amounts of arachidonic acid in the absence of any dietary arachidonic acid is that PUFA undergo β-oxidation and carbon recycling to differing extents (see Chapter 3; 8,73). Carbon recycling refers to the capture of part of the β-oxidized carbon skeleton and its incorporation into lipid (fatty acid and cholesterol) synthesis, a process that can consume 30 times more carbon from α-linolenic acid than goes into docosahexaenoic acid (32). Research on this topic is not yet extensive but shows that carbon recycling is a major fate of the 18 carbon PUFA—linoleic and α-linolenic acids—both in rodents and primates. It is both interesting and unexpected that docosahexaenoic acid also undergoes considerable carbon recycling (12) but, puzzlingly, arachidonic acid does not (74).

This is a reasonable, though perhaps unconventional, mechanism to explain how the brain retains arachidonic acid but cannot retain docosahexaenoic acid in the presence of inadequate levels of the precursors; it is not so much a question of their synthesis but rather a question of their *differential degradation* (β-oxidation). The teleological benefit of controlling the brain levels of these fatty acids by this mechanism is that, in early human evolution, their supply in the diet was not as limiting as it is now so mechanisms consequently developed to protect more against their *excess* than their deficiency.

Given the evidence that the shorter chain n-3 PUFA have actions in the brain that are not necessarily linked in any substantial way to producing more docosahexaenoic acid, "forcing" α-linolenic acid and eicosapentaenoic acid to function primarily as docosahexaenoic acid precursors (by using infant formulas without added docosahexaenoic acid) may compromise the ability of these precursors to perform their primary functions and could well be contributing to the functional brain deficits we currently, but perhaps mistakenly, attribute mostly to docosahexaenoic acid.

Perspective

Despite abundant evidence of multiple routes of α-linolenic acid metabolism (see Chapter 3), the bottom line is that it is not yet clear that α-linolenic acid does anything more important than act as a form of insurance for the synthesis of long chain n-3 PUFA when they are not present in milk or formulas in adequate amounts. In other words, if long chain n-3 PUFA are present in milk consumed by an infant, presently we have little reason to believe that α-linolenic acid itself has any role in infant nutrition. Thus, the importance of α-linolenic acid in human infant nutrition is really a product of suckling infants on milk formulas in which it is the only n-3 PUFA. Nevertheless, we should keep an open mind on this because we cannot say with certainty that we have discovered all the nutritional or health attributes of α-linolenic acid or even of n-3 PUFA as a whole.

If recent reports of abundant metabolism of α-linolenic acid outside its well-known conversion to long chain n-3 PUFA are any indication (i.e., its utilization in the synthesis of cholesterol and saturated fatty acids appearing in the brain; 7,73), it is certainly clear that we do not yet fully understand the metabolism of α-linolenic acid. Equally, and despite the apparent redundancy of α-linolenic acid in human milk or milk formulas containing long chain n-3 PUFA, we should be cautious about closing the door on exploring how α-linolenic acid itself may contribute to infant development and brain function.

Several of the studies reported here describe dietary supplementation with flaxseed oil but none discuss flaxseed itself. A theme running through this review is that, in my view, α-linolenic acid is not an adequate source of n-3 PUFA during early development, no matter how well it is titrated against the intake of n-6 PUFA and no matter whether we presently have tools to demonstrate a deficit of using α-linolenic acid as the sole n-3 PUFA in infant formulas. However, once beyond the infant period, α-linolenic acid is a key and, for many millions, the only n-3 PUFA. Its intake is marginally adequate in Western countries and raising its intake is associated with reduced risk of mortality and heart disease (see Chapter 3). Fish is still the best source of eicosapentaenoic acid and docosahexaenoic acid, but it is expensive, supplies are dwindling and often contaminated, and fish is not eaten by a large proportion of the population. Flaxseed offers benefits including improved laxation, reduction of serum cholesterol, and protection against cancer (see Chapters 11, 15) that extend beyond its content of α-linolenic acid. As we learn more about the attributes of supplemental n-3 PUFA intake for maintaining health, especially in old age, depending on one's lifestyle, either flaxseed is poised to be a valuable dietary ingredient or flaxseed oil a valuable dietary supplement.

Acknowledgments

CIHR and NSERC supported the author's research. Mary Ann Ryan provided excellent technical support.

References

1. Birch, E.E., D.R. Hoffman, R. Uauy, D.G. Birch, and F. Prestidge, Visual Acuity and the Essentiality of Docosahexaenoic Acid and Arachidonic Acid in the Diet of Term Infants, *Pediatr. Res. 44*:201–209 (1998).
2. Gibson, R.A., W. Chen, and M. Makrides, Randomized Trials with Polyunsaturated Fatty Acid Interventions in Preterm and Term Infants: Functional and Clinical Outcomes, *Lipids 36*:873–883 (2001).
3. Carlson, S.E., and M. Neuringer, Polyunsaturated Fatty Acid Status and Neuro-development: A Summary and Critical Analysis of the Literature, *Lipids 34*:171–178 (1999).
4. Willatts, P., J.S. Forsyth, M.K. DiModugno, S. Varma, and M. Colvin, Influence of Long-Chain Polyunsaturated Fatty Acids on Infant Cognitive Function, *Lipids 33*:973–980 (1998).
5. SanGiovanni, J.P., S. Parra-Cabrera, G.A. Colditz, C.S. Berkey, and J.T. Dwyer, Meta-Analysis of Dietary Essential Fatty Acids and Long Chain Polyunsaturated Fatty Acids as They Relate to Visual Resolution Acuity in Healthly Preterm Infants, *Pediatrics 105*:1292–1298 (2000).
6. Uauy, R., F. Calderon, and P. Mena, Essential Fatty Acids in Somatic Growth and Brain Development, *World Rev. Nutr. Diet 89*:134–160 (2001).
7. Cunnane, S.C., Carbon Recycling: An Important Pathway in α-Linolenate Metabolism in Fetuses and Neonates, in *Fatty Acids: Physiological and Behavioral Functions*, edited by D.I. Mostovsky, S. Yehuda, and N. Salem, Jr., Humana Press, Totowa, New Jersey, 2001, pp. 145–159.
8. Cunnane, S.C., The Contribution of α-Linolenic Acid in Flaxseed to Human Health, in *Flax—The Genus Linum*, edited by A. Muir and N. Westcott, Taylor and Francis, London, 2003, pp. 150–180.
9. Cunnane, S.C., The Conditional Nature of the Dietary Need for Polyunsaturates: A Proposal to Reclassify "Essential Fatty Acids" as "Conditionally Indispensable" or "Conditionally Dispensable" Fatty Acids, *Br. J. Nutr. 84*:803–812 (2000).
10. Brenna, J.T., Efficiency of Conversion of α-Linolenic Acid to Long Chain n-3 Fatty Acids in Man, *Curr. Opin. Nutr. Metabol. Care 5*:127–132 (2002).
11. Horrobin, D.F., The Phospholipid Concept of Psychiatric Disorders and Its Relationship to the Neurodevelopmental Concept of Schizophrenia, in *Phospholipid Spectrum Disorder in Psychiatry*, edited by M. Peet, I. Glen, and D.F. Horrobin, Marius Press, Carnford, Lancashire, United Kingdom, 1999, pp. 3–22.
12. Sheaff-Greiner, R.C., Q. Zhang, K.J. Goodman, D.A. Guissini, P.W. Nathanielsz, and J.T. Brenna, Linoleate, α-Linolenate and Docosahexaenoate Recycling into Saturated and Monounsaturated Fatty Acids Is a Major Pathway in Pregnant or Lactating Adults and Fetal or Infant Rhesus Monkeys, *J. Lipid Res. 37*:2675–2686 (1996).
13. Poumes-Ballihaut, C., B. Langelier, F. Houlier, J-M. Alessandri, G. Durand, C. Latge, and P. Guesnet, Comparative Bioavailability of Dietary α-Linolenic and Docosahexaenoic Acids in the Growing Rat, *Lipids 36*:793–800 (2001).
14. Yavin, E., S. Glozman, and P. Green, Docosahexaenoic Acid Accumulation in the Perinatal Brain: Prooxidant and Antioxidant Properties, *J. Molec. Neurosci. 16*:229–236 (2001).
15. Koletzko, B., M. Rodriguez-Palmero, H. Demmelmair, M. Fidler, R. Jensen, and T. Sauerwald, Physiological Aspects of Human Milk Lipids, *Early Hum. Dev. 65 (Suppl. 2)*:S3–18 (2001).

16. Cuthbertson, W.J.F., Evolution of Infant Nutrition, *Br. J. Nutr. 81*:359–371 (1999).
17. Sinclair, A.J., and M.A. Crawford, The Accumulation of Arachidonate and Docosa-hexaenoate in the Developing Rat Brain, *J. Neurochem. 19*:1753–1758 (1972).
18. Lamptey, M.S. and B.L. Walker, A Possible Essential Role for Dietary Linolenic Acid in the Development of the Young Rat, *J. Nutr. 106*:86–93 (1976).
19. Clandinin, M.T., J.E. Chappel,, S. Leong, T. Heim, P.R. Swyer, and G.W. Chance, Extrauterine Fatty Acid Accretion in Infant Brain: Implications for Fatty Acid Requirements, *Early Hum. Dev. 4*:131–138 (1980).
20. Neuringer, M., W.E. Connor, C. van Petten, and L. Barstad, Dietary Omega-3 Fatty Acid Deficiency and Visual Loss in Infant Rhesus Monkeys, *J. Clin. Invest. 73*:272–276 (1984).
21. Bourre, J.M., G. Pascal, G. Durand, M. Masson, O. Dumont, and M. Piciotti, Alterations in the Fatty Acid Composition of Rat Brain Cells (Neurons, Astrocytes, and Oligodendrocytes) and of Subcellular Fractions (Myelin, Synaptosomes) Induced by a Diet Devoid of n-3 Fatty Acids, *J. Neurochem. 43*:342–348 (1984).
22. Haggarty, P., K. Page, D.R. Abramovich, J. Ashton, D. and Brown, Long-Chain Polyunsaturated Fatty Acid Transport Across the Perfused Human Placenta, *Placenta. 18*:635–642 (1997).
23. Dutta-Roy, A.K., Fatty Acid Transport and Metabolism in the Feto-Placental Unit and the Role of Fatty Acid Binding Proteins, *Nutr. Biochem. 8*:548–557 (1997).
24. Uauy, R., P. Mena, B. Wegher, S. Nieto, and N. Salem, Jr., Long Chain Polyunsaturated Fatty Acid Formation in Neonates: Effect of Gestational Age and Intrauterine Growth, *Pediatr. Res. 47*:127–135 (2000).
25. Crawford, M.A., The Role of Essential Fatty Acids in Neural Development: Implications for Perinatal Nutrition, *Am. J. Clin. Nutr. 57 (Suppl.)*:703S–709S (1993).
26. Martinez, M., Tissue Levels of Polyunsaturated Fatty Acids During Early Human Development, *J. Pediatr. 120*:129S–138S (1992).
27. Farquharson, J., R. Cockburn, W.A. Patrick, E.C. Jamieson, and R.W. Logan, Infant Cerebral Cortex Phospholipid Fatty Acid Composition and Diet, *Lancet 340*:810–813 (1992).
28. Makrides, M., M.A. Neumann, R.W. Byard, K. Simmer, and R.A. Gibson, Fatty Acid Composition of Brain, Retina, and Erythrocytes in Breast- and Formula-Fed Infants, *Am. J. Clin. Nutr. 60*:189–194 (1994).
29. Edmond, J., Essential Polyunsaturated Fatty Acids and the Barrier to the Brain: The Components of a Model for Transport, *J. Molec. Neurosci. 16*:181–194 (2001).
30. Lin, Y., and N. Salem, Jr., A Technique for the *in vivo* Study of Multiple Stable Isotope-Labelled Essential Fatty Acids, *Prostagl. Leukotri. Essent. Fatty Acids 67*:141–146 (2002).
31. Sinclair, A.J., Incorporation of Radioactive Polyunsaturated Fatty Acids into Liver and Brain of the Developing Rat, *Lipids 10*:175–184 (1975).
32. Menard, C.R., K. Goodman, T. Corso, J.T. Brenna, and S.C. Cunnane, Recycling of Carbon into Lipids Synthesized *de novo* Is a Quantitatively Important Pathway of [U-^{13}C]-α-Linolenate Utilization in the Developing Rat Brain, *J. Neurochem. 7*:2151–2158 (1998).
33. Anderson, G.J., and W.E. Connor, Uptake of Fatty Acids by the Developing Brain, *Lipids 23*:286–290 (1988).
34. Emmison, N., P.A. Gallagher, and R.A. Coleman, Linoleic and α-Linolenic Acids Are Selectively Secreted in Triacylglycerol by Hepatocytes from Neonatal Rats, *Am. J. Physiol. 269*:R80–R86 (1995).

35. Cook, H.W., *In vitro* Formation of Polyunsaturated Fatty Acids by Desaturation in Rat Brain: Some Properties of the Enzymes in Developing Brain and Comparisons with Liver, *J. Neurochem. 30*:1327–1334 (1978).
36. Willard, D.E., S.D. Harman, T.L. Kaduce, M. Preuss, S.A. Moore, M.E.C. Robbins, and A.A. Spector, Docosahexaenoic Acid Synthesis from n-3 Polyunsaturated Fatty Acids in Differentiated Rat Brain Astrocytes, *J. Lipid Res. 42*:1368–1376 (2001).
37. Cunnane, S.C., V. Francescutti, J.T. Brenna, and M.A. Crawford, Breast-Fed Infants Achieve a Higher Rate of Brain and Whole Body Docosahexaenoic Acid Accumulation Than Formula-Fed Infants Not Consuming Dietary Docosahexaenoate, *Lipids 35*:105–111 (2000).
38. Carnielli, V.P., D.J.L. Wattimena, I.H.T. Luijendijk, A. Boerlage, H.J. Degenhart, and P.J.J. Sauer, The Very Low Birth Weight Premature Infant is Capable of Synthesizing Arachidonic and Docosahexaenoic Acid from Linoleic and α-Linolenic Acids, *Pediatr. Res. 40*:169–171 (1996).
39. Salem, N., B. Wegher, P. Mena, and R. Uauy, Arachidonic and Docosahexaenoic Acids Are Biosynthesized from Their 18 Carbon Precursors in Human Infants, *Proc. Natl. Acad. Sci. 93*:49–54 (1996).
40. Vermunt, S.H., R.P. Mensink, M.M. Simonis, and G. Hornstra, Effects of Dietary α-Linolenic Acid on the Conversion and Oxidation of ^{13}C-α-Linolenic Acid, *Lipids 35*:137–142 (2000).
41. Crawford, M.A., K. Costeloe, K. Gebremeskel, A. Phylactos, L. Skirvin, and F. Stacey, Are Deficits of Arachidonic and Docosahexaenoic Acids Responsible for the Neural and Vascular Complications of Preterm Babies? *Am. J. Clin. Nutr. 66 (Suppl.)*:1032S–1041S (1997).
42. Carlson, S.E., S. Werkman, and E. Tolley, Effect of Long Chain n-3 Supplementation on Visual Acuity and Growth in Preterm Infants with and without Bronchopulmonary Dysplasia, *Am. J. Clin. Nutr. 63*:687–697 (1996).
43. Lucas, A., M. Stafford, and R. Morley, Efficacy and Safety of Long-Chain Polyunsaturated Fatty Acid Supplementation of Infant-Formula Milk: A Randomized Trial, *Lancet 354*:1948–1954 (1999).
44. Jensen, C.L., T.C. Prager, J.K. Fraley, H. Chen, R.E. Anderson, and W.C. Heird, Effect of Dietary Linoleic/Alpha-Linolenic Acid Ratio on Growth and Visual Function of Term Infants, *J. Pediatr. 131*:200–209 (1997).
45. Life Sciences Research Office Report, Assessment of Nutrient Requirements for Infant Formulas, *J. Nutr. 128*:2059S–2078 (1998).
46. Tang, A.B., K.Y. Nishimura, and S.D. Phionney, Preferential Reduction in Adipose Tissue α-Linolenic Acid (18:3ω3) During Very Low Calorie Dieting Despite Supplementation with 18:3ω3, *Lipids 28*:987–993 (1993).
47. Chen, Z-Y., and S.C. Cunnane, Refeeding After Fasting Increases Apparent Oxidation of n-6 and n-3 Fatty Acids in Pregnant Rats, *Metabolism 42*:1206–1211 (1993).
48. Cunnane, S.C., and J. Yang, Zinc Deficiency Impairs Whole–Body Accumulation of Polyunsaturates and Increases the Utilization of [1-^{14}C]-Linoleate for *de novo* Lipid Synthesis in Pregnant Rats, *Can. J. Physiol. Pharmacol. 73*:1246–1252 (1995).
49. Cunnane, S.C., R. Ross, J.L. Bannister, and D.J.A. Jenkins, β-Oxidation of Linoleate in Obese Men Undergoing Weight Loss, *Am. J. Clin. Nutr. 73*:709–714 (2001).
50. Holman, R.T., S.B. Johnson, and T.F. Hatch, A Case of Human Linolenic Acid Deficiency Involving Neurological Abnormalities, *Am. J. Clin. Nutr. 35*:617–623 (1982).

51. Bjerve, K.S., L. Thoresen, and S. Borsting, Linseed and Cod Liver Oil Induce Rapid Growth in a 7-Year-Old Girl with n-3 Fatty Acid Deficiency, *J. Parent. Enter. Nutr. 12*:521–525 (1988).

52. Lucas, A., R. Morley, T.J. Cole, G. Lister, and C. Leeson-Payne, Breast Milk and Subsequent Intelligence Quotient in Children Born Preterm, *Lancet. 339*:261–264 (1992).

53. Auestad, N., M.B. Montalto, R.T. Hall, K.M. Fitzgerald, R.E. Wheeler, W.E. Connor, J.A. Taylor, and E.E. Hartmann, Visual Acuity, Erythrocyte Fatty Acid Composition, and Growth in Term Infants Fed Formulas with Long Chain Polyunsaturated Fatty Acids for One Year, *Pediatr. Res. 41*:1–10 (1997).

54. Auestad, N., R. Halter, R.T. Hall, M. Blatter, M. Bogle, W. Burks, J.R. Erickson, K.M. Fitzgerald, V. Dobson, S.M. Innis, L.T. Singer, M.B. Montalto, J.R. Jacobs, W. Qui, and M.H. Bornstein, Growth and Development in Term Infants Fed Long Chain Polyunsaturated Fatty Acids: A Double-Masked, Randomized, Parallel, Prospective, Multivariate Trial, *Pediatr. Res. 108*:372–381 (2001).

55. Martinez, M., Severe Deficiency of Docosahexaenoic Acid in Peroxisomal Disorders: A Defect of Δ-4 Desaturation, *J. Pediatr. 40*:1292–1298 (1990).

56. Martinez, M., Restoring the DHA Levels in the Brains of Zellweger Patients, *J. Molec. Neurosci. 16*:309–316 (2001).

57. Moser, H.W., and G.V. Raymond, Docosahexaenoic Acid Therapy for Disorders of Peroxisome Biogenesis, in *Fatty Acids. Physiological and Behavioral Functions*, edited by D.I. Mostovsky, S. Yehuda, and N. Salem Jr., Humana Press, Totowa, New Jersey, 2001, 257–271.

58. Faust, P.L., H-M. Su, A. Moser, and H. Moser, The Peroxisome Deficient PEX2 Zellweger Mouse: Pathologic and Biochemical Correlates of Lipid Dysfunction, *J. Molec. Neurosci. 16*:289–297 (2001).

59. Infante, J. and V. Huszagh, Zellweger Syndrome Knock-Out Mouse Models Challenge Putative Peroxisomal β-Oxidation Involvement in Docosahexaenoic Acid (22:6n-3) Biosynthesis, *Mol. Genet. Metab. 72*:1–7 (2001).

60. Agostoni, C., N. Massetto, G. Biasucci, A. Rottoli, M. Boinvissuto, M.G. Bruzzese, M. Giovannini, and E. Riva, Effects of Long Chain Polyunsaturated Fatty Acid Supplementation on Fatty Acid Status and Visual Function in Treated Children with Hyperphenylalaninemia, *137*:504–509 (2000).

61. Peet, M., New Strategies for the Treatment of Schizophrenia: Omega-3 Polyunsaturated Fatty Acids, in *Phospholipid Spectrum Disorder in Psychiatry*, edited by M. Peet, I. Glen and D.F. Horrobin, Marius Press, Carnford, Lancashire, United Kingdom, 1999, pp. 189–194.

62. Hamazaki, T., and Okuyama,H. (eds.), *Fatty Acids and Lipids—New Findings*, Karger, Basel, 2001. pp. 35–68.

63. Hibbeln, J.R., Membrane Lipids in Relation to Depression, in *Phospholipid Spectrum Disorder in Psychiatry*, edited by M. Peet, I. Glen and D.F. Horrobin, Marius Press, Carnford, Lancashire, United Kingdom, 1999, pp. 195–210.

64. Hamazaki, T., S. Sawazaki, M. Itomura, Y. Nagao, A. Theinprasert, T. Nagasawa, and S. Watanabe, Effect of Docosahexaenoic Acid on Hostility, *World Rev. Nutr. Dietet. 88*:53–57 (2001).

65. Stoll, A.L., K.E. Damico, B.P. Daly, W.E. Severus, and L.B. Marangell, Methodological Considerations in Clinical Studies of n-3 Fatty Acids in Major Depression and Bipolar Disorder, *World Rev. Nutr. Dietet. 88*:58–67 (2001).

66. Peet, M., and S. Ryles, Eicosapentaenoic Acid: A Potential New Treatment for Schizophrenia, in *Fatty Acids. Physiological and Behavioral Functions*, edited by D.I.

Mostovsky, S. Yehuda, and N. Salem, Jr., Humana Press, Totowa, New Jersey, 2001, pp. 345–356.

67. Bjerve, K.S., K.J. Fougner, K. Midthjell, and K. Bonaa, n-3 Fatty Acids in Old Age, *J. Intern. Med. Suppl. 225*:191–196 (1989).

68. Suzuki, H., Y. Morikawa, and H. Takahashi, Effect of DHA Oil Supplementation on Intelligence and Visual Acuity in the Elderly, *World Rev. Nutr. Diet. 88*:68–71 (2001).

69. Feller, S.E., K. Gawrisch, and A.D. MacKerell, Jr., Polyunsaturated Fatty Acids in Lipid Bilayers: Intrinsic and Environmental Contributions to Their Unique Physical Properties, *J. Am. Chem. Soc. 124*:318–326 (2002).

70. Applegate, K.R., and J.A. Glomset, Computer-Based Modelling of the Conformation and Packing of Docosahexaenoic Acid, *J. Lipid Res. 27*:658–680 (1986).

71. Ximenes da Silva, A., F. Lavialle, G. Gendrot, P. Guesnet, J-M. Alessandri, and M. Lavialle, Glucose Transport and Utilization Are Altered in the Brain of Rats Deficient in n-3 Polyunsaturated Fatty Acids, *J. Neurochem. 81*:1–10 (2002).

72. Horrobin, D.F., Disorders of Phospholipid Metabolism in Schizophrenia, Affective Disorders and Neurodegenerative Disorders, in *Fatty Acids: Physiological and Behavioral Functions*, edited by D.I. Mostovsky, S. Yehuda, and N. Salem, Jr., Humana Press, Totowa, New Jersey, 2001, pp. 331–344.

73. Cunnane, S.C., C.R. Menard, S.S. Likhodii, J.T. Brenna, and M.A. Crawford, Carbon Recycling into *de novo* Lipogenesis is a Major Pathway in Neonatal Metabolism of Linoleate and α-Linolenate, *Prostagl. Leukotri. Essential Fatty Acids 60*:387–392 (1999).

74. Wijendran, V., P. Lawrence, G.Y. Diau, G. Boehm, P.W. Nathanielsz, and J.T. Brenna, Significant Utilization of Dietary Arachidonic Acid Is for Brain Adrenic Acid in Baboon Neonates, *J. Lipid Res. 43*:762–767 (2002).

75. Hirsch, J., J.W. Farquhar, E.H. Ahrens, Jr., M.L. Peterson, and W. Stoffel, Studies of Adipose Tissue in Man: A Microtechnic for Sampling and Analysis, *Am. J. Clin. Nutr. 8*:499–511 (1960).

76. Sarda, P., G. Lepage, C.C. Roy, and P. Chessex, Storage of Medium Chain Triglycerides in Adipose Tissue of Orally Fed Infants, *Am. J. Clin. Nutr. 45*:399–405 (1987).

77. Farquharson, J., E.C. Jamieson, R.W. Logan, W.J. Patrick, A.G. Howatson, and F. Cockburn, Age- and Diet-Related Distributions of Hepatic Arachidonic and Docosahexaenoic Acid in Early Infancy, *Pediatr. Res. 38*:361–365 (1995).

78. Farquharson, J., F. Cockburn, W.A. Patrick, E.C. Jamieson, and R.W. Logan, Effect of Diet on Infant Subcutaneous Tissue Triglyceride Fatty Acids, *Arch. Dis. Child. 69*:589–593 (1993).

79. Cunnane, S.C., and M.J. Anderson, The Majority of Linoleate in the Rat is β-Oxidized or Stored in Visceral Fat, *J. Nutr. 127*:146–152 (1997).

80. Palowsky, R.J., J.R. Hibbeln, J.A. Novotny, and N. Salem, Jr., Physiological Compartmental Analysis of α-Linolenic Acid Metabolism in Adult Humans, *J. Lipid Res. 42*:1257–1265 (2001).

81. McCloy, U., Metabolism of ^{13}C Unsaturated Fatty Acids in Healthy Women, Ph.D. Thesis, University of Toronto, Toronto, 2002.

82. Holliday, M.A., Metabolic Rate and Organ Size During Growth from Infancy to Maturity and During Late Gestation and Early Infancy, *Pediatrics 47 (Suppl. 1)*:169–181 (1971).

83. Widdowson, E.M., Changes in Body Proportion and Composition During Growth, in *Scientific Foundations of Pediatrics*, edited by J.A. Davies and J. Dobbing, Heinemann, London, 1974, pp. 153–163.

Chapter 9

Flaxseed, Lignans, and Cancer

Lilian U. Thompson

Department of Nutritional Sciences, Faculty of Medicine, Toronto Ontario, M5S 3E2, Canada

Introduction

Flaxseed (FS) is an oilseed with exceptionally high concentration of the n-3 fatty acid α-linolenic acid (ALA; 53% of total fatty acids) and the phytoestrogen lignans (75–800 times that of 66 other plant foods) (1–3). Because these substances have been suggested to have anticancer effects (4,5), FS has been hypothesized to reduce the risk of cancer. This chapter describes the epidemiological and experimental studies, in animals and humans, on the effect of FS on carcinogenesis. The focus is on the effect of lignans, because the role of ALA on carcinogenesis has been discussed by others (6; Chapter 11).

Several plant lignans have been identified in FS including secoisolariciresinol diglycoside (SDG), matairesinol, isolariciresinol, and pinoresinol, but the major one is SDG (7,8). As previously reviewed (9; Chapter 4), ingested plant lignans are metabolized by the bacterial flora in the colon to the major mammalian lignans enterodiol (ED) and enterolactone (EL). They undergo enterohepatic circulation and some are excreted in the urine, mainly as glucuronide conjugates. Urinary excretion and blood levels of mammalian lignans in humans or animals correlate directly with the intake of plant food lignans (10,11). Hence, they have been used as markers of lignan intake or exposure in epidemiological and experimental studies. The similarity of ED and EL structures to the natural estrogen, 17β-estradiol, suggests that these lignans may have weak estrogenic/antiestrogenic properties and hence anticancer effects. However, evidence regarding these properties is still inconclusive (12; Chapter 10).

Epidemiological Studies

Breast Cancer

The relationship between the intake of lignan-rich FS and cancer risk has not been studied, because FS has not been used extensively for human consumption until recently. However, in the 1980s, Adlercreutz *et al.* (4) observed that urinary lignan excretion was lower in breast cancer patients and omnivores than in vegetarians, who have lower risk of breast cancer, suggesting that the lignans may have protective effects against breast cancer. Since then, several epidemiological studies have been conducted to determine the relationship between breast cancer risk and the urinary excretion or plasma levels of mammalian lignans, or intake of plant lignans. These are outlined in Table 9.1.

Three consecutive urine samples were collected from premenopausal and postmenopausal Australian women with newly diagnosed breast cancer and from matched controls, and analyzed for EL (13). After adjustments for confounding variables (age at menarche, parity, alcohol intake, and total fat intake), the urinary EL odds ratios showed a threefold reduction in risk of breast cancer for subjects with the highest *vs.* the lowest quartile of EL excretion. Separate analysis of premenopausal and postmenopausal women resulted in analogous trends.

In a case-control study from eastern Finland, single serum samples from 194 breast cancer patients and 208 controls were analyzed for EL (14). The mean serum EL concentration was lower in cancer cases (20 nM) than in controls (26 nM) ($P = 0.003$). The adjusted odds ratios showed about a threefold lower risk in the highest *vs.* the lowest quintile of serum EL. The mean serum EL concentration was 3.0 nM in the lowest and 54.0 nM in the highest quintile. The negative association between breast cancer risk and serum EL was observed among both premenopausal and postmenopausal women, but the relationship did not reach significance due to the reduced number of cases and controls upon subject subdivision.

The above studies were done in populations with low soy intake, but a recent Shanghai study (15) analyzed a population with high soyfood consumption. Soy is the richest source of the other major phytoestrogens, the isoflavones genistein and daidzein. Urine samples from 250 women with newly diagnosed breast cancer and 250 matched controls were collected after the initial cancer diagnosis and before cancer therapy, and analyzed for lignans. The median excretion rate of total lignans (nmol/mg creatinine) was significantly lower in cases (1.77) than in controls (4.16). The breast cancer risk was reduced as the excretion of total lignans increased, $P <$ 0.01 for trend. An adjusted odds ratio of 0.40 was observed for the highest *vs.* the lowest tertile of total lignan (i.e., ED + EL). The effects of lignans are more pronounced among premenopausal than postmenopausal women, suggesting that the effect of lignans may depend on factors related to endogenous hormone levels.

A weakness of the above retrospective studies, however, is the short urine collection time, and the single blood sampling time, which are assumed to represent the patients' lifetime lignan exposure. The stressful period after breast cancer diagnosis may also have caused changes in the patients' appetite and food selection, which then were reflected in the urine or blood lignan data. Because breast cancer takes a number of years to develop, it is more accurate to conduct a prospective study in which the development of breast cancer and the intake of lignan are monitored. Hence, in a study in the Netherlands (16), EL was analyzed in the urine collected from postmenopausal subjects 1–9 years before breast cancer was diagnosed. Two overnight urine samples with a time interval of about 1 yr were analyzed from 88 cases and 268 controls. The median urinary EL/creatinine (μmol/mol) of 576.2 for cases and 565.6 for controls did not differ significantly. The higher EL excretion was not significantly associated with increased breast cancer risk, although the urinary EL excretion (3280

TABLE 9.1
Epidemiological Studies on Lignans and Breast Cancer

Study design/site	Subjects/sampling	Group/lignan	Median levels		P	Adjusted OR (95% CI), P for trend	Reference
			Cases	Control			
Case-control Australia	144 cases, 144 controls; hospital-based cases, population control; one 72-h urine sample		nmol/24 h				13
		All/ED	282	316.5	—	0.73 (0.33–1.64) 0.288[a]	
		All/EL	1973.4	3097.7	—	0.36 (0.15–0.86) 0.013[a]	
Case-control Finland	194 cases, 208 controls; hospital-based cases, population control; one serum sample		nmol/L				14
		All/EL	20	26	0.003	0.38 (0.18–0.77) 0.03[b]	
		Premen./EL	17	21	0.10	0.42 (0.10–1.77) 0.18[b]	
		Postmen./EL	21	29	0.01	0.52 (0.19–1.28) 0.10[b]	
Case-control Shanghai	250 cases, 250 controls; population based; urine samples before cancer therapy		μmol/mg Cr				15
		All/ED + EL	1.77	4.16	<0.01		
		All/EL	1.34	3.40	<0.01		
		All/ED	0.19	0.31	<0.01		
		All/ED + EL				0.40 (0.24–0.64) <.01[c]	
		Premen./ED + EL				0.24 (0.12–0.50) <.01[c]	
		Postmen./ED + EL				0.62 (0.31–1.26) 0.16[c]	

Nested case-control Netherlands	88 cases, 268 controls; urine collected 1–9 yr before breast cancer	All/EL	µmol/mol Cr 565.6	576.2	0.81	1.43 (0.79–2.59) 0.25[c]	16
Nested case-referent Sweden	248 cases, 492 referents; baseline plasma sample	All/EL	nmol/L 19.3–26.8	22.9–20.4		1.6 (1.0–2.6)[d] 1.8 (1.1–2.8)[b]	17
Case-control San Francisco	1272 cases, 1610 controls; food frequency questionnaire	All/SECO+MAT	µg/d			1.3 (1.0–1.6) n.s.[e]	19
Case-control Western New York	301 cases, 316 controls, premen. Caucasian women; 439 cases, 494 controls, postmen. Caucasian women; food frequency questionnaire	ED + EL, mg/d Premen./ED + EL Postmen./ED + EL	0.35–0.64 0.57–0.77	0.51–0.67 0.67–0.76		0.49 (0.32–0.75) <0.05[f] 0.72 (0.51–1.02) <0.05[f]	21

Abbreviations: Cr, creatinine; ED, enterodiol; EL, enterolactone; MAT, matairesinol; n.s., not significant; Postmen., postmenopausal; Premen., premenopausal; SECO, secoisolariciresinol. [a]highest quartile of excretion, [b]highest quintile of plasma, [c]highest tertile of excretion, [d]lowest quintile of plasma, [e]highest quartile of intake, [f]highest tertile of intake.

nmol/24 h) in this study was comparable to those observed in the Australian study (3098 nmol/24 h) (13) among the controls.

Baseline blood samples were collected from subjects (248 cases and 492 referents) in three ongoing cohort studies in northern Sweden (The Vasterbotten Intervention Project; Swedish component of the World Health Organization's multinational study, Monitoring of Trends and Cardiovascular Disease; Mammary Screening Project) and analyzed for lignans (17). It is of interest that both low plasma EL (mean 2.9 nmol/L) and high plasma EL (mean 58.2 nmol/L) levels were significantly associated with increased breast cancer risk. The discrepancy between this study and that of the Finnish study, which showed lower risk with high serum EL levels (14), was suggested to be related to the higher intake of alcohol in Sweden. Alcohol from 0.5–1 drink per day has been shown to increase plasma EL levels by 131% (18).

With the rationale that urine or blood analysis for lignans may not reflect the period of cancer development or preclinical progression, a study in the San Francisco Bay area (U.S.A.) determined the phytoestrogen-breast cancer relationship in a multiethnic group of premenopausal and postmenopausal women including non-Asian, African American, Latina, and Caucasion women, using a food frequency questionnaire and a newly developed phytoestrogen database to estimate their usual phytoestrogen intake (19). No significant association between phytoestrogen intake and breast cancer risk was observed, and this conclusion did not vary substantially by race/ethnicity or menopausal status of the women. The average total phytoestrogen consumption by cases (3.174 mg/d) did not differ significantly from that of control (3.326 mg/d), with 87–88% as isoflavones and the rest as lignans (secoisolariciresinol and matairesinol) and coumestans. The lack of association may be related to the low level of phytoestrogen intake (3 mg/d) in the population in comparison with the average intake in Asian countries (about 15–30 mg/d). Because the ethnic groups did not differ in relative risk, the phytoestrogen intake did not account for the lower risk of breast cancer reported among Latina women (20).

In contrast to the above study (19), reduced risk of breast cancer in postmenopausal and premenopausal Caucasian women in western New York was significantly associated with high total lignan intake, also estimated from a food frequency questionaire (21). The lignan intakes ranged from 0.06–2.48 mg/d. Because the association of high urinary lignan with reduced breast cancer risk may be related to competitive inhibition of endogenous estrogens and because there is evidence that reproductive risk factors from breast cancer differ according to cytochrome P450c17α (CYP17) genotype (22), it was hypothesized that genetic variability in estrogen metabolism could affect lignan metabolism and therefore breast cancer risk (21). The A2 allele variant of CYP17 has been associated with elevated serum estrogen and progesterone levels (22). Women with at least one A2 allele had twice the risk of cancer compared with women without it (23). Indeed, in a subsample of the Caucasian women, the investigators observed a substantially greater reduction in breast cancer risk among premenopausal women with at least one A2 allele and in the highest tertile of lignans. However, the CYP17 genotype has lesser effect in postmenopausal women (21). Although these associations are interesting, this study did not measure the plasma levels of sex hormones or lignans to further prove the relationships.

Prostate Cancer

The relationship of circulating EL to prostate cancer risk has been examined in a longitudinal, nested case-control study linking three biobanks to the cancer registries in Finland, Norway, and Sweden (24; Table 9.2). A single measurement of EL concentration in serum samples collected an average of 14 yr before prostate cancer diagnosis showed no significant association with prostate cancer risk. This has been attributed in part to the low serum EL levels in the subjects, although evaluation of the very high levels of serum EL by analysis of the upper octile still showed no protective effect. The median serum EL level was 8.4 nmol/L for combined case and control subjects, with more than two-thirds of the subjects with serum or plasma levels below 15 nmol/L attributable to the low levels in the Norwegian group. The Finnish and Swedish groups had median EL concentrations of 15.6 and 13.6 nmol/L, respectively. Some of the Norwegian blood samples had been stored for more than 20 yr, and EL degradation may be a reason for their low EL levels. However, the cases and controls were matched for storage time so the differences between these two groups cannot be attributed to any degradation during storage.

Thyroid Cancer

It has been suggested that thyroid cancer is an estrogen-dependent disease (25) and could therefore be affected by the intake of phytoestrogens. Therefore, the association of thyroid cancer with phytoestrogen intake, based on a food frequency questionnaire during the year before cancer diagnosis, was detemined in a case-control study of a multiethnic population in the San Francisco Bay area (26; Table 9.2). The intake of genistein (which represents about half of total phytoestrogen consumption), daidzein, and the lignan secoisolariciresinol (2% of total consumption) was associated with a reduction in risk of thyroid cancer. The findings were similar for white and Asian women and for premenopausal and postmenoapusal women. These findings were weakened after adjustments for dietary and nondietary factors. However, the intake of isoflavones and lignans by this population was very low, with daily median intake of about 2 mg and 0.1 mg, respectively. Hence, further studies need to be done to confirm the association of thyroid cancer with the intake of phytoestrogens, particularly the lignans.

Experimental Studies

Breast Cancer

The carcinogenesis process generally consists of several stages: initiation, promotion, progression, and metastasis (Fig. 9.1). Because FS and its lignans may influ-

TABLE 9.2

Epidemiological Studies on Lignans and Prostate and Thyroid Cancer

Study/design/site	Subjects/sampling	Group	Median levels		P	Adjusted OR (95% CI), P for trend	Reference
			Cases	Control			
Prostate cancer							
Nested case-control Finland, Norway, Sweden	794 cases; 2550 controls; serum collected average 14 yr before cancer	EL	nmol/L 8.4	8.5		1.08 (0.83–1.39) n.s.[a]	24
Thyroid cancer			mg/d				
Case-control San Francisco Bay area	608 cases, 558 controls; multiethnic group; food frequency questionnaire	Total lignans MAT SECO	— — —	— — —		0.68 (0.43–1.1) 0.07[b] 0.72 (0.46–1.1) 0.49[b] 0.56 (0.35–0.89) 0.009[b]	26

Abbreviations: See Table 9.1. [a]highest quartile of plasma, [b]highest quintile of intake.

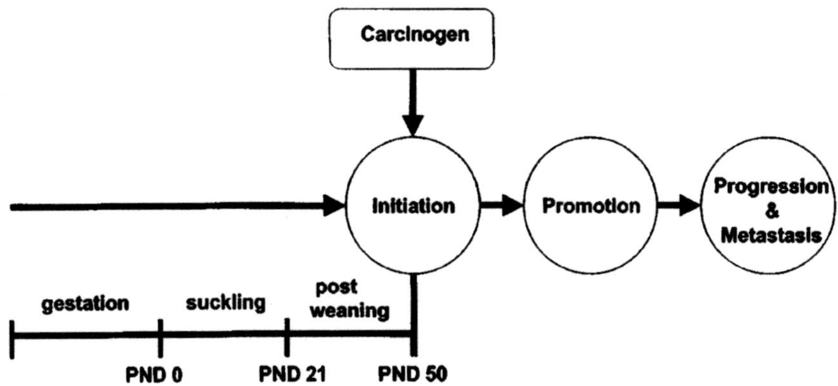

Fig. 1.9. Stages of carcinogenesis where flaxseed and lignans have been tested. Abbreviations: PND, postnatal day.

ence specific stages of the cancer process differently, our laboratory and those of others have been systematically studying their effects at the various stages of carcinogenesis. The effect of early exposure to FS and lignans, that is, during gestation, suckling, and post weaning (puberty), when mammary glands are developing, is also being investigated because it may also influence breast cancer risk later in life. These studies are summarized in Table 9.3.

Effect of Flaxseed and SDG: Carcinogen-Treated Rat Model. When 5 or 10% FS or defatted FS was fed to rats at the preinitiation stage, that is, for 4 wk before injection with the carcinogen dimethylbenzanthracene (DMBA) at 50 d old, significant reductions in cell proliferation and nuclear aberration were observed compared with the high fat (20% corn oil) basal diet control (BD) (27). A significant negative relationship between urinary lignan excretion and the cell nuclear aberration indicates that the lignans may in part be responsible for this effect. Because highly proliferating cells are more sensitive to the action of carcinogens and because nuclear aberrations are lethal events that have been associated with carcinogen exposure, the results indicate that FS may have a protective effect against breast cancer. Indeed, when rats were fed 5% FS at either the preinitiation stage (4 wk before DMBA injection) or the early promotion stage of carcinogenesis (for 20 wk after DMBA injection), a 21–32% reduction in tumor incidence was observed (28). The group fed the 5% FS at the preinitiation stage only or throughout the study had fewer tumors than the group fed the BD control. In contrast, the group fed 5% FS at the early promotion stage only or throughout the study had the smallest tumor size, with the former group significantly different from the control. This suggests that fewer tumors were initiated with FS treatment. The results are in agreement with the earlier observed reduction in cell proliferation and nuclear aberration in the short-term study (27); they also show that the subsequent removal of FS after carcinogen treatment led to enhanced

TABLE 9.3
Experimental Studies on Flaxseed, Lignans, and Breast Cancer

Model system	Treatment	Observations	Ref.
Sprague-Dawley rats	5 or 10% FS or defatted FS; fed at 21–50 d old; inject DMBA; test after 24 h	↓ Nuclear aberration; ↓ cell proliferation	27
Sprague-Dawley rats	5% FS or BD fed at either preinitiation stage (21–50 d before DMBA), or early promotion stage (1–21 wk post DMBA), or continuously	↓ Tumor incidence and number with FS at preinitiation stage and continuously; ↓ Tumor size with FS at promotion stage	28
Sprague-Dawley rats	1.5 mg SDG/d fed at early promotion stage (1–20 wk post DMBA)	↓ Number of tumors/group, multiplicity	29
Sprague-Dawley rats	2.5 or 5% FS, 1.82% FO, or 1.5 mg SDG/d at late promotion/early progression stage (13–20 wk post DMBA)	↓ Established tumor volume with FS, FO, and SDG; ↓ New tumor volume, number new tumors/group, and new tumor incidence	30
Sprague-Dawley rats	2.5 or 5% FS or equivalent amount of SDG (LSDG = 0.7 mg/d or HSDG = 1.4 mg/d) fed at promotion stage (1–22 wk post MNU)	Tumor multiplicity ↓ with HSDG, ↑ with LSDG; ↓ tumor invasiveness and grade with FS or SDG; trend to ↓ tumor volume with FS	31
Sprague-Dawley rats	Fed HMR 15 mg/kg b.w./d for 51 d starting 9 wk post DMBA	↓ Established tumor growth; ↓ New tumor number; HMR metabolized to EL	32
Sprague-Dawley rats	Fed HMR 4.7 mg/kg b.w./d starting either 1 wk before DMBA, or 9 wk post DMBA	↓ Established tumor growth; no effect on tumor multiplicity	33
Sprague-Dawley rats	Gavaged EL daily 1 mg/kg b.w. or 10 mg/kg b.w. for 7 wk starting 9 wk post DMBA	Both doses ↓ established tumor growth first 4 wk; at 7 wk, low-dose group tumor size similar to control, high-dose group tumor size lower than control	34

(Continued)

TABLE 9.3
(*Cont.*)

Model system	Treatment	Observations	Ref.
Athymic mice NCR nu/nu	Injected with human cancer cells (ER-negative MDA MB 435); fed 10% FS or BD for 7 wk starting 8 wk post cancer cell injection	↓ Tumor growth; ↓ tumor metastasis incidence; ↓ tumor cell proliferation, EGFR, IGF-1, VEGF	35
Athymic mice NCR nu/nu	Injected with human cancer cells (ER-negative MDA MB 435); fed 10% FS or BD for 7 wk starting 8 wk after cancer cell injection	↓ Tumor metastasis incidence; ↓ extracellular VEGF	36
Athymic mice NCR nu/nu	Injected with human cancer cells (ER-negative MDA MB 435); fed 10% FS, SDG, FO, SDG + FO, or BD for 7 wk starting 8 wk after cancer cell injection	↓ Tumor growth and/or metastasis incidence with FS, SDG, FO, or SDG + FO; SDG and FO together better than alone	38
Athymic mice Balb/c, Ovariectomized	Injected with human cancer cells (ER-positive MCF-7); fed BD or 10% FS with or without estrogen (E2) when tumors are established	With E2, FS ↓ tumor growth; without E2, FS did not increase tumor growth	Thompson, L.U. unpublished data
Sprague-Dawley rats	Dams fed 5 or 10% FS or SDG equivalent to 10% FS only during pregnancy and lactation; at PND 50, offspring mammary gland examined	↓ Terminal end buds, alveolar buds	40
Sprague-Dawley rats	Dams fed 10% FS or SDG equivalent to 10% FS only during lactation; offspring fed same only at post weaning; at PND 50, offspring mammary gland examined	Offspring suckling exposure ↓ terminal end buds, ↑ alveolar buds; exposure post weaning, no effect	41

(*Continued*)

TABLE 9.3
(*Cont.*)

Model system	Treatment	Observations	Ref.
Sprague-Dawley rats	Dams fed 10% FS or SDG equivalent to 10% FS only during lactation; at PND 21 and 50, offspring mammary gland examined	Offspring suckling exposure; at PND 21, ↑ terminal end buds , ↓ alveolar buds; at PND 50, ↓ terminal end buds, ↑ lobules	42
Sprague-Dawley rats	Dams fed 10% FS during lactation; offspring injected with DMBA at PND 50; fed BD post weaning until PND 200	↓ Tumor incidence, tumor number, tumor size	43
Postmenopausal breast cancer patients	Double-blind placebo controlled randomized design; fed muffins with 25 g flaxseed or placebo muffin, mean 39 d	↓ Tumor cell proliferation; ↓ tumor CerB2 labeling index; ↑ tumor cell apoptosis	44

Abbreviations: BD, basal diet; b.w., body weight; DMBA, dimethylbenzanthracene; EGFR, epidermal growth factor receptor; EL, enterolactone; ER, estrogen receptor; E2, estradiol; FO, flaxseed oil; FS, flaxseed; HMR, hydroxymatairesinol; HSDG, high amount of SDG; IGF, insulin-like growth factor; LSDG, low amount of SDG; MNU, *N*-methyl-*N*-nitrosourea; PND, postnatal day; SDG, secoisolariciresinol diglycoside; VEGF, vascular endothelial growth factor.

tumor growth. However, feeding FS after carcinogen treatment reduced the growth of the tumors that were already initiated.

To establish whether the effect of FS on mammary carcinogenesis was due to its lignans, SDG was isolated from FS (10) and fed to rats starting 1 wk after DMBA induction (29). A 46% reduction in the number of tumors per rat in the SDG group was observed. Although the tumor incidence and volume were lower in the SDG group than in the control, the difference did not reach significance. These results differed from those of the earlier study, which showed a reduction in tumor volume upon feeding 5% FS (28). This could be related to differences in the feeding regimen used before carcinogen administration. Whereas in the previous FS study (28), 21-d-old weanling rats were fed a high-fat BD until DMBA administration at 50 d, this study commenced feeding the high-fat BD when the rats were 43 d old. The consumption of the BD when the mammary gland is actively differentiating to its various structures may impact future tumor formation.

The effect of feeding the BD or BD supplemented with either 2.5% FS, 5% FS, SDG, or flaxseed oil (FO) at levels equivalent to the amount present in 5% FS was also tested at the late tumor promotion stage (i.e., starting 9 wk after injection with DMBA when tumors are already established) (30). After 7 wk of treatment, the established tumor volume was reduced in size by more than 50% in all treatment groups, whereas there was no change in the BD group. The SDG group developed the lowest number of new tumors. Moreover, new tumor volume was also lower in the SDG group compared with all other groups. Hence, the combined established and new tumor volumes were smaller in the SDG and the 2.5 and 5% FS groups compared with the FO and BD groups. There was a significant negative relationship between urinary lignan excretion and the established tumor volumes in the BD, SDG, 2.5 and 5% FS groups, indicating that the reduction in tumor size was in part due to the lignans derived from SDG in FS. However, the fact that FO also caused reduction in established tumor volume suggests that it may contribute as well to the tumor inhibitory effect of FS, perhaps through mechanisms involving its high ALA content.

In the above two studies (29,30), only one level of SDG (1.5 mg/d; equivalent to the level in 5% FS) was tested using the indirect acting carcinogen DMBA. Hence, investigation was made of the effect of feeding a high fat (20%) BD supplemented with 2.5 or 5% FS or SDG equivalent to the amount in 2.5% FS (0.7 mg/d; low SDG, LSDG) and 5% FS (1.4 mg/d; high SDG, HSDG) at the early promotion stage, that is 2 ds after injection with the direct-acting carcinogen *N*-methyl-*N*-nitrosourea (MNU) (31). The tumor size tended to be smallest in the 5% FS, but the FS diets did not significantly affect the tumor size, multiplicity, or incidence compared with the BD. The tumor multiplicity was highest in the LSDG group and lowest in the HSDG group suggesting that HSDG inhibited whereas LSDG promoted the tumor development. However, regardless of dose, the FS- and SDG-supplemented diets reduced the tumor invasiveness and grade, indicating that they delay the progression of the MNU-induced mammary tumorigenesis. Disparities in the results between this study and previous studies (28–30) have been related to differences in the experimental design including the use and amount of different carcinogens (5 mg DMBA/rat *vs.* about 9 mg MNU/rat) and the protective effect of the higher amount of ALA (1.60%) in the BD in this study, which used 20% soybean oil *vs.* 0.24% in the previous studies (28–30), which used 20% corn oil. The carcinogen level may be too high to see the subtle protective effect of FS, and the protective effect of the high level of ALA in the BD may have masked the inhibitory effect of FS on tumor development. Although in most cases, the effect of FS and SDG appears to be protective, it is evident that the nature of their effect is dependent on the experimental design.

Effect of Enterolactone: Carcinogen-Treated Rat Model. Matairesinol is a known precursor of EL (9). Recently, hydroxymatairesinol (HMR) isolated from Norway spruce (*Picea abios*) was also found to be metabolized primarily to EL (32). Hence, it has been used as a model to determine whether EL has anticancer effects. Rats gavaged with HMR, at a level of 15 mg/kg b.w. for 51 d starting 9 wk after DMBA induction, reduced the growth of established tumors and increased the proportion of regressing and stabilized tumors compared with the control (32). New tumor formation was also suppressed. When HMR was provided at 4.7 mg/kg b.w. starting 1 wk before DMBA induction, smaller tumor size was observed compared with the control, with a maximum reduction of 42% observed 13 wk after DMBA administration (33). A similar effect was observed when it was provided starting 9 wk after DMBA, with maximum reduction of 42% seen also at 13 wk post DMBA (after 4 wk HMR treatment). However, at the end of the study (i.e., 17 wk after DMBA), the tumor size with the longer HMR treatment was 22% smaller than the control and the tumors with the shorter treatment did not differ from the control.

Similar to the effect of HMR and SDG, daily gavage of EL, at 10 mg/kg b.w. for 7 wk starting 9 wk after DMBA induction, significantly reduced the growth of established rat tumors but reduced the growth of the new tumors that developed during the treatment period more (34). It increased the proportion of nongrowing (regressing, stabilized, or totally disappeared) tumors and decreased the proportion of the growing tumors. Tumor size reduction was also observed with daily gavage of 1 mg/kg b.w., but the difference did not reach statistical significance (34). The percentage tumor size reduction seen after treatment with HMR at 15 mg/kg b.w. (32) was close to that with EL at 10 mg/kg b.w. (34). Hence, the effect of HMR can be largely attributed to its EL metabolite.

Although the above studies provide direct evidence that the mammalian lignan EL and its precursors HMR and SDG have inhibitory effect on tumor growth, the optimum concentration of lignans for growth inhibition has yet to be established. The serum EL level was significantly increased by the effective 10 mg/kg b.w. EL dose but not by the 1 mg/kg b.w. dose, compared with the control (34). The serum EL level was 405 nM with the 10 mg/kg b.w., and 351 nM with the 4.7 mg/kg b.w. HMR treatment (33). These levels are much higher than the mean levels observed in breast cancer patients (20 nM) and controls (26 nM) in epidemiological studies (14). Perhaps the EL levels observed in population studies are just markers of a healthy diet rather than the actual EL level that can reduce tumor growth.

Effect of Flaxseed and SDG: Human Breast Cancer in the Athymic Mouse Model. All the above studies on the effect of FS, SDG, HMR, or EL on preinitiation, early, and/or late promotion stages of carcinogenesis have used carcinogen-treated rats, and it is unclear whether FS or its lignans can also influence the growth of human tumor cells. Also, tumors from carcinogen-treated rats do not readily metastasize. Hence, the growth and metastasis of the estrogen receptor (ER)-negative human breast cancer cells MDA MB 435 were investigated using the athymic mouse model (35), immunodeficient mice that can accept human tissues. The MDA MB 435 cells were tested because they are known to metastasize in the animal model, whereas the ER-positive cells such as MCF-7 cells do not. The cells were injected at the mammary fat pads and 8 weeks later, when tumors were already established, the BD or 10% FS was fed for 7 wk. Significant reductions in established tumor growth rate and 45% reduction in total incidence of metastasis in the FS-fed group were observed. Lung metastasis incidence was 55.6% in the BD *vs.* 22.2% in the FS group, and lymph node metastasis incidence was 88.9% in the BD group *vs.* 33.3% in the FS group. Metastatic lung tumor number was also lower by 82% in the FS group. Reductions in cell proliferation (as Ki67 labeling index) and expression of insulin-like growth factor-1 and epidermal growth factor receptor by FS were observed, suggesting that the effect of FS was in part due to the down regulation of these growth factors. A similar reduction in tumor growth rate and metastasis incidence (70% in the BD group *vs.* 10% in the 10% FS group) was also observed in another experiment with athymic mice using a similar experimental design (36). The level of vascular endothelial growth factor (VEGF) in the extracellular fluid was lower in the tumors, particularly the large tumors, in animals exposed to 10% FS compared with those in the BD group. Angiogenesis, the generation of new capillaries from preexisting vessels, is an important event in tumor growth, progression, and metastasis, and VEGF is known to be a potent stimulator of angiogenesis (37). Hence, the study suggests that the slower tumor growth rate and lower metastasis caused by FS may also be attributed to its effect on tumor angiogenesis.

To determine whether the lignan and/or the oil components are responsible for the inhibitory effect of FS on human tumor growth and metastasis, two studies were again conducted in athymic mice (38). As before, mice were injected with MDA MB 435 cells. In the first study, 8 wk after tumor cell injection when the tumors were already established, the mice were randomized into five groups such that the mean tumor sizes were similar. They were then fed either the BD, or BD supplemented with 10% FS, SDG, FO, or a mixture of SDG and FO. The SDG and FO levels were equivalent to the amount present in the 10% FS diet. Compared with the BD control, the primary tumor growth rate was significantly reduced in the FS, FO, and SDG + FO groups but not in the SDG group. Metastasis incidence in the lungs, distant lymph nodes, and other organs was reduced in all the treatment groups *vs.* the control, but significant difference was observed only with the FS and SDG + FO groups in lung metastasis, and with the FO group in lymph node metastasis. Metastasis incidence in other organs was significantly reduced by all treatments. Considering total metastasis, the greatest effect was seen in the SDG + FO group.

In the second study, the study design was the same as that in the first study except that, 8 wk after tumor cell injection, the tumors were excised prior to randomization of the mice to the five different diets (38). Again, all treatment diets reduced the metastasis incidence, with the treatment groups not differing significantly from each other. However, compared with the control, statistical significance was reached only by the difference in lung metastasis in the SDG + FO group, lymph node metastasis in the FS, FO, and SDG + FO groups, other organ metastasis in the FO group, and total metastasis in the FS, FO, and SDG + FO, groups.

The above two studies suggest that both SDG and FO can reduce human tumor growth and metastasis, but they have a greater protective effect in combination than individually. Because SDG in combination with FO, in most cases, had an effect similar to that of FS, the effect of FS on ER-negative human breast cancer cells may be attributed to both the SDG and FO.

It was hypothesized that the high amount of ALA in FS or FO may lead to increased lipid peroxidation in the primary tumors, which may be responsible for the reduced tumor growth and metastasis (38). However, no significant relationship was observed between the tumor growth and malonaldehyde concentration in the tumors, thus indicating that lipid peroxidation is not the main mechanism involved in this study.

Although FS may have a protective effect against ER-negative human breast cancer growth and metastasis, it was not known whether it would produce the same effect on ER-positive human breast cancer cells particularly in the presence of high or low levels of estrogen. If the lignans exert their effect through mechanisms involving the ER, then a different response might be expected in ER-positive cancer cells in the presence or absence of estrogen. Experiments were thus conducted where ovariectomized athymic mice were injected with the ER-positive MCF-7 cells into mammary fat pads, implanted with estradiol pellet (E2; 1.7 mg, 60-d release), and fed BD (Thompson, L.U., unpublished data). When the tumors became established (reached about 40 mm^2 size), the E2 pellet was removed and the mice were randomized to either the BD or 10% FS diet. In some mice, the E2 pellets were replaced, and mice were fed either the 10% FS diet or the BD. During the 6-wk treatment, in the absence of added E2 to simulate postmenopausal hormone conditions, the 10% FS did not increase the established tumor growth, but rather decreased the tumor size to the same level as the negative control. Conversely, in the presence of E2 to simulate premenopausal conditions, the established tumor continued to grow but at a rate significantly slower than that of the positive control (Thompson, L.U., unpublished data). Therefore, FS appears to have no estrogenic effect at low levels of circulating estrogen but is able to counteract the tumor-promoting effect of high levels of estrogen.

Effect of Flaxseed and SDG: Early Exposure and Carcinogenesis. A reduction in the number of the highly proliferative terminal end bud (TEB) structures in the developing mammary gland through differentiation to the less proliferative alveolar buds (AB) and lobules has been suggested to be protective against mammary cancer (39). Because estrogens are known mediators of mammary gland development and differentiation, the potential role of early exposure to phytoestrogens, such as the lignans in FS, on mammary gland development and differentiation and hence breast cancer risk, is of recent interest.

Rat offspring were exposed to 5 or 10% FS during either gestation and suckling, after weaning from postnatal day (PND) 21–50, or lifetime, that is, from gestation to PND 50 (40). Exposure to 5 or 10% FS during gestation and suckling or throughout lifetime significantly reduced the TEB density in the mammary gland, whereas exposure after weaning had no effect. This suggests that the reduction in TEB occurred during gestation and suckling and not after weaning. When SDG or FO were fed at levels equivalent to the amount found in the 5% FS diet during the same period as the FS, the same effect as that found in the 5% FS diet was observed with SDG but not with FO. Therefore, the effect of FS was primarily due to the SDG that it contains.

Because the reduction in TEB was observed when the rats were exposed to FS during gestation and suckling only, it was not clear whether the effect seen occurred during the gestation or suckling period. A repeat of the experiment was therefore conducted where the rats were exposed to 10% FS or SDG equivalent to the amount present in 10% FS only during gestation or gestation to PND 50 (41). Reductions in the number of TEB with corresponding increase in the number of AB, indicating enhanced differentiation, were observed in the FS- and SDG-treated groups compared with those fed the BD. However, continuous exposure to SDG from gestation to PND 50 resulted in the most differentiation of the mammary gland structure. Because the effect of SDG on TEB and AB did not differ from that of the FS-treated group, the effect of FS was again attributed to the SDG that it contains. The reduction in TEB density in this study (41) was similar to that observed upon exposure to FS during gestation and suckling (40). This indicates that the critical period for enhancing the mammary gland differentiation is during suckling and not during gestation.

To determine whether the reduction in the number of TEB at PND 50 seen in the above experiment will indeed reduce the risk of carcinogenesis in the offspring upon exposure to carcinogen, the suckling exposure to 10% FS or SDG was repeated in another experiment with offspring treated with the carcinogen DMBA when they reached PND 50 (42,43). The reduction in TEB with SDG or FS treatment was again observed (43). More important, the FS and SDG diets reduced the tumor incidence (by 31.3 and 42%, respectively), total tumor load (by 50.8 and 62.5%, respectively), and mean tumor size (by 43.9 and 67.7%, respectively) compared with the BD control (43) at 21 wk post-DMBA treatment. The FS group did not differ significantly from the SDG group. Hence, this study provided the first experimental evidence that exposure to FS during an early stage of mammary gland development (i.e., suckling) can enhance the mammary gland differentiation, which then can reduce the risk of breast cancer at adulthood, and that this effect is mediated primarily by the lignans.

Effect of Flaxseed: Clinical Study. Encouraged by the results in the animal models, which showed the ability of FS and lignans to reduce established tumor growth (30,32–36), Thompson *et al.* (44) conducted a randomized double-blind, placebo-controlled, prospective clinical trial to determine the effect of FS in patients with newly diagnosed breast cancer. Patients were randomized to consume either a muffin containing 25 g FS or a placebo control muffin once daily from the time of diagnosis to time of surgery (mean duration = 39 d). There was a significant increase in urinary lignan excretion, a marker of lignan intake, in the FS-muffin group. FS significantly reduced the tumor cell proliferation (measured as Ki67 labeling index) and c-erB-2 expression, and increased the apoptosis index when compared with the control, suggesting a beneficial effect. These results agreed with the observations in animal studies (28–30,35,36,38). Although feeding SDG in clinical trials still needs to be evaluated, based on the results in animal studies, the effect of FS on breast cancer patients may in part be due to its lignans.

Prostate Cancer

Studies on flaxseed, lignans, and prostate cancer are summarized in Table 9.4.

Effect of Lignans: Observational Study. The incidence of prostate cancer is lower in populations whose diets are Asian or Mediterranean than in those whose diets

L.U Thompson

TABLE 9.4
Experimental Studies on Flaxseed, Lignans, and Prostate Cancer

Model system	Treatment	Observations	Ref.

50 Portuguese
53 Hong Kong
  Chinese
36 British
	Measured EL, ED, daidzein,
	  and equol in plasma and
	  prostatic fluid
		↑ EL in prostatic fluid in Portuguese,
		↑ daidzein in Hong Kong men;
		  higher levels in prostatic fluid than
		  in plasma
			45

15 BPH patients;
  10 controls
	Measured ED, EL in plasma
	  and prostatic tissues
		BPH patients and controls, similar
		  lignan levels
			46

Nude mice
  Balb/c
	LNCaP cells transplanted;
	  fed 30% rye bran; 9 wk
		↓ Tumor number and growth, PSA;
		↑ apoptosis; ↑ EL not related
		  to antitumor effect
			51

Sprague-Dawley
  rats
	Dams fed 5 or 10% FS during
	  pregnancy and lactation;
	  male offspring prostate
	  examined
		↓ Cell proliferation with 5% FS;
		↑ cell proliferation with 10% FS
			50,52

TRAMP
  mice
	Fed 5% FS; 20 or 30 wk	At 20 wk, no difference from control;
		  at 30 wk, ↓ tumor aggressiveness;
		↓ cell proliferation; ↑ apoptosis
			48

Prostate cancer
  patients
	25 patients; on low-fat diet
	  fed 30 g/d FS for average
	  of 34 d
		↓ Serum testosterone, free androgen
		  index, mean tumor cell proliferation,
		↑ apoptotic index; ↓ PSA in men with
		  Gleason sum score ≤ 6
			47

Abbreviatons: See Tables 9.1, 9.3. BPH, benign prostatic hyperplasia; PSA, prostate specific antigen; TRAMP, transgenic adenocarcinoma mouse prostate model.

are of the Western type (46). Hence, Morton *et al.* (45) determined the concentration of the isoflavones daidzein and equol and the lignans EL and ED in the plasma and prostatic fluids of men of Asian (Hong Kong), Western (Britain), and Mediterranean (Portugal) culture and dietary habits to examine the relationship between prostate cancer and lignan exposure. The concentrations of daidzein and equol were higher in the plasma and prostatic fluid of men from Hong Kong compared with those from Britain and Portugal. In contrast, the levels of ED and EL were much higher in the prostatic fluid of Portuguese men than in the British or Hong Kong men. Hence, the isoflavone or lignan concentrations in the prostatic fluid are higher in populations with lower risk of prostate cancer, indicating a potential prostate cancer protective effect of phytoestrogens. It is of interest that the phytoestrogen concentration in the prostatic fluid was severalfold higher than that in the plasma, indicating the ability of the prostate to concentrate these phyto-estrogens. However, although the mean lignan concentration was 40 times greater in the prostatic fluid than in the plasma, the EL levels in the plasma were not significantly associated with the prostatic fluid levels.

The relationship between prostatic hyperplasia and the various phytoestrogens, including isoflavones and lignans, in the plasma and prostatic tissues has also been determined in 25 men with similar dietary habits (46). Tissue samples were obtained from 15 patients with benign prostatic hyperplasia (BPH) and 10 patients with bladder cancer and with prostate volume <25 mL, who were used as controls. Prostatic tissue concentrations of ED, EL, equol, and daidzein were similar in the BPH and control groups. Conversely, the prostatic tissue concentrations of genistein were lower in the BPH group than in the control, thus suggesting that isoflavones, specifically genistein, but not lignans, influence the growth of BPH. However, the plasma levels of isoflavones and lignans were similar in the BPH and control groups, which the authors suggest could be due to differences in rates of absorption and metabolism of the phytoestrogens.

Effect of Flaxseed: Clinical Study. To determine whether the lignan-rich FS can protect against prostate cancer, 25 patients with prostate cancer who were awaiting prostatectomy were instructed to consume daily a low fat (20% kcal or less) diet, supplemented with 30 g FS (47). After an average duration of 34 d (range 21–77 d), significant reductions in total serum cholesterol, total testosterone, and free androgen index were observed when compared with historic cases matched by age, race, prostate specific antigen (PSA) level at diagnosis, and biopsy Gleason sum. The PSA level decreased among men with Gleason sums of 6 or less, but it continued to increase in men with higher Gleason sum scores. The reason for this is not clear, although the authors speculated that, at higher Gleason sum, PSA may be a marker of tumor volume or that one mechanism for release of PSA from tumor is apoptosis. Significantly lower mean tumor cell proliferation and higher apoptotic index were observed in the FS-fed group than the historic control. These data suggest that an FS-supplemented low fat diet may affect prostate cancer biology, but it can not be conclusively stated that the effect is due solely to the FS or its lignans because the study used historic controls instead of a low fat diet treatment group or a placebo group as controls.

Effect of Flaxseed: Animal Models. Studies on the role of diet on prostate cancer are limited by the lack of suitable animal models that are independent of exogenous hormones or transplanted tumors. The development of the transgenic adenocarcinoma mouse prostate (TRAMP) model helped solve the problem. Because the TRAMP mice harbor a rat probasin promoter that drives prostate-specific epithelial expression of the SV40 T-antigen in the dorsolateral and ventral lobes of the prostate, this model can mimic heterogeneous human prostate cancer (48,49). Thus, to establish that FS can protect against prostate cancer, the same research group as above (47) did an animal study using this model (48). At 5–6 wk of age, the TRAMP mice were fed the 5% FS diet or control diet for 20 or 30 wk. All control mice developed prostate cancer *vs.* only 97% of the mice in the FS group. No significant difference in the tumor grade was observed at 20 wk, but at 30 wk less aggresive tumors were observed in the FS group compared with the control. The control mice had a higher prevalence of lung and lymph node metastasis (13 and 16%, respectively) than the FS group (5 and 12%, respectively), although these differences were not statistically significant. However, at 20 wk, control mice had significantly higher cellular proliferation (Ki67 labeling index) (38.1 *vs.* 26.2) and lower apoptotic index (1.45 *vs.* 3.3) of prostatic tissues than did the FS group. Similar differences were observed at 30 wk. Overall, these results indicate that the lignan-rich FS is protective against prostate cancer and supports the earlier observations in prostate cancer patients (47). The TRAMP model is an androgen-dependent animal model. Thus, it was anticipated that the effect seen was due to suppression of testosterone formation by FS supplementation, particularly around puberty, a hormone sensitive time, as has been observed in earlier studies (47,50). However, because hormone levels were not measured in this study, the mechanism for the protective effect of FS cannot be concluded.

Rye bran contains lignans, and the low incidence of prostate cancer in Finland, where the intake of rye bran is high, may be related to the high intake of lignans from this food (51). Hence, the effect of feeding 30% rye bran or ethyl acetate extract of rye bran for 9 wk was determined using Balb/c nude mice injected with the LNCaP androgen-sensitive human prostate adenocarcinoma cell line (51). Of the animals fed the control diet, 75% developed tumors, but only 30% of the animals fed rye bran developed tumors. The tumors in the rye bran group were also smaller, secreted less PSA, and had a higher apoptotic index but similar cell proliferation rate compared with the control group. However, the urinary lignan excretion, which was high in the rye bran group and low in the ethyl acetate extract group, did not correlate with the antitumor effect. Therefore, although rye bran appears to be protective against prostate cancer, compounds other than lignans may be primarily responsible.

Exposure to 5 or 10% FS or equivalent levels of SDG at an early age has been shown to be protective against breast cancer (40–43), but its effect on prostate cancer remains to be determined. The offspring of dams who were fed 10% FS during pregnancy and lactation had heavier prostates than controls (50,52). Whether this will lead to increased or decreased prostates cancer risk is unknown.

Effect of Lignans: In Vitro *Study.* In support of the hypothesis that lignans are protective aganst prostate cancer, an *in vitro* study has shown that EL and ED (10–100 µM) can inhibit the growth of the androgen-sensitive LNCAP and androgen-insensitive (DU-145 and PC3) prostate cancer cell lines (53). The androgen-sensitive cell line was more responsive than the androgen-insensitive cell lines, and EL was more effective than ED. The dose for 50% growth inhibition of LNCAP cells (IC50) by

EL was 57 µM, whereas it was 100 µM for ED. Both are weaker than genistein, which has an IC50 of 25 µM. Finnish men who consumed diets containing high amounts of lignans had a median serum concentration of 13.8 µM (range 0–95.6 µM) (14), levels much lower than the effective levels observed *in vitro*. However, reports that the lignan levels in prostatic fluids are manyfold higher than in the blood (45) may explain why the protective effect of lignan-rich foods on prostate cancer has been demonstrated *in vivo* (48,51).

Colon Cancer

Several studies have been conducted on the effect of FS and lignans on colon cancer (Table 9.5).

Effect of lignans: In Vitro *Study*. ED and EL at 10–100 µM levels have been shown to significantly decrease the cell proliferation of four cancer cell lines (LS174T, T84, Caco-2, and HCT-15) (54) . These cell lines are not estrogen-dependent, because estradiol at 0.5, 1, and 10 µM did not enhance their cell proliferation. Hence the reduction in colon cancer cell proliferation by ED and EL may be due to nonhormone-related mechanisms. The inhibitory levels are physiological considering that the human colon may produce 665 µM lignans assuming an intake of 50 g FS/d with 2.93 µmol SDG/g FS, and a colonic content of about 220 g (54).

Effect of Flaxseed and SDG: Animal Studies. To test the *in vivo* effect of FS and its lignans on colon cancer, aberrant crypts (AC) were used as early biomarkers of cancer risk. AC are putative precursor lesions, which differ from normal colon crypts by their larger size, thicker epithelial lining, and increased pericryptal zone. Multiple AC in localized regions are referred to as aberrant crypt foci (ACF), and the higher the number of AC per ACF, the higher is the possibility that the AC may develop to tumors. A high fat (20%) BD supplemented with either 5 or 10% FS or defatted FS fed to rats for 4 wk starting 1 wk after induction with the colon carcinogen azoxymethane (AOM) reduced the number of AC and ACF by about 50% (55). This is indicative of reduced colon cancer risk. The colonic epithelial cell proliferation, expressed as labeling index, was also significantly reduced by all diets except the 5% defatted FS. Because the effect of defatted FS did not differ significantly from that of the full fat FS, it was suggested that the effect of FS on the colon is not due to its oil component but rather to its other components such as the lignans. Furthermore, the similar effect of 5 and 10% FS suggests that a high level of FS is not necessary to produce a colon cancer protective effect.

Hence, in a follow-up study, high fat diets supplemented with either 2.5 or 5% FS, or defatted FS, or SDG (1.5 mg/d; equivalent to the amount present in 5% FS) were fed to rats for 100 d, again starting 1 wk after AOM induction (56). All treatment diets reduced the number of AC per ACF. The control group already had four microadenomas and two polyps, whereas none were observed in the other groups, indicating that the control group progressed faster through the tumorigenesis process

TABLE 9.5
Experimental Studies on Flaxseed, Lignans, and Colon Cancer

Model system	Treatment	Observations	Ref.
In vitro: LS174T	1–100 µM ED, EL; T84, Caco-2, HCT-15 colon cancer cells	↓ Cell proliferation, 20–85% at 100 µM	54
Sprague-Dawley rats	AOM-treated; fed 5 or 10% FS or defatted FS; 4 wk	↓ AC and ACF; ↓ cell proliferation except with 5% defatted FS	55
Sprague-Dawley rats	AOM-treated; fed 2.5 or 5% FS, or defatted FS, or 1.5 mg/d SDG; 100 d	↓ ACF multiplicity with FS and defatted FS; ↓ AC and ACF multiplicity with SDG; negative relationship between lignans and ACF multiplicity; positive relationship with β-glucuronidase activity	56
Sprague-Dawley rats	Fed 2.5, 5.0, or 10% FS; 4 wk	Positive relationship between urinary lignans and β-glucuronidase activity	57
F-344 rats	AOM-treated; fed 30% rye bran with lignans; 31 wk	↓ Large ACF, tumor numbers	59
Rats	DMH-treated; fed flaxseed (2 g/d); 34 wk	No effect on tumors	60
Rats	AOM-treated; fed 16.8% FS	No effect on tumors	61
APC[MIN] mice	0.02% HMR (10 mg/kg b.w.); 5–6 wk	↓ Adenomas in small intestine; normalize β-catenin in adenoma tissue	62
	10% rye bran; 5–6 wk	No effect	

Abbreviations: See Tables 9.1, 9.3. AC, aberrant crypt; ACF, aberrant crypt focus; AOM, azoxymethane; APC, adenomatous polyposis coli; DMH, dimethylhydrazine.

than the other treatment groups. The 2.5% FS diet group did not differ significantly from the 5% FS diet group, suggesting that an even lower level of FS is sufficient to produce a protective effect. Because SDG produced an effect that did not differ significantly from that of the 5% FS diet, and a negative correlation was observed between the number of AC per ACF and urinary lignan excretion, it was again suggested that the effect of FS is primarily due to the lignans that it contains.

The urinary lignan excretion had a positive relationship with β-glucuronidase activity, indicating a possible involvement of this enzyme in the lignan effect (56). This was confirmed in another study where 2.5, 5, and 10% FS were fed (57). β-glucuronidase, produced by the colonic bacteria, assists the enterohepatic circulation of many compounds including estrogen, lignans, and other compounds, some of which are toxic. Although it has been suggested that increased β-glucuronidase activity relates to increased risk of colon cancer (58), this does not appear to be the case in high lignan diets. It may just be an indicator of bacterial activity, and whether it is a marker of good or adverse effects depends on what compound it helps circulate. Because lignans appear to be protective, β-glucuronidase activity is an indicator of protective effects in this study.

Although early biomarkers of colon cancer risk suggest a protective effect of FS, results of studies with tumors as the endpoint are ambivalent. Feeding 30% rye bran diet, which contains lignans, to AOM-treated rats for 31 wk, lowered the number of large ACF (>6 AC per ACF) and the number of tumors compared with the control (59). The effect was thought to be due to the lignans rather than the dietary fiber in rye bran because the fiber contents of the rye bran and control diets were the same. In contrast, FS (2 g/d) in the chow fed to rats given multiple injections of the carcinogen dimethylhydrazine resulted in no significant effect on number or histopathological characteristics of tumors compared with the control (7,60). The subtle effect of FS may have been masked by the large amount of carcinogen used for tumor induction. Only a single injection was used in the studies described earlier (55,56), whereas multiple injections were used in this study (7,60). The protective effect of any phytoestrogens present in the rat chow used as a control diet may also have masked any protective effect of the FS. It is also possible that a high intake of FS (2 g/d) is not protective; another study, which also fed high levels of FS (16.8%) to rats, did not observe a reduction in colon tumorigenesis compared with the control (61).

HMR, as a major precursor of EL, was also tested for its effect on colon cancer risk using the adenomatous polyposis coli (APC^{Min}) mouse, an animal model of human familial adenomatous polyposis (62). The BD containing 2.5% inulin and supplemented with 0.02% HMR (provided at 10 mg HMR/kg b.w.) significantly reduced the number of adenomas in the small intestine compared with the control group. The β-catenin levels in the adenoma tissue of APC^{Min} mouse was also normalized, indicating that the protective effect of HMR or its EL metabolite is mediated through the β-catenin pathway. Supplementation with a 10% rye bran diet, which also contains lignans, did not produce the same effect (62), perhaps because the lignan intake level was not high enough (10–15 µg/kg b.w.) when fed in the presence of inulin.

Skin Cancer

The effect of FS and lignans on skin cancer was tested primarily at the metastasis stage. Male C57BL/6 mice were fed 2.5, 5, or 10% FS from 2 wk before to 2 wk after the melanoma cell line B16BL6 was injected directly into the blood stream (63). Respective reductions of 32, 54, and 63% in the number of lung tumors and 16.1, 29.0, and 58.1% in lung tumor area were observed. When the experiment was repeated by feeding 73, 147, and 293 µmol SDG/kg diet, equivalent to the amount present in 2.5, 5, and 10% FS, similar reductions were observed (i.e., 38.7, 41.9, and 53.2% reductions in number of tumors, and 55, 60, and 70% reductions in tumor size) (64). This indicates that the reduction in lung metastasis by FS was in part due to the lignans that it contains. The metastasis process involves adhesion, invasion, and migration of cancer cells. Because this study injected the tumor cells directly into the blood stream instead of allowing the cells to detach from growing tumors and invade the basement membrane before reaching the circulation, it remains unclear if the FS and its lignans can reduce all the steps involved in melanoma cell metastasis.

Summary, Conclusions, and Future Work

Epidemiological studies are not all in agreement, but a majority support a protective effect of lignans on breast cancer. Studies conducted in Australia, Finland, Shanghai, and New York showed reduction in breast cancer risk with high lignan intake, particularly EL, but studies in the Netherlands, Sweden, and San Francisco showed no relationship. The lack of agreement among studies may be related to methodological differences including the biomarker of lignan exposure used (i.e., urine or blood levels of lignans) or intake levels based on food frequency questionnaire, menopausal status, ethnicity, genetic susceptibility, and whether the study is retrospective or prospective. In contrast, experimental studies in animal and human models showed a promising protective effect of FS and lignans on breast cancer. In carcinogen-treated rats, FS reduced the tumor incidence, number, and size depending on whether it was fed at the preinitiation, early or late promotion, or progression stages of carcinogenesis. It reduced the growth of established human tumors in athymic mice injected with ER-negative (MDA MB 435) or ER-positive (MCF-7) human breast cancer cells and in postmenopausal patients with newly diagnosed breast cancer. It can reduce the metastasis of ER-negative human breast cancer cells. Early exposure to FS during the suckling period enhanced mammary gland development, which then reduced tumor development at adulthood. The effect of SDG on breast cancer is similar to that of FS. Hence, the mammalian lignans ED and EL derived from SDG appear to be in part responsible for the effect of FS. Further support for the anticancer effect of lignans comes from similar results obtained with EL itself and with HMR, its direct precursor.

The only epidemiological study on prostate cancer and urinary lignan excretion showed no significant relationship, whereas one study on thyroid cancer showed a protective effect of lignans. It is, however, premature to make a conclusion on the effect of lignans on prostate and thyroid cancer based on limited epidemiological studies. Small observational studies showed conflicting results. Nevertheless, a clinical study on prostate cancer patients demonstrated a beneficial effect of 30 g FS in a low-fat diet regimen. Although such a study did not conclusively show that the effect was due to FS because of lack of appropriate controls, a follow-up feeding study using the TRAMP mouse model demonstrated the ability of

FS to reduce prostate tumor development. An *in vitro* study also showed the ability of ED and EL to reduce prostate tumor cell growth.

No epidemiological study has addressed specifically the effect of lignans on colon cancer. An *in vitro* study showed an antiproliferative effect of ED or EL. In AOM-treated rats, reductions in early biomarkers of colon cancer such as number of AC, ACF, and AC per ACF have been reported with the feeding of 2.5–10% FS and SDG equivalent to the level in 5% FS. However, rat studies with tumors as the endpoint have shown no protective effect of feeding FS at an intake level of 2 g/d or at l6.8% level in the diet. It is unclear, in these experiments, whether the effect is due to the high level of intake or the multiple carcinogen injections used in these studies. An experiment using the APC^{Min} mouse model showed that 0.02% HMR in an inulin-containing diet can reduce the number of intestinal tumors. The effects of SDG or its mammalian lignan metabolites have yet to be tested in this model.

It is clear that a 2.5–10% FS diet can reduce melanoma cell metastasis and that the SDG is in part responsible for this effect. However, the cancer cells were injected directly into the bloodstream in this study, and therefore it is not known whether the lignans in flaxseed can inhibit all the stages of the metastasis process.

Overall, studies to date indicate that FS has cancer protective effects and that the lignans are in part responsible for this effect. However, more research needs to be done before definitive recommendations can be made regarding their use for cancer prevention or treatment because their effects differ depending on the dose and stage of the cancer process in which they are consumed. More clinical trials have to be conducted to confirm the observations in epidemiological and animal studies. The optimum intake level of FS and lignan that will produce the effect remains to be established. Several mechanisms whereby the lignans influence the cancer process have been discussed by others (12,65; Chapters 6,10). These include the ability of lignans to act as antiestrogens, antioxidants, and inhibitors of growth factor expression and enzymes involved in estrogen synthesis and metabolism, but none of them are conclusive and further exploration is needed. The effects of lignans appear to differ depending on the hormonal milieu, suggesting the need to differentiate their effects in men and premenopausal and postmenopausal women. The interaction of FS or its lignans with other phytoestrogens as well as cancer drugs has to be studied. This is particularly important because, with increasing interest on the use of alternative medicine and dietary approaches to reducing cancer risk, consumers may consume FS in combination with other phytoestrogen-rich foods such as soy or cancer drugs such as tamoxifen without knowing whether the effect will be more beneficial or adverse. There is certainly more work ahead for flaxseed and lignan researchers in the cancer area.

Acknowledgments

Work of the author was funded by the Natural Sciences and Engineering Research Council of Canada, American Institute for Cancer Research, Flax Council of Canada, Saskatchewan

Flax Development Commission, Cancer Research Society, Canadian Breast Cancer Research Foundation, and National Cancer Institute (United States), and conducted with the help of many graduate students, postdoctoral fellows (funded by National Institute of Nutrition, Canada), technicians, and collaborators.

References

1. Bhatty, R.S., Nutrient Composition of Whole Flaxseed and Flaxseed Meal, in *Flaxseed in Human Nutrition*, edited by S.C. Cunnane and L.U. Thompson, 1st edn., AOCS Press, Champaign, Illinois, 1995, pp. 22–42.

2. Thompson, L.U., P. Robb, M. Serraino, and F. Cheung, Mammalian Lignan Production from Various Foods, *Nutr. Cancer 16*:43–52 (1991).

3. Mazur, W., Phytoestrogen Content in Foods, *Bailliere's Clin. Endocrin. Metab. 12*:729–742 (1998).

4. Adlercreutz, H., T. Fotsis, B. Heikkinen, J.R. Dwyer, M. Wood, B.R. Goldin, and S.L. Gorbach, Excretion of the Lignans Enterolactone and Enterodiol and of Equol in Omnivorous and Vegetarian Women and in Women with Breast Cancer, *Lancet 2*: 1295–1299 (1982).

5. Johnston, P.V., Flaxseed Oil and Cancer: α-Linolenic Acid and Carcinogenesis, in *Flaxseed in Human Nutrition*, edited by S.C. Cunnane and L.U. Thompson, 1st edn., AOCS Press, Champaign, Illinois, 1995, pp. 207–218.

6. Bougnoux, P., α-Linolenic Acid and Cancer, in *Flaxseed in Human Nutrition*, edited by L.U. Thompson and S.C. Cunnane, 2nd edn., AOCS Press, Champaign, Illinois, 2003, pp. 232–244.

7. Setchell, K.D.R., Discovery and Potential Clinical Importance of Mammalian Lignans, in *Flaxseed in Human Nutrition*, edited by S.C. Cunnane and L.U. Thompson, 1st edn., AOCS Press, Champaign, Illinois, 1995, pp. 82–98.

8. Meagher, L.P., G.R. Beecher, V.P. Flanagan, and B.W. Li, Isolation and Characterization of the Lignans, Isolariciresinol and Pinoresinol, in Flaxseed Meal, *J. Agric. Food Chem. 47*:3173–3180 (1999).

9. Thompson, L.U., Analysis and Bioavailability of Lignans, in *Flaxseed in Human Nutrition*, edited by L.U. Thompson and S.C. Cunnane, 2nd edn., AOCS Press, Champaign, Illinois, 2003, pp. 92–116.

10. Rickard, S.E., L.J. Orcheson, M.M. Seidl, L. Luyengi, H.H. Fong, and L.U. Thompson, Dose-Dependent Production of Mammalian Lignans in Rats and *in-vitro* from the Purified Precursor Secoisolariciresinol Diglycoside from Flaxseed, *J. Nutr. 126*:2012–2019 (1996).

11. Nesbitt, P.D., Y. Lam, and L.U. Thompson, Human Metabolism of Mammalian Lignan Precursors in Raw and Processed Flaxseed, *Am. J. Clin. Nutr. 69*:549–555 (1999).

12. Saarinen, N., S. Makela, and R. Santti, Mechanisms of Anticancer Effects of Lignans with a Special Emphasis on Breast Cancer, in *Flaxseed in Human Nutrition*, edited by L.U. Thompson and S.C. Cunnane, 2nd edn., AOCS Press, Champaign, Illinois, 2003, pp. 223–231.

13. Ingram, D., K. Sanders, M. Kolybaba, and D. Lopez, Case-Control Study of Phyto-Oestrogens and Breast Cancer, *Lancet 350*:990–994 (1997).

14. Pietienen, P., K. Stumpf, S. Mannisto, V. Kataja, M. Uusitupa, and H. Adlercreutz, Serum Enterolactone and Risk of Breast Cancer: A Case-Control Study in Eastern Finland, *Cancer Epidemiol. Biomarkers Prev. 10*:339–344 (2001).

15. Dai Q., A.A. Franke, F. Jin, X. Shu, J.R. Hebert, L.J. Custer, J. Cheng, Y.T. Gao, and W.Zheng, Urinary Excretion of Phytoestrogens and Risk of Breast Cancer Among Chinese Women in Shanghai, *Cancer Epidemiol. Biomarkers, Prev. 11*:815–821 (2002).

16. Tonkelaar, I., L. Keinan-Boker, P. Van't Veer, C. Arts, H. Adlercreutz, J.H. Thijssen, and P. Peeters, Urinary Phytoestrogens and Postmenopausal Breast Cancer Risk, *Cancer Epidemiol. Biomarkers Prev. 10*:223–228 (2001).

17. Hulten, K., A. Winkvist, P. Lenner, R. Johansson, H. Adlercreutz, and G. Hallmans, An Incident Case-Referent Study on Plasma Enterolactone and Breast Cancer Risk, *Eur. J. Nutr. 41*:168–176 (2002).

18. Horner, N.K, A.R. Kristal, J. Prunty, H.E. Skor, J.D. Potter, and J.W. Lampe, Dietary Determinants of Plasma Enterolactone, *Cancer Epidemiol. Biomarker Prev. 11*:121–126 (2002).

19. Horn-Ross, P.L., E.M. John, M. Lee, S.L. Stewart, J. Koo, L.C. Sakoda, A.C. Shau, J. Goldstein, P. Davis, and E.J. Perez-Stable, Phytoestrogen Consumption and Breast Cancer Risk in a Multiethnic Population, *Am. J. Epidemiol. 154*:434-41 (2001).

20. Lum, R., C. O'Malley, and S. Lui, *California Incidence and Mortality in the San Francisco Bay Area*, 1988–1997, Northern California Cancer Center (2000).

21. McCann, S.E., K.B. Moysich, J.L. Freudenheim, C.B. Ambrosone, and P.G. Shields, The Risk of Breast Cancer Associated with Dietary Lignans Differs by CYP17 Genotype in Women, *J. Nutr. 132*:3036–3041 (2002).

22. Felgelson, H.S., L.S. Shames, M.C. Pike, G.A. Coetzee, F.Z. Stanczyk, and B.E. Henderson, Cytochrome P450c17α gene (CYP17) Polymorphism Is Associated with Serum Estrogen and Progesterone Concentrations, *Cancer Res. 58*:585–587 (1998).

23. Bergman-Jungestrom, M., M. Gentile, A.C. Lundin, and S. Wingren, Association Between CYP17 Gene Polymorphism and Risk of Breast Cancer in Young Women, *Int. J. Cancer 84*:350–353 (1999).

24. Stattin, P., H. Adlercreutz, L. Tenkanen, E. Jellum, S. Lumme, G. Hallmans, S. Harvei, L. Teppo, K. Stumpf, T. Luostarinen, M. Lehtinen, J. Dillner, and M. Hakama, Circulating Enterolactone and Prostate Cancer Risk: A Nordic Nested Case-Control Study, *Int. J. Cancer 99*:124–129 (2002).

25. Rossing, M.A., L.F. Voigt, K.G. Wicklund, and J.R. Daling, Reproductive Factors and Risk of Papillary Thyroid Cancer in Women, *Am. J. Epidemiol. 151*:765–772 (2000).

26. Horn Ross, P.L, K.J. Hogatt, and M.M. Lee, Phytoestrogens and Thyroid Cancer Risk: The San Francisco Bay Area Thyroid Cancer Study, *Cancer Epidemiol. Biomarkers Prev. 11*:43–49 (2002).

27. Serraino, M., and L.U. Thompson, The Effect of Flaxseed Supplementation on Early Risk Markers for Mammary Carcinogenesis, *Cancer Lett. 60*:135–142 (1991).

28. Serraino, M., and L.U. Thompson, The Effect of Flaxseed Supplementation on the Initiation and Promotional Stages of Mammary Tumorigenesis, *Nutr. Cancer 17*:153–159 (1992).

29. Thompson, L.U., M. Seidl, S. Rickard, L. Orcheson, and H. Fong, Antitumorigenic Effect of Mammalian Lignan Precursor from Flaxseed, *Nutr. Cancer 26*:159–165 (1996).

30. Thompson, L.U., S. Rickard, L. Orcheson, and M. Seidl, Flaxseed and Its Lignan and Oil Components Reduce Mammary Tumor Growth at a Late Stage of Carcinogenesis, *Carcinogenesis 17*:1373–1376 (1996).

31. Rickard, S.E., Y.V. Yuan, J. Chen, and L.U. Thompson, Dose Effects of Flaxseed and Its Lignan on Methyl-*N*-nitrosourea-Induced Mammary Tumorigenesis in Rats, *Nutr. Cancer 35*:50–57 (1999).

32. Saarinen, N.M., A. Warri, S.I. Makela, C. Eckerman, M. Reunanen, M. Ahotupa, S.M. Salmi, A.A. Franke, L. Kangas, and R. Santti, Hydroxymatairesinol, a Novel Enterolactone Precursor with Antitumor Properties from Coniferous Tree (*Picea abies*), *Nutr. Cancer 36:*207–216 (2000).

33. Saarinen, N.M., R. Huovinen, A. Warri, S.I. Makela, L.V. Valentin-Blasini, L. Needham, C. Eckerman, Y.U. Collan, and R. Santti, Uptake and Metabolism of Hydroxymatairesinol in Relation to Its Anticarcinogenicity in DMBA-Induced Rat Mammary Carcinoma Model, *Nutr. Cancer 41:*82–90 (2001).

34. Saarinen, N.M., R. Huovinen, A. Warri, S.I. Makela, L.V. Valentin-Blasini, R. Sjoholm, J. Ammala, R. Lehtila, C. Eckerman, Y.U. Collan, and R.S. Santti, Enterolactone Inhibits the Growth of 7,12-Dimethylbenzanthracene-Induced Mammary Carcinomas in the Rat, *Molec. Cancer Therapeut. 1:*869–876 (2002).

35. Chen, J.M., M. Stavro, and L.U. Thompson, Dietary Flaxseed Inhibits Breast Cancer Growth and Metastasis and Downregulates Expression of Epidermal Growth Factor Receptor and Insulin-like Growth Factor, *Nutr. Cancer 43:*187–192 (2002).

36. Dabrosin, C., J. Chen, L. Wang, and L.U. Thompson, Flaxseed Inhibits Metastasis and Decreases Extracellular Vascular Endothelial Growth Factor in Human Breast Cancer Xenografts, *Cancer Lett. 186:*31–37 (2002).

37. Yoshiji, H., D.E Gomez, M. Shibuya, and U.P. Thorgeirsson, Expression of Vascular Endothelial Growth Factor, Its Receptor, and Other Angiogenic Factors in Human Breast Cancer, *Cancer Res. 56:*2013–2016 (1996).

38. Wang, L., *The Effect of Dietary Flaxseed on the Growth, Recurrence and Metastasis of Human Breast Cancer Cells*, M.Sc. Thesis, University of Toronto, Ontario, Canada (2002).

39. Russo, J., and I.H. Russo, DNA Labeling Index and Structure of the Rat Mammary Gland as Determinants of Susceptibility to Carcinogenesis, *J. Natl. Cancer Inst. 61:*1451–1495 (1978).

40. Tou, J.C.L., and L.U. Thompson, Exposure to Flaxseed or Its Lignan Component During Different Developmental Stages Influences Rat Mammary Gland Structures, *Carcinogenesis 20:*1831–1835 (1999).

41. Ward, W.E, F.O. Jiang, and L.U. Thompson, Exposure to Flaxseed or Purified Lignan During Lactation Influences Rat Mammary Gland Structures, *Nutr. Cancer 37:*187–192 (2000).

42. Tan, K.P., J. Chen, and L.U. Thompson, Early Mammary Morphogenesis, Reproductive Indices, and Splenocyte Blastogenesis in Rats Exposed to Flaxseed and Secoisolariciresinol Diglycoside (SDG) During Suckling Period, *FASEB J. 16:*A742 (2002).

43. Chen, J., K.P. Tan, and L.U. Thompson, Exposure to Flaxseed and Its Lignans During Suckling Inhibits Rat Mammary Tumorigenesis, *FASEB J. 16:*A1005 (2002).

44. Thompson, L.U., T. Li, J. Chen, and P.E. Goss, Biological Effects of Dietary Flaxseed in Patients with Breast Cancer, *Breast Cancer Res. Treat. 64:*50 (2000).

45. Morton, M.S., P.S.F. Chan, C. Cheng, N. Blacklock, A. Matos-Ferreira, L. Abranches-Montero, R. Correia, S. Lloyd, and K. Griffiths, Lignans and Isoflavonoids in Plasma and Prostatic Fluid in Men: Samples from Portugal, Hong Kong, and the United Kingdom, *The Prostate 32:*122–128 (1997).

46. Hong, S.J., S.I. Kim, S.M. Kwon, J.R. Lee, and B.C. Chung, Comparative Study of Concentration of Isoflavones and Lignans in Plasma and Prostatic Tissues of Normal Control and Benign Prostatic Hyperplasia, *Yonsei Med. J. 43:*236–241 (2002).

47. Demark-Wahnefried, W., D.T. Price, T.J. Polascik, C.N. Robertson, E.E. Anderson, D.F. Paulson, P.J. Walther, M. Gannon, and R.T. Vollmer, Pilot Study of Dietary Fat Restriction and Flaxseed Supplementation in Men with Prostate Cancer Before Surgery: Exploring the Effects on Hormonal Levels, Prostate-Specific Antigen, and Histopathologic Features, *Urology 58*:47–52 (2001).
48. Lin, X., J.R. Gingrich, J. Li, Z. Haroon, B. Peterson, and W. Demark-Wahnefried, Effect of Flaxseed Supplementation on the Development of Prostatic Neoplasia and Metastasis in Transgenic (TRAMP) Mice, *J. Nutr. 131*:3136S–3137S (2001).
49. Greenberg, N.M., F.J. De Mayo, P.C. Sheppard, R. Barrios, R. Lebovitz, M. Finegold, R. Angelopolou, J.G. Dodd, M.L. Duckworth, and J.M. Rosen, The Rat Probasin Gene Promoter Directs Hormonally and Developmentally Regulated Expression of a Heterologous Gene Specifically to the Prostate in Transgenic Mice, *Mol. Endocrinol. 8:* 230–239 (1994).
50. Tou, J.C.L., J. Chen, and L.U. Thompson, Flaxseed and Its Lignan Precursor, Secoisolariciresinol Diglycoside, Affect Pregnancy Outcome and Reproductive Development in Rats, *J. Nutr. 128*:1861–1868 (1998).
51. Byland A., J.X. Zhang, A. Bergh, J.E. Damber, A. Widmark, A. Johansson, H. Adlercreutz, P. Aman, M.J. Shepherd, and G. Hallmans, Rye Bran and Soy Protein Delay Growth and Increase Apoptosis of Human LNCaP Prostate Adenocarcinoma in Nude Mice, *The Prostate 42*:304–314 (2000).
52. Tou, J.C., J. Chen, and L.U. Thompson, Dose, Timing, and Duration of Flaxseed Exposure Affect Reproductive Indices and Sex Hormone Levels in Rats, *J. Toxicol. Environ. Health 56*:555–570 (1999).
53. Lin, X, B.R. Switzer, and W. Demark-Wahnefried, Effect of Mammalian Lignans on the Growth of Prostate Cancer Cell Lines, *J. Nutr. 131*:3136S (2001).
54. Sung, M., M. Lautens, and L.U. Thompson, Mammalian Lignans Inhibit the Growth of Estrogen-Independent Human Colon Cancer Cells, *AntiCancer Res. 18*:1405–1408 (1998).
55. Serraino, M., and L.U. Thompson, Flaxseed Supplementation and Early Markers of Colon Carcinogenesis, *Cancer Lett. 63*:159–165 (1992).
56. Jenab, M., and L.U. Thompson, The Influence of Flaxseed and Lignans on Colon Carcinogenesis and β-Glucuronidase Activity, *Carcinogenesis 17*:1343–1348 (1996).
57. Jenab, M., S.E. Rickard, L.J. Orcheson, and L.U. Thompson, Flaxseed and Lignans Increase Cecal β-Glucuronidase Activity in Rats, *Nutr. Cancer 33*:154–158 (1999).
58. Reddy, B.S., A. Engle, B. Simi, and M. Goldman, Effect of Dietary Fiber on Colonic Bacterial Enzymes and Bile Acids in Relation to Colon Cancer, *Gastroenterology 102*:1475–1482 (1992).
59. Davies, M.J., E.A. Bowey, H. Adlercretuz, I.R. Rowland, and P.C. Rumsby, Effects of Soy or Rye Supplementation of High Fat Diets on Colon Tumor Development in Azoxymethane-Treated Rats, *Carcinogenesis 20*:927–931 (1999).
60. Gilbert, J.M., Experimental Colorectal Cancer as a Model of Human Disease, *Ann. R. Coll. Surg. Engl. 69*:48–53 (1987).
61. Bennink, M.R, and E.A. Rondini, Phytoestrogens, Estrogens and Risk of Colon Cancer, in *Phytoestrogens and Health*, edited by G.S. Gilani and J. Anderson, AOCS Press, Champaign, Illinois, 2002, pp. 427–439.
62. Oikarinen, S.I., A.M. Pajari, and M. Mutanen, Chemopreventive Activity of Crude Hydroxymatairesinol (HMR) Extract in APC Min Mice, *Cancer Lett. 161*:253–258 (2000).

63. Yan, L., J. Yee, M. McGuire, and L.U. Thompson, Dietary Flaxseed Supplementation and Experimental Metastasis of Melanoma Cells in Mice, *Cancer Lett. 124*:181–186 (1998).
64. Li, D., J.A.Yee, L.U. Thompson, and L. Yan, Dietary Supplementation with Secoiso-lariciresinol Diglycoside (SDG) Reduces Experimental Metastasis of Melanoma Cells in Mice, *Cancer Lett. 142*:91–96 (1999).
65. Hutchins, A.M., J.L. Slavin, Effects of Flaxseed on Sex Hormone Metabolism, in *Flaxseed in Human Nutrition*, edited by L.U. Thompson and S.C. Cunnane, 2nd edn., AOCS Press, Champaign, Illinois, 2003, pp. 126–149.

Chapter 10

Mechanism of Anticancer Effects of Lignans with a Special Emphasis on Breast Cancer

Niina Saarinen, Sari Mäkelä, and Risto Santti

Institute of Biomedicine, Department of Anatomy, and Functional Foods Forum, University of Turku, Turku, Finland

Introduction

Both epidemiological and experimental evidence is accumulating to show that a lignan-rich diet may reduce the risk of human breast cancer (BC). Low serum and urine concentrations of enterolactone (EL) have been correlated with an increased risk of BC (1–3). Further, the anticarcinogenic effects of lignans have been demonstrated in animal models. Thompson *et al.* (4,5) were the first to demonstrate the anticancer effects of dietary plant lignan, secoisolariciresinol diglycoside (SDG), in the dimethylbenzanthracene (DMBA)-induced mammary carcinoma in rats. We have shown that 7-hydroxymatairesinol (HMR), which is converted to EL in rats, also inhibited the progression of DMBA-induced rat mammary carcinoma (6,7) and, further, that EL in part mediates the anticancer effects of HMR, seen in the experimental mammary cancer models (8). The mean serum EL concentration associated with the inhibition of tumor growth was 0.35–0.41 µM after daily administration of EL. These EL concentrations are 10-fold when compared with those measured in the general population. Because of the limited supply of pure lignans and lack of toxicological data, however, no clinical trials have so far been conducted in an effort to assess the anticarcinogenic effects of lignans *per se* in humans. Moreover, the mechanisms of their possible anticarcinogenic effects remain obscure.

Estrogenicity and Antiestrogenicity of EL and Enterodiol (ED)

Lignans are often classified as phytoestrogens. Many lignans have an aromatic hydroxylated phenyl ring in their structure, which is in accordance with the chemical structure considered to be important for the estrogenicity of the compound (9). In examining the potential mechanisms of action of lignans and particularly their role in the chemoprevention of BC, many investigators have concentrated on the ability of lignans to antagonize the estrogen metabolism and the biological action of 17β-estradiol (E_2). This is understandable because the anticancer effects of EL in DMBA-induced breast cancer are in parallel with the hypothesis that EL acts as an antiestrogen by inhibiting the estrogen receptor (ER)-mediated action of estrogen. The possible estrogenic and antiestrogenic activities of the mammalian lignans EL and ED and some plant lignans have been studied both *in vitro* and *in vivo*.

Both EL and ED have been reported to have weak estrogenic activity in the ER-positive MCF-7 cells (10–12). Relatively high concentrations of EL (1–10 μM) increased the MCF-7 cell proliferation and DNA synthesis (10–14). The antiestrogen tamoxifen blocked the EL-induced cell proliferation measured as DNA synthesis (10,14), suggesting a mechanism associated with ER. Furthermore, 1 μM EL increased the expression of both estrogen-responsive pS2 protein (11) and progesterone receptors in MCF-7 cells (10). ED was less estrogenic in MCF-7 cells, having approximately 0.1 the activity of EL (10), or it was devoid of estrogenic activity (11).

Mousavi and Adlercreutz (13) reported that the combination of EL (0.5–2 μM) and E_2 (1 nM) resulted in lower MCF-7 cell proliferation than EL or E_2 alone. This is the only study showing antiestrogenic activity for EL *in vitro* at realistic concentrations. High concentrations (>10–100 μM) of EL have repeatedly been shown to inhibit the proliferation of breast cancer cells (10,12,13,15). The possibility that the observed antiproliferative effects are based on other than receptor-mediated effects has not been excluded in these *in vitro* studies.

We have studied the transcriptional action of EL and ED in ERα- and ERβ-positive cells (6). Stable subtype-specific estrogen-receptor cell lines (293 human embryonal kidney epithelial cells) (16) were used in our study. The transactivation of an estrogen response element (ERE)-containing estrogen-responsive reporter gene *via* human ER was measured as the activity of reporter protein alkaline phosphatase (ALP) in the cell culture medium. The cells were exposed to the test compounds for 72 h. EL or ED did not increase the ALP activity *via* ERα or ERβ in concentrations below 1.0 μM. In addition, mammalian lignans ED and EL showed no antagonistic activity against 0.5 nM E_2 action at concentrations below 1.0 μM. We concluded that EL had no significant estrogenic or antiestrogenic activity *via* ERα or ERβ *in vitro* at concentrations found in human organisms without special diets. However, it should be noted that the assay we used is based on a reporter gene that is controlled by ERE, and thus measures action *via* the classical ER-mediated pathway only. It is well known that ER-ligand complexes may act *via* other hormone-responsive elements, such as AP-1 sites, and regulate the expression of genes devoid of ERE. If EL (or other lignans) bind to ER, it is thus possible that they may exert actions *via* these pathways. At present there is no clear evidence to indicate that EL (or other lignans) bind to ER at any significant affinity. Further studies are warranted to clarify this question.

The estrogenicity or antiestrogenicity of lignans has rarely been studied *in vivo*. In contrast to the suggestions based on the *in vitro* findings, no convincing evidence for the estrogenic activity of EL has been obtained when uterine growth has been used as an indicator of estrogenic activity. Setchell *et al.* (17) showed that EL administered subcutaneously (s.c.) to immature female mice (6.7 and 20 mg/kg b.w.) did not increase uterine weight, suggesting the lack of an estrogenic effect. We have confirmed the findings of Setchell *et al.* HMR or EL had no significant estrogen-like effect in the immature rat uterine growth test (8). Feeding of SDG to rats during pregnancy and lactation increased the uterine weight of offspring at weaning, but the uterotropic effect was not evident later (18). In a later study by Ward *et al.* (19), SDG given to dams during lactation or continuously to offspring up to postnatal day 132 had no effects on anogenital distance, puberty onset, or estrous cycle length in female offspring or the weight of reproductive organs (uterus, ovaries, prostate). Waters and Knowler (20) demonstrated that EL given in a dose of 1 mg/kg b.w. (s.c.) reduced the E_2-stimulated RNA synthesis in uterus (EL administered 22 h before E_2). When E_2 was administered at the same time or up to 12 h before EL, no significant effect was seen on the stimulation by E_2. The authors pointed to the structural resemblance between EL and antiestrogens such as tamoxifen and suggested competition with E_2 for the receptor binding as a possible mechanism of action. In adult female rats, supplementation of the diet with purified SDG (0.75, 1.5, or 3.0 mg/d/rat) lengthened the estrous cycling by 18–39% (21). The highest dose of SDG (3.0 mg/d) caused cessation of estrous cycling or irregular cycles in 66% of the rats. A similar type of response was observed with the antiestrogen tamoxifen (1 mg/kg b.w.), suggesting an antiestrogenic effect. It is pertinent to mention that a 4-wk or lifelong oral administration of HMR to adult rats did not show any detectable antiandrogenic responses of the weights of the testis or accessory sex glands or changes in androgen or luteinizing hormone (LH) concentration (6). This means that HMR or its metabolites are not able to cause feedback inhibition of LH and androgen secretion, which is the primary antiandrogenic effect of estrogen in the male. These studies also lessen the likelihood that lignans would act as inhibitors of 5α-reductase (22).

Altogether, there is no firm evidence for the estrogenic effects of lignans *in vivo*. However, the possibility of a weak antiestrogenic effect by lignans cannot be ruled out. It is surprising that the estrogenicity or antiestrogenicity of lignans has never been studied in detail in breast tissue or bone. In theory, they could act as selective estrogen receptor modulators (SERM). It would be important to broaden our understanding of the mechanism of lignan action on breast cancer cell proliferation with particular attention to the effects at the concentrations likely to be found in humans (23).

Inhibition of Estrogen Biosynthesis

Aromatase is a cytochrome P-450 enzyme converting testosterone and androstenedione to 17β-estradiol and estrone, respectively. The inhibition of aromatase enzyme activity by EL resulting in reduced estrogen formation has been suggested to be a mechanism by which consumption of lignan-rich plant food contributes to a reduction of estrogen-dependent diseases such as BC (24,25). Inhibition of aromatase by pharmaceuticals has been demonstrated to inhibit mammary cancer growth in humans even more efficiently than the antiestrogen tamoxifen (26–28). In the DMBA-induced mammary carcinoma model, aromatase inhibitors are also effective (29–32). Aromatase inhibition would, therefore, be an attractive explanation for EL action as well. The aromatase inhibition activities of several lignans have been tested *in vitro*. The tritiated water method, which shows the aromatization of the A ring, has been applied. Adlercreutz *et al.* (24) were the first to demonstrate that EL inhibits aromatase in human placental microsomes and in JEG-3 choriocarcinoma cells. The IC_{50} concentration reported for EL in placental microsomes was 14 μM. Wang *et al.* (25) demonstrated that EL and its suggested intermediate metabolites (3-demethoxy-3'-O-demethylmatairesinol and 3,3'-didemethoxymatairesinol) were weak aromatase inhibitors in human preadipocytes (i.e., the determined IC_{50} values were 74 μM, 84 μM, and 60 μM, respectively). ED and its suggested intermediates (3'-O-demethylsecoisolariciresinol, 3'-demethoxysecoisolariceresinol, and 3,3'-O,O-didemethyl-secoisolariciresinol) were even weaker inhibitors (i.e., the IC_{50} concentrations were more than 100 μM; 25).

We have used the aromatase assay based on the identification and quantitation of the end product, ³H-estrone (8). The potential aromatase-inhibiting capacity of several lignans was tested in HEK-293 cells permanently transfected with human aromatase gene (Arom+HEK 293) (33). Of the tested lignans, EL was the most potent aromatase inhibitor with the IC_{50} concentration of 8.9 μM. The IC_{50} concentrations for secoisolariciresinol (SECO), ED, matairesinol (MR), and HMR were over 10 μM which was the highest tested concentration.

The aromatase inhibiting capacity of lignans has been studied *in vivo,* but the data achieved so far should be regarded as preliminary. The complete study would require the measurements of estrogen concentrations in serum or estrogen formation in ovary after inhibitor treatment, but the methods currently available for estrogen measurements are not sensitive enough, and further, the variation of the concentrations particularly in serum makes single-point determinations unreliable. The statistically significant decrease in uterine weight seen among the DMBA-treated female rats after the long-term administration of EL could be due to aromatase inhibition (8). However, we could not demonstrate aromatase inhibition by EL in a short-term test *in vivo* (i.e., an androstenedione-induced uterine growth assay). Because of the methodological limitations, no solid evidence was found to support the hypothesis that EL inhibits aromatization, and through this mechanism the tumor growth in the DMBA-induced mammary cancer model. Furthermore, the inhibition of aromatase could occur locally in tissues with relatively low aromatase activity, such as mammary gland or mammary tumors. Such an effect would not be observable in serum estrogen concentrations or in the inhibition of uterine growth, but might play a significant role in the aromatase-expressing target tissues (34).

Reduction of Growth Factors by Lignans

In addition to the transcription factors such as ER, alterations in the complex signal-transduction cascades could also contribute to the carcinogenic process. Controlling these events could offer specific molecular targets for the prevention of end-stage disease (35). High EL concentrations may have antiproliferative effects that are not associated with ER as mentioned earlier. The mechanism of this inhibition is not clear in MDA-MB-468 ER-negative human BC cells, because neither 100 μM EL nor ED affected epidermal growth factor-induced gene transcription, epidermal growth factor receptor tyrosine kinase activity, protein kinase C activity, or TPA-induced *c-fos* mRNA levels (36). However, high mammalian lignan concentrations did inhibit the colony formation (i.e., proliferation) of MDA-MB-468 cells, with IC_{50} values of 41 μM and 19.8 μM for EL and ED, respectively (36). The relevance of these findings for the understanding of lignan action *in vivo* is unclear.

The importance of insulin-like growth factor I receptor (IGF-IR) signaling in neoplastic transformation is clearly evident from a variety of studies. Insulin-like growth factor I (IGF-I) has well-characterized mitogenic and antiapoptotic effects in both normal and neoplastic mammary epithelial cells. These are mediated through the IGF-IR expressed on the surface of many cell types, including mammary gland epithelial cells. Strong positive associations have been reported between serum IGF-I levels and mammary cancer risks.

The response of DMBA-induced mammary tumors to estrogen-deprivation therapies involves reduction of IGF bioactivity secondary to upregulation of expression of IGF binding proteins. A decrease in the IGF-I gene expression of DMBA-induced tumors and concomitant regression of tumor size was seen after treatment of the rats with tamoxifen (37) or a potent and specific nonsteroidal aromatase inhibitor (vorozole) (32). Further, the gene expression of IGF binding proteins 2, 3, 4, and 5 increased rapidly in DMBA-induced mammary tumors following ovariectomy (38). On the other hand, estrogen has been shown to increase IGF-I mRNA, tyrosine phosphorylation of IGF-IR, and the formation of IGF-IR/IRS-1/p85 complex in target tissue (39). Unlike many other estrogen-regulated genes, the IGF-I gene apparently is not regulated through the classical mechanism that involves the direct binding of an E_2-ER complex to an estrogen response element in the regulatory regions of target genes. Rather, E_2 regulation of IGF-I gene transcription was shown to be mediated through an activating protein-1 (AP-1) site in the IGF-I gene promoter (40). SDG was shown to reduce plasma IGF-I in rats, suggesting that a reduction in the availability of IGF-I to bind to receptors may be one mechanism by which lignans decrease breast cancer risk (41). The possibility that lignans would control IGF-I gene expression through an AP-1 site needs to be explored.

Antioxidative Effects of Lignans

The chemopreventive action of plant lignans SDG, HMR, and EL was evident when the lignan administration was started prior to or after the tumor induction with DMBA. This suggests that mechanisms other than DMBA detoxification or anti-initiation action may be involved. Antioxidative activity of the dietary components is one suggested mechanism of action in mammary carcinogenesis. McCormick *et al.* (42) documented that synthetic antioxidants such as butylated hydroxy toluene (BHT) and butylated hydroxy anisole (BHA) showed antitumor activity when administered to rats *via* an undefined laboratory rodent diet. The doses of BHT and BHA needed for the antitumorigenic effect were 2500 mg/kg of diet or 5000 mg/kg of diet, respectively. The anticarcinogenic activity was evident when the administration had started 2 wk prior to or 1 wk after tumor induction, suggesting that mechanisms other than carcinogen detoxification may also be involved. Cohen *et al.* (43) also reported a significant inhibitory effect of BHT in a dose range of 300–6000 mg/kg of diet. We showed a strong antioxidative capacity for HMR *in vitro* when compared with mammalian lignans ED and EL (6). HMR was a 12-fold greater inhibitor of *t*-butylperoxide-induced lipid peroxidation in rat liver microsomes compared with BHA, and even 170-fold greater compared with BHT. In this assay, HMR was at least a 140-fold stronger antioxidant than the mammalian lignans ED and EL. Our recent analysis of rat urine after a single-dose administration of lignans clearly indicated that SECO and HMR may be absorbed as such from the alimentary tract (44). It may very well be that some of the anticarcinogenic properties of HMR may be due to the strong antioxidative activity of the compound itself. The antioxidative activity of EL is comparable to that of BHT, but much higher doses of BHT (300–6000 mg/kg of diet) were required for tumor inhibition (43) as compared with the EL doses (1 mg/kg and 10 mg/kg of b.w.). It is thus unlikely that EL action would be based solely on its antioxidative activity.

Conclusions

Delay in the clinical appearance is desirable strategy for controlling cancer. Prevention of cancer *sensu stricto* is not a realistic goal. Inhibition of the carcinogenesis through diverse mechanisms by structurally diverse compounds is most probably needed. Epidemiological data and chemoprevention experiments in animal models suggest that diets rich in lignans reduce the susceptibility to mammary cancer. Lignans are among the candidates to be added to future diets designed for chemoprevention of mammary cancer. The minimum or optimal dose of lignans associated with a reduced risk of cancer remains to be established. EL mediates in part the anticancer effects of plant lignans. When plant lignans such as SECO and HMR are absorbed from the alimentary tract, the possibility that they would also be anticarcinogenic, even through a mechanism different from that of EL, cannot be excluded. The possible mechanisms of the action of lignans could include (i) aromatase inhibition, (ii) reduction of bioactive IGF-I, and (iii) antioxidation. In general, lignans have been well tolerated in all animal studies performed so far.

Acknowledgments

Funded by the National Technology Agency of Finland (TEKES) and Hormos Nutraceutical Ltd.

References

1. Ingram, D., K. Sanders, M. Kolybaba, and D. Lopez, Case-Control Study of Phyto-Oestrogens and Breast Cancer, *Lancet 350*:990–994 (1997).
2. Pietinen, P., K. Stumpf, S. Männisto, B. Kataja, M. Uusitupa, and H. Adlercreutz, Serum Enterolactone and Risk of Breast Cancer: A Case-Control Study in Eastern Finland, *Cancer Epidemiol. Biomarkers Prev. 10*:339–344 (2001).

3. Hulten, K., A. Winkvist, P. Lenner, R. Johansson, H. Adlercreutz, and G. Hallmans, An Incident Case-Referent Study on Plasma Enterolactone and Breast Cancer Risk, *Eur. J. Nutr. 41*:168–176 (2002).

4. Thompson, L.U., M.M. Seidl, S.E. Rickard, L.J. Orcheson, and H.H. Fong, Antitumorigenic Effect of a Mammalian Lignan Precursor from Flaxseed, *Nutr. Cancer 26(2)*:159–165 (1996a).

5. Thompson, L.U., S.E. Rickard, L.J. Orcheson, and M.M. Seidl, Flaxseed and Its Lignan and Oil Components Reduce Mammary Tumor Growth at a Late Stage of Carcinogenesis, *Carcinogenesis 17(6)*:1373–1376 (1996b).

6. Saarinen, N.M., A. Wärri, S.I. Mäkelä, C. Eckerman, M. Reunanen, M. Ahotupa, S.M. Salmi, A.A. Franke, L. Kangas, and R. Santti, Hydroxymatairesinol, a Novel Enterolactone Precursor with Antitumor Properties from Coniferous Tree (*Picea abies*), *Nutr. Cancer 36*:207–216 (2000).

7. Saarinen, N.M., R. Huovinen, A. Wärri, S.I. Mäkelä, L. Valentín-Blasini, L. Needham, C. Eckerman, Y.U. Collan, and R. Santti, Uptake and Metabolism of Hydroxymatairesinol in Relation to Its Anticarcinogenicity in DMBA-Induced Rat Mammary Carcinoma Model, *Nutr. Cancer 41*(1,2):82–90 (2001).

8. Saarinen, N.M., R. Huovinen, A. Wärri, S.I. Mäkelä, L. Valentín-Blasini, R. Sjöholm, J. Ämmälä, R. Lehtilä, C. Eckerman, Y.U. Collan, and R.S. Santti, Enterolactone Inhibits the Growth of 7,12-Dimethylbenz[a]anthracene-Induced Mammary Carcinomas in the Rat, *Mol. Cancer Ther. 1*:869–876 (2002).

9. Jordan, V.C., S. Mittal, B. Gosden, R. Koch, and M.E. Lieberman, Structure-Activity Relationship of Estrogens, *Environ. Health Perspec. 61*:97–110 (1985).

10. Welshons, W.V., C.S. Murphy, R. Koch, G. Calaf, and V.C. Jordan, Stimulation of Breast Cancer Cells *in vitro* by the Environmental Estrogen Enterolactone and the Phytoestrogen Equol, *Breast Cancer Res. Treat. 10*:169–175 (1987).

11. Sathyamoorthy, N., T.T.Y. Wang, and J.M. Phang, Stimulation of pS2 Expression by Diet-Derived Compounds, *Cancer Res. 54*:957–961 (1994).

12. Wang, C., and M.S. Kurzer, Phytoestrogen Concentration Determines Effects on DNA Synthesis in Human Breast Cancer Cells, *Nutr. Cancer 28*:236–247 (1997).

13. Mousavi, Y., and H. Adlercreutz, Enterolactone and Estradiol Inhibit Each Other's Proliferative Effect on MCF-7 Breast Cancer Cells in Culture, *J. Steroid Biochem. Mol. Biol. 41*(3–8):615–619 (1992).

14. Wang, C., and M.S. Kurzer, Effects of Phytoestrogens on DNA Synthesis in MCF-7 Cells in the Presence of Estradiol or Growth Factors, *Nutr. Cancer 31*:90–100 (1998).

15. Hirano, T., K. Fukuoka, K. Oka, T. Naito, K. Hosaka, H. Mitsuhashi, and Y. Matsumoto, Antiproliferative Activity of Mammalian Lignan Derivatives Against the Human Breast Carcinoma Cell Line ZR-75-1, *Cancer Invest. 8*:595–602 (1990).

16. Barkhem, T., B. Carlsson, Y. Nilsson, E. Enmark, J. Gustafsson, and S. Nilsson, Differential Response of Estrogen Receptor Alpha and Estrogen Receptor Beta to Partial Estrogen Agonists/Antagonists, *Mol. Pharmacol. 54*:105–112 (1998).

17. Setchell, K.D., A.M. Lawson, E. Conway, N.F. Taylor, D.N. Kirk, G. Cooley, R.D. Farrant, S. Wynn, and M. Axelson, The Definitive Identification of the Lignans *trans*-2,3-*bis*(3-hydroxybenzyl)-gamma-butyrolactone and 2,3-*bis*(3-hydroxybenzyl)butane-1,4-diol in Human and Animal Urine, *Biochem. J. 197*:447–458 (1981).

18. Tou, J.C., J. Chen, and L.U. Thompson, Flaxseed and Its Lignan Precursor, Secoisolariciresinol Diglycoside, Affect Pregnancy Outcome and Reproductive Development in Rats, *J. Nutr. 128*:1861–1868 (1998).

19. Ward, W.E., J. Chen, and L.U. Thompson, Exposure to Flaxseed or Its Purified Lignan During Suckling Only or Continuously Does not Alter Reproductive Indices in Male and Female Offspring, *J. Toxicol. Environ. Health 64*:567–577 (2001).

20. Waters, A.P., and J.T. Knowler, Effect of a Lignan (HPMF) on RNA Synthesis in the Rat Uterus, *J. Reprod. Fert. 66*:379–381 (1982).

21. Orcheson, L.J., S.E. Rickard, M.M. Seidl, and L.U. Thompson, Flaxseed and Its Mammalian Lignan Precursor Cause a Lengthening or Cessation of Estrous Cycling in Rats, *Cancer Lett. 125*(1,2):69–76 (1998).

22. Evans, B.J., K. Griffiths, and M.S. Morton, Inhibition of 5α-Reductase in Genital Skin Fibroblasts and Prostate Tissue by Dietary Lignans and Isoflavonoids, *J. Endocr. 147*:295–302 (1995).

23. Wang, L-Q., Mammalian Phytoestrogens: Enterodiol and Enterolactone, *J. Chromatog. 777*:289–309 (2002).

24. Adlercreutz, H., C. Bannwart, K. Wähälä, T. Mäkelä, G. Brunow, T. Hase, P.J. Arosemena, J.T. Kellis, and L.E. Vickery, Inhibition of Human Aromatase by Mammalian Lignans and Isoflavonoid Phytoestrogens, *J. Steroid Biochem. Mol. Biol. 44*(2):147–153 (1993b).

25. Wang, C., T. Mäkelä, T. Hase, H. Adlercreutz, and M.S. Kurzer, Lignans and Flavonoids Inhibit Aromatase Enzyme in Human Preadipocytes, *J. Steroid Biochem. Mol. Biol. 50*(3,4):205–212 (1994).

26. Bonneterre, J., B. Thurlimann, J.F. Robertson, M. Krzakowski, L. Mauriac, P. Koralewski, I. Vergote, A. Webster, M. Steinberg, and M. von Euler, Anastrozole Versus Tamoxifen as First-Line Therapy for Advanced Breast Cancer in 668 Postmenopausal Women: Results of the Tamoxifen or Arimidex Randomized Group Efficacy and Tolerability Study, *J. Clin. Oncol. 18*:3748–3757 (2000).

27. Bonneterre, J., A. Buzdar, J.M. Nabholtz, J.F. Robertson, B. Thurlimann, M. von Euler, T. Sahmoud, A. Webster, and M. Steinberg, Anastrozole is Superior to Tamoxifen as First-Line Therapy in Hormone Receptor Positive Advanced Breast Carcinoma, *Cancer 92*:2247–2258 (2001).

28. Mouridsen, H., M. Gershanovich, Y. Sun, R. Perez-Carrion, C. Boni, A. Monnier, J. Apffelstaedt, R. Smith, H.P. Sleeboom, F. Janicke, A. Pluzanska, M. Dank, D. Becquart, P.P. Bapsy, E. Salminen, R. Snyder, M. Lassus, J.A. Verbeek, B. Staffler, H.A. Chaudri-Ross, and M. Dugan, Superior Efficacy of Letrozole Versus Tamoxifen as First-Line Therapy for Postmenopausal Women with Advanced Breast Cancer: Results of a Phase III Study of the International Letrozole Breast Cancer Group, *J. Clin. Oncol. 19*:2596–2606 (2001).

29. Houjou, T., T. Wada, and M. Yasutomi, Antitumor and Endocrine Effects of an Aromatase Inhibitor (CGS 16949A) on DMBA-Induced Rat Mammary Tumor, *Clin. Ther. 15*:137–147 (1993).

30. Dukes, M., The Relevance of Preclinical Models to the Treatment of Postmenopausal Breast Cancer, *Oncology 54(Suppl. 2)*:6–10 (1997).

31. Iino, Y., T. Karakida, N. Sugamata, T. Andoh, H. Takei, M. Takahashi, S. Yaguchi, T. Matsuno, M. Takehara, M. Sakato, S. Kawashima, and Y. Morishita, Antitumor Effects of SEF19, a New Nonsteroidal Aromatase Inhibitor, on 7,12-Dimethylbenz[a]anthracene-Induced Mammary Tumors in Rats, *Anticancer Res. 18*(1A):171–176 (1998).

32. Sugamata, N., Y. Koibuchi, Y. Iino, and Y. Morishita, A Novel Aromatase Inhibitor, Vorozole, Shows Antitumor Activity and a Decrease of Tissue Insulin-Like Growth

Factor-I Level in 7,12-Dimethylbenz[a]anthracene-Induced Rat Mammary Tumors, *Int. J. Oncol. 14*:259–263 (1999).

33. Li, X., E. Nokkala, W. Yan, T. Streng, N. Saarinen, A. Wärri, I. Huhtaniemi, R. Santti, S. Mäkelä, and M. Poutanen, Altered Structure Function of Reproductive Organs in Transgenic Male Mice Overexpressing Human Aromatase, *Endocrinol. 142*:2435–2442 (2001).

34. Simpson, E.R., and S.R. Davis, Minireview: Aromatase and the Regulation of Estrogen Biosynthesis—Some New Perspectives, *Endocrinol. 142*:4589–4594 (2001).

35. Sporn, M.B., and N. Suh, Chemoprevention: An Essential Approach to Controlling Cancer, *Nat. Rev. Cancer 2*:537–543 (2002).

36. Schultze-Mosgau, M.H., I.L. Dale, T.W. Gant, J.K. Chipman, D.J. Kerr, and A. Gescher, Regulation of c-fos Transcription by Chemopreventive Isoflavonoids and Lignans in MDA-MB-468 Breast Cancer Cells, *Eur. J. Cancer 34*:1425–1431 (1998).

37. Huynh, H.T., E. Tetenes, L. Wallace, and M. Pollak, *In vivo* Inhibition of Insulin-Like Growth Factor I Gene Expression by Tamoxifen, *Cancer Res. 53*:1727–1730 (1993).

38. Nickerson, T., J. Zhang, and M. Pollak, Regression of DMBA-Induced Breast Carcinoma Following Ovariectomy is Associated with Increased Expression of Genes Encoding Insulin-Like Growth Factor Binding Proteins, *Int. J. Oncol. 14*:987–990 (1999).

39. Salerno, M., D. Sisci, L. Mauro, M.A. Guvakova, S. Ando, and E. Surmacz, Insulin Receptor Substrate 1 is a Target for the Pure Antiestrogen ICI 182,780 in Breast Cancer Cells, *Int. J. Cancer 81*:299–304 (1999).

40. Umayahara, Y., R. Kawamori, H. Watada, E. Imano, N. Iwama, T. Morishima, Y. Yamasaki, Y. Kajimoto, and T. Kamada, Estrogen Regulation of the Insulin-Like Growth Factor I Gene Transcription Involves an AP-1 Enhancer, *J. Biol. Chem. 269*:16433–16442 (1994).

41. Rickard, S.E., Y.V. Yuan, and L.U. Thompson, Plasma Insulin-Like Growth Factor I Levels in Rats Are Reduced by Dietary Supplementation of Flaxseed or Its Lignan Secoisolariciresinol Diglycoside, *Cancer Lett. 161*:47–55 (2000).

42. McCormick, D.L., N. Major, and R.C. Moon, Inhibition of 7,12-Dimethyl-benz(a)anthracene-Induced Rat Mammary Carcinogenesis by Concomitant or Postcarcinogen Antioxidant Exposure, *Cancer Res. 44*:2858–2863 (1984).

43. Cohen, L.A., M. Polansky, K. Furuya, M. Reddy, B. Berke, and J.H. Weisburger, Inhibition of Chemically Induced Mammary Carcinogenesis in Rats by Short-Term Exposure to Butylated Hydroxytoluene (BHT): Interrelationships Among BHT Concentration, Carcinogen Dose, and Diet, *J. Natl. Cancer Inst. 72*:165–174 (1984).

44. Saarinen, N., A. Smeds, S. Mäkelä, J. Ämmälä, K. Hakala, J. Pihlava, E. Ryhänen, R. Sjöholm, and R. Santti, Structural Determinants of Plant Lignans for the Formation of Enterolactone *in vivo*, *J. Chromatogr. B 777*(1,2):311–319 (2002).

Chapter 11

α-Linolenic Acid and Cancer

Philippe Bougnoux and Véronique Chajès

Nutrition, Croissance, Cancer, INSERM E-0211, University François-Rabelais, 2, Boulevard Tonnellé, F-37032 Tours, France

Introduction

α-Linolenic acid (ALA, 18:3 n-3) is the major n-3 polyunsaturated fatty acid (PUFA) in the human diet. It is derived mainly from some vegetable oils (e.g., flaxseed, rapeseed) and is found in moderate concentrations in plant leaves (e.g., spinach, mustard, red leaf lettuce, buttercrunch lettuce, and in citrus fruits). Purslane, a commonly eaten plant in Crete, is rich in ALA (4.5 mg/g) (1). The importance of a Cretan-type diet, rich in ALA from vegetables and in docosahexaenoic acid (DHA) from fish, was demonstrated in several studies in relation to the risk of coronary heart disease, and it could account for the low rate of cancer in Crete (1). However, few epidemiological and experimental data are available on the association between dietary intake of ALA and cancer risk, and results are conflicting.

Epidemiological Studies

Breast Cancer

Few data are available on the specific association between dietary intake of ALA and breast cancer. In a case-control study conducted in Italy, specific food and fatty acid intakes were compared between 2569 women with breast cancer and 2588 control women (2). A high dietary intake of ALA was associated with a decreased risk of breast cancer (odds ratio for highest *vs.* lowest quintile 0.69). In the Netherlands Cohort Study on Diet and Cancer, a significant association was found between a high dietary intake of ALA and a decreased risk of breast cancer (odds ratio for highest *vs.* lowest quintile 0.70, 95% CI 0.51–0.97) (3). In contrast, in a case-control study conducted in Uruguay, in a population of 365 women with breast cancer and 397 controls, a high dietary intake of ALA was associated with an increased risk of breast cancer (odds ratio for highest *vs.* lowest quartile 3.79, 95% CI 1.53–9.40) (4). However, meat consumption is high in the Uruguayan population, one of the highest in the world, and the main source of ALA in Uruguay is red meat (53.8%) and not vegetables (10.7%). In a case-control study conducted in Uruguay, a high consumption of meat (total meat, red meat) was associated with an increased risk of breast cancer (odds ratio for highest *vs.* lowest quartile 4.2) (5). Thus, the positive association between dietary ALA and breast cancer risk could be the consequence of a high dietary intake of meat, source of carcinogenic heterocyclic amines. The role of ALA in the promotion of breast cancer needs to be further explored.

Several population-based studies were undertaken to investigate the relation between n-3 fatty acid composition, including ALA, and breast cancer risk, using biomarkers of past dietary intake of n-3 PUFA.

Several prospective cohort studies based on n-3 PUFA levels of phospholipids in prediagnostic sera, conducted in Norway (6), Sweden (7), and the United States (the New York University Women's Health Study) (8) found no significant association between ALA level and breast cancer risk (Table 11.1).

Using erythrocyte membrane fatty acid composition as a biomarker of past dietary intake of fatty acids, a prospective study conducted in northern Italy (the ORDET study) (9) found no significant association between ALA level and breast cancer risk (Table 11.1).

Because adipose tissue has been shown to best reflect dietary exposures to the essential fatty acids (with the linoleic acid, ALA, and long-chain n-3 PUFA) (10,11), several studies used the fatty acid composition of adipose tissue samples as a biomarker to investigate the relation between exposure to n-3 PUFA and breast cancer. The determination of fatty acid profiles in adipose tissue offers the advantages over questionnaire methods of dietary assessment in case-control studies of being free from recall bias and, unlike serum levels of fatty acids, of not being potentially altered by recent changes in diet that may occur due to the disease.

In a case-control study conducted in the United States (12), no significant association was found between ALA level and breast cancer risk (Table 11.1). In the European Community Multicenter Study on Antioxidants, Myocardial Infarction, and Cancer of the breast (EURAMIC), the fatty acid content of adipose tissue from the subcutaneous buttock in postmenopausal breast cancer cases and controls from five European countries (Germany, Switzerland, the Netherlands, Northern Ireland, and Spain) was determined (13). ALA alone showed a nonsignificant inverse association with breast cancer risk in three centers, Coleraine (odds ratio for highest *vs.* lowest tertile 0.66, CI 0.31–1.39), Berlin (odds ratio 0.88, 95% CI 0.15–5.18), and Zurich (odds ratio 0.54, 95% CI 0.22–1.30), and a non-significant positive association in the two other centers, Zeist (odds ratio 1.29, 95% CI 0.53–3.12) and Malaga (odds ratio 1.92, 95% CI 0.76–4.85). Pooling all centers gave little evidence of an inverse association with breast cancer for ALA (Table 11.1). In pooled analyses, the effect estimates for ALA were higher when considering the relation to n-6 fatty acids rather than absolute levels (Table 11.1). In our case-control studies conducted in central France, we found significant inverse associations between ALA level in breast adipose tissue and the relative risk of breast cancer (14,15; Table 11.1).

In conclusion, except for the case-control studies conducted in Europe and based on adipose tissue ALA level which found an inverse association between ALA or ALA/n-6 PUFA ratio and breast cancer risk, the other studies conducted in the United States or in Europe and based on serum or erythrocyte membrane ALA level found no

TABLE 11.1
Biomarkers of α-Linolenic Acid (ALA) Intake and Breast Cancer Risk

Study		Population				Odds ratio	
Reference	Type (country)	Cases	Controls	Biomarker	Comparison	(CI 95%)	P
(6)	Case-control study (Norway)	87	235	Serum phospholipids ALA level (mg/mL)	High vs. low quartile	0.6 (0.3–1.4)[a]	0.15
(7)	Prospective study (Sweden)	196	388	Serum phospholipids ALA level (% of total fatty acids)	High vs. low quartile	1.36 (0.63–2.96)[b]	0.42
(8)	Prospective study (New York, USA)	197	197	Serum phospholipids ALA level (% of total fatty acids)	High vs. low quartile	0.80 (0.44–1.46)c	0.48
(9)	Prospective study (Italy)	71	141	Erythrocyte membrane phospholipids ALA level (% of total fatty acids)	High vs. low tertile	1.38 (0.70–2.70)[d]	0.35
(12)	Case-control study (USA, Boston)	(Postmenopausal women) 380	176	Subcutaneous adipose tissue ALA level (% of total fatty acids)	High vs. low quintile	0.9 (0.6–1.5)[e]	0.59
(13)	Multicentric case-control study (Europe)	(All centers pooled, postmenopausal) 291	351	Subcutaneous adipose tissue ALA (% of total fatty acids) ALA/total n-6 fatty acids	All centers pooled High vs. low tertile	0.92 (0.53–1.60)[f] 0.66 (0.36–1.21)[f]	0.78 0.17
(14)	Case-control study (France)	123	59	Breast adipose tissue ALA level (% of total fatty acids)	High vs. low quartile	0.36 (0.12–1.02)[g]	0.026

| (15) | Case-control study (France) | 241 | 88 | Breast adipose tissue ALA level (% of total fatty acids) | High *vs.* low tertile | 0.39 (0.19–0.78)[h] | 0.01 |
| | | | | ALA/linoleic acid | | 0.41 (0.20–0.81)[h] | 0.0004 |

[a]The study was based on blood samples provided by the Janus serum bank, initiated in 1973, comprising nearly 500,000 samples from approximately 170,000 donors who had no diagnosed cancer at the time of blood donation. The case sera were obtained from women who developed breast cancer up to several years subsequent to blood donation. Three controls, free of any diagnosed cancer, born within 1 yr of a case, and who had donated blood within 6 mon of a case donation, were randomly selected for each case. Analysis was restricted to women who were 55 yr and younger (65 cases; 195 controls).

[b]The study was conducted in three cohort studies in northern Sweden. Two referents were randomly selected for each case from the cohort and matched for age, age of blood sample, and sampling center. Results were adjusted for age at menarche, age at full-term pregnancy, number of children, use of hormone-replacement therapy, height, and weight.

[c]The study is a part of the prospective New York University Women's Health Study on hormones, diet, and cancer. Controls were cohort members, free of cancer, randomly selected from among those who matched a case by age at recruitment, menopausal status at baseline, date of baseline blood samplings, and number of blood samplings before a case's date of diagnosis. Only the baseline blood samples were used for the analysis of fatty acids. Results were adjusted for age at first full-term birth, family history of breast cancer, history of benign breast disease, and total cholesterol.

[d]The study is part of a prospective study of hormones, diet, and breast cancer risk (ORDET) conducted in northern Italy. For each case, two matched control subjects were randomly selected from the cohort. Because body mass index, waist-to-hip ratio, age at menarche, age at first childbirth, age at menopause, months of lactation, parity, and educational level exerted a major confounding effect on the relationship of fatty acids to breast cancer risk, an unadjusted odds ratio is presented.

[e]The control group was composed of women with nonproliferative benign breast disease. Results were adjusted for age, alcohol intake, age at first birth, parity, family history of breast cancer, age at menopause, age at menarche, prior history of benign breast disease, and weight 5 yr before study entry.

[f]Controls were randomly selected from population registries in Germany and Switzerland. Other centers (the Netherlands, Northern Ireland, and Spain) drew controls from patient lists of the cases of general practitioners. Results were adjusted for age, body mass index, nulliparity, family history of breast cancer, age exceeding 35 yr at first childbirth, and study center.

[g]The control group was composed of women with benign breast pathologies. Results were adjusted for age at diagnosis, body mass index, and menopausal status.

[h]The control group was composed of women with benign breast pathologies. Results were adjusted for age at diagnosis, height, body mass index, menopause, and menopausal status-body mass index interaction.

association, or a trend to an increased risk of breast cancer with increased ALA level. Before any association can be concluded, other epidemiological studies integrating both dietary intakes of ALA (animal, vegetables sources) and biomarkers should be carried out.

Prostate Cancer

Several epidemiological studies examined the association between dietary intake of ALA and prostate cancer risk. Data from the prospective Health Professionals Follow-up Study conducted in the United States showed a positive association between dietary intake of ALA and prostate cancer risk (odds ratio for highest *vs.* lowest quintile of intake 3.43, 95% CI 1.67–7.04) (16). In this study, red meat was the food group with the strongest positive association with prostate cancer (odds ratio for highest *vs.* lowest quintile intake 2.64, 95% CI 1.21–5.77). However, the red meat and ALA associations were independent, suggesting that other sources of ALA, such as some vegetable oils or dairy foods, could be responsible for any real effects. Similar results were found in a case-control study conducted in an Uruguayan population of 217 men with advanced prostate cancer and 431 controls: a high intake of ALA was associated with an increased risk of prostate cancer (odds ratio for highest vs. lowest quartile 3.91; 95% CI 1.5–10.1) (17). Furthermore, in this study, both ALA from animal (odds ratio for highest *vs.* lowest quartile 2.98, 95% CI 1.02–8.68) and vegetable (odds ratio for highest *vs.* lowest quartile 2.03, 95% CI 1.01–4.07) sources displayed increased risks of prostate cancer. In a case-control study conducted in Spain in a population of 217 incident cases and 434 matched controls, the relative risk of prostate cancer increased in relation to increased dietary intake of ALA (odds ratio for highest *vs.* lowest quartile 3.1, 95% CI 1.1–3.8) (18). In contrast, in a cohort study conducted in the Netherlands, a (nonsignificant) decreased risk of prostate cancer was associated with increasing quintile of ALA (odds ratio for highest *vs.* lowest quintile 0.76, 95% CI 0.66–1.04) (19). Thus, most of the studies on the relationship between dietary ALA and prostate cancer showed positive associations, suggesting that ALA could enhance the risk of developing prostate cancer. Alternatively, ALA could be a marker of a more complex dietary pattern.

Few epidemiological studies based on biomarkers of past dietary intake of ALA are available, but results are strongly consistent (Table 11.2). In the prospective Physicians' Health Study conducted in the United States, a nested case-control study was used to compare fatty acid composition in plasma from 120 men who developed prostate cancer and 120 matched controls who did not: men with elevated levels of plasma ALA had a two- to threefold increase in risk of prostate cancer compared with those with low levels (Table 11.2) (20). These results were consistent with findings in the larger and independent prospective analysis from the Health Professionals Follow-up Study based on dietary histories (16). In a case-control study conducted in the United States on 89 cases and 38 controls and based on erythrocyte membrane and adipose tissue fatty acid composition as a biomarker, nonsignificant positive associations were found between ALA level, either in erythrocyte membrane or in

TABLE 11.2

Biomarkers of α-Linolenic Acid (ALA) Intake and Prostate Cancer Risk

| Study | | Population | | | | Odds ratio | |
Reference	Type (country)	Cases	Controls	Biomarker	Comparison	(CI 95%)	P
(20)	Nested case-control study (USA)	120	120	Plasma cholesterol esters ALA level (% of total fatty acids)	High *vs.* low quartile	2.05 (0.88–4.79)[a]	0.04
(22)	Nested case-control study (Norway)	141	282	Serum phospholipids ALA level (mg/L)	High *vs.* low quartile	2.0 (1.1–3.6)[b]	0.03
(21)	Case-control study (USA)	89	38	Erythrocytes membranes ALA level (%)	High *vs.* low quartile	1.69 (0.54–5.26)[c]	0.23
(21)	Case-control study (USA)	89	38	Adipose tissue	High *vs.* low quartile	2.73 (0.70–10.61)[c]	0.18
(23)	Case-control study (USA)	67	156	Erythrocyte membrane ALA level (% of total fatty acids)	Highest *vs.* lowest quartile	2.6 (1.1–5.8)[d]	0.01

[a]Part of the Physician's Health Study, on a population of 14,916 male physicians in the United States who provided plasma samples from 120 men who later developed prostate cancer and 120 controls who did not. Unadjusted estimated relative risks of prostate cancer were based on level of baseline plasma cholesterol fatty acids.

[b]The study was carried out as a nested case-control study among men with no known prostate cancer at the time of blood sampling. The population contributed serum to the Janus serum bank in Norway. Controls were matched to cases by country, age, and date of blood sample.

[c]Cases were recruited from a university-based urology outpatient clinic and had confirmation of a prostate cancer diagnosis within 1 yr of entry into the study. Controls were free of prostate cancer, were recruited from the same clinic over the same period, and had a prostate biopsy or a prostatectomy specimen that was free of prostate cancer. The base model included race and age as covariates.

[d]Cases were newly diagnosed with primary adenocarcinoma of the prostate. Controls were selected in the population and had a similar age distribution as the cases. Results were adjusted for age.

adipose tissue and prostate cancer risk (Table 11.2) (21). A similar finding was reported in a nested case-control study conducted in Norway, based on prediagnostic levels of fatty acids in serum: an increased risk of prostate cancer was associated with an increased level of ALA (22) (Table 11.2). Similarly, in a case-control study conducted in the United States in a population of 67 incident prostate cancer cases and 156 population-based controls, erythrocyte membrane n-3 PUFA levels were determined: prostate cancer risk was increased with an increased level of ALA (Table 2) (23). In a study conducted in Korea based on serum fatty acid levels as biomarker of dietary intakes, the authors reported higher levels of ALA in patients with prostate cancer (mean: 2.88% of total fatty acids) than in patients with benign prostatic hyperplasia (mean: 2.21%) or subjects without evidence of benign or prostatic disease (mean: 1.97%) (24).

In conclusion, few epidemiological data are yet available on the association between dietary intake of ALA and cancer, and findings on the association with breast cancer are conflicting. A consistent positive association between dietary intake of ALA or biomarkers of ALA intake and prostate cancer risk was found and remains to be clarified.

Experimental Studies

A few studies have specifically addressed the role of ALA-enriched diets on tumorigenesis in experimental animal systems. It was concluded from most of the experimental systems investigated that ALA can inhibit tumor growth and development at several steps.

In a model of mouse syngenic tumors, linseed (flaxseed) oil (10%) in the diet reduced the growth of grafted mammary tumors as well as metastases, but 10% corn oil or a mixture of fish oil and corn oil did not (25). It is of interest that tumor growth inhibition was not correlated to inhibition of prostaglandin E synthesis, suggesting a specific effect of linseed oil independent of n-3 PUFA-induced prostaglandin inhibition. Spontaneous mammary tumorigenesis in mice was significantly inhibited by dietary ALA (26). Mistol seed oil, rich in ALA, had protective effects on survival and metastases in another mouse syngenic grafted mammary tumor system (27). Hirose *et al.* (28) used a rat model of carcinogen-induced mammary and colon tumors, and provided diets containing 10% of either perilla oil (containing more than 45% ALA) or safflower oil (rich in n-6 PUFA, less than 1% ALA), or soybean oil (rich in n-6 PUFA, 5–7% ALA). They found that the number of mammary tumors per rat was significantly lower in the group of rats given perilla oil diet than in the soybean oil diet group. They also found that colon tumor incidence was lower in animals receiving the perilla oil supplement than in those given safflower oil diet, and the numbers of colon tumors per rat tended to be lowest in rats administered perilla oil. Also, the incidence of nephroblastomas in rats receiving perilla oil diet was significantly lower than that for the soybean oil diet group. The results thus indicate that the ALA (n-3 PUFA)-rich perilla oil diet inhibits development of mammary gland, colon, and kidney tumors as compared with the linoleic acid (n-6 PUFA)-rich safflower or soybean oil diet (28). Similar results were reported using other experimental animal systems. Narisawa *et al.* (29) found that perilla oil had an inhibitory effect on rat colon carcinogenesis induced by intrarectal *N*-methyl-*N*-nitrosourea through a decreased sensitivity of colonic mucosa to tumor promoters. They further found that a small proportion of perilla oil in the diet was able to reduce the risk of colon cancer in the same experimental system (30). Onogi *et al.* (31) found that perilla oil suppressed the early stages of colon carcinogenesis as assessed in azoxymethane-induced foci of colonic aberrant crypts in rats in a dose-dependent fashion. In contrast, ALA did not reduce tumorigenesis in Apc(Min/+) mice, whereas long chain n-3 PUFA did, although ALA like n-3 PUFA led to a decrease in prostaglandin synthesis (32).

In a diethylnitrosamine-induced hepatocarcinogenesis rat model, Okuno *et al.* (33) found that perilla oil and also oils containing n-6 PUFA inhibited multiplicity of liver adenoma but not carcinoma.

With respect to prostate cancer, experimental investigations into the actions of dietary n-3 PUFA remain at an early stage and have been hampered by a lack of satisfactory animal models. A diet enriched in long chain n-3 PUFA from fish oil (not ALA) reduced the growth of tumors from DU-145 transplantable human prostatic cells in nude mice (34). Using a rat system of prostate carcinogenesis induced by 3,2'-dimethyl-4-aminobiphenyl along with testosterone propionate as a prostate tumor promoter, Mori *et al.* (35) found perilla oil (20%) reduced statistically the incidence of prostatic intraepithelial neoplasia in the ventral lobe from 70 to 10%, compared with the same amount of beef tallow, indicating that perilla oil, which contains a high level of ALA, is not a promoter of prostate carcinogenesis under these experimental conditions. In summary, epidemiological studies combined with animal model and *in vitro* experiments indicate that natural components of the diet such as n-3 PUFA may serve as chemopreventive agents that suppress the growth and dissemination of neoplastic prostate cells (36). However, experiments specifically designed to investigate the action of ALA rather than long chain n-3 PUFA are still missing.

A protective effect of ALA on carcinogenesis has not been consistently reported. Dietary ALA increased liver metastases in a model of BOP-induced pancreatic ductular carcinoma in the Syrian golden hamster (37). ALA had no effect on tumor growth or metastasis to the liver in mice implanted intrasplenically with highly metastatic Lewis lung carcinoma tumor cells (38). In a mouse transplantable lung carcinoma cell line, a diet containing n-3 PUFA as fish oil or linseed oil elicited more liver metastases than a diet with n-6 PUFA or saturated fatty acids (39). In a two-stage urinary bladder carcinogenesis male rat model, ALA was not protective (40). The reasons for these discrepancies are not known and may need further experimentation in other *in vivo* models.

ALA is never provided experimentally as a single dietary component, but rather as a component of food or of oil added to the diet. Whether ALA is solely responsible for tumor inhibition, or whether there is a role for the other components present in oils remains unanswered. To address this issue, Thompson *et al.* (41) investigated the effect of flaxseed oil, which is rich in ALA (more than 40%), along with other flaxseed components such as the lignan precursor secoisolariciresinol-diglycoside. Using a rat model of carcinogen-induced mammary tumors, they investigated the inhibitory effect of the lignan precursor compared with that of the oil during the late phase of tumor development. They concluded that the oil had an inhibitory effect on tumor growth independent of the lignan, and that ALA present in high amounts in flaxseed oil had protective effects (41), a conclusion that was also provided by the comparison of flaxseed with soybean oils (42).

Is the antitumor effect of ALA influenced by its interaction with n-6 PUFA? This possibility has already been suggested in experimental mammary carcinogenesis, where tumor growth was suppressed only when equal parts of n-3 and n-6 PUFA-rich oils were fed (43,44). The role of PUFA relative to other fatty acids was further investigated. Sasaki *et al.* (45) used a 10% fat diet in a rat model of mammary carcinogenesis, and modified the relative proportion of n-3 and n-6 PUFA, through mixing coconut oil (rich in saturates), safflower oil (rich in linoleate), and fish oil [from sardine, rich in eicosapentaenoic acid (EPA) and DHA] in such a way that the ratio of total PUFA over saturated fatty acid was kept constant. They found that increasing the n-3/n-6 PUFA ratio (from 0.01 to 7.8) did not suppress the incidence or reduce the latency of mammary tumor development, but even promoted development of tumors. The potential confounding effect of antioxidants was not examined. Although long chain n-3 PUFA were specifically investigated, a similar effect is likely to exist for ALA. Using a model of transgenic mice with the c-neu breast cancer oncogene under the control of a MMTV promoter, Rao *et al.* (46) found that flaxseed oil along with melatonin decreased the number of tumors and tumor weight per mouse, and concluded that flaxseed oil may delay the growth of mammary tumors if the n-6: n-3 PUFA ratio of fat consumed is close to 1. Thus the antitumor action of ALA seems to depend on the amount of other PUFA present in the diet.

Another mechanism involved in the tumor growth inhibition by n-3 PUFA is the promotion of apoptosis of tumor cells. Oxidation products of n-3 PUFA have been known to contribute to their cytotoxicity. Lipoperoxides are directly cytotoxic, and deficiencies in antioxidant defense mechanisms have been reported to enhance the cytotoxic effect of n-3 PUFA. Recent data from several studies underline this emerging important role of PUFA in cancer. Ramesh and Das (47) studied methylcholanthrene-induced sarcoma ascitic tumor cells *in vitro* and *in vivo*. They found that n-3 PUFA (DHA>ALA>EPA) were the most potent in inhibiting the growth of tumor cells. Vitamin E partially blocked the cytotoxicity of these fatty acids, and lipid peroxidation was enhanced by all fatty acids tested. Using a rat mammary carcinogenesis model with diet rich in PUFA, Lhuillery *et al.* (48) reported that tumor incidence and growth was increased in rats receiving additional vitamin E in their diet, suggesting that even in this model of late stages of carcinogenesis, oxidized PUFA have an inhibiting role on tumor growth. Using diets rich in PUFA and containing a high level of ALA (linseed oil) in an identical experimental system, Cognault *et al.* (49) found that tumor growth increased in the presence of vitamin E and decreased when prooxidant agents were added to the diet. When they used a diet low in PUFA and devoid of ALA (hydrogenated palm/sunflower oil), no difference was observed in the presence or absence of vitamin E, suggesting that sensitivity of PUFA to peroxidation may interfere with tumor growth. Growth kinetic parameter analysis indicated that tumor growth resulted from variations in cell loss but not from changes in cell proliferation. These data show that, *in vivo*, PUFA effects on tumor growth are highly dependent on diet oxidative status.

Conclusion

Dietary ALA inhibits tumor growth and development in several experimental systems and therefore belongs to a family of promising molecules for cancer prevention. Recent studies highlight the importance of the interaction with other food components. Presence or absence of effects on tumor growth, depending on background level of n-6 PUFA or antioxidants, may account for previously inconsistent results in experimental carcinogenesis. Recognition of the role of lipid peroxidation in the antitumor effects of n-3 PUFA, apparent in a variety of *in vitro* or *in vivo* systems, represents a major advance in the field.

References

1. Simopoulos, A.P., The Mediterranean Diets: What Is So Special About the Diet of Greece? The Scientific Evidence, *J. Nutr. 131*:3065S–3073S (2001).
2. Franceschi, S., A. Favero, A. Decarli, E. Negri, C. La Vecchia, M. Ferraroni, A. Russo, S. Salvini, D. Amadori, E. Conti, M. Montella, and A. Giacosa, Intake of Macronutrients and Risk of Breast Cancer, *Lancet 347*:1351–1356 (1996).
3. Voorrips, L., H. Brants, A. Kardinaal, G. Hiddink, and P. van den Brandt, Intake of Conjugated Linoleic Acid, Fat, and Other Fatty Acids in Relation to Postmenopausal Breast Cancer: The Netherlands Cohort Study on Diet and Cancer, *Am. J. Clin. Nutr. 76*:873–882 (2002).
4. De Stéfani, E., H. Deneo-Pellegrini, M. Mendilaharsu, and A. Ronco, Essential Fatty Acids and Breast Cancer: A Case-Control Study in Uruguay, *Int. J. Cancer 76*:491–494 (1998).
5. Ronco, A., E. De Stefani, M. Mendilaharsu, and H. Deneo-Pellegrini, Meat, Fat, and Risk of Breast Cancer: A Case-Control Study from Uruguay, *Int. J. Cancer 65*:328–331 (1996).
6. Vatten, L.J., K.S. Bjerve, A. Andersen, and E. Jellum, Polyunsaturated Fatty Acids in Serum Phospholipids and Risk of Breast Cancer: A Case-Control Study from the Janus Bank in Norway, *Eur. J. Cancer 29A*:532–538 (1993).
7. Chajès, V., K. Hultén, A.L. Van Kappel, A. Winkvist, R. Kaaks, G. Hallmans, P. Lenner, and E. Riboli, Fatty Acid Composition in Serum Phospholipids and Risk of Breast Cancer: An Incident Case-Control Study in Sweden, *Int. J. Cancer 83*:585–590 (1999).
8. Saadatian-Elahi, M., P. Toniolo, P. Ferrari, J. Goudable, A. Akhmedkhanov, A. Zeleniuch-Jacquotte, and E. Riboli, Serum Fatty Acids and Risk of Breast Cancer in a Nested Case-Control Study of the New York University Women's Health Study, *Cancer Epidemiol. Bio. Prev. 11*:1353–1360 (2002).

9. Pala, V., V. Krogh, P. Muti, V. Chajès, E. Riboli, A. Micheli, M. Saadatian, S. Sieri, and F. Berrino, Erythrocyte Membrane Fatty Acids and Subsequent Breast Cancer: A Prospective Italian Study, *J. Natl. Cancer Inst. 93*:1088–1095 (2001).

10. London, S.J., F.M. Sacks, J. Caesar, M.J. Stampfer, E. Siguel, and W.C. Willett, Fatty Acid Composition of Subcutaneous Adipose Tissue and Diet in Postmenopausal US Women, *Am. J. Clin. Nutr. 54*:340–345 (1991).

11. Kohlmeier, L., and M. Kohlmeier, Adipose Tissue as a Medium for Epidemiologic Exposure Assessment, *Environ. Health Perspect. 103*:99–106 (1995).

12. London, S.J., F.M. Sacks, M.J. Stampfer, I.C. Henderson, H. Maclure, A. Tomita, W.C. Wood, S. Remine, N.J. Robert, and J.R. Dmochowski, Fatty Acid Composition of the Subcutaneous Adipose Tissue and Risk of Proliferative Benign Breast Disease and Breast Cancer, *J. Natl. Cancer Inst. 85*:785–793 (1993).

13. Simonsen, N., P. Van't Veer, J.J. Strain, J.M. Martin-Moreno, J.K. Huttunen, J.F. Navajas, B.C. Martin, M. Thamm, A.R. Kardinaal, F.J. Kok, and L. Kohlmeier, Adipose Tissue Omega-3 and Omega-6 Fatty Acid Content and Breast Cancer in the EURAMIC Study, *Am. J. Epidemiol. 147*:342–352 (1998).

14. Klein, V., V. Chajès, E. Germain, G. Schulgen, M. Pinault, D. Malvy, T. Lefrancq, A. Fignon, O. Le Floch, C. Lhuillery, and P. Bougnoux, Low Alpha-Linolenic Acid Content of Adipose Breast Tissue Is Associated with an Increased Risk of Breast Cancer, *Eur. J. Cancer 36*:335–340 (2000).

15. Maillard, V., P. Bougnoux, P. Ferrari, M.L. Jourdan, M. Pinault, F. Lavillonnière, G. Body, O. Le Floch, and V. Chajès, N-3 and n-6 Fatty Acids in Breast Adipose Tissue and Relative Risk of Breast Cancer in a Case-Control Study in Tours, France, *Int. J. Cancer 98*:78–83 (2002).

16. Giovannucci, E., E.B. Rimm, G.A. Colditz, M.J. Stampfer, A. Ascherio, C.C. Chute, and W.C. Willett, A Prospective Study of Dietary Fat and Risk of Prostate Cancer, *J. Natl. Cancer Inst. 85*:1571–1579, 1993.

17. De Stéfani, E., H. Deneo-Pellegrini, P. Boffetta, A. Ronco, and M. Mendilaharsu, α-Linolenic Acid and Risk of Prostate Cancer: A Case-Control Study in Uruguay, *Cancer Epidemiol. Bio. Prev. 9*:335–338, 2000.

18. Ramon, J.M., R. Bou, S. Romea, M.E. Alkiza, M. Jacas, J. Ribes, and J. Oromi, Dietary Fat Intake and Prostate Cancer Risk: A Case-Control Study in Spain, *Cancer Causes Control 11*:679–685 (2000).

19. Schuurman, A.G., P.A. Van den Brandt, E. Dorant, H.A.M. Brants, and R.A. Goldbohm, Association of Energy and Fat Intake with Prostate Carcinoma Risk: Results from the Netherlands Cohort Study, *Cancer 86*:1019–1027 (1999).

20. Gann, P.H., C.H. Hennekens, F.M. Sacks, F. Grodstein, E.L. Giovannucci, and M.J. Stampfer, Prospective Study of Plasma Fatty Acids and Risk of Prostate Cancer, *J. Natl. Cancer Inst. 86*:281–286 (1994).

21. Godley, P., M. Campbell, P. Gallagher, F. Martinson, J. Mohler, and R. Sandler, Biomarkers of Essential Fatty Acid Consumption and Risk of Prostatic Carcinoma, *Cancer Epidemiol. Biol. Prev. 5*:889–895 (1996).

22. Harvei, S., K.S. Bjerve, S. Tretli, E. Jellum, T.E. Robsahm, and L. Vatten, Prediagnostic Level of Fatty Acids in Serum Phospholipids: ω-3 and ω-6 Fatty Acids and the Risk of Prostate Cancer, *Int. J. Cancer 71*:545–551 (1997).

23. Newcomer, L.M., I.B. King, K.G. Wicklund, and J.L. Standford, The Association of Fatty Acids with Prostate Cancer Risk, *The Prostate 47*:262–268 (2001).

24. Yang, Y.J., S.H. Lee, S.J. Hong, and B.C. Chung, Comparison of Fatty Acid Profiles in the Serum of Patients with Prostate Cancer and Benign Prostatic Hyperplasia, *Clin. Biochem. 32*:405–409 (1999).

25. Fritsche, K.L., and P.V. Johnston, Effect of Dietary α-Linolenic Acid on Growth, Metastasis, Fatty Acid Profile and Prostaglandin Production of Two Murine Mammary Adenocarcinomas, *J. Nutr. 120*:1601–1609 (1990).

26. Kamano, K., H. Okuyama, R. Konishi, and H. Nagasawa, Effects of a High-Linoleate and a High-α-Linolenate Diet on Spontaneous Mammary Tumourigenesis in Mice, *Anticancer Res. 9*:1903–1908 (1989).

27. Munoz, S.F., R.A. Silva, A. Lamarque, C.A. Guzman, and A.R. Eynard, Protective Capability of Dietary Zizyphus Mistol Seed Oil, Rich in 18:3, n-3, on the Development of Two Murine Mammary Gland Adenocarcinomas with High or Low Metastatic Potential, *Prostaglandins Leukot. Essent. Fatty Acids 53*:135–138 (1995).

28. Hirose, M., A. Masuda, N. Ito, K. Kamano, and H. Okuyama Effects of Dietary Perilla Oil, Soybean Oil and Safflower Oil on 7,12-Dimethylbenz[a]anthracene (DMBA) and 1,2-Dimethyl-hydrazine (DMH)-Induced Mammary Gland and Colon Carcinogenesis in Female SD Rats, *Carcinogenesis 11*:731–735 (1990).

29. Narisawa, T., M. Takahashi, H. Kotanagi, H. Kusaka, Y. Yamazaki, H. Koyama, Y. Fukaura, Y. Nishizawa, M. Kotsugai, and Y. Isoda *et al.*, Inhibitory Effect of Dietary Perilla Oil Rich in the ω-3 Polyunsaturated Fatty Acid α-Linolenic Acid on Colon Carcinogenesis in Rats, *Jpn. J. Cancer Res. 82*:1089–1096 (1991).

30. Narisawa, T., Y. Fukaura, K. Yazawa, C. Ishikawa, Y. Isoda, and Y. Nishizawa, Colon Cancer Prevention with a Small Amount of Dietary Perilla Oil High in α-Linolenic Acid in an Animal Model, *Cancer 73*:2069–2075 (1994).

31. Onogi, N., M. Okuno, C. Komaki, H. Moriwaki, T. Kawamori, T. Tanaka, H. Mori., and Y. Muto, Suppressing Effect of Perilla Oil on Azoxymethane-Induced Foci of Colonic Aberrant Crypts in Rats, *Carcinogenesis 17*:1291–1296 (1996).

32. Petrik, M.B., M.F. McEntee, B.T. Johnson, M.G. Obukowicz, and J. Whelan, Highly Unsaturated (n-3) Fatty Acids, but Not α-Linolenic, Conjugated Linoleic or γ-Linolenic Acids, Reduce Tumorigenesis in Apc(Min/+) Mice, *J. Nutr. 130*:2434–2443 (2000).

33. Okuno, M., T. Tanaka, C. Komaki, S. Nagase., Y. Shiratori, Y. Muto, K. Kajiwara, T. Maki, and H. Moriwaki, Suppressive Effect of Low Amounts of Safflower and Perilla Oils on Diethylnitrosamine-Induced Hepatocarcinogenesis in Male F344 Rats, *Nutr. Cancer 30*:186–193 (1998).

34. Karmali, R.A., P. Reichel, L.A. Cohen, T. Terano, A. Hirai, Y. Tamura, and S. Yoshida, The Effects of Dietary Omega-3 Fatty Acids on the DU-145 Transplantable Human Prostatic Tumor, *Anticancer Res. 7*:1173–1179 (1987).

35. Mori, T., K. Imaida, S. Tamano, M. Sano, S. Takahashi, M. Asamoto, M. Takeshita, H. Ueda, and T. Shirai, Beef Tallow, but Not Perilla or Corn Oil, Promotion of Rat Prostate and Intestinal Carcinogenesis by 3,2'-Dimethyl-4-aminobiphenyl, *Jpn. J. Cancer Res. 92*:1026–1033 (2001).

36. Cohen, L.A., Nutrition and Prostate Cancer: A Review, *Ann. N.Y. Acad. Sci. 963*:148–155 (2002).

37. Wenger, F.A., C.A. Jacobi, M. Kilian, J. Zieren, H.U. Zieren, and J.M. Muller, Does Dietary α-Linolenic Acid Promote Liver Metastases in Pancreatic Carcinoma Initiated by BOP in Syrian Hamster? *Ann. Nutr. Metab. 43(2)*:121–126 (1999).

38. Kimura, Y., Carp Oil or Oleic Acid, but Not Linoleic Acid or Linolenic Acid, Inhibits Tumor Growth and Metastasis in Lewis Lung Carcinoma-Bearing Mice, *J. Nutr. 132*:2069–2075 (2002).

39. Coulombe, J., G. Pelletier, P. Tremblay, G. Mercier, and D. Oth, Influence of Lipid Diets on the Number of Metastases and Ganglioside Content of H59 Variant Tumors, *Clin. Exp. Metastasis 15*:410–417 (1997).

40. Kitano, M., S. Mori, T. Chen, T. Murai, and S. Fukushima, Lack of Promoting Effects of α-Linolenic, Linoleic or Palmitic Acid on Urinary Bladder Carcinogenesis in Rats, *Jpn. J. Cancer Res. 86*:530–534 (1995).

41. Thompson, L.U., S.E. Rickard, L.J. Orcheson, and M.M. Seidl, Flaxseed and Its Lignan and Oil Components Reduce Mammary Tumor Growth at a Late Stage of Carcinogenesis, *Carcinogenesis 17*:1373–1376 (1996).

42. Rickard, S.E., Y.V. Yuan, J. Chen, and L.U. Thompson, Dose Effects of Flaxseed and Its Lignan on *N*-Methyl-*N*-Nitrosourea-Induced Mammary Tumorigenesis in Rats, *Nutr. Cancer 35*:50–57 (1999).

43. Ip, C., M.M. Ip, and P. Sylvester, Relevance of *trans* Fatty Acids and Fish Oil in Animal Tumorigenesis Studies, *Prog. Clin. Biol. Res. 222*:283–294 (1986).

44. Cohen, L.A., J.Y. Chen-Backlund, D.W. Sepkovic, and S. Sugie, Effect of Varying Proportions of Dietary Menhaden and Corn Oil on Experimental Rat Mammary Tumor Promotion, *Lipids 28*:449–456 (1993).

45. Sasaki, T., Y. Kobayashi, J. Shimizu, M. Wada, S. In'nami, Y. Kanke, and T. Takita, Effects of Dietary n-3-to-n-6 Polyunsaturated Fatty Acid Ratio on Mammary Carcinogenesis in Rats, *Nutr. Cancer 30*:137–143 (1998).

46. Rao, G.N., E. Ney, and R.A. Herbert, Effect of Melatonin and Linolenic Acid on Mammary Cancer in Transgenic Mice with c-neu Breast Cancer Oncogene, *Breast Cancer Res. Treat. 64*:287–296 (2000).

47. Ramesh, G., and D.N. Das, Effect of *cis*-Unsaturated Fatty Acids on Meth-A Ascitic Tumour Cells *in vitro* and *in vivo*, *Cancer Lett. 123*:207–214 (1998).

48. Lhuillery, C., S. Cognault, E. Germain, M.L. Jourdan, and P. Bougnoux, Suppression of the Promoter Effect of Polyunsaturated Fatty Acids by the Absence of Dietary Vitamin E Experimental Mammary Carcinoma, *Cancer Lett. 114*:233–234 (1997).

49. Cognault, S., M.L. Jourdan, E. Germain, R. Pitavy, E. Morel, G. Durand, P. Bougnoux, and C. Lhuillery, Effect of an α-Linolenic Acid-Rich Diet on Rat Mammary Tumor Growth Depends on the Dietary Oxidative Status, *Nutr. Cancer 36*:33–41 (2000).

Chapter 12

α-Linolenic Acid and Heart Disease

Duo Li[a,b], Nadia Attar-Bashi[b], and Andrew J. Sinclair[b]

[a]Department of Food Science, Hangzhou University of Commerce, Hangzhou, China 310035; and [b]Department of Food Science, Royal Melbourne Institute of Technology University, Melbourne, Australia 3000

Introduction

Dietary fat intake is known to play a critical role in influencing coronary heart disease (CHD) risk factors (1–3). Dietary saturated fatty acid (SFA) intake, especially myristic and palmitic acids, increases plasma cholesterol and low-density lipoprotein cholesterol (LDL-C) and is associated with CHD mortality (2). Polyunsaturated fatty acid (PUFA) has generally been associated with lowered plasma cholesterol and LDL-C and CHD risk (4). Recent research suggests that the diet-CHD relationship is more complex than previously recognized. The role of n-3 PUFA in the diet and tissues is of increasing interest in the medical research literature. α-Linolenic acid (ALA, 18:3n-3) is an essential fatty acid that can be used to synthesize longer chain n-3 PUFA such as eicosapentaenoic acid (EPA, 20:5n-3), docosapentaenoic acid (DPA, 22:5n-3), and docosahexaenoic acid (DHA, 22:6n-3) by desaturation and elongation steps. In the past three decades, a substantial number of studies have examined the effect of n-3 PUFA from marine sources on CHD mortality and risk factors (5). However, the relationship between ALA from plant sources and CHD mortality and risk factors is relatively less studied. In this chapter, we consider the effect of ALA on CHD mortality and risk factors, based on publications from the most recent international literature.

One issue to consider in any report on ALA is whether ALA itself has a biological role, or whether ALA is merely important as a substrate for the production of EPA, DPA, and DHA. In plants, ALA clearly plays an important role because lipoxygenase products of ALA include jasmonic acid and related phytohormones produced via the oxylipin cascade (6). The evidence in mammals is less clear. In humans, feeding high ALA diets (up to 15 g/d) in 4-wk studies led to significant increases in ALA, EPA, and DPA in plasma triacylglycerols (TAG) and phospholipids, and very little if any detectable increase in DHA in plasma, platelets, and white and red blood cells (7–9). Deuterium-labeling studies in humans with ALA showed that the percentage conversion of ALA to EPA and other long-chain n-3 PUFA was between 11–19% of the dose, and that the conversion rate was reduced by 40–54% when the diet was rich in linoleic acid (10). In a recent study, physiological compartmental analysis of ALA metabolism in adult humans was carried out. Subjects received a 1-g oral dose of an isotope tracer of ALA and only about 0.2% of the plasma ALA was destined for synthesis of EPA; approximately 63% of the plasma EPA was accessible for production of DPA, and 37% of DPA was available for synthesis of DHA (11). The very limited conversion of ALA to EPA indicates that the biosynthesis of long-chain n-3 PUFA from ALA is limited in healthy individuals.

α-Linolenic Acid Protection Against Coronary Heart Disease

There is some evidence from recent prospective, cross-sectional and intervention studies that the dietary intake of ALA is protective against CHD.

Prospective Studies

Two prospective cohort investigations from the USA Health Professionals study and Nurses' Health study examined the effect of the intake of dietary fat and n-3 PUFA from plants on CHD in humans. In the first study, 43,757 healthy male professionals aged 40–75 yr, free of diagnosed cardiovascular disease (CVD) or diabetes were followed up for 6 yr from 1986 (12). Each subject completed a food-frequency questionnaire at the beginning of the study. The subjects returned food-frequency questionnaires in each 2-yr follow-up cycle. During the follow-up, 505 nonfatal myocardial infarctions and 229 deaths were documented. After adjustment for nondietary risk factors and total fat intake, intake of ALA was significantly negatively correlated with risk of myocardial infarction (relative risk 0.41 for a 1% increase in energy from ALA, $P < 0.01$). In the second study, the dietary intake of ALA was calculated from a food-frequency questionnaire completed in 1984 by 76,283 nurses aged 38–63 yr, free from previously diagnosed CVD and cancer. There were 597 cases of nonfatal myocardial infarction and 232 cases of fatal ischemic heart disease documented during 10 yr of follow-up. After the adjustment of confounding factors, such as age, standard coronary risk factors, and dietary intake of linoleic acid (LA) and other nutrients, the results showed that women who had a higher intake of ALA ($\geq$ 5–6 times/wk) were significantly associated with reduced risk of fatal ischemic heart disease compared with women who consumed ALA less than once per month in this study population ($P < 0.001$) (13). The major foods contributing to ALA in this study were mayonnaise or other creamy salad dressings and oil and vinegar salad dressings.

Cross-sectional Studies

The Family Heart Study, a cross-sectional study by the USA National Heart, Lung, and Blood Institute, found that higher intakes of either ALA or LA were inversely related to the prevalence of coronary artery disease (CAD) (14). Dietary intakes of 4584 volunteers were assessed with a semi-quantitative food-frequency questionnaire. After adjustment for confounding factors such as age, LA, and anthropometric, lifestyle, and metabolic factors, the prevalence odds ratios of CAD from the lowest to the highest quintile of ALA intake were 1.0, 0.77, 0.61, 0.58, and 0.60 for the men (P = 0.012) and 1.0, 0.57, 0.52, 0.30, and 0.42 for the women (P = 0.014). LA was also inversely related to the prevalence odds ratios of CAD in the multivariate model (0.60

and 0.61 in the second and third quintiles, respectively) after adjustment for ALA. It was noted that the combined effects of LA and ALA were stronger than the effect of either of the fatty acids individually.

Dietary Intervention Studies

The Lyon diet-heart study concluded that ALA prevented secondary CHD (15). In this study, the diet chosen was associated with a low mortality rate from CHD and all causes in the Seven-Countries Study (16). The Cretan diet had a high intake of ALA and was rich in antioxidants because it was rich in fruits and vegetables. Crete had a lower mortality rate from CHD compared with similar cohorts in other countries. Cretan participants had threefold higher serum concentrations of ALA compared with a similar cohort from the Netherlands (17). In the Lyon study, 605 patients who had suffered a first myocardial infarction were randomly divided into two groups, experimental ($n = 302$) and control ($n = 303$). Patients in the experimental group received a Mediterranean style diet (rapeseed oil and rapeseed oil based margarine). This diet was rich in ALA (ALA/LA ratio of 1:4), and total fat provided 30.5% of energy with S/M/P ratio of 0.9:1.4:1. The control group consumed a habitual diet which was poor in ALA (ALA/LA was 1:20) and in which total fat contributed 32.7% of energy with S/M/P ratio of 1.2:1:1. After 27 mon of follow-up, there were 16 cardiac deaths in the control and 3 in the experimental groups, 17 nonfatal myocardial infarctions in the control and 5 in the experimental groups, and the relative risk ratio for cardiac deaths and nonfatal myocardial infarction in the ALA-rich group was 0.27 ($P = 0.001$).

This study was continued after the original conclusion because there was a high adherence of the experimental group to the program over the total 46 mon of mean follow-up for each patient (18). It was found that the three composite outcomes [(i) cardiac death and nonfatal myocardial infarction; (ii) the preceding plus major secondary endpoints including unstable angina, stroke, heart failure, pulmonary or peripheral embolism; or (iii) the preceding plus minor events requiring hospital admission] were significantly reduced in the Mediterranean diet group compared with the Western diet group. The traditional risk factors such as high blood cholesterol and raised blood pressure were significantly and independently associated with recurrence of events. Plasma ALA, measured at 2 mon after randomization, was the only fatty acid that was significantly negatively associated with myocardial infarction plus cardiac death after adjustment for age, sex, smoking, total cholesterol, blood pressure, leukocyte count, and aspirin use.

Another study investigated the secondary prevention of CHD by ALA as mustard oil. In this study, 360 patients less than 1 d after acute myocardial infarction were randomized to one of three dietary groups: fish oil capsules (EPA, 1.08 g/d, and DHA, 0.72 g/d); mustard seed oil, 20 g/d (ALA, 2.9 g/d); and a control group (aluminum hydroxide, 100 mg/d) (19). After 1 yr, this study showed that there was a significant reduction in cardiac events in the fish oil and mustard oil groups compared with the control group (24.5% and 28.2%, respectively, *vs.* 34.7%; $P < 0.01$). Nonfatal infarctions were also significantly lower in the fish oil and mustard oil groups compared with the placebo group (13.0% and 15.0% *vs.* 25.4%, $P < 0.05$). Total cardiac deaths were significantly reduced in the fish oil group but not in the mustard oil group compared with the placebo group. The fish oil and mustard oil groups also showed significant reductions in total cardiac arrhythmias, left ventricular enlargement, and angina pectoris compared with the placebo group.

The Multiple Risk Factor Intervention Trial (MRFIT) was a study of 12,866 men randomly assigned to either a usual care or special intervention group. The latter received advice and programs regarding reduction in smoking, blood pressure, and blood cholesterol. In this report, multivariate regression analysis was used to determine the effect of dietary PUFA intakes on 10-yr mortality rates in 6,250 usual care men (20). Dietary PUFA intake was calculated from four dietary recall interviews at baseline and at 1-, 2-, and 3-yr follow-up. Dietary intake of ALA was significantly negatively associated with CHD mortality rates ($P < 0.04$), total CVD ($P < 0.03$), and all-cause mortality ($P < 0.02$). There were also significant inverse associations for long chain n-3 PUFA on CHD mortality ($P < 0.02$), CVD mortality ($P < 0.006$), and all-cause mortality ($P < 0.02$).

The mechanisms whereby ALA could prevent CHD include reduction of blood pressure and levels of plasma/serum TAG and lipoprotein lipids, and support of antithrombotic and fibrinolytic activities, and of antiarrhythmia, anti-inflammatory, and anti-immunity actions.

Blood Pressure

The n-3 PUFA from fish have been found to reduce both systolic and diastolic blood pressure (BP), and this has been evaluated in a meta-analysis of 31 placebo-controlled trials in 1,356 subjects. The results indicated that systolic BP fell 3.4 mm Hg and diastolic BP fell 2.0 mm Hg following ingestion of 5.6 g/d of fish oil in hypertensive subjects (21).

Epidemiological, prospective, and cross-sectional studies found that vegetarians have a lower mortality from heart disease (22) and a lower diastolic blood pressure than omnivores in general populations (23,24). Berry and Hirsch (25) studied the relationship between adipose tissue fatty acids and blood pressure in 399 free-living male subjects (average age 37 yr). Stepwise regression analysis was performed to assess the separate contributions of age, body mass index, and adipose tissue fatty acids composition to the variance in BP. The analysis showed that adipose tissue LA was not associated with BP; however, they reported that an absolute 1% increase in adipose tissue ALA was associated with a decrease of 5 mm Hg in the systolic, diastolic, and composite mean arterial blood pressure. It was reported that ALA concentration had a disproportionate influence on the mean arterial blood pressure because it comprised 1/8 the amount of LA in adipose tissue.

The MARGARIN study was a prevention of CHD project. It investigated the association between dietary intake of ALA and LA, as assessed by a food-frequency questionnaire and levels of plasma cholesterol ester (CE), with CHD risk factors. The study was a double blind, randomized placebo-controlled trial, which involved 266 subjects with hypercholesterolemia (6.0–8.0 mmol/L) and at least two other CHD risk factors (26). In multivariate analysis, CE ALA was inversely associated with diastolic blood pressure ($r = -0.13$; $P < 0.05$) and positively with serum TAG levels ($P < 0.01$), whereas the CE LA was inversely associated with serum TAG ($P < 0.01$). In the lowest quintile of CE ALA, mean dietary intake was 0.4% energy of ALA (1.2 g/d), 8.4% energy of LA, and an LA/ALA ratio of 21, and in the highest quintile 0.6% energy of ALA (1.7 g/d), 6.8% energy of LA, and an LA/ALA ratio of 12. In the highest quintile of CE ALA, the diastolic BP was 4 mm Hg lower and the serum TAG 0.3 mmol/L higher compared with the top quintile, suggesting that replacing LA with ALA might decrease diastolic blood pressure.

A recent animal study found that ALA deficiency in the perinatal period resulted in an increase in BP later in life in Sprague-Dawley rats (27). The authors suggested that although the study was carried out on rats, the implication of these findings is that an adequate ALA intake at an early age may help prevent increased BP in later life in humans.

Plasma and Lipoprotein Lipids

LA reduced plasma/serum total and LDL-C when substituted for carbohydrate in the diet, whereas saturated fatty acids were cholesterolemic and *cis* monounsaturated fatty acid was neutral in early studies (28,29). Recent dietary intervention studies found that like LA, ALA from plant sources is able to decrease plasma/serum total and LDL-C levels (30,31). McDonald *et al.* (30) reported that canola oil and sunflower oil had an equivalent hypercholesterolemic and antithrombotic effect in eight healthy young hypocholesterolemic men aged 19–32 yr. Approximately 75% of the fat in the diet was provided by a mixture of fats (tallow, lard, corn oil, butter, and vegetable shortening) during the 6 d pre- and postexperimental periods, and either canola oil or sunflower oil during the two 18-d experimental periods with 1.2, 7.9, and 0.8% of total fatty acid as ALA in the mixture of fats, canola oil, and sunflower oil, respectively. The canola oil and sunflower oil diets produced similar decreases in plasma total cholesterol of 20 and 15%, and LDL-C of 25 and 21%, respectively. It cannot be ruled out that the monounsaturated fatty acid and LA content of the canola diet contributed to the cholesterol-lowering effects seen.

Chan *et al.* (31) have compared the effect of dietary oleic acid (OA), LA, and ALA on plasma lipid metabolism in eight normolipidemic men aged 20–34 yr. A mixed-fat diet composed of conventional foods was fed for 6 d pre- and postexperimental periods. The subjects consumed four different experimental diets during four 18-d experimental periods. The four diets were identical in the proportions of nutrients: the diets provided 53% of energy from carbohydrate, 13% from protein, and 34% from fat. Diet 1 provided the 75% fat from a mixture of sunflower and olive oils; diet 2 from canola oil; diet 3 from soybean oil; and diet 4 from a mixture of sunflower, olive, and flaxseed oils. There were significant reductions in the mean concentrations of plasma total cholesterol (TC) (–18%), LDL-C (–22%), and VLDL-C (–41%) after the experimental diets than after the pre- and postexperimental mixed-fat diet ($P < 0.004$). Mean serum apolipoprotein B (–19%) and apolipoprotein A-I (–9%) concentrations were also significantly lower after the experimental diets ($P < 0.0007$). This study demonstrated that dietary ALA was as effective as OA and LA in lowering TC, LDL-C, VLDL-C, and serum apo B and apo A-I concentrations.

In a randomized, crossover feeding trial, 10 men with polygenic hypercholesterolemia were fed a Mediterranean-type cholesterol-lowering diet (control) and a diet of similar composition in which walnuts replaced approximately 35% of energy from monounsaturated fat for 6 wk for each diet (32). Compared with the control diet, the walnut diet reduced serum TC and LDL-C by 4.2% ($P = 0.176$) and 6% ($P = 0.087$), respectively. There was also a reduction in the apolipoprotein B level (6% reduction) in parallel with LDL-C reduction. Whole LDL was enriched with ALA and LA from walnuts. Also, LDL obtained during the walnut diet showed a 50% increase in association rates to the LDL receptor in human hepatoma HepG2 cells compared with LDL obtained during the control diet. There was a positive correlation between LDL uptake by the HepG2 cells and the ALA content of the TAG and cholesterol ester fraction of the LDL particles ($r^2 = 0.42$, $P < 0.05$). This data suggests that a walnut-enriched diet can increase the receptor-mediated LDL clearance, which might be responsible for the reduced LDL cholesterol levels on this diet.

Kelley *et al.* (33) studied the effect of a diet rich in flaxseed oil (6.3% energy from ALA) on blood lipids and coagulation status in 10 volunteers in a study lasting 126 d. This study was a comparison between a moderate fat diet rich in LA compared with a moderate fat diet rich in ALA. Subjects consumed a baseline diet (23.4% energy from fat, polyunsaturated/saturated (P/S) = 0.89, LA/ALA = 66) or the flax oil diet (28.8% energy from fat, P/S = 1.5, LA/ALA = 0.7) for 56 d and then changed to the other diet. The flax oil diet did not significantly alter serum TAG, TC, HDL-C, or LDL-C compared with the values for the baseline diet period. There was no difference between the diets on the bleeding time, prothrombin time, and partial prothrombin time in this study.

The n-3 PUFA, especially those from marine oils, have been shown to reduce serum TAG levels (5,34,35). In contrast, increased ALA intakes were associated with raised serum TAG compared with LA in a cross-sectional analysis (26) and with a significantly raised serum TAG compared with a high LA diet in a double blind, randomized study (36).

We have compared the effect of low, moderate, and high ALA/LA diets on plasma and lipoprotein lipids in vegetarian men (8). Three dietary periods in this study were (i) a low ALA (safflower oil and safflower oil-based margarine), (ii) a moderate ALA (canola oil and canola oil-based margarine), and (iii) a high ALA (linseed oil and linseed oil-based margarine) diet. Seventeen healthy male vegetarians (34 ± 8 yr of age) completed the study. All subjects consumed a low ALA diet for 14 d, and then were randomly divided into moderate ALA or high ALA diet groups for the following 28 d. Subjects were requested to refrain from consuming fish during the 42 d of the intervention. There were no significant differences in plasma TAG, TC, LDL-C, or HDL-C concentrations between subjects on the diets rich in LA and low in ALA (31-g LA/d, 2-g ALA/d) and those on the diet containing almost equal amounts of LA and ALA (17-g LA/d, 15-g ALA/d), indicating that diets rich in either ALA or LA do not differ in their effect on plasma TAG levels. This confirms previous reports (37,38) suggesting that the plasma lipid-lowering effect of flaxseed (as opposed to flaxseed oil) is not due to its oil or ALA content but rather to its soluble fiber.

Vermunt *et al.* (39) reported that *trans* ALA may increase plasma ratios of LDL-C/HDL-C and TC/HDL-C. In this study, 88 healthy male volunteers from four European nations (France, Scotland, UK, and the Netherlands) consumed a *trans* ALA-free diet for 6 wk, followed by either high or low *trans* ALA for another 6 wk. Daily total *trans* ALA intake in the high *trans* group was 1.41 g. Experimental oils were delivered via margarines, cheeses, muffins, and biscuits. Compared with the low *trans* ALA diet, the high *trans* ALA diet significantly increased the plasma LDL-C, ratios of LDL-C/HDL-C by 8.1% (95% CI 1.4, 15.3), and the TC/HDL-C ratio by 5.1% (95% CI 0.4, 9.9). No effects were found on plasma concentrations of TC and HDL-C, TAG, apolipoprotein B and A-1, and lipoprotein(a).

Anti-Thrombotic and Fibrinolytic Activities

Arterial thrombosis is generally recognized to play a major role in the transition from stable to acute ischaemic heart and cerebral diseases, manifested by unstable angina, acute thrombotic infarction, and sudden death. Platelet aggregation is an early event in the development of thrombosis. It is initiated by thromboxane A_2 (TXA_2), a potent platelet aggregation agent and vascular contractor, produced from arachidonic acid (AA), a long chain n-6 PUFA in the platelet membrane (40,41). EPA is released from phospholipids of the platelet membrane and as a "false" substrate competes with AA for access to cyclooxygenase and produces an alternative form of thromboxane A_3 (TXA_3), which is relatively inactive in promoting platelet aggregation and vasoconstriction (42). This situation can lead to a reduced TXA_2 production and thus a lower thrombosis tendency (43,44). A diet with a high n-6/n-3 PUFA ratio can cause a high tissue n-6/n-3 PUFA ratio (i.e., increased AA/EPA ratio), which may promote production of TXA_2, leading to an increased thrombosis tendency (45,46). A number of other studies using diets rich in ALA have shown reductions in TXA_2 production (47–49).

Platelet EPA level doubled and collagen-induced platelet aggregation significantly declined ($P < 0.05$) after healthy male subjects consumed diets supplemented with 40-g of flaxseed oil ($n = 5$) for 23 d compared with a group which consumed an identical quantity of sunflower seed oil ($n = 6$) (50). Freese *et al.* (51) investigated the effect of dietary LA/ALA ratio on platelet aggregation in 20 male subjects using low-erucic acid rapeseed oil and high-oleic acid sunflower oil as the major fat source in a crossover design. LA/ALA ratio was 2.8 and 28 in the rapeseed oil and sunflower oil diets, respectively. Platelet aggregation induced by collagen was significantly ($P < 0.05$) reduced after using rapeseed oil compared with sunflower oil.

Chan *et al.* (52) investigated the effect of dietary ALA and its ratio to LA on platelet and plasma phospholipid fatty acids and prostanoid production in eight normolipidemic male subjects. The study consisted of two 42-d phases. Each was divided into a 6-d pre-experimental period, during which a mixed fat diet was fed, and two 18-d experimental periods, during which a mixture of sunflower and olive (low ALA, high LA/ALA ratio [LO-HI diet]), soybean (intermediate ALA, intermediate LA/ALA ratio), and canola oils (intermediate ALA, low LA/ALA ratio); and a mixture of sunflower, olive, and flax oil (high ALA, low LA/ALA ratio [HI-LO diet]) provided 77% of the fat in the diet. Each subject consumed each of the 4 experimental diets for 18 d per diet. The ALA content and the LA/ALA ratio of the four experimental diets were 0.8%, 27.4; 6.5%, 6.9; 6.6%, 3.0; and 13.4%, 2.7 respectively. There was a significant ($P < 0.05$) increase in platelet EPA level following the HI-LO diet compared with the LO-HI diet. Also, production of 6-keto-PGF$_1$ was significantly higher ($P < 0.05$) following the HI-LO diet than the LO-HI diet.

Evidence from dietary intervention studies has found that production of TXA$_2$ was decreased by LC n-3 PUFA in humans (53,54) and by plant ALA in humans (55,56). In one dietary intervention study, 15 healthy men (aged 37.7 ± 6.5 yr) were provided with foods enriched in ALA (cooking oil, margarine, salad dressing, and mayonnaise) and EPA and DHA (sausages and savory dip) and with foods naturally rich in n-3 fatty acids, such as flaxseed meal and fish. Subjects incorporated these foods into their diets at home for 4 wk. The average intake of EPA plus DHA was 1.8 g/d and of ALA was 9.0 g/d. EPA levels increased threefold in plasma, platelet, and mononuclear cell phospholipids. TXB$_2$, prostaglandin E$_2$, and interleukin 1β synthesis decreased by 36, 26, and 20% ($P < 0.05$), respectively (56).

In a study of one healthy 64-yr-old male subject, in which a 7-wk intervention involved a diet rich in ALA from canola and flaxseed oils, it was found that the urinary excretion of 11-dehydrothromboxane B$_2$ declined by 34% from baseline level 7 wk after the n-6/n-3 ratio of dietary PUFA was reduced from 28 to 1. Return to the baseline diet brought about a rapid return of this metabolite to baseline levels. The excretion of 2,3-dinor-6-oxo-prostaglandin F1-α was also reduced by approximately 32%; however, the levels remained low throughout the entire study. The dietary adjustment was brought about by substituting measured amounts of canola and flaxseed oils (3:1) for measured amounts of olive and corn oils (3:1) in an otherwise fat-free basal diet. This pilot study indicates that dietary ALA may be an effective modulator of thromboxane and prostacyclin biosynthesis (55).

The MARGARIN study from the Netherlands evaluated the effect of an increased intake of ALA on CVD risk factors after 2 yr. Subjects with multiple CVD risk factors (124 men, 158 women) participated in the double-blind intervention study (36), where they consumed margarine rich in either ALA ($n = 114$) or LA ($n = 110$). The average ALA intake was 6.3 g/d in the ALA group and 1.0 g/d in the LA groups. After 2 yr, the ALA group had a higher ratio of total to HDL cholesterol (+0.34; 95% CI: 0.12, 0.56), lower HDL cholesterol (–0.05 mmol/L; 95% CI: –0.10, 0), and higher serum TAG (+0.24 mmol/L; 95% CI: 0.02, 0.46), and after 1 yr, the ALA group had lower plasma fibrinogen (–0.18 g/L; 95% CI: –0.31, –0.04) than did the LA group adjusted for baseline values, gender, and lipid-lowering drugs (36).

Arterial Compliance

Arterial compliance or elasticity is an important index of circulatory function, which decreases with increasing CVD risk. Fish oil supplementation has been shown to improve arterial compliance in diabetic subjects (57), and to inhibit norepinephrine-mediated vasoconstriction, which also influences compliance (58).

A dietary intervention study found that ALA from flaxseed oil raised arterial compliance (59). Fifteen obese subjects (eight men, seven postmenopausal women) aged between 45–63 yr with markers of insulin resistance and with an absence of known metabolic disorders, consumed four diets for 4 wk each: two control diet periods, and two experimental periods (high ALA and low ALA) with flaxseed and sunola oils being the basic oils for the high and low ALA periods, respectively. In the two control periods, fat intake was 35% of energy; during the two experiment periods, fat intake fell to 26% of energy. Systemic arterial compliance was calculated from aortic flow velocity and aortic root driving pressure. Systemic arterial compliance during the first and last control periods was 0.42 ± 0.12 and 0.56 ± 0.21 units (mL/mm Hg). The arterial compliance rose significantly to 0.78 ± 0.28 ($P < 0.0001$) following the high ALA diet, and it was 0.62 ± 0.19 on the low ALA diet. The significant increase in arterial compliance ($P < 0.05$) with ALA reflected rapid functional improvement in the systemic arterial circulation despite the fact that insulin sensitivity and HDL-C decreased and LDL oxidizability increased with the high ALA diet.

Cardiac Arrhythmia

Results from animal models and cultured cells studies have indicated that marine n-3 PUFA (60–62), purified EPA (63), and plant n-3 PUFA (64,65) reduce cardiac arrhythmias. This effect may be at least partly responsible for the protective action of ALA on CHD mortality, because postinfarction patients assigned to a Mediterranean ALA-enriched diet had significantly reduced CHD mortality and morbidity compared with patients receiving no advised traditional diet controls (15). The protective effects in that study were attributed to ALA; however, there were no differences in primary recognized CVD risk factors such as blood pressure or plasma and lipoprotein lipids (15).

ALA from canola oil inhibits cardiac arrhythmias in rats when compared with oleic acid from olive oil and LA from soybean and sunflower oils (64). In that study, male Sprague-Dawley rats were randomly assigned to one of four experimental diet groups for 12 wk. The fat source in the diets was 12% olive (63% oleic acid), canola (55% oleic, 8% ALA), soybean (50% LA, 7% ALA), or sunflower seed oil (64% LA). Arrhythmias were induced by coronary artery occlusion and reperfusion. The rats fed the diet containing canola oil had a significantly lower incidence of ventricular fibrillation, mortality, and arrhythmia score during reperfusion than those fed the olive, soybean, and sunflower oil diets. The proportion of n-3 PUFA in myocardial phospholipids of canola oil group was significantly increased compared with the other diets.

In the dog, the antiarrhythmic effect of ALA was similar to that of EPA and DHA (62). Surgical myocardial infarction was produced by ligating the left anterior coronary artery and placing an inflatable cuff around the left circumflex artery. The dogs were trained to run on a treadmill and were screened for susceptibility to ventricular fibrillation when the cuff was inflated. In the control exercise, ischemia tests were conducted 1 wk before and 1 wk after infusion of free fatty acids of EPA, DHA, and ALA. ALA showed an antiarrhythmic activity similar to that of EPA and DHA. The mechanism of the antiarrhythmic action of n-3 PUFA is thought to be due to the n-3 PUFA stabilizing the electrical activity of cardiac myocytes by modulating the sodium and calcium currents in the myocytes, resulting in a prolonged relative refractory period (65).

Inflammation and Immunity

The n-3 PUFA have a regulatory influence on different processes of inflammatory and immune cell activation, and can provide positive effects on various states of immune diseases with a hyperinflammatory nature (66–68). Increased tissue levels of AA increase the eicosanoid family of inflammatory mediators such as prostaglandins, leukotrienes, and related metabolites, and through these regulate the activities of inflammatory cells, the production of cytokines, and the various balances within the immune system (69). The n-3 PUFA act as AA antagonists, decrease the levels of AA in cell membranes, and modulate the amount and types of eicosanoids (69,70). The n-3 PUFA might improve conditions of inflammation and immunity by eicosanoid-independent mechanisms. The n-3 PUFA can down-regulate the T-helper 1-type response, which is associated with chronic inflammatory disease. Components of both acquired and natural immunity, including the production of key inflammatory cytokines, can be influenced by n-3 PUFA. The n-3 PUFA decreased cytokine-induced adhesion molecule expression through mediated mechanisms, reducing inflammatory leucocyte-endothelium interactions and modifying lipid mediator synthesis, thus affecting the transendothelial migration of leucocytes and leucocyte trafficking in general. The n-3 fatty acids influence inflammatory cell activation processes from signal transduction to protein expression even involving effects at the genomic level.

Kelley *et al.* (71) studied the effect of a diet rich in flaxseed oil (6.3% energy from ALA) on immunocompetence in 10 volunteers in a study lasting 126 d. This study compared a moderate fat diet rich in LA with a moderate fat diet rich in ALA. Subjects consumed a baseline diet (23.4% energy from fat, P/S = 0.89, LA/ALA = 66) or the flax oil diet (28.8% energy from fat, P/S = 1.5, LA/ALA = 0.7) for 56 d and then changed to the other diet. On the flax oil diet it was found that the proliferation of peripheral blood mononuclear cells was suppressed when they were cultured with phytohemagglutinin-P and concanavalin A; the flax oil diet also suppressed the delayed hypersensitivity response to seven recall antigens (71). There were no changes in the concentration of immunoglobulins in the serum or the number of helper cells, suppressor cells, and total T and B cells in the peripheral blood between the control and flax oil diet.

Summary

Results from prospective/epidemiological and dietary intervention studies indicate that ALA reduces cardiac mortality. The beneficial effects of ALA may operate via various mechanisms. ALA has been reported to prevent secondary CHD, reduce platelet aggregability, and decrease TXA_2 production. ALA has also been reported to have anti-inflammatory and anti-immunity activities through influence on eicosanoid-dependent and eicosanoid-independent mechanisms. Unlike LA, ALA from plant sources inconsistently decreases plasma/serum total and LDL cholesterol levels. Animal studies suggest that ALA can prevent cardiac arrhythmia and ventricular fibrillation, and it is as effective as EPA and DHA in the prevention of ventricular fibrillation. Whether the effects reported here are due to ALA itself or to longer chain n-3 PUFA formed from ALA is not known. Because most studies with ALA feeding show very little formation or accumulation of EPA in plasma lipids and almost no accumulation of DHA, this suggests that the reported benefits are from ALA itself.

References

1. Willett, W.C., Diet and Health: What Should We Eat? *Science 264*:532–537 (1994).
2. Kromhout, D., A. Menotti, B. Bloemberg, C. Aravanis, H. Blackburn, R. Buzina, A.S. Dontas, F. Fidanza, S. Giaipaoli, A. Jansen, M. Karvonen, M. Katan, A. Nissinen, S. Nedeljkovic, J. Pekkanen, M. Pekkarinen, S. Punsar, L. Rasanen, B. Simic, and H. Toshima, Dietary Saturated and *trans* Fatty Acids and Cholesterol and 25-Year Mortality from Coronary Heart Disease—The Seven Countries Study, *Prev. Med.* 24:308–315 (1995).
3. Key, T.J., M. Thorogood, P.N. Appleby, and M.L. Burr, Dietary Habits and Mortality in 11,000 Vegetarians and Health Conscious People: Results of a 17 Year Follow-Up, *Brit. Med. J. 313*:775–779 (1996).
4. Hegsted, D.M., M.A. Ausman, J.A. Jonson, and G.E. Dallal, Dietary Fat and Blood Lipids: An Evaluation of the Experimental Data, *Am. J. Clin. Nutr. 57*:875–883 (1993).
5. Bucher, H.C., P. Hengstler, C. Schindler, and G. Meier, n-3 Polyunsaturated Fatty Acids in Coronary Heart Disease: A Meta-Analysis of Randomized Controlled Trials, *Am. J. Med. 112*:298–304 (2002).
6. Koch, T., T. Krumm, V. Jung, J. Engelbert, and W. Boland, Differential Induction of Plant Volatile Biosynthesis in the Lima Bean by Early and Late Intermediates of the Octadecanoid-Signalling Pathway, *Plant Physiol. 121*:153–162 (1999).
7. Gerster, H., Can Adults Adequately Convert α-Linolenic Acid to Eicosapentaenoic Acid and Docosahexaenoic Acid? *Internat. J. Vit. Nutr. Res. 68*:159–173 (1998).

8. Li, D., A. Sinclair, A. Wilson, S. Nakkote, F. Kelly, L. Abedin, N. Mann, and A. Turner, Effect of Dietary Alpha-Linolenic Acid on Thrombotic Risk Factors in Vegetarian Men, *Am. J. Clin. Nutr. 69*:872–882 (1999).

9. Mantzioris, E., M.J. James, R.A. Gibson, and L.G. Cleland, Dietary Substitution with an α-Linolenic Acid-Rich Vegetable Oil Increases Eicosapentaenoic Acid Concentrations in Tissues, *Am. J. Clin. Nutr. 59*:1304–1309 (1994).

10. Emken, E.A., R.O. Adlof, and G.M. Gulley, Dietary Linoleic Acid Influences the Desaturation and Acylation of Deuterium-Labelled Linoleic and Alpha-Linolenic Acid in Young Adult Males, *Biochim. Biophys. Acta 1213*:277–288 (1994).

11. Pawlosky, R.J., J.R. Hibbeln, J.A. Novotny, and N. Salem, Jr., Physiological Compartmental Analysis of α-Linolenic Acid Metabolism in Adult Humans, *J. Lipid Res. 42*:1257–1265 (2001)

12. Ascherio, A., E.B. Rimm, E.L. Giovannucci, D. Spiegelman, M. Stampfer, and W.C. Willett, Dietary Fat and Risk of Coronary Heart Disease in Men: Cohort Follow-Up Study in the United States, *Brit. Med. J. 313*:84–90 (1996).

13. Hu, F.B., M.J. Stampfer, J.E. Manson, E.B. Rimm, A. Wolk, G.A. Colditz, C.H. Hennekens, and W.C.Willett, Dietary Intake of Alpha-Linolenic Acid and Risk of Fatal Ischemic Heart Disease Among Women, *Am. J. Clin. Nutr. 69*:890–897 (1999).

14. Djousse, L., J.S. Pankow, J.H. Eckfeldt, A.R. Folsom, P.N. Hopkins, M.A. Province, Y. Hong, and R.C. Ellison, Relation Between Dietary Linolenic Acid and Coronary Artery Disease in the National Heart, Lung, and Blood Institute Family Heart Study, *Am. J. Clin. Nutr. 74*:612–619 (2001).

15. de Lorgeril, M., S. Renaud, N. Mamelle, P. Salen, J.L. Martin, I. Monjaud, J. Guidollet, P. Touboul, and J. Delaye, Mediterranean Alpha-Linolenic Acid-Rich Diet in Secondary Prevention of Coronary Heart Disease, *Lancet 343*:1454–1459 (1994).

16. Keys, A., Wine, Garlic, and CHD in Seven Countries, *Lancet 1*:145–146 (1980).

17. Sandker, G.W., D. Kromhout, C. Aravanis, B.P. Bloemberg, R.P. Mensink, N. Karalias, and M.B. Katan, Serum Cholesteryl Ester Fatty Acids and Their Relation with Serum Lipids in Elderly Men in Crete and The Netherlands, *Eur. J. Clin. Nutr. 47*:201–208 (1993).

18. de Lorgeril, M., P. Salen, J.L. Martin, I. Monjaud, J. Delaye, and N. Mamelle, Mediterranean Diet, Traditional Risk Factors, and the Rate of Cardiovascular Complications after Myocardial Infarction: Final Report of the Lyon Diet Heart Study, *Circulation 99*:779–785 (1999).

19. Singh, R.B., M.A. Niaz, J.P. Sharma, R. Kumar, V. Rastogi, and M. Moshiri, Randomized, Double-Blind, Placebo-Controlled Trial of Fish Oil and Mustard Oil in Patients with Suspected Acute Myocardial Infarction: The Indian Experiment of Infarct Survival-4, *Cardiovasc. Drugs Ther. 11*:485–491 (1997).

20. Dolecek, T.A., Epidemiological Evidence of Relationships Between Dietary Polyunsaturated Fatty Acids and Mortality in the Multiple Risk Factor Intervention Trial, *Proc. Soc. Exp. Biol. Med. 200*:177–182 (1992).

21. Morris, M.C., F. Sack, and B. Rosner, Does Fish Oil Lower Blood Pressure? A Meta-Analysis of Controlled Trials, *Circulation 88*:523–533 (1993).

22. Key, T.J., G.E. Fraser, M. Thorogood, P.N. Appleby, V. Beral, G. Reeves, M.L. Burr, J. Chang-Claude, R. Frentzel-Beyme, J.W. Kuzma, J. Mann, and K. McPherson, Mortality in Vegetarians and Non-Vegetarians: A Collaborative Analysis of 8300 Deaths Among 76,000 Men and Women in Five Prospective Studies, *Public Health Nutr. 1*:33–41 (1998).

23. Rouse, I.L., L.J. Beilin, B.K. Armstrong, and R. Vandongen, Blood-Pressure-Lowering Effect of a Vegetarian Diet: Controlled Trial in Normotensive Subjects, *Lancet 1*:5–10 (1983).

24. Beilin, L.J., I.L. Rouse, B.K. Armstrong, B.M. Margetts, and R. Vandongen, Vegetarian Diet and Blood Pressure Levels: Incidental or Causal Association? *Am. J. Clin. Nutr. 48*:806–810 (1988).

25. Berry, E.M., and J. Hirsch, Does Dietary Linolenic Acid Influence Blood Pressure? *Am. J. Clin. Nutr. 44*:336–340 (1986).

26. Bemelmans, W.J., F.A. Muskiet, E.J. Feskens, J.H. de Vries, J. Broer, J.F. May, and B.M. Jong, Associations of Alpha-Linolenic Acid and Linoleic Acid with Risk Factors for Coronary Heart Disease, *Eur. J. Clin. Nutr. 54*:865–871 (2000).

27. Weisinger, H.S., J.A. Armitage, A.J. Sinclair, P.B. Vingrys, and R.S. Weisinger, Perinatal Omega-3 Fatty Acid Deficiency Affects Blood Pressure Later in Life, *Nature Med. 7*:258–259 (2001).

28. Keys, A., J.T. Anderson, and F. Grande, Serum Cholesterol Response to Changes in the Diet. IV. Particular Saturated Fatty Acids in Diet, *Metabolism 14*:776–787 (1965).

29. Hegsted, D.M., R.B. McGandy, M.L. Myers, and F.L. Stare, Quantitative Effects of Dietary Fat on Serum Cholesterol in Man, *Am. J. Clin. Nutr. 17*:281–295 (1965).

30. McDonald, B.E., J.M. Gerrard, V.M. Bruce, and E.J. Corner, Comparison of the Effect of Canola Oil and Sunflower Oil on Plasma Lipids and Lipoproteins and on *in vivo* Thromboxane A_2 and Prostacyclin Production in Healthy Young Men, *Am. J. Clin. Nutr. 50*:1382–1388 (1989).

31. Chan, J.K., V.M. Bruce, and B.E. McDonald, Dietary Alpha-Linolenic Acid is as Effective as Oleic Acid and Linoleic Acid in Lowering Blood Cholesterol in Normolipidemic Men, *Am. J. Clin. Nutr. 53*:1230–1234 (1991).

32. Munoz, S., M. Merlos, D. Zambon, C. Rodriguez, J. Sabate, E. Ros, and J.C. Laguna, Walnut-Enriched Diet Increases the Association of LDL from Hypercholesterolemic Men with Human He G2 Cells, *J. Lipid Res. 42*:2069–2076 (2001).

33. Kelley, D.S., G.J. Nelson, J.E. Love, L.B. Branch, P.C. Taylor, P.C. Schmidt, B.E. Mackey, and J.M. Iacono, Dietary Alpha-Linolenic Acid Alters Tissue Fatty Acid Composition, but Not Blood Lipids, Lipoproteins or Coagulation Status in Humans, *Lipids 28*:533–537 (1993).

34. Svaneborg, N., J.M. Moller, E.B. Schmidt, K. Varming, H.H. Lervang, and J. Dyerberg, The Acute Effects of a Single Very High Dose of n-3 Fatty Acids on Plasma Lipids and Lipoproteins in Healthy Subjects, *Lipids 29*:145–147 (1994).

35. Zampelas, A., A.S. Peel, B.J. Gould, J. Wright, and C.M. Williams, Polyunsaturated Fatty Acids of the n-6 and n-3 Series: Effects on Postprandial Lipid and Apolipoprotein Levels in Healthy Men, *Eur. J. Clin. Nutr. 48*:842–848 (1994).

36. Bemelmans, W.J., J. Broer, E.J. Feskens, A.J. Smit, F.A. Muskiet, J.D. Lefrandt, V.J. Bom, J.F. May, and B. Meyboom de-Jong, Effect of an Increased Intake of Alpha-Linolenic Acid and Group Nutritional Education on Cardiovascular Risk Factors: The Mediterranean Alpha-linolenic Enriched Groningen Dietary Intervention (MARGARIN) Study, *Am. J. Clin. Nutr. 75*:221–227 (2002).

37. Cunnane, S.C., S. Ganguli, C.R. Menard, A.C. Liede, M.J. Hamadeh, Z-Y. Chen, T.M.S. Wolever, and D.J.A. Jenkins, High α-Linolenic Acid Flaxseed (*Linum usitatissimum*): Some Nutritional Properties in Humans, *Br. J. Nutr. 69*: 443–453 (1993).

38. Cunnane, S.C., M.J. Hamadeh, A.C. Liede, L.U. Thompson, T.M.S., Wolever, and D.J.A. Jenkins, Nutritional Attributes of Traditional Flaxseed in Healthy Young Adults, *Am. J. Clin. Nutr. 61*:62–68 (1995).

39. Vermunt, S.H., B. Beaufrere, R.A. Riemersma, J.L. Sébédio, J.M. Chardigny, and R.P. Mensink, Dietary Trans Alpha-Linolenic Acid from Deodorised Rapeseed Oil and Plasma Lipids and Lipoproteins in Healthy Men: The TransLinE Study, *Br. J. Nutr. 85*:387–392 (2001).

40. Hamberg, M., J. Svensson, and B. Samuelsson, Thromboxanes: A New Group of Biologically Active Compounds Derived from Prostaglandin Endoperoxides, *Proc. Natl. Acad. Sci. 72*:2994–2998 (1975).

41. Moncada, S., and J.R. Vane, Arachidonic Acid Metabolites and the Interactions Between Platelets and Blood Vessel Walls, *N. Engl. J. Med. 300*:1142–1148 (1979).

42. Raz, A., M.S. Minkes, and P. Needlemen, Endoperoxides and Thromboxanes: Structural Determinants for Platelet Aggregation and Vasoconstriction, *Biochim. Biophys. Acta 488*:305–311 (1977).

43. Leaf, A., and P.C. Weber, Cardiovascular Effects of n-3 Fatty Acids, *N. Engl. J. Med. 318*:549–557 (1988).

44. Dyerberg. J., Linolenate Derived Polyunsaturated Fatty Acids and Prevention of Atherosclerosis, *Nutr. Rev. 44*:125–134 (1986).

45. Kinsella, J.E., B. Lokesh, and R.A. Stone, Dietary n-3 Polyunsaturated Fatty Acids and Amelioration of Cardiovascular Disease: Possible Mechanisms, *Am. J. Clin. Nutr. 52*:1–28 (1990).

46. Okuyama, H., T. Kobayashi, and S. Watanabe, Dietary Fatty Acids—the n-6/n-3 Balance and Chronic Elderly Diseases: Excess Linoleic Acid and Relative n-3 Deficiency Syndrome Seen in Japan, *Prog. Lipid Res. 35*:409–457 (1996).

47. Abeywardena, M.Y., P.L. McLennan, and J.S. Charnock, Differential Effects of Dietary Fish Oil on Myocardial Prostaglandin I2 and Thromboxane A2 Production, *Am. Physiol. 260*:379–385 (1991).

48. Henry, M.M., J.N. Moore, E.B. Feldman, J.K. Fischer, and B. Russell, Effect of Dietary Alpha Linolenic Acid on Equine Monocyte Procoagulant Activity and Eicosanoid Synthesis, *Circ. Shock 32*:173–188 (1990).

49. Ikeda, I., H. Yoshida, M. Tomooka, A. Yosef, K. Imaizumi, H. Tsuji, and A. Seto, Effects of Long-Term Feeding of Marine Oils with Different Positional Distribution of Eicosapentaenoic and Docosahexaenoic Acids on Lipid Metabolism, Eicosanoids Production, and Platelet Aggregation in Hypercholesterolemic Rats, *Lipids 33*:897–904 (1998).

50. Allman, M.A., M.M. Pena, and D. Peng, Supplementation with Flaxseed Oil Versus Sunflower Seed Oil in Healthy Young Men Consuming a Low Fat Diet: Effects on Platelet Composition and Function, *Eur. J. Clin. Nutr. 49*:169–178 (1995).

51. Freese, R., M. Mutanen, L.M. Valsta, and I. Salminen, Comparison of the Effects of Two Diets Rich in Monounsaturated Fatty Acids Differing in Their Linoleic/α-Linolenic Acid Ratio on Platelet Aggregation, *Thromb. Haemost. 71*:73–77 (1994).

52. Chan, J.K., B.E. McDonald, J.M. Gerrard, V.M. Bruce, B.J. Weaver, and B.J. Holub, Effect of Dietary α-Linolenic Acid and Its Ratio to Linoleic Acid on Platelet and Plasma Fatty Acids and Thrombogenesis, *Lipids 28*:811–817 (1993).

53. Ferretti, A., G.J. Nelson, P.C. Schmidt, G. Bartolini, D.S. Kelley, and V.P. Flanagan, Dietary Docosahexaenoic Acid Reduces the Thromboxane/Prostacyclin Synthetic Ratio in Humans, *J. Nutr. Biochem. 9*:88–92 (1998).

54. Schacky, C.V., W. Siess, S. Fischer, and P.C. Weber, A Comparative Study of Eicosapentaenoic Acid Metabolism by Human Platelets *in vivo* and *in vitro*, *J. Lipid Res. 26*:457–464 (1985).

55. Ferretti, A., and V.P. Flanagan, Antithromboxane Activity of Dietary Alpha-Linolenic Acid: A Pilot Study, *Prostaglandins Leukot. Essent. Fatty Acids 54*:451–455 (1996).

56. Mantzioris, E., L.G. Cleland, R.A. Gibson, M.A. Neumann, M. Demasi, and M.J. James, Biochemical Effects of a Diet Containing Foods Enriched with n-3 Fatty Acids, *Am. J. Clin. Nutr. 72*:42–48 (2000).

57. McVeigh, G.E., G.M. Brennan, J.N. Cohn, S.M. Finkelstein, R.J. Hayes, and G.D. Johnston, Fish Oil Improves Arterial Compliance in Non–Insulin-Dependent Diabetes Mellitus, *Arterioscler. Thromb. 14*:1425–1429 (1994).

58. Chin, J.F.P., A.P. Gust, P.J. Nestel, and A.M. Dart, Marine Oils Dose-Dependently Inhibit Vasoconstriction of Forearm Resistance Vessels in Humans, *Hypertension 21*:22–28 (1993).

59. Nestel, P.J., S.E. Pomeroy, T. Sasahara, T. Yamashita, Y.L. Liang, A.M. Dart, G.L. Jennings, M. Abbey, and J.D. Cameron, Arterial Compliance in Obese Subjects Is Improved with Dietary Plant n-3 Fatty Acid from Flaxseed Oil Despite Increased LDL Oxidizability, *Arterioscler. Thromb. Vasc. Biol. 17*:1163–1670 (1997).

60. McLennan, P.L., M.Y. Abeywardena, and J.S. Charnock, Dietary Fish Oil Prevents Ventricular Fibrillation Following Coronary Artery Occlusion and Reperfusion, *Am. Heart J. 116*:709–717 (1988).

61. McLennan, P.L., T.M. Bridle, M.Y. Abeywardena, and J.S. Charnock, Comparative Efficacy of n-3 and n-6 Polyunsaturated Fatty Acids in Modulating Ventricular Fibrillation Threshold in Marmoset Monkeys, *Am. J. Clin. Nutr. 58*:666–669 (1993).

62. Billman, G.E., H. Hallaq, and A. Leaf, Prevention of Ischemia-Induced Ventricular Fibrillation by Omega 3 Fatty Acids, *Proc. Natl. Acad. Sci. USA 91*:4427–4430 (1994).

63. Hallaq, H., A. Sellmayer, T.W. Smith, and A. Leaf, Protective Effect of Eicosapentaenoic Acid on Ouabain Toxicity in Neonatal Rat Cardiac Myocytes, *Proc. Natl. Acad. Sci. USA 87*:7834–7838 (1990).

64. McLennan, P.L., and J.A. Dallimore, Dietary Canola Oil Modifies Myocardial Fatty Acids and Inhibits Cardiac Arrhythmias in Rats, *J. Nutr. 125*:1003–1009 (1995).

65. Kang, J.X., and A. Leaf, Prevention of Fatal Cardiac Arrhythmias by Polyunsaturated Fatty Acids, *Am. J. Clin. Nutr. 71*:202S–207S (2000).

66. Jeffery, N.M., P. Sanderson, E.J. Sherrington, E.A. Newsholme, and P.C. Calder, The Ratio of n-6 to n-3 Polyunsaturated Fatty Acids in the Rat Diet Alters Serum Lipid Levels and Lymphocyte Functions, *Lipids 31*:737–745 (1996).

67. Inui, K., Y. Fukuta, A. Ikeda, H. Kameda, Y. Kokuba, and M. Sato, The Effect of Alpha-Linolenic Acid-Rich Emulsion on Fatty Acid Metabolism and Leukotriene Generation of the Colon in a Rat Model with Inflammatory Bowel Disease, *Ann. Nutr. Metab. 40*:175–182 (1996).

68. Benquet, C., K. Krzystyniak, R. Savard, F. Guertin, D. Oth, and M. Fournier, Modulation of Exercise-Induced Immunosuppression by Dietary Polyunsaturated Fatty Acids in Mice, *J. Toxicol. Environ. Health 43*:225–237 (1994).

69. Calder, P.C., P. Yaqoob, F. Thies, F.A. Wallace, and E.A. Miles, Fatty Acids and Lymphocyte Functions, *Br. J. Nutr. 87*:S31–48 (2002).

70. Grimm, H., K. Mayer, P. Mayser, and E. Eigenbrodt, Regulatory Potential of n-3 Fatty Acids in Immunological and Inflammatory Processes, *Br. J. Nutr. 87*:S59–67 (2002).

71. Kelley, D.S., L.B. Branch, J.E. Love, P.C. Taylor, Y.M. Rivera, and J.M. Iacono, Dietary Alpha-Linolenic Acid and Immunocompetence in Humans, *Am. J. Clin. Nutr. 53*:40–46 (1991).

Chapter 13

Flaxseed and Prevention of Experimental Hypercholesterolemic Atherosclerosis

Kailash Prasad

Department of Physiology, College of Medicine, University of Saskatchewan, Saskatoon, Saskatchewan, Canada

Introduction

Atherosclerosis is a disease of large- and medium-sized arteries characterized by focal thickening of intima associated with fatty deposits. Atherosclerosis can progress and can abruptly interfere with blood flow, particularly through the heart and brain, causing heart attack and stroke. Atherosclerosis and its complications, such as myocardial ischemia and infarction, stroke and peripheral vascular disease, remain major causes of mortality and morbidity in the Western world. Heart disease is the number one killer in the West. Major risk factors for atherosclerosis are hypercholesterolemia, diabetes, hypertension, and smoking. Other risk factors include hyperhomocysteinemia, infection, and obesity. Hypercholesterolemia is a major factor for the development of atherosclerosis. Every 1% increase in serum cholesterol increases the risk of coronary artery disease (CAD) by 2–3% (1). Lowering blood cholesterol by 10% reduces CAD risk over 5 yr by half for men 40 years of age and by a quarter for men 60 yr of age. Hypercholesterolemia is a major risk factor for the development of CAD, as evidenced from epidemiological data and clinical trial data (2–4). Numerous strategies have been used to reduce the serum levels of cholesterol to prevent the complications of hypercholesterolemia.

In this chapter, the effectiveness of flaxseed and its components in the prevention/reduction of hypercholesterolemic atherosclerosis will be discussed. This chapter deals with oxidative stress and hypercholesterolemic atherosclerosis, and the effects of flaxseed and its components on serum lipids, hypercholesterolemic atherosclerosis, and oxidative stress parameters.

Oxidative Stress and Hypercholesterolemic Atherosclerosis

Various theories have been formulated for the pathogenesis of hypercholesterolemic atherosclerosis. The response to injury hypothesis has gained prominence. This hypothesis states that some form of injury to the endothelial cells occurs at particular sites in the arterial wall. Endothelial cell injury is the basic mechanism for initiation and maintenance of atherosclerosis (5). Hypercholesterolemia produces endothelial cell injury (6,7). The mechanism by which cholesterol produces endothelial cell injury was not known until some investigators (8–11) showed that hypercholesterolemia increases the levels of oxygen radicals (OR). Hypercholesterolemia increases the production of superoxide anion (O_2^-) by endothelial cells (12). Prasad and Kalra (8) and Mantha *et al.* (13) have shown that hypercholesterolemia increases the production of oxygen radicals by polymorphonuclear leukocytes (PMNL). Hypercholesterolemia increases the lipid peroxidation product malondialdehyde (MDA), an indirect index of levels of OR, in blood (8–11) and aortic tissue (8,9,11), and decreases the antioxidant reserve in the aorta (9–11). Antioxidant enzymes are also affected adversely in hypercholesterolemia (11,13,14). The evidence that levels of oxygen radicals are elevated in hypercholesterolemia also comes from the observation that oxidized-low-density lipoprotein (OX-LDL) is present in the blood (15,16) and atherosclerotic plaques (17).

These data suggest that there is oxidative stress, defined as a serious imbalance between oxidants and antioxidants in favor of oxidants, in hypercholesterolemia. Oxygen radicals are known to produce endothelial cell damage (18,19). Thus, increases in the levels of oxygen radicals would damage endothelial cells and set the stage for development and maintenance of hypercholesterolemic atherosclerosis. It has also been suggested that LDL is readily modified by oxidation and produces cholesterol deposition in tissue macrophages, leading to formation of foam cells characteristic of early atherosclerotic fatty streak lesions (20–23).

Hypercholesterolemic atherosclerosis is associated with an increase in the production of oxygen radicals by PMNL (8), an increase in MDA levels in blood (8,11) and aortic tissue (8,9,11), and a decrease in the antioxidant reserve (8,10,11). Antioxidant-induced reduction in atherosclerosis is associated with a decrease in oxidative stress (8–11). Oxidative stress during hypercholesterolemic atherosclerosis, and the protection afforded by antioxidants which was associated with a decrease in oxidative stress, supports the hypothesis that oxygen radicals are involved in the development of hypercholesterolemic atherosclerosis.

Pharmacological Activity of Flaxseed

Flaxseed contains 32–45% of its mass as oil, of which 51–55% is α-linolenic acid (n-3 fatty acids) (24,25). Flaxseed also contains secoisolariciresinol diglucoside (SDG), a lignan (26), the levels of which in flaxseed vary from 0.6–1.8 g/100 g (27). α-Linolenic acid suppresses production of interleukin-1 (IL-1) and tumor necrosis factor (TNF) (28,29). Dietary n-3 fatty acid reduces production of inflammatory mediators and leukotrienes by monocytes (30) and suppresses respiratory bursts of PMNL (31). n-3 Fatty acid also reduces the production of superoxide anion (O_2^-) by monocytes (32). SDG has been shown to have antioxidant activity (33,34).

Flaxseed, Atherosclerosis, and Serum Lipids

Because hypercholesterolemic atherosclerosis is associated with oxidative stress and flaxseed components have the ability to reduce oxygen radical production by monocytes and PMNL, flaxseed should be able to reduce hypercholesterolemic atherosclerosis. Prasad (35) studied the effect of flaxseed on high cholesterol diet-induced atherosclerosis, serum lipids, and OR-producing activity of PMNL in rabbits. The rabbits were on a high cholesterol diet with or without flaxseed for 8 wk. In this study, the rabbits were divided into four groups: Group I, control (on regular diet); Group II, flaxseed (regular diet supplemented with flaxseed in the dose of 7.5 g/kg b.w., orally, daily); Group III, 1% cholesterol (regular diet supplemented with 1% cholesterol); and Group IV, cholesterol + flaxseed (diet similar to that in Group III plus flaxseed in the dose of flaxseed similar to that in Group II).

The high cholesterol diet produced an increase in the serum cholesterol and OR-producing activity of PMNLS [PMNL-chemiluminescence (PMNL-CL)] (Fig. 13.1) without affecting serum triglycerides. These changes were associated with develop-

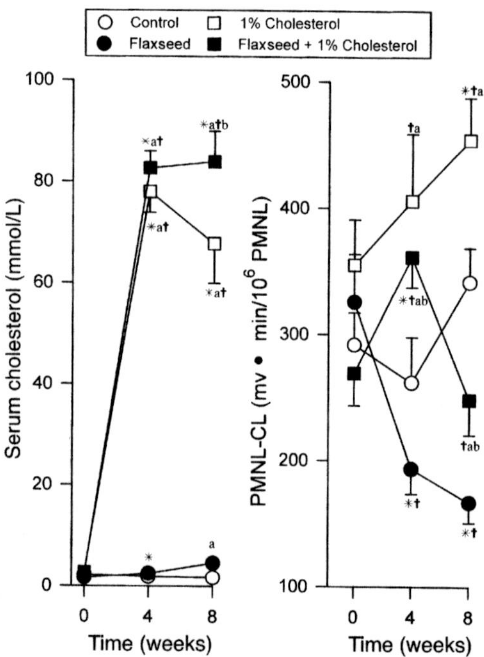

Fig. 13.1. Changes in the serum total cholesterol and PMNL-CL in the four experimental groups. Results are expressed as mean ± SE. *P < 0.05, comparison of values at various times with respect to time 0 in the respective groups. [a]P < 0.05, control vs. other groups for serum cholesterol while flaxseed vs. cholesterol or cholesterol + flaxseed for PMNL-CL. Control and flaxseed groups had no atherosclerotic changes. [†]P < 0.05, flaxseed vs. cholesterol with or without flaxseed for serum cholesterol while control vs. other groups for PMNL-CL. [b]P < 0.05, cholesterol vs. cholesterol + flaxseed. Abbreviation: PMNL-CL, polymorphonuclear leukocyte chemiluminescence. Source: Reproduced with permission from Prasad (35).

ment of atherosclerosis. Flaxseed diet reduced the development of atherosclerosis by 46% (Fig. 13.2) without affecting serum triglycerides. However, the serum cholesterol with flaxseed treatment was higher than without flaxseed treatment (Fig. 13.1). Flaxseed treatment reduced the PMNL-CL in hypercholesterolemic rabbits. These results suggest that flaxseed is effective in reducing hypercholesterolemic atherosclerosis without lowering serum cholesterol. The antiatherogenic activity of flaxseed may be due to suppression of OR production by PMNL by α-linolenic acid and/or due to the antioxidant activity of SDG.

It has been reported that 10% flaxseed diet in rats did not affect serum lipids, but 20% and 30% flaxseed diets lowered plasma triglycerides (TG), total cholesterol (TC), and low-density lipoprotein (LDL-C) by 23% and 23%; 21% and 33%; and 33.7% and 67%, respectively (36). Flaxseed diet in rats did not change serum TG

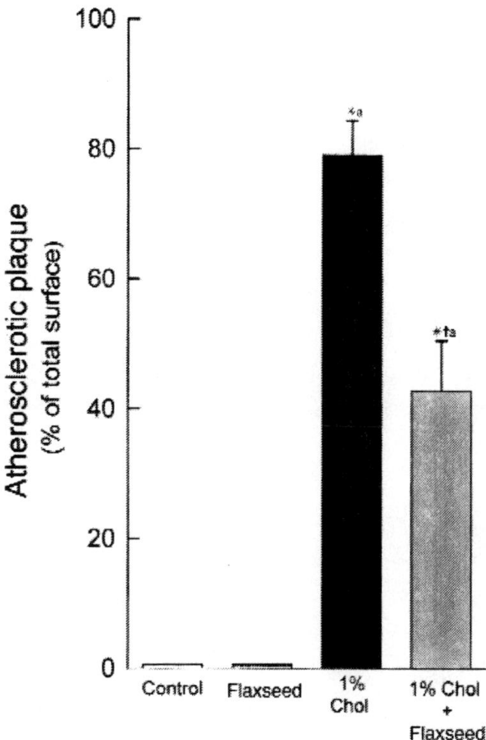

Fig. 13.2. Extent of atherosclerotic area on the intimal surface of aortae of the four groups. Results are expressed as mean ± SE. *P < 0.05, control *vs.* other groups. ^{a}P < 0.05, flaxseed *vs.* cholesterol with or without flaxseed. †P < 0.05, cholesterol *vs.* cholesterol + flaxseed. *Abbreviation:* Chol, cholesterol. *Source:* Reproduced with permission from Prasad (35).

levels but decreased serum TC and LDL-C (37). Flaxseed and flax oil (15% in diet) prevented the rise in TC and TG in a renal ablation rat model, the flaxseed being more effective than flax oil (38). These results suggest that the lipid-lowering effect of flaxseed mostly resides in the flaxmeal. Full-fat flaxseed negated the adverse effects of hard fat on TC and high-density lipoprotein (HDL-C) but did not lower serum TG. Sanders and Roshani (39) reported that flax oil does not lower plasma TG, TC, or HDL-C.

The effects of flaxseed and flax oil on serum lipids have been reported in humans. Flaxseed consumption of 50 g daily for 4 wk reduced serum TC by 9% and LDL-C by 18% (40). Consumption of 15 g of flaxseed daily for 3 mon reduced TC (18 mg/dL) and LDL-C (19 mg/dL) in hyperlipidemic individuals (41). HDL-C levels remained unaltered, whereas TG levels fell slightly in these individuals. Flaxseed consumption of 50 g daily for 4 wk reduced plasma TC by 6% and LDL-C by 9% (42). Flaxseed at a dose of 30 g daily for 4 wk in humans reduced serum TC by 11% and LDL-C by 12% (43). At a higher dose of flaxseed, the reductions in TC and LDL-C were low (43). Flax oil given to healthy individuals for 56 d had no effect on plasma TG, TC, LDL-C, or HDL-C (44). These results suggest that the effect of flaxseed on the serum lipids in animals and humans is variable. Also, it appears that flax oil does not have effects on serum lipids (45).

Flaxseed with Very Low α-Linolenic Acid and Atherosclerosis

As stated earlier, the antiatherogenic activity of flaxseed could be due to α-linolenic acid and/or SDG. An investigation was made by Prasad *et al.* (46) to determine whether the antiatherogenic activity of flaxseed is due to α-linolenic acid. Their assumption was that if the antiatherogenic effect of flaxseed is due to α-linolenic acid, then flaxseed with very low α-linolenic acid would have no antiatherogenic activity. Such type of flaxseed has been developed by Crop Development Center (CDC). This CDC flaxseed (Type II flaxseed) has an oil content (35% of total mass) similar to that of ordinary flaxseed but has only 2–3% (*vs.* 55%) of α-linolenic acid content. However, both types of flaxseed have similar concentrations of SDG (16.4 mg/g *vs.* 15.4 mg/g defatted flaxmeal) (46). Type II flaxseed at a dose of 7.5 g/kg b.w. (similar to doses of ordinary flaxseed used in an earlier study) were fed to rabbits for 8 wk. Hypercholesterolemia was produced by feeding the rabbits a high cholesterol diet. Prasad *et al.* (46) investigated the effects of Type II flaxseed on high cholesterol diet-induced atherosclerosis and serum lipids in rabbits. Rabbits in this study were assigned to four groups: Group I, control (regular diet); Group II, Flax II (regular diet supplemented with CDC flaxseed with very low α-linolenic acid in a dose of 7.5 g/kg b.w., orally, daily); Group III, cholesterol (1% cholesterol in regular diet); Group IV, Flax II + 1% cholesterol (1% cholesterol diet supplemented with Flax II in a dose similar to that in Group II). Type-II flaxseed reduced the development of atherosclerosis by 69% (Fig. 13.3). Reduction in atherosclerosis was associ-

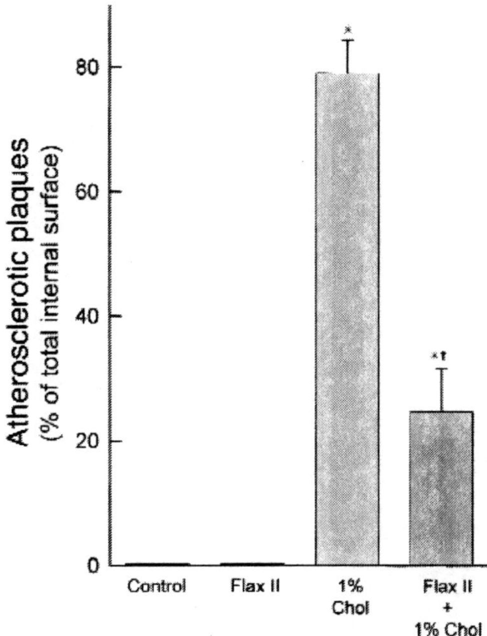

Fig. 13.3. Extent of atherosclerotic plaques on the intimal surface of the aortae of the four groups. Results are expressed as mean ± SE. Control and Flax II groups had no atherosclerosis. *$P < 0.05$, control *vs.* other groups. †$P < 0.05$, cholesterol *vs.* cholesterol + Flax II. *Abbreviation:* See Figure 13.2. *Source:* Reprinted with permission from Prasad *et al.* (46).

ated with a decrease in serum TC by 14% (Fig. 13.4), LDL-C by 17% (Fig. 13.4), and TC/HDL-C by 28% (Fig. 13.5). HDL-C remained unaltered, but serum TG was elevated with flaxseed treatment.

These results suggest that the antiatherogenic effect of Type II flaxseed is not due to α-linolenic acid, but may be due to the SDG component of flaxseed, which may have a lipid-lowering effect in addition to antioxidant activity. The decrease in the TC and LDL-C, and the increase in the TG could be due to low levels of α-linolenic acid because n-3 fatty acids from fish oil are known to decrease serum TG (47,48) and HDL-C (48) and increase serum LDL-C (48) and TC (49). The soluble fiber in flaxseed may also have a lipid-lowering effect synergistic with or independent of SDG (45).

Secoisolariciresinol Diglucoside and Atherosclerosis

The results of the studies on Type II flaxseed suggested that antiatherogenic activity of flaxseed may reside in flaxmeal which contains SDG. Prasad (27) studied the

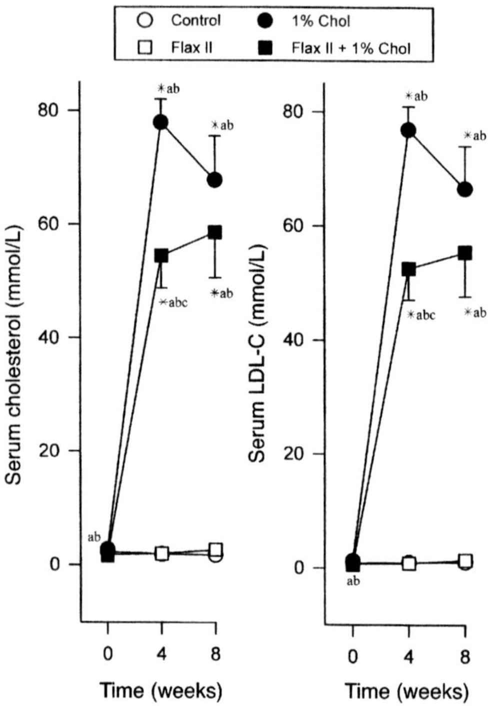

Fig. 13.4. Changes in serum cholesterol and low-density-lipoprotein cholesterol (LDL-C) in the four groups. Results are expressed as mean ± SE. *P < 0.05, comparison of values at various times with respect to time 0 in the respective groups. [a]P < 0.05, control *vs.* other groups. [b]P < 0.05, Flax II *vs.* Chol with or without Flax II. [c]P < 0.05, Chol *vs.* Chol + Flax II. *Abbreviation:* see Figure 13.2. *Source:* Reproduced with permission from Prasad (46).

effect of SDG on hypercholesterolemic atherosclerosis to determine if SDG prevents/retards its development and whether the antiatherogenic effect of SDG is associated with decrease in oxidative stress and serum lipids. Rabbits in this study were assigned to four groups: Group I, control (regular diet); Group II, SDG (regular diet supplemented with SDG in the dose of 15 mg/kg b.w., orally, daily); Group III, 1% cholesterol (regular diet containing 1% cholesterol); Group IV, 1% cholesterol + SDG (1% cholesterol diet supplemented with SDG in the dose similar to that in Group II). SDG reduced serum TC by 33%, serum LDL-C by 35%, and TC/HDL-C by approximately 64% (Fig. 13.6). SDG initially raised serum HDL-C in the rabbits on high cholesterol diet but it remained unchanged at the end of the protocol. Although SDG decreased serum TG in the control diet group, it did not change the levels in the high cholesterol-fed group. Atherosclerosis in the high cholesterol-fed group was associated with an increase in MDA levels in the aortic tissue and a

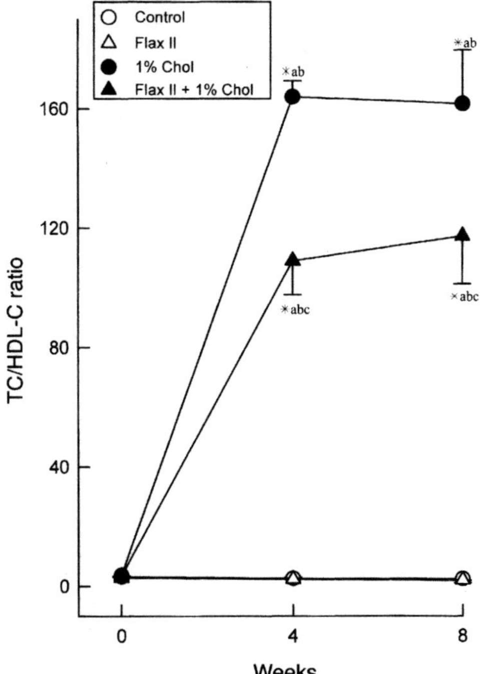

Fig. 13.5. Changes in the ratio of total cholesterol (TC) to high-density-lipoprotein cholesterol (HDL-C) in the four groups. This ratio provides the index of risk. High ratio indicates increased risk and vice versa. Results are expressed as mean ± SE. *P < 0.05, comparison of values at various times with respect to time 0 in the respective groups. [a]P < 0.05, control *vs.* other groups. [b]P < 0.05, Flax II *vs.* cholesterol with or without Flax II. [c]P < 0.05, cholesterol *vs.* cholesterol + Flax II. *Abbreviation:* See Figure 13.2. *Source:* Reprinted with permission from Prasad *et al.* (46).

decrease in the antioxidant reserve as indicated by an increase in the aortic tissue chemiluminescence (Aortic-CL) (Fig. 13.7). SDG reduced the development of atherosclerosis by 73% (Fig. 13.8). This reduction was associated with a decrease in the aortic tissue MDA and an increase in the antioxidant reserve.

These results suggest that hypercholesterolemic atherosclerosis is associated with oxidative stress, and that reduction in hypercholesterolemic atherosclerosis by flaxseed and its components is associated with reduction in oxidative stress. Reduction in atherosclerosis could also be due to a decrease in the serum levels of cholesterol by flaxseed and its components. Reduction in cholesterol may ultimately lead to reduction in oxidative stress. As such, hypercholesterolemia is associated with oxidative stress (8–11,13,20,21).

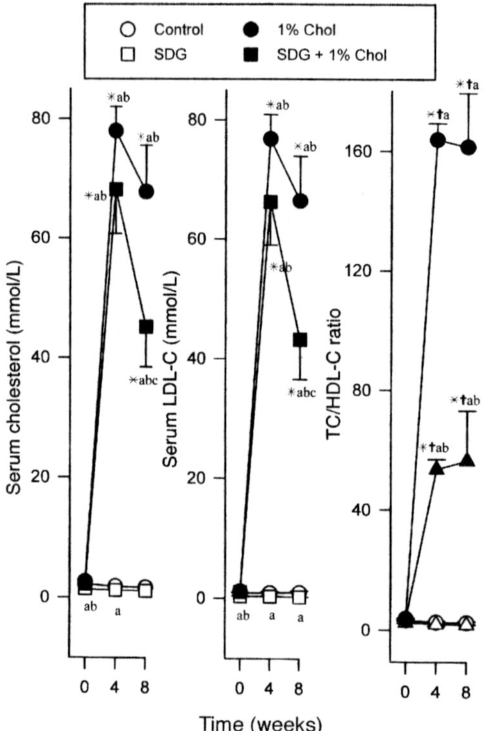

Fig. 13.6. Changes in the serum cholesterol, LDL-C, and TC/HDL-C ratio in the four groups. Results are expressed as mean ± SE. *P < 0.05, comparison of values at different times with respect to time 0 in the respective groups. [a]P < 0.05, control *vs.* other groups for Chol and LDL-C, and SDG *vs.* Chol with or without SDG for TC/HDL-C. [b]P < 0.05, SDG *vs.* Chol with or without SDG for Chol and LDL-C, and Chol *vs.* Chol plus SDG for TC/HDL-C. [c]P < 0.05, Chol *vs.* Chol + SDG. [†]P < 0.05, control *vs.* other groups. *Abbreviations:* See Figures 13.2, 13.4, and 13.5; SDG, secoisolariciresinol diglucoside. *Source:* Reproduced with permission from Prasad (27).

Clinical Implications

Flaxseed and several of its components, including SDG and fiber, may be helpful in reducing the development of hypercholesterolemic athersclerosis and might reduce the risk of heart attack, stroke, and other peripheral vascular disease. SDG may also be beneficial in patients with hypercholesterolemia because it reduces the serum total cholesterol and LDL-C and risk ratio. SDG has the potential to be an effective, safe, and inexpensive therapeutic agent for prevention of ischemic heart disease and stroke. Beneficial effects of flaxseed and SDG against serum cholesterol and hypercholesterolemic atherosclerosis would change the strategy for management of hypercholes-

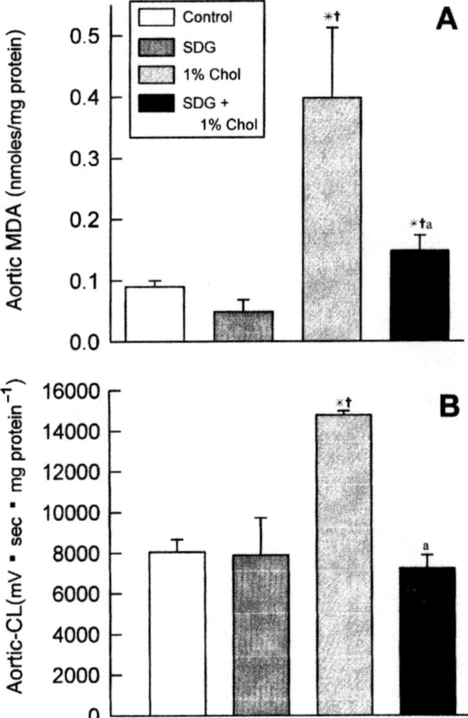

Fig. 13.7. Aortic tissue MDA and chemiluminescence (CL) in four experimental groups. Results are expressed as mean ± SE. *$P < 0.05$, control *vs.* other groups. †$P < 0.05$, SDG *vs.* Chol with or without SDG. [a]$P < 0.05$, Chol *vs.* Chol ± SDG. *Abbreviations:* See Figures 13.2 and 13.6. *Source:* Reprinted with permission from Prasad (27).

terolemic atherosclerosis and its sequelae. Prevention of ischemic heart disease and stroke will markedly reduce the health care costs, and social and economic burden to society because of morbidity and mortality associated with heart attack and stroke.

Summary

Flaxseed reduces development of atherosclerosis by 46% without lowering serum total cholesterol. However, flaxseed with very low α-linolenic acid reduces the development of atherosclerosis by 69%, and this reduction is associated with a reduction in serum TC and LDL-C and a rise in serum TG. Serum HDL-C remained unaffected. Effects of flaxseed on serum lipids are very variable, from no change to a slight reduction. Flax oil has no effect on serum lipids. SDG reduces hypercholesterolemic atherosclerosis by 73% and serum TC and LDL-C and the ratio of TC/HDL-C by 33%, 35%, and 64%, respectively. HDL-C, although elevated initially, remained unchanged

     K. Prasad

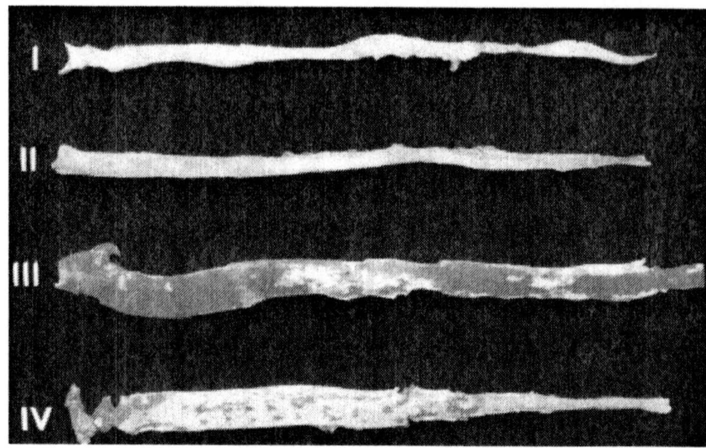

Fig. 13.8. Intimal surface of aortae from four experimental groups showing Sudan IV-stainable lipid deposits. Note marked dark lipid deposits in Groups III and IV, the deposits being less in Group IV than in Group III. Group I, control; Group II, control diet + secoisolariciresinol diglucoside (SDG); Group III, 0.5% cholesterol diet; Group IV, 0.5% cholesterol diet with SDG. *Source:* Reproduced with permission from Prasad (27).

later on. SDG reduced the oxidative stress induced by hypercholesterolemia. SDG-induced reduction in atherosclerosis was associated with reduction in oxidative stress and serum TC and LDL-C and risk ratio (TC/HDL-C). These results suggest that the effectiveness of flaxseed in reduction of atherosclerosis is not due to α-linolenic acid but may be due to its SDG or fiber content.

Acknowledgments

The author acknowledges the technical assistance of P.K. Chattopadhyay and Barbara Raney. This study was supported by grants from CIHR Regional Partnership Program, and Heart and Stroke Foundation of Saskatchewan, Canada. The author is also thankful to Linda Steckler for excellent assistance in preparation of this manuscript.

References

1. Davis, C., B. Rifkind, H. Brenner, and D. Gordon, A Single Cholesterol Measurement Underestimates the Risk of CHD: An Empirical Example from Lipid Research Clinic's Mortality Follow-up Study, *JAMA 264*:3044–3046 (1990).
2. Anderson, K.M., W.P. Castelli, and D. Levy, Cholesterol and Mortality 30 Years Follow-up from Framingham Study, *JAMA 257*:2176–2180 (1987).
3. Lipid Research Clinic Program, Lipid Research Clinic's Coronary Primary Prevention Trial Results, *JAMA 251*:351–374 (1954).
4. Castelli, W.P., Cholesterol and Lipids in the Risk of Coronary Artery Disease: The Framingham Heart Study, *Canad. J. Cardiol. 4(Suppl. A)*:5A–10A (1988).

5. Ross, R., The Pathogenesis of Atherosclerosis—an Update, *N. Engl. J. Med. 314*: 488–500 (1986).
6. Ross, R., and L. Harker, Hyperlipidemia and Atherosclerosis: Chronic Hyperlipidemia Initiates and Maintains Lesions by Endothelial Cell Desquamation and Lipid Accumulation, *Science 193*:1094–1100 (1976).
7. Nelson, E., S.D. Gertz, M.S. Forbes, M.I. Rennels, F.P. Heald, M.A. Kahn, T.M. Farber, E. Miller, M.M. Hussain, and F.L. Earl, Endothelial Lesion in the Aorta of Egg Yolk-Fed Miniature Swine: A Study by Scanning and Transmission Electron Microscopy, *Exp. Mol. Pathol. 25*:208–220 (1976).
8. Prasad, K., and J. Kalra, Oxygen Free Radicals and Hypercholesterolemic Atherosclerosis: Effect of Vitamin E, *Am. Heart J. 125*: 958–973 (1993).
9. Prasad, K., J. Kalra, and P. Lee, Oxygen Radicals as a Mechanism of Hypercholesterolemic Atherosclerosis: Effects of Probucol, *Intl. J. Angiol. 3*:100–112 (1994).
10. Prasad, K., S.V. Mantha, J. Kalra, and P. Lee, Prevention of Hypercholesterolemic Atherosclerosis by Garlic, an Antioxidant, *J. Cardiovasc. Pharmacol. Ther. 2*:309–320 (1997).
11. Prasad, K., S.V. Mantha, J. Kalra, R. Kapoor, and B.R.C. Kamalarajan, Purpurogallin in the Prevention of Hypercholesterolemic Atherosclerosis, *Intl. J. Angiol. 6*:157–166 (1997).
12. Rosen, G.M., and B.A. Freeman, Detection of Superoxide Generated by Endothelial Cells, *Proc. Natl. Acad. Sci. USA 81*:7269–7273 (1984).
13. Mantha, S.V., M. Prasad, J. Kalra, and K. Prasad, Antioxidant Enzymes in Hypercholesterolemia and Effects of Vitamin E in Rabbits, *Atherosclerosis 101*:135–144 (1993).
14. Mantha, S.V., J. Kalra, and K. Prasad, Effects of Probucol on Hypercholesterolemia-Induced Changes in Antioxidant Enzymes, *Life Sci. 58*:503–509 (1996).
15. Demuth, K., I. Myara, B. Chappey, B. Vedie, M.A. Pech-Amsellem, M.E. Haberland, and N. Moatti, A Cytotoxic Electronegative LDL Subfraction Is Present in Human Plasma, *Arteroscl. Thromb. Vasc. Biol. 16*:773–783 (1996).
16. Sevanian, A., J. Hwang, H. Hodis, G. Cazzolato, P. Avagaro, and G. Bittolo-Bon, Contribution of an *in vivo* Oxidized LDL to LDL Oxidation and Its Association with Dense LDL Subpopulations, *Arterioscl. Thromb. Vasc. Biol. 16*:784–793 (1996).
17. Jull, K., L.B. Nielsen, K. Munkholm, S. Stender, and B.G. Nordestgaard, Oxidation of Plasma Low-Density Lipoprotein Accelerates Its Accumulation and Degradation in the Arterial Wall *in vivo*, *Circulation 94*:1698–1704 (1996).
18. Hiebert, L.M., and J.M. Lu, Heparin Protects Cultured Arterial Endothelial Cells from Damage by Toxic Oxygen Metabolites, *Atherosclerosis 83*:47–51 (1990).
19. Hiebert, L.M., and J.M. Liu, Dextran Sulphates Protect Porcine Arterial Endothelial Cells from Free Radical Injury, *Human Exptl. Toxicol. 13*:233–239 (1994).
20. Henriksen, T., E.M. Mahoney, and D. Steinberg, Enhanced Macrophage Degradation of Low-Density Lipoprotein Previously Incubated with Cultured Endothelial Cells: Recognition by Receptors for Acetylated Low-Density Lipoproteins, *Proc. Natl. Acad. Sci. USA 78*:6499–6503 (1981).
21. Keaney, J.F., Jr., and J.A. Vita, Atherosclerosis, Oxidative Stress, and Antioxidant Protection in Endothelial-Derived Relaxing Factor Action, *Prog. Cardiovasc. Dis. 38*:129–154 (1995).
22. Schwartz, C.J., A.J. Valente, and E.A. Sprague, A Modern View of Atherogenesis, *Am. J. Cardiol. 71*:9B–14B (1993).

23. Steinberg, D., Antioxidants and Atherosclerosis: A Current Assessment, *Circulation* *84*:1420–1425 (1991).

24. Oomah, B.D., and G. Mazza, Flaxseed Proteins: A Review, *Food Chem.* *48*:109–114, (1993).

25. Hettiarachchy, N.S., G.A. Hareland, A. Ostenson, and G. Blander-Shank, Chemical Composition of 11 Flaxseed Varieties Grown in North Dakota, *Proc. Flax. Institute 53*:36 (1990).

26. Westcott, N.D., and A.D. Muir, U.S. Patent 5,705,618 (1988).

27. Prasad, K., Reduction of Serum Cholesterol and Hypercholesterolemic Atherosclerosis in Rabbits by Secoisolariciresinol Diglucoside (SDG) Isolated from Flaxseed, *Circulation* *99*:1355–1362 (1999).

28. Endres, S., R. Ghorbani, V.E. Kelley, K. Georgilis, G. Lonnemann, J.W. van der Meer, J.G. Cannon, T.S. Rogers, M.S. Klempner, P.C. Weber, et al., The Effect of Dietary Supplementation with n-3 Polyunsaturated Fatty Acids on the Synthesis of Interleukin-1 and Tumor Necrosis Factor by Mononuclear Cells, *N. Engl. J. Med. 320*:265–271 (1989).

29. Chandrasekar, B., and G. Fernandes, Decreased Pro-Inflammatory Cytokines and Increased Antioxidant Enzyme Gene Expression by Omega-3 Lipids in Murine Lupus Nephritis, *Biochem. Biophys. Res. Commun. 200*:893–898 (1994).

30. Lee, T.H., R.L. Hoover, J.D. Williams, R.I. Sperling, J. Ravalese, III, B.W. Spur, D.R. Robinson, E.J. Corey, R.A. Lewis, and K.F. Austen, Effect of Dietary Enrichment with Eicosapentaenoic and Docosahexaenoic Acids on *in vitro* Neutrophil and Monocyte Leukotriene Generation and Neutrophil Function, *N. Engl. J. Med. 312*:1217–1224 (1985).

31. Fisher, M., P.H. Levine, B.H. Weiner, M.H. Johnson, E.M. Doyle, P.A. Ellis, and J.J. Hoogasian, Dietary n-3 Fatty Acid Supplementation Reduces Superoxide Production and Chemiluminescence in a Monocyte-Enriched Preparation of Leukocytes, *Am. J. Clin. Nutr. 51*:804–808 (1990).

32. Fisher, M., K.S. Upchurch, and P.H. Levine, Effects of Dietary Fish Oil Supplementation on Polymorphonuclear Leukocyte Inflammatory Potential, *Inflammation 10*:387–392 (1986).

33. Prasad, K., Hydroxyl Radical-Scavenging Property of Secoisolariciresinol Diglucoside (SDG) Isolated from Flaxseed, *Mol. Cell. Biochem. 168*:117–123 (1997).

34. Prasad, K., Antioxidant Activity of Secoisolariciresinol Diglucoside-Derived Metabolites, Secoisolariciresinol, Enterodiol, and Enterolactone, *Intl. J. Angiol. 9*:220–225 (2000).

35. Prasad, K., Dietary Flaxseed in Prevention of Hypercholesterolemic Atherosclerosis, *Atherosclerosis 132*:69–76 (1997).

36. Ratnayake, W.M.N., W.A. Behrens, P.W.F. Fischer, M.R. L'Abbé, R. Mongeau, and J.L. Beare-Rogers, Chemical and Nutritional Studies of Flaxseed (Variety Linott) in Rats, *J. Nutr. Biochem. 3*:232–240 (1992).

37. Kritchevsky, D., S.A. Tepper, and D.M. Klurfeld, Influence of Flaxseed on Serum and Liver Lipids in Rats, *J. Nutr. Biochem. 2*:133–134 (1991).

38. Ingram, A.J., A. Parbtani, W.F. Clark, E. Spanner, M.W. Huff, D.J. Philbrick, and B.J. Holub, Effects of Flaxseed and Flax Oil Diets in a Rat—5/6 Renal Ablation Model, *Am. J. Kidney Dis. 25*:320–329 (1995).

39. Sanders, T.A.B., and F. Roshani, The Influence of Different Types of Omega 3 Polyunsaturated Fatty Acids on Blood Lipids and Platelet Function in Healthy Volunteers, *Clin. Sci. 64*:91–99 (1983).

40. Cunnane, S.C., S. Ganguli, C. Menard, A.C. Liede, M.J. Hamadeh, Z.Y. Chen, T.M. Wolever, and D.J. Jenkins, High-α-Linolenic Acid Flaxseed (*Linum usitatissimum*): Some Nutritional Properties in Humans, *Br. J. Nutr. 69*:443–453 (1993).
41. Bierenbaum, M.L., R. Reichstein, and T.R. Watkins, Reducing Atherogenic Risk in Hyperlipedemic Humans with Flaxseed Supplementation: A Preliminary Report, *J. Am. Coll. Nutr. 12*:501–504 (1993).
42. Cunnane, S.C., M.J. Hamadeh, A.C. Liede, L.U. Thompson, T.M. Wolever, and D.J. Jenkins, Nutritional Attributes of Traditional Flaxseed in Healthy Young Adults, *Am. J. Clin. Nutr. 61*:62–68 (1995).
43. Clark, W.F., A. Parbtani, M.W. Huff, E. Spanner, H. de Salis, I. Chin-Yee, D.J. Philbrick, and B.J. Holub, Flaxseed: A Potential Treatment for Lupus Nephritis, *Kidney Int. 48*:475–480 (1995).
44. Kelley, D.S., G.J. Nelson, J.E. Love, L.B. Branch, P.C. Taylor, P.C. Schmidt, B.E. Mackey, and J.M. Iacono, Dietary α-Linolenic Acid Alters Tissue Fatty Acid Composition but Not Blood Lipids, Lipoproteins or Coagulation Status in Humans, *Lipids 28*:533–537 (1993).
45. Cunnane, S.C., Metabolism and Function of α-Linolenic Acid in Humans, in *Flaxseed in Human Nutrition*, edited by S.C. Cunnane and L.U. Thompson, AOCS, Champaign, Illinois, 1995, pp. 99–127.
46. Prasad, K., S.V. Mantha, A.D. Muir, and N.D. Westcott, Reduction of Hypercholesterolemic Atherosclerosis by CDC-Flaxseed with Very Low Alpha-Linolenic Acid, *Atherosclerosis 136*:367–375 (1998).
47. Kestin, M., P. Clifton, G.B. Belling, and P.J. Nestel, n-3 Fatty Acids of Marine Origin Lower Systolic Blood Pressure and Triglycerides but Raise LDL Cholesterol Compared with n-3 and n-6 Fatty Acids from Plants, *Am. J. Clin. Nutr. 51*:1028–1034 (1990).
48. Phillipson, B.E., D.W. Rothrock, W.E. Connor, W.S. Harris, and D.R. Illingworth, Reduction of Plasma Lipids, Lipoproteins and Apoproteins by Dietary Fish Oils in Patients with Hypertriglyceridemia, *N. Eng. J. Med. 312*:1210–1216 (1985).
49. Nestel, P.J., W.E. Connor, M.F. Reardon, S. Connor, S.Wong, and R. Boston, Suppression by Diets Rich in Fish Oil of Very Low Density Lipoprotein Production in Man, *J. Clin. Invest. 74*:82–89 (1984).

Chapter 14

Flaxseed Components in the Prevention of Experimental Diabetes

Kailash Prasad

Department of Physiology, College of Medicine, University of Saskatchewan, Saskatoon, Saskatchewan, Canada

Introduction

Diabetes is characterized by hyperglycemia, glucosuria, polyurea, polydipsia, polyphagia, impaired glucose tolerance, insulin resistance, hyperlipidemia, and diabetic coma. There are two types of diabetes, type-1 and type-2. Type-1 is also known as insulin-dependent diabetes mellitus or juvenile onset diabetes. It occurs suddenly and is of early age onset. In this type of diabetes, the pancreas is unable to produce insulin and the individuals require daily insulin injections for survival. Type-2 diabetes, also called noninsulin-dependent diabetes mellitus, appears slowly and is of adult onset (peak age 50–60 yr). In type-2 diabetes, the pancreas is unable to produce enough insulin to meet the body's requirements or the body is unable to properly use the insulin produced.

Type-2 diabetes is 10 times more common than type-1 diabetes. In Germany, type-2 diabetes is almost 20 times more common than type-1 diabetes (1). Diabetes mellitus is the seventh leading cause of mortality in the United States. Prevalence of diagnosed diabetes between 1988–1994 in the United States was estimated to be 5.1% for adults ≥ 20 yr of age (2). According to the National Institute of Diabetes and Digestive and Kidney Disease, Diabetes Statistics (3), type-2 diabetes affects more than 16 million people in the United States. The prevalence of diabetes is two to three times higher in aboriginal people of Canada than in the rest of the Canadian population (4). Overall prevalence of type-2 diabetes in Canada was 7.2% in 1995 and is estimated to be 9.2% in the year 2025 (5). It is estimated that the number of persons with diabetes in the world will rise from 135 million in 1995 to 300 million in 2025 (5). Diabetes is diagnosed when the fasting plasma glucose level is >7.8 mM/L or the oral glucose tolerance test shows the plasma glucose level higher than 11.1 mM/L (6). The time between onset and diagnosis is usually 9–12 yr.

Diabetes and Its Complications

Complications of diabetes are numerous including coronary artery disease, strokes, polyneuropathy, nephropathy, and retinopathy. Diabetes is responsible for 50% of all nontraumatic limb amputations, 40% of kidney failures, 25% of cardiac surgeries, and is the leading cause of blindness and death. The treatment of diabetes has involved the use of insulin, insulin secretagogues (sulfonylureas, biguanides) and the insulin-sensitizing drug, thiazolidinedione (triglitazone, rosiglitazone, and pioglitazone) for glycemic control (7). However, despite intensive use of antidiabetic drugs, including insulin, many type-2 diabetic patients exhibit poor glycemic control and some develop serious complications. The main mechanism of microvascular complications is suggested to be hyperglycemia, but the current treatment, reducing blood sugar levels, does not completely prevent complications. This suggests that factor(s) other than glucose may be involved in diabetes and its complications.

In this chapter the potential use of secoisolariciresinol diglucoside (SDG) isolated from flaxseed will be discussed in type-1 and type-2 diabetes.

Pharmacological Properties of SDG

Flaxseed contains 32–45% of its mass as oil, of which 51–55% is α-linolenic acid (n-3 fatty acid) (8–10). SDG, a lignan, has been isolated from defatted flaxseed in 98% pure form (11). Flaxseed is the richest source of plant lignan SDG (12). The levels of SDG in flaxseed vary between 0.6–1.8 g/100 g (13). HPLC has shown that SDG scavenges hydroxyl radical (•OH) generated by photolysis of hydrogen peroxide (H_2O_2) with ultraviolet light and trapped with salicylic acid (14). This effect was concentration dependent. Also, SDG prevented the •OH-induced lipid peroxidation of liver homogenate in a concentration-dependent manner (14).

Prasad (15) measured the antioxidant activity of SDG using the ability of this compound to reduce the chemiluminescence (CL) of activated polymorphonuclear leukocytes ([PMNL]; PMNL-CL). Activated PMNLs produce superoxide anion (O_2^-), H_2O_2, •OH, and singlet oxygen ($^1O_2$) (16–18). SDG produced a concentration-dependent reduction in PMNL-CL, suggesting that SDG has antioxidant activity. Antioxidant activity of SDG was 1.27 times greater than that of vitamin E. These results suggest that SDG is a potent antioxidant.

Increased hepatic glucose production has been reported to be due to increased hepatic gluconeogenesis (19). Phosphoenolpyruvate carboxykinase (PEPCK) is a rate-limiting enzyme for gluconeogenesis in the liver (20) and is elevated in all models of diabetes (21). Regulation of the activity of PEPCK is primarily controlled through gene expression. SDG suppresses the expression of the PEPCK gene (22).

Oxidative Stress and Diabetes

Reactive oxygen species (ROS) have been implicated in the pathophysiology of both type-1 and type-2 diabetes. Lipid peroxidation product malondialdehyde (MDA) have been shown to increase in the pancreas and serum of streptozotocin-induced type-1 diabetes (23–26) and in patients with diabetes (27). Dimethylthiourea, a free-radical scavenger, inhibits the cytotoxic effects of *tert*-butylhydroperoxide, interleukin-1, tumor necrosis factor, interferon gamma, and alloxan on the islet cells (28). Cytokines increase MDA content of islet cells and cause islet cell necrosis (29). Alloxan generates ROS in the presence of islets of pancreas (30). The fact that antioxidants (superoxide dismutase, dimethylthiourea) prevent the effect of alloxan on pancreas suggests that the cytotoxic effects of alloxan are through ROS (30–32). The hypothesis that streptozotocin-induced diabetes is through generation of ROS is supported by the fact that dimethylthiourea protects beta cells of the pancreas against the diabetogenic effect of streptozotocin (33–35). Recently, serum and pancreatic MDA levels were shown to be elevated in streptozotocin-induced diabetes (36). MDA levels are elevated in the serum and pancreas of the Bio-Breed diabetic-prone (BBdp) rat, which is a genetic model of type-1 diabetes (37). These reports suggest that ROS may be involved in the pathogenesis of type-1 diabetes.

ROS may also be involved in the pathogenesis of type-2 diabetes. As such, increased serum MDA levels have been observed in patients with type-2 diabetes (38–42). Plasma free-radical concentrations are positively correlated to fasting plasma insulin in patients with type-2 diabetes (43). Increased oxidation of low density lipoprotein has been reported in type-2 diabetic patients (44,45). Recently, Prasad has reported an increase in the oxidative stress in the animal model of type-2 diabetes (46).

SDG and Type-1 Diabetes

Because ROS have been implicated in the pathogenesis of type-1 diabetes and SDG has antioxidant activity, SDG might be effective in the prevention of development of type-1 diabetes. Studies have been conducted on the effects of SDG in prevention of two animal models of type-1 diabetes (36,37). In one model, diabetes is produced by injecting streptozotocin in rats (36). The other is found in genetic model BBdp rats that develop diabetes spontaneously (47). The incidence of diabetes in BBdp rats is 40–70%; more than 85% develop this condition between 60–120 d of age, and fewer than 0.5% develop it before 60 d of age.

In streptozotocin-induced diabetes, the purpose of the study was twofold: To determine whether (i) SDG prevents/reduces the development of diabetes, and (ii) prevention/reduction is associated with reduction in oxidative stress (36). The rats were assigned to four groups: Group I, Control; Group II, Control + SDG (22 mg/kg b.w. daily, orally); Group III, streptozotocin (STZ) treatment; and Group IV, STZ + SDG (dose similar to that in Group II). Diagnosis of diabetes was made on the basis of glucosuria and was confirmed at the time of sacrifice by measuring serum glucose. Oxidative stress was assessed by measuring serum and pancreatic MDA, pancreatic antioxidant reserve [pancreatic tissue chemiluminescence (pancreatic-CL)], and ROS-producing activity of white blood cells (WBC-CL). The rats in Groups I and II did not develop diabetes (Fig. 14.1). However, all the rats in Group III and 25% in Group IV developed diabetes.

SDG prevented the development of diabetes by 75%. Serum glucose increased markedly in the rats that developed diabetes (Fig. 14.2). Serum glucose levels in

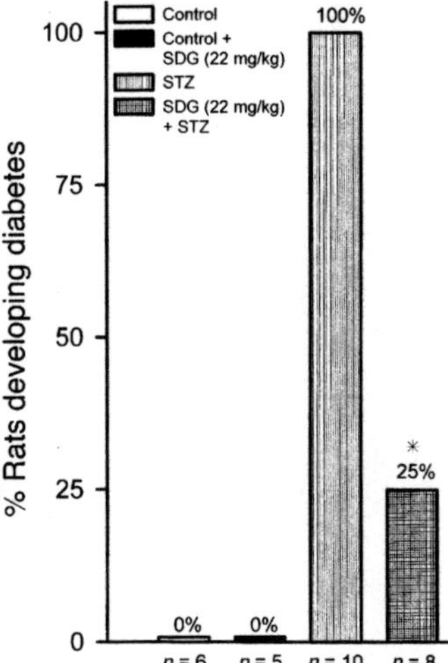

Fig. 14.1. Incidence of development of diabetes (glucosuria) in the four groups. Percentage of rats developing diabetes in each group is given over the top of the bar. *$P < 0.05$, STZ *vs.* STZ + SDG. *Abbreviations:* SDG, secoisolariciresinol diglucoside; STZ, streptozotocin; *n*, number of rats in each group. *Source:* Reproduced with permission from Prasad *et al.* (36).

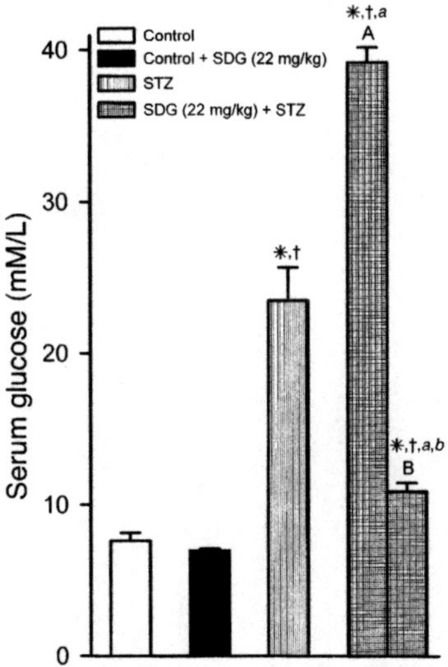

Fig. 14.2. Levels of serum glucose in the four groups. Results are expressed as mean ± SE. "A" represents the rats in Group IV that developed glucosuria, and "B" represents the rats in Group IV that did not develop glucosuria. *$P < 0.05$, control *vs.* other groups. †$P < 0.05$, control + SDG *vs.* other groups. a$P < 0.05$, STZ *vs.* SDG + STZ. b$P < 0.05$, A *vs.* B. *Abbreviations:* See Figure 14.1. *Source:* Reproduced with permission from Prasad *et al.* (36).

Group IV that did not develop diabetes were lower than in those that developed diabetes. Serum, and pancreatic MDA (Fig. 14.3) levels were elevated in rats that developed diabetes. Antioxidant reserve was lower in the rats that developed diabetes compared with those that did not develop diabetes. The WBC-CL activity was greater in the rats that developed diabetes than in those that did not develop diabetes (Fig. 14.4). These results suggest that STZ-induced diabetes is associated with oxidative stress and that SDG is effective in reducing the STZ-induced diabetes mellitus with concomitant reduction in oxidative stress.

Reports are also available on the effects of SDG in BBdp rats (37). In this study the rats were divided into three groups: Group I, Control, BBn rats that do not develop diabetes; Group II, BBdp rats; and Group III, BBdp rats that received SDG (22 mg/kg b.w. orally, daily). The SDG treatment started at the age of 5 wk. Approximately 73% of untreated BBdp rats developed diabetes whereas only 21.4% of SDG-treated BBdp rats developed diabetes (Fig. 14.5) suggesting that SDG prevented the development of diabetes by 71%. Serum glucose levels were higher in the rats that developed diabetes compared with those that did not develop diabetes (Fig. 14.6). Serum and pancreatic MDA levels were elevated in rats that developed diabetes. Antioxidant reserve in the pancreas was diminished. The data suggest that the development of diabetes is associated with an increase in the serum and pancreatic MDA and a decrease in the antioxiant reserve. Prevention of development of diabetes by SDG is associated with a decrease in serum and pancreatic MDA and an increase in the antioxidant reserve.

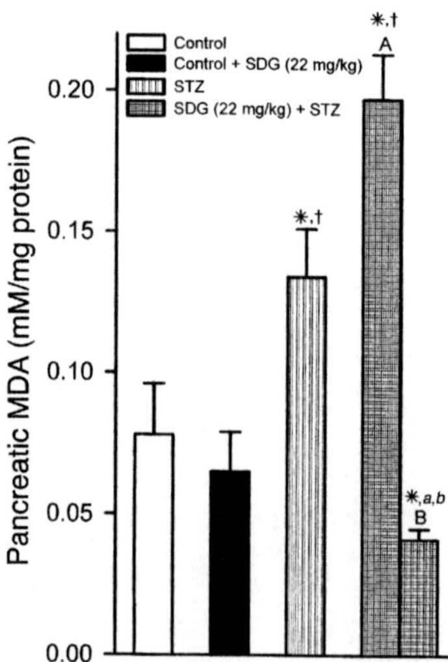

Fig. 14.3. MDA content of pancreas in various groups. Results are expressed as mean ± SE. *$P < 0.05$, control *vs.* other groups. †$P < 0.05$, control + SDG *vs.* other groups. [a]$P < 0.05$, STZ *vs.* STZ + SDG. [b]$P < 0.05$, A *vs.* B. *Abbreviations:* See Figure 14.1; MDA, malondialdehyde. *Source:* Reproduced with permission from Prasad *et al.* (36).

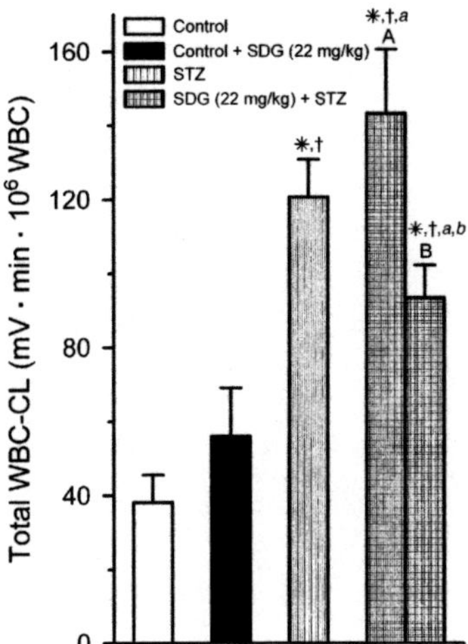

Fig. 14.4. WBC-CL activity in the four groups. "A" represents the rats in Group IV that developed diabetes, and "B" represents the rats in Group IV that were nondiabetic. Results expressed as mean ± SE. *$P < 0.05$, Group I *vs.* other groups. [†]$P < 0.05$, Group II *vs.* other groups. [a]$P < 0.05$, Group III *vs.* A and B of Group IV. [b]$P < 0.05$, A *vs.* B. *Abbreviations:* See Figure 14.1. WBC-CL, white blood cells chemiluminescence. *Source:* Reproduced with permission from Prasad *et al.* (36).

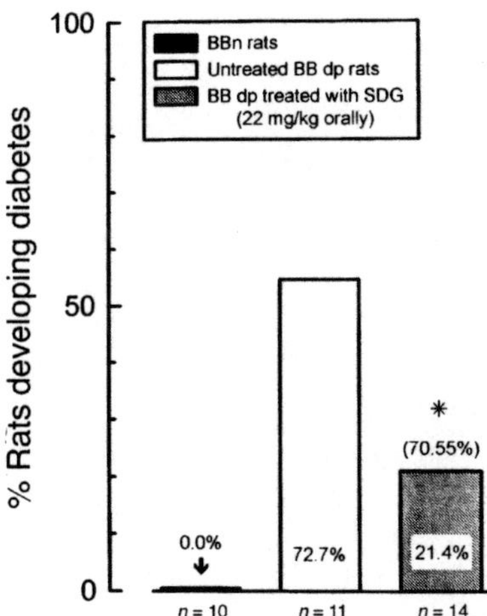

Fig. 14.5. Incidence of diabetes in the three groups. Numerals in the bar show the percentage of rats developing diabetes (n = number of rats). *$P < 0.05$, untreated *vs.* SDG-treated BBdp rats. *Abbreviations:* See Figure 14.1. *Source:* Reproduced with permission from Prasad, K. (37).

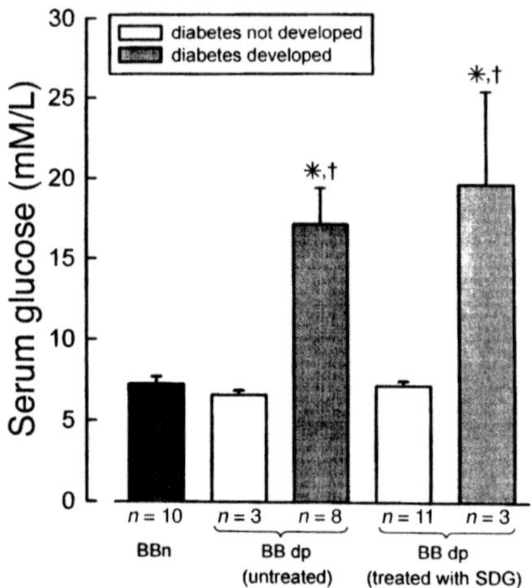

Fig. 14.6. Levels of serum glucose in various groups. Results are expressed as mean ± SE. *n* = number of rats. *$P < 0.05$, BB$_n$ *vs.* other groups. †$P < 0.05$, rats in the groups that did not develop glucosuria *vs.* those that did develop glucosuria. *Abbreviation:* See Figure 14.1. *Source:* Reproduced with permission from Prasad, K. (37).

SDG and Type-2 Diabetes

An extensive study was conducted on the effect of SDG in the prevention of type-2 diabetes (46). This study was carried out to determine whether (i) type-2 diabetes was associated with oxidative stress, (ii) SDG can prevent/reduce the development of type-2 diabetes, and (iii) SDG-induced prevention/reduction is associated with prevention/reduction in oxidative stress. This study was conducted in ZDF/gmi-fa/fa female rats (Genetic Models Inc., Indianapolis, IN), an animal model of type-2 diabetes. These rats become glucose intolerant at about 70 d of age on a special diet. The study included five groups of rats. Group I, ZLC (Zucker lean control) rats that do not develop diabetes; Group II, ZLC + SDG (40 mg/kg b.w. orally, daily); Group III, ZDF diabetic-prone rats; Group IV, ZDF + SDG in a dose similar to that in Group II; and Group V, ZDF + SDGB (SDG dose similar to that in Group II). The rats in Groups III and IV were sacrificed 1 wk after the development of diabetes or at the age of 101 d. The rats in Group V were killed at the age of 70 d to determine whether SDG treatment prevents the development of diabetes by that age. The purpose was also to determine whether the prevention is associated with a decrease in oxidative stress. Groups I–IV were used to investigate whether (i) SDG prevents/retards the development of diabetes, (ii) development of diabetes is associated with oxidative stress, and (iii) prevention is associated with decrease in oxidative stress. A diagnosis of diabetes was made on the basis of glucosuria and fasting blood glucose levels greater than 11.1 mM/L.

None of the rats in Groups I, II, and V had glucosuria by the age of 72 d. All of the rats in Group III and 20% of the rats in Group IV developed diabetes by the age of 72 d (Figs. 14.7,14.8). The results suggest that SDG prevented the development of diabetes by 80% by the age of 72 d. However, 80% of ZDF rats treated with SDG that did not develop diabetes by the age of 72 d developed diabetes later. The onset of diabetes in SDG-treated rats was significantly delayed (Fig. 14.8). Blood glucose was measured at the age of 42 d and at the time of sacrifice. It was noted that the blood glucose levels were similar at both times in Groups I, II, and V but were elevated when rats in Groups III and IV developed diabetes (Fig. 14.9). The results also suggest that SDG prevented the rise in blood glucose by the time these rats were 72 d of age, the time when untreated they all had elevated blood glucose. The glucose tolerance test was positive in Groups III and IV but negative in Groups I and II (46). Levels of glycated hemoglobin (HbA$_1$) and serum MDA (Fig. 14.10) were elevated in diabetic rats; however, they were lower in SDG-treated rats that did not develop diabetes (46).

These results suggest that type-2 diabetes is associated with an increase in oxidative stress and that SDG is effective in retarding the development of diabetes. Reduction in diabetes is associated with reduction in oxidative stress. Also, SDG prevented the development of diabetes by 80% by the age at which all the untreated diabetic-prone rats developed diabetes.

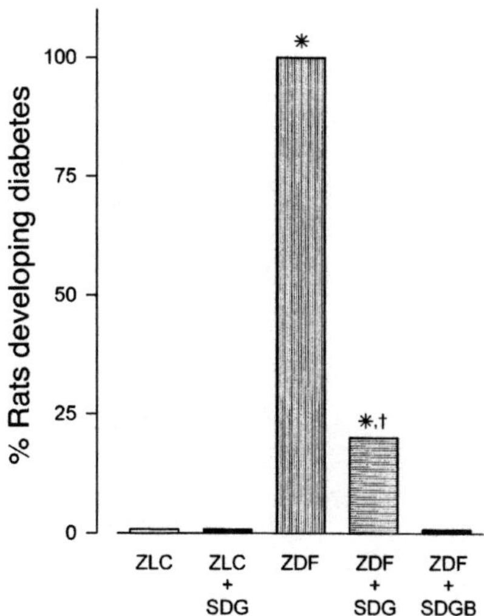

Fig. 14.7. Incidence of glucosuria (diabetes) in the five groups of rats at the age of 72 d. SDGB, SDG given until 68–70 d of age. *$P < 0.05$, ZLC or ZLC + SDG *vs.* ZDF, ZDF + SDG or ZDF + SDGB. †$P < 0.05$, ZDF *vs.* ZDF + SDG. *Abbreviations:* See Figure 14.1; SDGB, SDg-treated ZDF rats at age 70 day before developing diabetes; ZLC, Zucker lean controls; ZDF, Zucker diabetic-prone fatty rats. *Source:* Reproduced with permission from Prasad, K. (46).

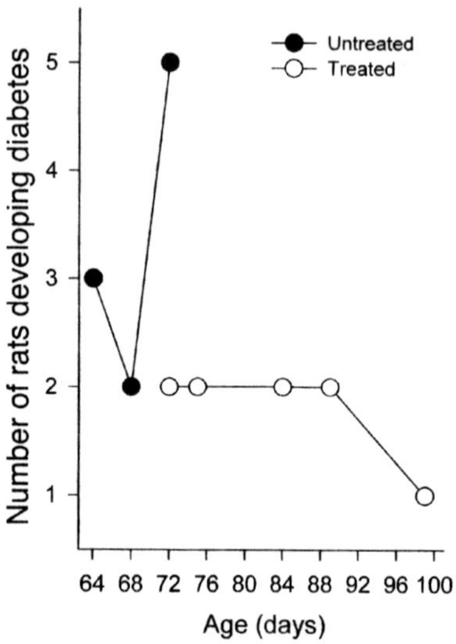

Fig. 14.8. Number of rats developing diabetes in untreated and SDG-treated ZDF rats with time. *Abbreviations:* See Figures 14.1, 14.7. *Source:* Reproduced with permission from Prasad, K. (46).

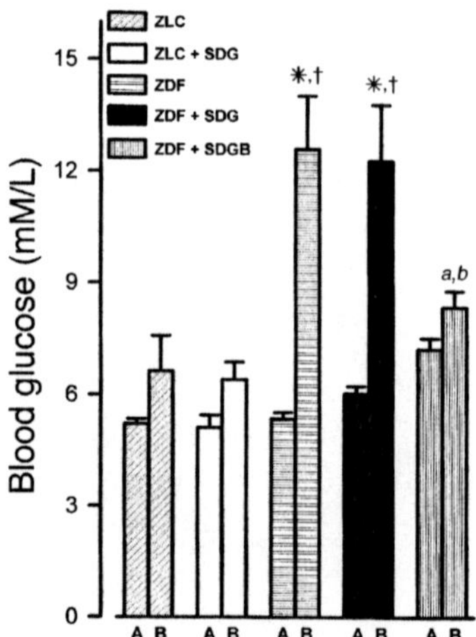

Fig. 14.9. Fasting blood glucose levels in various groups. Results are expressed as mean ± SE. A, initial blood glucose (42 d of age); B, blood glucose levels at the time of sacrifice in the respective groups. *$P < 0.05$, ZLC or ZLC + SDG *vs.* ZDF or ZDF + SDG or ZDF + SDGB. [†]$P < 0.05$, A *vs.* B in various groups. [a]$P < 0.05$, ZDF *vs.* ZDF + SDGB. [b]$P < 0.05$, ZDF + SDG *vs.* ZDF + SDGB. *Abbreviations:* See Figures 14.1, 14.7. *Source:* Reproduced with permission from Prasad, K. (46).

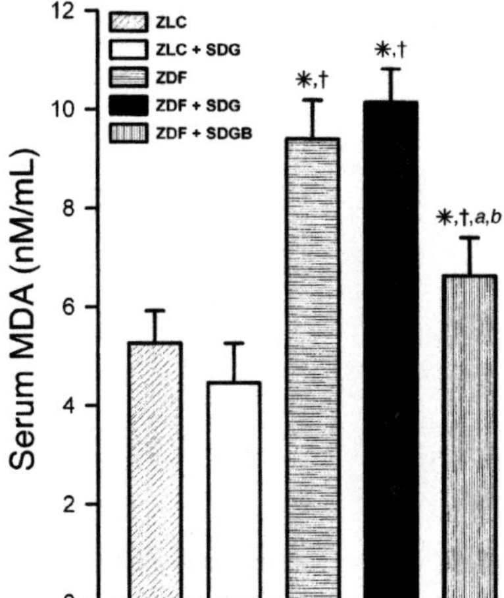

Fig. 14.10. Serum MDA levels in the five groups at the time of sacrifice. Results are expressed as mean ± SE. *$P < 0.05$, ZLC *vs.* other groups. †$P < 0.05$, ZLC + SDG *vs.* other groups. [a]$P < 0.05$, ZDF *vs.* ZDF + SDG or ZDF + SDGB. [b]$P < 0.05$, ZDF + SDG *vs.* ZDF + SDGB. *Abbreviations:* See Figures 14.1, 14.3, 14.7. *Source:* Reproduced with permission from Prasad, K. (46).

The prevention and delay in the development of type-1 and type-2 diabetes by SDG isolated from flaxseed could be due to its antioxidant activity (14) and/or due to the antioxidant activity of its metabolites (secoisolariciresinol, enterolactone, and enterodiol) (15). The antidiabetic effect of SDG could also be due to its suppressant effect on expression of the PEPCK gene (22).

Flaxseed and Oil in Diabetes

Effects of flaxseed in diabetes are not known. However, there are reports which suggest that flaxseed decreases postprandial glucose responses (10,48). Flaxseed oil has been shown to slow the rise in blood glucose levels due to a ketogenic diet which had a 4:1 ratio by weight of fat to protein plus carbohydrate (49). About 23 studies have been conducted on the effects of n-3 fatty acids in patients with type-2 diabetes (50). In most of these studies fish-oil consumption lowered serum triacylglycerol concentrations significantly, but in some studies plasma glucose concentrations increased. In a randomized, double-blind, placebo-controlled, crossover trial in patients with type-2 diabetes, consumption of 6 g of n-3 fatty acid produced no significant change in serum glucose (51). Flaxseed oil, in 15 obese patients with insulin resistance, did not affect glucose tolerance; however, it decreased insulin sensitivity and increased ability of LDL to oxidize (52). In a recent study, Kleemann *et al.* (53) showed that flaxseed oil did not significantly affect the insulitis in BBdp rats.

Summary

This chapter describes the effect of flaxseed and its components in type-1 and type-2 diabetes. Efficacy of flaxseed in diabetes is not known. However, reports suggest that flaxseed in normal individuals decreases postprandial glucose responses. Flaxseed oil does not appear to affect blood glucose, glucose tolerance, or insulitis. In some studies flaxseed oil was shown to increase blood glucose. SDG isolated from defatted flaxseed showed antioxidant activity and suppressed the PEPCK gene expression. SDG reduced the development of type-1 diabetes by 73–75% and retarded the development of type-2 diabetes. Both type-1 and type-2 diabetes were associated with an increase in oxidative stress. The effectiveness of SDG in reducing/retarding the development of diabetes was associated with a reduction in the oxidative stress.

In conclusion, diabetes is associated with oxidative stress. SDG isolated from flaxmeal is effective in reducing the development of diabetes (type-1 and type-2), and this effect of SDG in diabetes is associated with a decrease in oxidative stress.

Acknowledgments

The author acknowledges the technical assistance of P.K. Chattopadhya, Barbara Raney, and S.V. Mantha. The study on type-2 diabetes was supported by a grant from the Saskatchewan Flax Development Commission and Canadian Adaptation and Rural Development in Saskatchewan.

References

1. Janka, H.U., and D. Michaelis, Epidemiology of Diabetes Mellitus: Prevalence, Incidence, Pathogenesis, and Prognosis, *Z. Arztl. Fortbild. Qualitatssich. 96*:159–165 (2002).

2. Harris, M.I., K.M.F. Flegal, C.C. Cowie, M.S. Eberhardt, D.E. Goldstein, R.R. Little, H.M. Wiedmeyer, and D.D. Byrd-Holt, Prevalence of Diabetes, Impaired Fasting Glucose, and Impaired Glucose Tolerance in U.S. Adults, The Third National Health and Nutrition Examination Survey, 1988–1994, *Diabetes Care 21*:518–524 (1998).

3. National Institutes of Health, National Institute of Diabetes and Digestive and Kidney Diseases: Diabetes Statistics, NIH Publication No. 96-3926, National Institutes of Health. Bethesda, Maryland, 1995, pp. 1–4.

4. Young, T.K., J.M. Kaufert, J.K. McKenzie, A. Hawkins, and J. O'Neil, Excessive Burden of End-State Renal Disease Among Canadian Indians: A National Study, *Am. J. Public Health 79*:756–758 (1989).

5. King, H., R.E. Aubert, and W.H. Herman, Global Burden of Diabetes, 1995–2025: Prevalence, Numerical Estimates and Projections, *Diabetes Care 21*:1414–1431 (1998).

6. Davidson, J.K., *Diagnosis of Diabetes Mellitus. Clinical Diabetes Mellitus: A Problem Oriented Approach*, 3rd edn., Thieme, New York, 2000, pp. 169–188.

7. Olefsky, J.M., Treatment of Insulin Resistance with Peroxisome Proliferator-Activated Receptor Gamma Agonists, *J. Clin. Invest. 106*:467–472 (2000).

8. Oomah, B.D., and G. Mazza, Flaxseed Proteins: A Review, *Food Chem. 48*:109–114, (1993).

9. Hettiarachchy, N.S., G.A. Hareland, A. Ostenson, and G. Blander-Shank, Chemical Composition of 11 Flaxseed Varieties Grown in North Dakota, *Proc. Flax Institute 53*:36 (1990).

10. Cunnane, S.C., S. Ganguli, C. Menard, A.C. Liede, M.J. Hamadeh, Z.-Y. Chen, T.M.S. Wolever, and D.J.A. Jenkins, High Alpha-Linolenic Acid Flaxseed (*Linum usitatissimum*): Some Nutritional Properties in Humans, *Br. J. Nutr. 69*:443–453 (1993).

11. Westcott, N.D., and A.D. Muir, U.S. Patent 5,705,618 (1988).

12. Bakke, J.E., and H.J. Klosterman, A New Diglucoside from Flaxseed, *Proc. North Dakota Acad. Sci. 10*:18–22 (1956).

13. Prasad, K., Reduction of Serum Cholesterol and Hypercholesterolemic Atherosclerosis in Rabbits by Secoisolariciresinol Diglucoside (SDG) Isolated from Flaxseed, *Circulation 99*:1355–1362 (1999).

14. Prasad, K., Hydroxyl Radical-Scavenging Property of Secoisolariciresinol Diglucoside (SDG) Isolated from Flaxseed, *Mol. Cell Biochem. 168*:117–123 (1997).

15. Prasad, K., Antioxidant Activity of Secoisolariciresinol Diglucoside-Derived Metabolites, Secoisolariciresinol, Enterodiol, and Enterolactone, *Intl. J. Angiol. 9*:220–225 (2000).

16. Babior, B.M., The Respiratory Burst of Phagocytes, *J. Clin. Invest. 73*:599–601 (1984).

17. Fantone, J.C., and P.A. Ward, Role of Oxygen Derived Free Radicals and Metabolites in Leukocyte-Dependent Inflammatory Reactions, *Am. J. Pathol. 107*:397–418 (1982).

18. Prasad, K., J. Kalra, A.K. Chaudhary, and D. Debnath, Effect of Polymorphonuclear Leukocyte-Derived Oxygen Free Radicals and Hypochloric Acid on Cardiac Function and Some Biochemical Parameters, *Am. Heart J. 119*:538–550 (1990).

19. Consoli, A., N. Nurjhan, F. Capani, and J. Gerich, Predominant Role of Gluconeo-Genesis in Increased Hepatic Glucose Production in NIDDM, *Diabetes 38*:550–557 (1989).

20. Rognstad, R., Rate-Limiting Steps in Metabolic Pathways, *J. Biol. Chem. 254*:1875–1878 (1979).

21. Veneziale, C.M., J.C. Donofrio, and H. Nishimura, The Concentration of p-Enolpyruvate Carboxykinase Protein in Murine Tissues in Diabetes of Chemical and Genetic Origin, *J. Biol. Chem. 258*:14257–14262 (1983).

22. Prasad, K., Suppression of Phosphoenolpyruvate Carboxykinase Gene Expression by Secoisolariciresinol Diglucoside (SDG), a New Antidiabetic Agent, *Intl. J. Angiol. 11*:107–109 (2002).

23. Jain, S.K., S.N. Levine, J. Duett, and B. Hollier, Elevated Lipid Peroxidation Levels in Red Blood Cells of Streptozotocin-Treated Diabetic Rats, *Metabolism 39*:971–975 (1990).

24. Kakkar, R., J. Kalra, S.V. Mantha, and K. Prasad, Lipid Peroxidation and Activity of Antioxidant Enzymes in Diabetic Rats, *Mol. Cell Biochem. 151*:113–119 (1995).

25. Kakkar, R., S.V. Mantha, J. Radhi, K. Prasad, and J. Kalra, Increased Oxidative Stress in Rat Liver and Pancreas During Progression of Streptozotocin-Induced Diabetes, *Clin. Sci. 94*:623–632 (1998).

26. Kakkar, R., S.V. Mantha, J. Kalra, and K. Prasad, Time Course Study of Oxidative Stress in Aorta and Heart of Diabetic Rat, *Clin. Sci. 91*:441–448 (1996).

27. Jennings, P.E., A.F. Jones, C.M. Florkowski, J. Lunec, and A.H. Barnett, Increased Diene Conjugates in Diabetic Subjects with Microangiopathy, *Diabet. Med. 4*:452–456 (1987).

28. Sumoski, W., H. Baquerizo, and A. Rabinovitch, Oxygen Radical Scavengers Protect Rat Islet Cells from Damage by Cytokines, *Diabetologia 32*:792–796 (1989).

29. Rabinovitch, A., W.L. Suarez, P.D. Thomas, K. Strynadka, and I. Simpson, Cytotoxic Effects of Cytokines on Rat Islets: Evidence for Involvement of Free Radicals and Lipid Peroxidation, *Diabetologia 35*:409–413 (1992).

30. Grankvist, K., S. Marklund, J. Sehlin, and I.B. Taljedal, Superoxide Dismutase, Catalase and Scavengers of Hydroxyl Radical Protect Against the Toxic Action of Alloxan on Pancreatic Islet Cells *in vitro*, *Biochem. J. 182*:17–25 (1979).

31. Fischer, L.J., and S.A. Hamburger, Dimethylthiourea: A Radical Scavenger That Protects Isolated Pancreatic Islets from the Effects of Alloxan and Dihydroxyfumarate Exposure, *Life Sci.* 26:1405–1409 (1980).

32. Tibaldi, J., J. Benjamin, F.S. Cabbat, and R.E. Heikkila, Protection Against Alloxan-Induced Diabetes by Various Urea Derivatives: Relationship Between Protective Effects and Reactivity with the Hydroxyl Radical, *J. Pharmacol. Exp. Ther.* 211:415–418 (1979).

33. Like, A.A., and A.A. Rossini, Streptozotocin-Induced Pancreatic Insulitis: New Model of Diabetes Mellitus, *Science* 193:415–417 (1976).

34. Robbins, M.J., R.A. Sharp, A.E. Slonim, and I.M. Burr, Protection Against Streptozotocin-Induced Diabetes by Superoxide Dismutase, *Diabetologia* 18:55–58 (1980).

35. Wilson, G.L., N.J. Patton, J.M. McCord, D.W. Mullins, and B.T. Mossman, Mechanism of Streptozotocin- and Alloxan-Induced Damage in Rat β-Cells, *Diabetologia* 27:587–591 (1984).

36. Prasad, K., S.V. Mantha, A.D. Muir, and N.D. Westcott, Protective Effect of Secoisolariciresinol Diglucoside Against Streptozotocin-Induced Diabetes and Its Mechanism, *Mol. Cell Biochem.* 206:141–150 (2000).

37. Prasad, K., Oxidative Stress as a Mechanism of Diabetes in Diabetic Prone BB Rats: Effect of Secoisolariciresinol Diglucoside (SDG), *Mol. Cell Biochem.* 209:89–96 (2000).

38. Armstrong, A.M., J.E. Chestnutt, M.J. Gormley, and I.S. Young, The Effect of Dietary Treatment on Lipid Peroxidation and Antioxidant Status in Newly Diagnosed Noninsulin Dependent Diabetes, *Free Radic. Biol. Med.* 21:719–726 (1996).

39. Sundaram, R.K., A. Bhaskar, S. Vijayalingam, M. Viswanathan, R. Mohan, and K.R. Shanmugasundaram, Antioxidant Status and Lipid Peroxidation in Type-II Diabetes with and without Complications, *Clin. Sci.* 90:255–260 (1996).

40. Sato, Y., N. Hotta, S. Sakamoto, S. Matsuoka, N. Ohishi, and K. Yagi, Lipid Peroxide Level in Plasma of Diabetic Patients, *Biochem. Med.* 21:104–107 (1979).

41. Griesmacher, A., M. Kindhauser, S.E. Andert, W. Schreiner, C. Toma, P. Knoebl, P. Pietschmann, R. Prager, C. Schnack, G. Schernthaner, and M.M. Mueller, Enhanced Serum Levels of Thiobarbituric Acid Reactive Substances in Diabetes Mellitus, *Am. J. Med.* 98:469–475 (1995).

42. Nourooz-Zadeh, J., J. Tajaddini-Sarmadi, S. McCarthy, D.J. Betteridge, and S.P. Wolff, Elevated Levels of Authentic Plasma Hydroperoxides in NIDDM, *Diabetes* 44:1054–1058 (1995).

43. Paolisso, G., and D. Giugliano, Oxidative Stress and Insulin Action: Is There a Relationship? *Diabetologia* 39:357–363 (1996).

44. Bellomo, G., E. Maggi, M. Poli, F.G. Agosta, P. Bollati, and G. Finardi, Autoantibodies Against Oxidatively Modified Low-Density Lipoproteins in NIDDM, *Diabetes* 44:60–66 (1995).

45. Reaven, P., Dietary and Pharmacologic Regimens to Reduce Lipid Peroxidation in Non-insulin-Dependent Diabetes Mellitus, *Am. J. Clin. Nutr.* 62 (Suppl. 6):1483S–1489S (1995).

46. Prasad, K., Secoisolariciresinol Diglucoside from Flaxseed Delays the Development of Type-2 Diabetes in Zucker Rat, *J. Lab. Clin. Med.* 138:32–39 (2001).

47. Mordes, J.P., J. Desemone, and A.A. Rossini, The BB Rat, *Diabetes Metab. Rev.* 3:725–750 (1987).

48. Cunnane, S.C., M.J. Hamadeh, A.C. Liede, L.U. Thompson, T.M.S. Wolever, and D.J.A. Jenkins, Nutritional Attributes of Traditional Flaxseed in Healthy Young Adults, *Am. J. Clin. Nutr.* 61:62–68 (1995).

49. Likhodii, S.S., K. Musa, A. Mendonca, C. Dell, W.M. Burnham, and S.C. Cunnane, Dietary Fat, Ketosis, and Seizure Resistance in Rats on the Ketogenic Diet, *Epilepsia 41*:1400–1410 (2000).
50. Harris, W.S., Do ω-3 Fatty Acids Worsen Glycemic Control in NIDDM? *ISSFAL Newsletter 3*:6–9 (1996).
51. Connor, W.E., M.J. Prince, D. Ullmann, M. Riddle, L. Hatcher, F.E. Smith, and D. Wilson, The Hypotriglyceridemic Effect of Fish Oil in Adult-Onset Diabetes Without Adverse Glucose Control, *Ann. NY Acad. Sci. 683*:337–340 (1993).
52. Nestel, P.J., S.E. Pomeroy, T. Sasahara, T. Yamashita, Y.L. Liang, A.M. Dart, G.L. Jennings, M. Abbey, and J.D. Cameron, Arterial Compliance in Obese Subjects Is Improved with Dietary Fatty Acid from Flaxseed Oil Despite Increased LDL Oxidizability, *Arterioscl. Thromb. Vasc. Biol. 17*:1163–1170 (1997).
53. Kleemann, R., F.W. Scott, U. Wörz-Pagenstert, W.M.N. Ratnayake, and H. Kolb, Impact of Dietary Fat on Th_1/Th_2 Cytokine Gene Expression in the Pancreas and Gut of Diabetes-Prone BB Rats, *J. Autoimmun. 11*:97–103 (1998).

Chapter 15

Flaxseed, Fiber, and Coronary Heart Disease: Clinical Studies

P. Mark Stavro, Augustine L. Marchie, Cyril W.C. Kendall, Vladimir Vuksan, and David J.A. Jenkins

Department of Nutritional Sciences, University of Toronto, Toronto, Ontario M5S 3E2, Canada

Introduction

Interest is developing in the use of flaxseed as a dietary agent in intervention trials to reduce risk of coronary heart disease (CHD). Some flaxseed components, including viscous fiber and α-linolenic acid (ALA), in addition to a product of lignan metabolism, enterolactone, show potential in the prevention of CHD (1–13). Also, a limited number of clinical studies show that consumption of flaxseed causes a reduction in blood levels of various CHD risk factors, including total cholesterol (TC) and low-density lipoprotein cholesterol (LDL-C) (14–18). For these reasons, flaxseed is an ideal dietary candidate for CHD risk reduction, and its incorporation into foods may help to lower the burden of this disease in North America.

Dietary Recommendations and Favored Foods

The introduction of a new food is likely to be most successful if the food has attributes allowing it to fit well with current dietary recommendations. Traditional flaxseed contains oil that is low in saturated fat and high in ALA. It is a rich source of protein and the seed is covered in a viscous fiber with beneficial effects on glucose and cholesterol metabolism (14,15,19–21). The content of viscous fiber is particularly relevant now that the latest recommendations of the National Cholesterol Education Program (NCEP) Adult Treatment Panel III (ATP III) has included the addition of 15–25 g/d of viscous fiber to the diet to facilitate blood cholesterol reduction (22). Overall, flaxseed supports the current international nutritional recommendations to increase the consumption of plant foods and reduce the intake of saturated fat and dietary cholesterol in order to reduce the risk of chronic diseases, especially CHD. In view of current trends in dietary guidelines, the nutritional environment is favorable for the introduction of foods containing flaxseed.

Components of Flaxseed Relevant to CHD

The components of interest in traditional flaxseed include fiber, oil, protein, and lignan precursors. A high amount of dietary fiber is provided by flaxseed, with 28% present by weight, of which 75% is insoluble and 25% is viscous fiber (flaxseed mucilage) (19). The content of oil by weight is approximately 41%, and the most abundant fatty acid is the n-3 polyunsaturated fatty acid, ALA, which comprises about 57% of the oil (23). Flaxseed is thus one of the richest dietary sources of ALA, along with other rich sources such as soybeans, canola seeds, walnuts, mustard oil, and salba seeds (24,25). The protein constitutes 20% of the seed's weight. Flaxseed is also the richest dietary source of the lignan precursors secoisolariceresinol diglycoside (SDG) and materesinol, which are converted to enterolactone and enterodiol, respectively (26,27) and which possess potential antioxidant activity (28). Each of these components present in flaxseed has demonstrated the potential to be protective against CHD.

Flaxseed Fiber

The outer coat of flaxseed contains the seed's fiber. Both viscous soluble and insoluble fiber are present. Findings from clinical studies indicate that viscous fibers, as opposed to insoluble fibers, reduce serum cholesterol—a risk factor for CHD (21). However, most of the association with fiber and CHD events is with the insoluble fraction, especially with the insoluble cereal fiber from wheat (1,3,5). An early prospective study by Morris *et al.* (5) first demonstrated this relationship in 337 middle-aged men from London and southeast England. These men provided a 7-d individual weighed record of their diet during the period 1956–1966. Follow-up continued until 1976, by which time 45 had suffered a CHD event. When the group of 337 men was stratified into tertiles of daily total dietary fiber intake, it was shown that in the upper tertile of intake (16.9–56.1 g/d) 7 CHD events occurred and in the lower tertile of intake (5.6–15.4 g/d) 22 such events occurred. Further, when the group was divided into tertiles of daily cereal fiber intake, 5 men in the upper tertile of intake (8.4–34.2 g/d) suffered a CHD event compared with 25 men in the lower tertile of intake (2.0–7.1 g/d). This study provided the first indication of the inverse relationship between dietary fiber intake, namely cereal fiber intake, and CHD risk.

Since then, a number of prospective studies in both men and women have also suggested that an inverse relationship exists between dietary fiber intake (1–4,6,29), especially cereal fiber intake, and CHD risk (1–3,6; Table 15.1). In the Health Professionals' Follow-up Study (1), 43,757 adult men were followed for 6 yr, and the number of CHD events according to cereal fiber intake was determined. The findings showed that the total number of myocardial infarction cases significantly decreased across quintiles of daily cereal fiber intake, such that men in the highest quintile (median intake = 9.7 g/d) compared with men in the lowest quintile (median intake = 2.2 g/d) had a 29% lower risk for myocardial infarction. Supporting this finding is a Finnish study (2), in which 21,930 adult men were monitored for the incidence of major coronary events over a 6.1-yr period. When the sample of men was divided into quintiles of daily cereal fiber intake, it was found that for men in the highest quintile (median intake = 26.3 g/d), risk for fatal CHD was 26% lower than for men in the lowest quintile (median intake = 16.1 g/d). Also, further analysis of the results showed that increasing daily soluble fiber intake by 3 g reduced the risk of coronary death by 27%.

TABLE 15.1
Components of Flaxseed and Coronary Heart Disease (CHD)

Dietary component (reference)	Variable	Comparison	Outcome	Effect (%)	P
Dietary fiber					
Prospective studies					
HPFS (1)	Cereal fiber	Q5 *vs.* Q1	Total MI	−29	= 0.007
CPS (2)	Cereal fiber	Q5 *vs.* Q1	Fatal CHD	−28	= 0.01
NHS (3)	Cereal fiber	Q5 *vs.* Q1	CHD events[a]	−37	< 0.001
IWHS (4)	Whole grain	Q5 *vs.* Q1	IHD events	−30	= 0.02
LCS (5)	Cereal fiber	T3 *vs.* T1	CHD events	−25	NR
α-Linolenic acid					
Prospective studies					
MRFIT (6)	α-Linolenic acid	Q5 *vs.* Q1	Fatal CHD	−42	< 0.05
NHS (7)	α-Linolenic acid	Q5 *vs.* Q1	Fatal CHD	−45	= 0.01
HPFS (8)	α-Linolenic acid	Q5 *vs.* Q1	CHD events	−20	= 0.07
Intervention trials					
IMDHS (10)	α-Linolenic acid	E *vs.* C	Total cardiac events	−52	NR
LHS (11)	α-Linolenic acid	E *vs.* C	CVD events[b]	−73	= 0.001
IEIS (12)	α-Linolenic acid	E *vs.* C	Total cardiac events	−19	< 0.05
Vegetable protein					
Prospective study					
NHS (44)	Vegetable protein	Q5 *vs.* Q1	Acute CHD events	−11	= 0.005
Lignan					
Prospective Study					
KIHDRFS (13)	Plasma enterolactone	Case *vs.* control	Acute CHD events	−58.8	= 0.005

[a]Fatal plus nonfatal CHD.
[b]Death from cardiovascular causes plus nonfatal myocardial infarction.
Abbreviations: C, control group; CPS, Cancer Prevention Study; E, experimental group; HPFS, Health Professionals' Follow-up Study; IEIS, Indian Experiment of Infarct Survival; IHD, Ischemic heart disease; IMDHS, Indo-Mediterranean Diet Heart Study; IWHS, Iowa Women's Health Study; KIHDRFS, Kuopio Ischemic Heart Disease Risk Factor Study; LCS, London Civil Study; LHS, Lyon Heart Study; MI, myocardial infarction; MRFIT, Mister Fit; NHS, Nurses' Health Study; NR, not reported; Q, quintile; T, tertile.

The findings from prospective trials with women parallel those in men. In the Nurses' Health Study (3), dietary fiber intake was assessed in 39,876 adult females with no previous history of CHD; the women were subsequently followed for an average of 6 yr for incidence of CHD events. The results showed that women in the highest quintile of cereal fiber intake (median = 7.7 g/d) had a 37% reduced risk for total CHD events as compared with women in the lowest quintile of cereal fiber intake (median = 2.2 g/d). These findings are also supported by more recent findings from the Nurses' Health Study (6). In addition, Jacobs *et al.* (4) showed in 34,492 adult women free of ischemic heart disease (IHD) at baseline, that whole-grain intake was inversely related to the risk of death from IHD during a 9-yr follow-up period. Across quintiles of whole-grain intake (0.2, 0.9, 1.2, 1.9, and 3.2 servings/d) there was a significant decrease in the adjusted risk for death from IHD (1.0, 0.96, 0.71, 0.64, and 0.70, respectively).

Overall, the prospective trials in both men and women show that an inverse relationship is present between dietary fiber intake, namely cereal fiber intake, and the risk of CHD. Thus, because flaxseed is high in both insoluble and viscous fiber, consuming it may contribute to preventing CHD in both men and women. In addition to the findings from the mentioned prospective studies, findings from clinical trials with dietary fiber should also be considered.

Many clinical studies show that consumption of viscous soluble fiber can effectively reduce serum cholesterol. A meta-analysis by Brown *et al.* (21) showed that 2–10 g/d of viscous soluble fiber was associated with small but significant decreases in TC and in LDL-C (–0.045 and –0.057 mmol/L per g/soluble fiber, respectively). Extrapolating from these findings, increasing daily soluble fiber intake by 3 g can decrease total cholesterol by $\approx$0.129 mmol/L (5 mg/dL). However, these data are very conservative, and the analysis omitted a significant number of studies with larger reductions in serum cholesterol (30). Thus, a 20 g serving of flaxseed consumed twice a day can provide approximately 3 g of viscous fiber, and consumption of it on a daily basis can help individuals regulate their blood cholesterol level.

Flaxseed Oil

For over 30 years flaxseed (linseed) oil has been of considerable interest in the prevention of CHD. This stems from early findings showing ALA to be the most potent fatty acid in preventing platelet adhesiveness (31,32). Recently, findings from major prospective and interventional studies have demonstrated ALA to be protective against CHD (Table 15.1).

An inverse relationship between dietary ALA intake and risk for CHD in both men and women has been shown in three prospective studies (7–9). The Multiple Risk Factor Intervention Trial initially demonstrated this relationship in 6250 middle-aged men (7). After a 10.5-yr follow-up period, CHD fatality was shown to decrease significantly across quintiles of dietary ALA intake, as determined from five dietary recall interviews. Between the highest and lowest quintiles, daily ALA intake differed by

                                    P.M. Stavro and D.J.A. Jenkins

0.56 g, and risk for fatal CHD decreased by 42%. More recently, the Nurses' Health Study (8) and the Health Professionals' Follow-up Study (9) showed that ALA was the only fatty acid that protected against fatal CHD in women and nonfatal CHD in men, respectively. Between the highest and lowest quintiles in each study the daily intake of ALA differed by 0.7 g, and the risk for the mentioned CHD events was reduced by 25 and 20%, respectively. Overall, the findings from these prospective studies show that significantly fewer CHD events occur in population groups consuming high amounts of ALA compared with those consuming low amounts of ALA.

Large interventional studies also show that increased ALA intake is associated with a decreased risk for CHD events. To date, three randomized, controlled, clinical trials have demonstrated that dietary ALA can reduce the risk of CHD (10–12). The most recent one, a 2-yr follow-up called the Indo-Mediterranean Diet Heart Study (10), tested an ALA-rich diet from walnuts and mustard oil, also high in fruits, vegetables, and legumes, against a control diet on CHD events in 1000 patients with one or more risk factors for CHD. The patients randomized to the ALA-rich diet (4.1 g/d of total n-3 fatty acids) compared with the control diet (1.9 g/d of total n-3 fatty acids) suffered 50% fewer nonfatal myocardial infarctions and total cardiac events. The earlier findings of the Lyon Diet Heart Study were similar (11). This study compared an ALA-rich Mediterranean diet (2.0 g/d of ALA) with a control diet (0.6 g/d of ALA) on recurrent CHD events in 600 patients who had previously suffered a myocardial infarction. After 2-yr, both fatal and nonfatal myocardial infarctions were lowered by over 70% with the ALA-rich diet. Statistical analysis has since indicated that most of the beneficial effects were attributable to plasma ALA concentrations (33,34).

Although other components of the Indo-Mediterranean and Lyon intervention diets may have contributed to the reduction in CHD risk, the Indian Experiment of Infarct Survival (12) indicates that ALA itself can reduce CHD risk. In this study, patients with suspected acute myocardial infarction consumed mustard oil or placebo capsules daily for a 1-yr period. Those in the mustard oil group (2.9 g/d ALA) *vs.* the control group (0.5 g/d ALA) experienced 10% fewer total cardiac events and 7% fewer nonfatal infarctions.

The reduction in the CHD events in the ALA interventional studies could be due to ALA's antiarrhythmic effect or beneficial effect on serum lipid profile. In the Indian Experiment of Infarct Survival, there was a 54% reduction in total arrhythmias (12). Arrhythmias are a major contribution to sudden cardiac death (35–37). In the case of serum lipids, a recent clinical study showed that camelina oil, a good source of ALA, reduced LDL-C by 12.2% in hypercholesterolemic individuals (38).

Flaxseed Protein

The high vegetable protein content of flaxseed is of interest for reducing CHD risk because plant protein tends to reduce serum cholesterol. The cholesterol-lowering effect of plant protein has been most clearly demonstrated using soy (39–41).

Numerous human studies to date have documented soy protein's efficacy in lowering serum cholesterol (42). Thus, the protein fraction of soy, or other legumes, may contribute to the observed inverse relationship between legume consumption and CHD risk (43). Bazzano *et al.* (43) showed a significant reduction in total (fatal plus nonfatal) CHD events across quartiles of legume consumption in a cohort of 9632 men and women. Those individuals in the highest quartile of legume consumption (≥4 times/wk) *vs.* those in the lowest quartile of legume consumption (<1 time/wk) suffered 21% fewer total CHD events. In addition, findings from the Nurses' Health Study (44) showed a significant inverse relationship between total protein intake and total CHD events. When the vegetable protein intake was examined separately, a nonsignificant inverse association between it and risk of total CHD events was found. Still, the estimated risk for total CHD events was 11% lower in the highest quintile of daily vegetable protein intake (median = 5.0% of total energy) compared with the lowest quintile of intake (median = 2.4% of total energy). The effect of flaxseed protein itself on CHD risk or cholesterol lowering remains to be documented but is probably contributing to the lipid changes seen when whole flaxseed is fed.

Flaxseed Lignans

Flaxseed is the richest dietary source of lignan precursors. Its main lignan precursors include SDG and materesinol, which can be converted *in vivo* into the mammalian lignans enterolactone and enterodiol (26,27). Lignans are part of a broad class of compounds called phytoestrogens, which also include the isoflavones found in soy. These compounds tend to show antioxidant activity. A part of their cardioprotective effects has been proposed to result from their ability to protect LDL from oxidation and prevent the production of the more atherogenic LDL particles (45–48).

A paucity of data is available on the role of lignans in human CHD. In a recent analysis of a 7.7-yr follow-up study in Finland called the the Kuopio Ischaemic Heart Disease Risk Factor Study (13), enterolactone was measured in the blood of the first 167 men (from a cohort of 2005 men) to suffer an initial CHD event and compared with 167 matched controls. The mean baseline serum enterolactone concentration was significantly lower among the cases than the controls. As well, the men in the highest quarter of the enterolactone distribution (>30.1 nmol/L) had a 58.5% lower risk of acute coronary events than men in the lowest quarter (<7.21 nmol/L).

The literature regarding the association of dietary lignan intake and risk factors for CHD supports the Finnish experience. A recent observational study showed that increased dietary intake of lignans is associated with decreased aortic stiffness. Aortic stiffness is an independent risk factor for CHD (49). Thus, preliminary evidence from prospective studies shows that dietary lignan intake may reduce the risk for CHD.

Milled Flaxseed

The efficacy of flaxseed and flaxseed oil in lowering blood cholesterol has been investigated in several clinical trials (14–18,50–56; Table 15.2). These trials have included

TABLE 15.2

Clinical Trials with Whole Ground Flaxseed

Reference	Flaxseed dose (g/d)	Treatment period (wk)	Study design	Sex (M)	Sex (F)	Age (yr; Mean ± SD)	Baseline cholesterol[a] (mmol/L)	Effect of flaxseed[b] (%)	P
(15)	50	4	Crossover	5	5	25 ± 3	TC = 4.20 ± 0.11	−5.7 TC	< 0.05
							LDL-C = 2.70 ± 0.46	−8.9 LDL-C	< 0.05
(14)	50	4	Crossover	—	9	24 ± 3	TC = 4.80 ± 1.00[c]	−9.0 TC	< 0.01
							LDL-C = 3.75 ± 1.00[c]	−18 LDL-C	< 0.01
(16)	40	12	Parallel	—	36	54 ± 8	TC = 5.76 ± 0.25	−5.6 TC	= 0.01
							LDL-C = 3.21 ± 0.25		
(50)	1.3 g/100 g[d]	4	Crossover	16	62	44 ± 11	TC = 5.80 ± 0.29	No change	
							LDL = 3.37 ± 0.27	No change	
(17)	38	6	Crossover	—	38	56 ± 7	TC = 5.95 ± 1.44	−6.9 TC	< 0.05
							LDL-C = 4.12 ± 1.39	−15.8 LDL-C	< 0.05
(18)	50[e]	3	Crossover	22	7	57 ± 2	TC = 6.42 ± 0.16	−5.5 TC	< 0.001
							LDL-C = 4.36 ± 0.13	−9.7 LDL-C	< 0.001

[a]Serum cholesterol level at baseline prior to flaxseed treatment.

[b]Percentage change is calculated from serum TC or LDL-C level at the end of flaxseed treatment relative to baseline.

[c]Values estimated from a figure.

[d]Flaxseed was incorporated into foods.

[e]Defatted flaxseed

Abbreviations: LDL-C, low density lipoprotein cholesterol; TC, total cholesterol.

both healthy and hypercholesterolemic men and women of different ages and have reported various lipid parameters such as TC, LDL-C, high density lipoprotein cholesterol (HDL-C), and triglyceride.

In healthy individuals, two clinical trials have been investigated thus far (14–16,50). An initial study in healthy young females by Cunnane *et al.* (14) showed that consumption of 50 g/d of ground flaxseed for 4 wk reduced serum TC and LDL-C by 9 and 18%, respectively. The same investigators later showed in a study with healthy young men and women that LDL-C decreased by 8% after the participants consumed 50 g/d of ground flaxseed in muffins for 4 wk (15). Thus, in healthy individuals, whole or milled flaxseed lowers serum cholesterol.

In hypercholesterolemic individuals, most but not all studies support a hypocholesterolemic effect of flaxseed. Jenkins *et al.* (18) tested a low fat diet supplemented with either wheat bran or flaxseed muffins (providing 50 g/d of partially defatted flaxseed) daily for 3 wk on serum lipids in hyperlipidemic men and women (postmenopausal). Consumption of the flaxseed muffin diet decreased serum TC, LDL-C, and apolipoprotein B by 5.5, 9.7, and 5.9%, respectively. However, serum triglyceride increased by 10.2% in the same study. In hypercholesterolemic postmenopausal women, Arjmandi *et al.* (17) showed that the consumption of 38 g/d ground flaxseed in bread or muffins for 6 wk lowered serum TC by 6.9% and LDL-C by 15.8%. More recently, in postmenopausal women with moderately elevated serum cholesterol who consumed 40 g/d of raw ground flaxseed for 3 mon, serum TC and non-HDL-C decreased by 6%, compared with controls (16). Also, serum apolipoproteins A-1 and B were significantly reduced by 6 and 7.5%, respectively. In contrast, Tarpila *et al.* (50) showed no effect on blood lipids after healthy middle-aged men and women ate a flaxseed-supplemented food diet (1.3 g ground flaxseed/100 g test foods plus 5 g flaxseed oil/100 g test foods) for 4 wk that provided 20% of total energy from flaxseed-supplemented foods. Thus far, however, the cholesterol-lowering efficacy of flaxseed in both healthy and hypercholesterolemic men and women has been demonstrated, and supports the role for this oilseed in the area of CHD risk reduction.

Flaxseed Oil

A small number of clinical trials have investigated the efficacy of flaxseed oil in cholesterol reduction in both hyper- and normocholesterolemic men and women (51–56; Table 15.2). Abbey *et al.* (51) showed that feeding hypercholesterolemic men (baseline TC range = 5.3–7.9 mmol/L) 28 g/d flaxseed oil for 6-wk had no effect on any blood lipid parameters. A similar lack of effect was shown in healthy men (mean ± SD baseline TC = 3.87 ± 0.70 mmol/L) who consumed a diet supplemented with a mean of 31.7 g/d flaxseed oil for 56 d (52) and in men with mildly elevated TC on average (mean ± SD baseline TC ≈5.5 ± 1.2 mmol/L) who consumed 14 g/d ALA for 4 wk (53). Layne *et al.* (54) also showed that ALA consumed at a level of 35 mg/kg b.w. for 3 mon by healthy young adults (mean baseline TC ≈4.15 mmol/L) had no effect on blood lipids. However, at the 60 g/d dose, Singer *et al.* (55) showed that consumption of flaxseed oil for 2 wk significantly decreased triglyceride by 24.5, 21.8, and 34.8% in healthy subjects, subjects with hypertension, and subjects with hypercholesterolemia, respectively. In the subjects with hypertension and in those with hypercholesterolemia, TC was reduced by 18 and 14%, respectively. The subjects with hypertension had 22.6% lower LDL-C after flaxseed supplementation. Thus, flaxseed oil does not appear to alter blood lipids except when consumed at very high dietary intakes.

The reduction of TC and LDL-C seen with ground flaxseed can therefore be attributed to a number of components in flaxseed including the viscous fiber or the lignans, such as SDG, which have also been independently shown to lower TC and LDL-C in cholesterol-fed rabbits (57). Further, the flaxseed protein may also contribute to the cholesterol-lowering effect of this oilseed.

Conclusions

Flaxseed contains components that can be expected to confer clear advantages for CHD risk reduction. Its viscous fiber and possibly its lignans may help to reduce serum cholesterol. These effects may also be supported by its protein content. Flaxseed is rich in ALA, which has been shown to reduce CHD risk in large cohort studies and to stabilize the contractile activity of the heart, thereby reducing the likelihood of fatal cardiac arrhythmias and sudden death. Hence, adding flaxseed to the Western diet has a potentially important role in reducing CHD risk.

References

1. Rimm, E.B., A. Ascherio, E. Giovannucci, D. Spiegelman, M.J. Stampfer, and W.C. Willett, Vegetable, Fruit, and Cereal Fiber Intake and Risk of Coronary Heart Disease Among Men, *JAMA 275*:447–451 (1996).
2. Pietinen, P., E.B. Rimm, and P. Korhonen, Intake of Dietary Fiber and Risk of Coronary Heart Disease in a Cohort of Finnish Men. The Alpha-Tocopherol, Beta-Carotene Cancer Prevention Study, *Circulation 94*:2720–2727 (1996).
3. Wolk, A., J.E. Manson, and M.J. Stampfer, Long-Term Intake of Dietary Fiber and Decreased Risk of Coronary Heart Disease Among Women, *JAMA 281*:1998–2004 (1999).
4. Jacobs, Jr., D.R., K.A. Meyer, L.H. Kushi, and A.R. Folsom, Whole-Grain Intake May Reduce the Risk of Ischemic Heart Disease Death in Postmenopausal Women: The Iowa Women's Health Study, *Am. J. Clin. Nutr. 68*:248–257 (1998).
5. Morris, J.N., J.W. Marr, and D.G. Clayton, Diet and Heart: A Postscript, *Br. Med. J. 2*:1307–1314 (1977).
6. Liu, S., J.E. Buring, H.D. Sesso, E.B. Rimm, W.C. Willett, and J.E. Manson, A Prospective Study of Dietary Fiber Intake and Risk of Cardiovascular Disease Among Women, *J. Am. Coll. Cardiol. 39*:49–56 (2002).
7. Dolecek, T.A., Epidemiological Evidence of Relationships Between Dietary Polyunsaturated Fatty Acids and Mortality in the Multiple Risk Factor Intervention Trial, *Proc. Soc. Exp. Biol. Med. 200*:177–182 (1992).

8. Hu, F.B., M.J. Stampfer, J.E. Manson, E.B. Rimm, A. Wolk, G.A. Colditz, C.H. Hennekins, and W.C. Willett, Dietary Intake of Alpha-Linolenic Acid and Risk of Fatal Ischemic Heart Disease Among Women, *Am. J. Clin. Nutr.* 69:890–897 (1999).

9. Ascherio, A., E.B. Rimm, E.L. Giovannucci, D. Spiegelman, M. Stampfer, and W.C. Willett, Dietary Fat and Risk of Coronary Heart Disease in Men: Cohort Follow-up Study in the United States, *Br. Med. J. 313*:84–90 (1996).

10. Singh, R.B., G. Dubnov, and M.A. Niaz, Effect of an Indo-Mediterranean Diet on Progression of Coronary Artery Disease in High Risk Patients (Indo-Mediterranean Diet Heart Study): A Randomised Single-Blind Trial, *Lancet 360*:1455–1461 (2002).

11. de Lorgeril, M., S. Renaud, N. Mamelle, P. Salen, J.L. Martin, I. Monjaud, J. Guidollet, P. Touboul, and J. Delaye, Mediterranean Alpha-Linolenic Acid-Rich Diet in Secondary Prevention of Coronary Heart Disease, *Lancet 343*:1454–1459 (1994).

12. Singh, R.B., M.A. Niaz, J.P. Sharma, R. Kumar, V. Rastogi, and M. Moshiri, Randomized, Double-Blind, Placebo-Controlled Trial of Fish Oil and Mustard Oil in Patients with Suspected Acute Myocardial Infarction: The Indian Experiment of Infarct Survival—4, *Cardiovasc. Drugs Ther. 11*:485–491 (1997).

13. Vanharanta, M., S. Voutilainen, T.A. Lakka, M. van der Lee, H. Adlercreutz, and J.T. Salonen, Risk of Acute Coronary Events According to Serum Concentrations of Enterolactone: A Prospective Population-Based Case-Control Study, *Lancet 354*:2112–2115 (1999).

14. Cunnane, S.C., S. Ganguli, C. Menard, A.C. Liede, M.J. Hamadeh, Z.Y. Chen, T.M. Wolever, and D.J. Jenkins, High Alpha-Linolenic Acid Flaxseed (*Linum usitatissimum*): Some Nutritional Properties in Humans, *Br. J. Nutr. 69*:443–453 (1993).

15. Cunnane, S.C., M.J. Hamadeh, A.C. Liede, L.U. Thompson, T.M. Wolever, and D.J. Jenkins, Nutritional Attributes of Traditional Flaxseed in Healthy Young Adults, *Am. J. Clin. Nutr. 61*:62–68 (1995).

16. Lucas, E.A., R.D. Wild, L.J. Hammond, D.A. Khalil, S. Juma, B.P. Daggy, B.J. Stoecker, and B.H. Arjmandi, Flaxseed Improves Lipid Profile Without Altering Biomarkers of Bone Metabolism in Postmenopausal Women, *J. Clin. Endocrinol. Metab. 87*:1527–1532 (2002).

17. Arjmandi, B.H., D.A. Khan, S. Juma, M.L. Drum, S. Venkatesh, E. Sohn, L. Wei, and R. Derman, Whole Flaxseed Consumption Lowers Serum LDL-Cholesterol and Lipoprotein(a) Concentrations in Postmenopausal Women, *Nutr. Res. 18*:1203–1214 (1998).

18. Jenkins, D.J., C.W. Kendall, and E. Vidgen, Health Aspects of Partially Defatted Flaxseed, Including Effects on Serum Lipids, Oxidative Measures, and *ex vivo* Androgen and Progestin Activity: A Controlled Crossover Trial, *Am. J. Clin. Nutr. 69*:395–402 (1999).

19. Carter, J.F., Potential of Flaxseed and Flaxseed Oil in Baked Goods and Other Products in Human Nutrition, *Cereal Foods World 38*:753–759 (1993).

20. Chandalia, M., A. Garg, D. Lutjohann, K. von Bergmann, S.M. Grundy, and L.J. Brinkley, Beneficial Effects of High Dietary Fiber Intake in Patients with Type 2 Diabetes Mellitus, *N. Engl. J. Med. 342*:1392–1398 (2000).

21. Brown, L., B. Rosner, W.W. Willett, and F.M. Sacks, Cholesterol-Lowering Effects of Dietary Fiber: A Meta-Analysis, *Am. J. Clin. Nutr. 69*:30–42 (1999).

22. Third Report of the National Cholesterol Education Program (NCEP) Expert Panel on Detection, Evaluation, and Treatment of High Blood Cholesterol in Adults (Adult Treatment Panel III) final report, *Circulation 106*:3143–3421 (2002).

23. Bhatty, R.S., Nutrient Composition of Whole Flaxseed and Flaxseed Meal, in *Flaxseed in Human Nutrition*, 1st edn., edited by S.C. Cunnane and L.U. Thompson, AOCS Press, Champaign, Illinois, 1995, pp. 22–42.

24. Harper, C.R., and T.A. Jacobson, The Fats of Life: The Role of Omega-3 Fatty Acids in the Prevention of Coronary Heart Disease, *Arch. Intern. Med. 161*:2185–2192 (2001).

25. Whitham, D., M. DiBuono, M. Stavro, J. Sievenpiper, R. Bazinet, and V. Vuksan, A Novel Dietary Supplement High in Omega-3 Fatty Acids Lowers Blood Pressure in Individuals with Type 2 Diabetes, *Diabetes 51*:A403 (2002).

26. Adlercreutz, H., and W. Mazur, Phyto-Oestrogens and Western Diseases, *Ann. Med. 29*:95–120 (1997).

27. Axelson, M., J. Sjovall, B.E. Gustafsson, and K.D. Setchell, Origin of Lignans in Mammals and Identification of a Precursor from Plants, *Nature 298*:659–660 (1982).

28. Kitts, D.D., Y.V. Yuan, A.N. Wijewickreme, and L.U. Thompson, Antioxidant Activity of the Flaxseed Lignan Secoisolariciresinol Diglycoside and Its Mammalian Lignan Metabolites Enterodiol and Enterolactone, *Mol. Cell. Biochem. 202*:91–100 (1999).

29. Kushi, L.H., R.A. Lew, F.J. Stare, C.R. Ellison, M. el Lozy, G. Bourke, L. Daly, I. Graham, N. Hickey, and R. Mulcahy, Diet and 20-Year Mortality from Coronary Heart Disease. The Ireland-Boston Diet-Heart Study, *N. Engl. J. Med. 312*:811–818 (1985).

30. Olson, B.H., S.M. Anderson, M.P. Becker, J.W. Anderson, D.B. Hunninghake, D.J. Jenkins, J.C. LaRosa, J.M. Rippe, D.C. Roberts, D.B. Stoy, C.D. Summerbell, A.S. Truswell, T.M. Wolever, D.H. Morris, and V.L. Fulgoni, III, Psyllium-Enriched Cereals Lower Blood Total Cholesterol and LDL Cholesterol, but Not HDL Cholesterol, in Hypercholesterolemic Adults: Results of a Meta-Analysis, *J. Nutr. 127*:1973–1980 (1997).

31. Owren, P.A., The Effect of Dietary Linolenic Acid on Thrombosis and Haemostasis, *Thromb. Diath. Haemorrh. Suppl. 17*:223–229 (1965).

32. Owren, P.A., Linolenic Acid and Coronary Thrombosis, *Ann. Intern. Med 63*:1160–1161 (1965).

33. De Lorgeril, M., P. Salen, J.L. Martin, N. Mamelle, I. Monjaud, P. Touboul, and J. Delaye, Effect of a Mediterranean Type of Diet on the Rate of Cardiovascular Complications in Patients with Coronary Artery Disease: Insights into the Cardioprotective Effect of Certain Nutriments, *J. Am. Coll. Cardiol. 28*:1103–1108 (1996).

34. de Lorgeril, M., P. Salen, J.L. Martin, I. Monjaud, J. Delaye, and N. Mamelle, Mediterranean Diet, Traditional Risk Factors, and the Rate of Cardiovascular Complications after Myocardial Infarction: Final Report of the Lyon Diet Heart Study, *Circulation 99*:779–785 (1999).

35. Spielman, S.R., A.M. Greenspan, H.R. Kay, K.F. Discigil, C.R. Webb, N.M. Sokoloff, A.P. Rae, J. Morganroth, and L.N. Horowitz, Electrophysiologic Testing in Patients at High Risk for Sudden Cardiac Death. I. Nonsustained Ventricular Tachycardia and Abnormal Ventricular Function, *J. Am. Coll. Cardiol. 6*:31–40 (1985).

36. Rabkin, S.W., F.L. Mathewson, and R.B. Tate, The Electrocardiogram in Apparently Healthy Men and the Risk of Sudden Death, *Br. Heart J. 47*:546–552 (1982).

37. Greene, H.L., Definition of Patients at High Risk of Sudden Arrhythmic Cardiac Death, *Clin. Cardiol. 11*:115–116 (1988).

38. Karvonen, H.M., A. Aro, N.S. Tapola, I. Salminen, M.I. Uusitupa., and E.S. Sarkkinen, Effect of Alpha-Linolenic Acid-Rich *Camelina sativa* Oil on Serum Fatty Acid Composition and Serum Lipids in Hypercholesterolemic Subjects, *Metabolism 51*:1253–1260 (2002).

39. Kritchevsky, D., Vegetable Protein and Atherosclerosis, *J. Am. Oil Chem. Soc. 56*:135–140 (1979).

40. Carroll, K.K., P.M. Giovannetti, M.W. Huff, O. Moase, D.C. Roberts, and B.M. Wolfe, Hypocholesterolemic Effect of Substituting Soybean Protein for Animal Protein in the Diet of Healthy Young Women, *Am. J. Clin. Nutr. 31*:1312–1321 (1978).

41. Sirtori, C.R., E. Agradi, F. Conti, O. Mantero, and E. Gatti, Soybean-Protein Diet in the Treatment of Type-II Hyperlipoproteinaemia, *Lancet 1*:275–277 (1977).

42. Anderson, J.W., B.M. Johnstone, and M.E. Cook-Newell, Meta-Analysis of the Effects of Soy Protein Intake on Serum Lipids, *N. Engl. J. Med. 333*:276–282 (1995).

43. Bazzano, L.A., J. He, L.G. Ogden, C. Loria, S. Vupputuri, L. Meyers, and P.K. Whelton, Legume Consumption and Risk of Coronary Heart Disease in US Men and Women: NHANES I Epidemiologic Follow-up Study, *Arch. Intern. Med. 161*:2573–2578 (2001).

44. Hu, F.B., M.J. Stampfer, J.E. Manson, E. Rimm, G.A. Colditz, F.E. Speizer, C.H. Hennekens, and W.C. Willett, Dietary Protein and Risk of Ischemic Heart Disease in Women, *Am. J. Clin. Nutr. 70*:221–227 (1999).

45. Wilson, T., H. March, W.J. Ban, Y. Hou, S. Adler, C.Y. Meyers, T.A. Winters, and M.A. Maher, Antioxidant Effects of Phyto- and Synthetic-Estrogens on Cupric Ion-Induced Oxidation of Human Low-Density Lipoproteins *in vitro*, *Life Sci 70*:2287–2297 (2002).

46. Wiseman, H., J.D. O'Reilly, H. Adlercreutz, A.I. Mallet, E.A. Bowey, I.R. Rowland, and T.A. Sanders, Isoflavone Phytoestrogens Consumed in Soy Decrease F(2)-Isoprostane Concentrations and Increase Resistance of Low-Density Lipoprotein to Oxidation in Humans, *Am. J. Clin. Nutr. 72*:395–400 (2000).

47. Tikkanen, M.J., K. Wahala, S. Ojala, V. Vihma, and H. Adlercreutz, Effect of Soybean Phytoestrogen Intake on Low Density Lipoprotein Oxidation Resistance, *Proc. Natl. Acad. Sci. USA 95*:3106–3110 (1998).

48. Kang, M.H., M. Naito, K. Sakai, K. Uchida, and T. Osawa, Mode of Action of Sesame Lignans in Protecting Low-Density Lipoprotein Against Oxidative Damage *in vitro*, *Life Sci. 66*:161–171 (2000).

49. van der Schouw, Y.T., A. Pijpe, C.E. Lebrun, M.L. Bots, P.H. Peeters, W.A. van Staveren, S.W. Lamberts, and D.E. Grobbee, Higher Usual Dietary Intake of Phytoestrogens Is Associated with Lower Aortic Stiffness in Postmenopausal Women, *Arterioscler. Thromb. Vasc. Biol. 22*:1316–1322 (2002).

50. Tarpila, S., A. Aro, I. Salminen, A. Tarpila, P. Kleemola, J. Akkila, and H. Adlercreutz, The Effect of Flaxseed Supplementation in Processed Foods on Serum Fatty Acids and Enterolactone, *Eur. J. Clin. Nutr. 56*:157–165 (2002).

51. Abbey, M., P. Clifton, M. Kestin, B. Belling, and P. Nestel, Effect of Fish Oil on Lipoproteins, Lecithin:Cholesterol Acyltransferase, and Lipid Transfer Protein Activity in Humans, *Arteriosclerosis 10*:85–94 (1990).

52. Kelley, D.S., G.J. Nelson, J.E. Love, L.B. Branch, P.C. Taylor, P.C. Schmidt, B.E. Mackey, and J.M. Iacono, Dietary Alpha-Linolenic Acid Alters Tissue Fatty Acid Composition, but Not Blood Lipids, Lipoproteins or Coagulation Status in Humans, *Lipids 28*:533–537 (1993).

53. Mantzioris, E., M.J. James, R.A. Gibson, and L.G. Cleland, Dietary Substitution with an Alpha-Linolenic Acid-Rich Vegetable Oil Increases Eicosapentaenoic Acid Concentrations in Tissues, *Am. J. Clin. Nutr. 59*:1304–1309 (1994).

54. Layne, K.S., Y.K. Goh, J.A. Jumpsen, E.A. Ryan, P. Chow, and M.T. Clandinin, Normal Subjects Consuming Physiological Levels of 18:3(n-3) and 20:5(n-3) from Flaxseed or

Fish Oils Have Characteristic Differences in Plasma Lipid and Lipoprotein Fatty Acid Levels, *J. Nutr. 126*:2130–2140 (1996).

55. Singer, P., I. Berger, M. Wirth, W. Godicke, W. Jaeger, and S. Voigt, Slow Desaturation and Elongation of Linoleic and Alpha-Linolenic Acids as a Rationale of Eicosapentaenoic Acid-Rich Diet to Lower Blood Pressure and Serum Lipids in Normal, Hypertensive and Hyperlipidemic Subjects, *Prostaglandins Leukot. Med. 24*:173–193 (1986).

56. Chen, W.J.L., and W.J. Anderson, Hypercholesterolemic Effects of Soluble Fibres, in *Dietary Fiber: Basic and Clinical Aspects*, edited by G.V. Vahouny and D. Kritchevsky, Plenum Press, New York, 1986, pp. 275–286.

57. Prasad, K., Reduction of Serum Cholesterol and Hypercholesterolemic Atherosclerosis in Rabbits by Secoisolariciresinol Diglucoside Isolated from Flaxseed, *Circulation 99*:1355–1362 (1999).

Chapter 16

Flaxseed and Flaxseed Products in Kidney Disease

Malcolm R. Ogborn

Department of Pediatrics and Child Health and Department of Human Nutritional Sciences, University of Manitoba, Winnipeg, Manitoba, R3E 3P4, Canada

Introduction

Controversy about the role of diet as a means by which the symptoms and progress of renal disease can be modified is a constant feature of modern nephrology (1,2). Studies in animal models of renal failure, particularly the 5/6 nephrectomy rat (3), and more recently the *pcy* mouse (4–7) and Han:SPRD-*cy* rat (8–11), have shown very promising results. In these models, diet changes involving various sources and amounts of protein and sources and amounts of lipids have had significant functional and histologic effects. Flaxseed and derivatives of flaxseed feature quite prominently in this area.

To appreciate the potential role of flaxseed products in kidney disease, it is necessary to appreciate the pathology of the chronically failing kidney, which involves patterns of renal damage that are common to many diverse forms of renal insult. The operation of the mammalian kidney relies not only on an adequate mass of functional renal parenchymal cells, but also on a tightly regulated arrangement of those cells. The healthy kidney is characterized by a closely packed arrangement of renal tubules, blood vessels, and glomeruli where filtration occurs. The renal interstitial space between the tubules is a small volume almost completely occupied by vessels and lymphatics. Glomeruli and tubules communicate and regulate electrolyte balance and vascular tone in part through the close association of renal tubules with juxtaglomerular apparatus that modulates the release of renin.

The cardinal distinguishing feature of the failing kidney is expansion of the tubulo-interstitial space (12). This expansion occurs both from the loss of volume as tubules of nonfunctional nephrons atrophy and from the expansion of other organ elements, including interstitial collagen and the development of interstitial infiltrates (13). Tubular epithelial cells may demonstrate both apoptosis and unregulated proliferation; cysts and even low-grade malignancies are commonly found in the failed kidneys of human dialysis patients (14). These processes may continue even when the original disease process, such as an immunologically mediated glomerulonephritis or a toxic agent, such as lithium, is no longer active (13). As functioning nephrons are lost, hyperfiltration occurs in those remaining, eventually progressing to glomerulosclerosis as intraglomerular collagen obliterates the vascular loops.

The biochemical cytokine and the hormonal mediation of these events are extremely complex and incompletely understood. Chronic renal injury is invariably associated with biochemical and histologic evidence of oxidant damage, irrespective of the model studied (15). Renal epithelial and mesangial cells are capable of elaborating many cytokines and peptide growth factors under different conditions (16). The kidney has complex regulation of fatty acid release and metabolism, a potential target of dietary therapy, with constitutive expression on phospholipase A_2, cyclooxygenases 1 and 2, and prostaglandin synthases (17). Like the physiological functions of the kidney, the expression of this complex family of enzymes varies from segment to segment of the nephron. Epidemiologic data suggest that estrogen may favorably influence disease progression, including effects on peptide growth factor dependent mechanisms (18).

Secondary effects of renal disease may also accelerate injury. Such effects include hypertension (19), which will accelerate arteriolar injury, hyperparathyroidism, which promotes cell damage through increasing intracellular calcium (20), and hyperlipidemias, which have been associated with increased rates of oxidant injury to the kidneys (21).

Flaxseed and its products are logical subjects of study in the dietary modification of renal disease. There is considerable evidence that modification of dietary fatty acid profile, particularly involving increased n-3 polyunsaturated fatt acids (n-3 PUFA) content, or reduced n-6/n-3 ratio, has significant biochemical and histologic effects in the kidney (3,4,7,9,11), possibly influencing both inflammatory and proliferative pathways. Flax lignan secoisolariciresinol diglucoside (SDG) has antioxidant activity, generates the estrogenic compounds enterodiol and enterolactone under the action of intestinal flora, and antagonizes at least one peptide growth and activation factor, platelet-activating factor (PAF). Flaxseed protein has not been studied independently due to technical difficulties in its purification, but may also play a role in a similar to that seen with soy protein that has substantial effects on both normal renal physiology and pathology.

Renal injury may be ameliorated in three ways: (i) improvement in the renal disease that is responsible for renal injury; (ii) modification of inflammatory, fibrotic, or proliferative processes that contribute to the progression of chronic renal injury; and (iii) modification of extrarenal manifestations of renal failure that may themselves accelerate renal damage. This review will focus mainly on the second category, because it is in this area that most work has been done with flaxseed products. In addition, the possibility of flax-based products to supplement the care of patients with renal failure and perhaps modify their risk of nonrenal morbidity will also be explored.

Flaxseed and Flaxseed Components in the 5/6 Nephrectomy Model of Chronic Renal Failure

The 5/6 nephrectomy rat is the most studied animal model of chronic renal failure. The most common technique is to ablate two of the three branches of the renal artery at one surgical procedure and then, either simultaneously or at a later date, remove the other kidney (22). Because the majority of studies in this model were conducted in an era in which the glomerulus was the main focus of experimental pathology, most studies emphasize glomerular change, but this model does demonstrate the interstitial changes that are the most important predictors of renal doom (23). Clinically, animals develop progressive uremia, with increasing proteinuria as a marker of ongoing glomerular and tubular injury, and hypertension (24).

Ingram *et al.* (3) studied the effect of 15% ground flaxseed or 15% flax oil diets on the progression of renal injury in the 5/6 nephrectomy rat. Standard rat chow was used for the control diet and as the base for the experimental diets. The study compared a range of renal outcomes, including blood pressure, glomerular filtration rate, and a semiquantitative morphometric assessment of the degree of mesangial expansion and glomerulosclerosis. They also studied the biochemical consequences of the dietary intervention, measuring serum cholesterol, triglycerides, renal phospholipids, and urinary thromboxane B_2 and 6-keto $PGF_{1\alpha}$. Both of the flax-based interventions were protective against the rise in blood pressure that was seen after destruction of renal tissue in control animals. Over a 20-wk period, all animals showed a substantial decline in glomerular filtration rate, but residual function in flaxseed-fed rats was 82% better than in controls, and in flax oil-fed rats, residual function was 59% better than in controls.

Translated to human disease, changes of this magnitude would make the difference between dialysis and management without high technology support for extended periods. Corresponding improvements were seen in proteinuria, mesangial expansion, and glomerulosclerosis. Both diets also moderated, but did not eliminate, the hypercholesterolemia and hypertriglyceridemia that develops in uremic animals. These diets also enriched renal tissue in n-3 PUFA and abrogated the rise in urinary excretion of the markers of increased prostanoid production. The authors concluded that because the dietary flax oil content differed significantly between the interventions, both n-3 PUFA and non–oil-based flax components had beneficial effects. The design of their study did not permit isolation of the extent to which lignan, oil, or other components of flaxseed contributed to clinical benefit. Although demonstrating significant effects, the magnitude of the interventions used in this study is difficult to extrapolate to human diets.

Flaxseed and Flaxseed Components in Polycystic Kidney Disease

Polycystic kidney disease (PKD) is the fourth most common cause of end-stage renal disease, with treatment costs exceeding $1 billion/yr in the United States with proportionately similar costs in other developed countries (25). Tubular dilatation is associated with increased rates of both epithelial proliferation (26) and apoptosis (27). These components may be seen in many forms of renal injury, and PKD is now generally considered to be a disorder caused by dysregulation of renal repair and regeneration processes (28). Disruption of architecture by cystic change occurs only in a minority of nephrons. Human cystic diseases demonstrate interstitial inflammation, fibrosis, and nephron loss through apoptosis similar to that seen in many forms of chronic renal disease (25).

The Han:SPRD-*cy* rat is a model of PKD that shares autosomal dominant inheritance, progression through early adult life and sexual dimorphism with human disease (29). The disease is characterized by progressive dilatation of nephrons in young animals, associated with marked interstitial inflammation and fibrosis with associated nephron loss in older animals (30). Unlike human PKD, Han:SPRD-*cy* rat PKD has proved amenable to treatment. Modification of this model of PKD has been achieved with a variety of environmental manipulations including dietary protein restriction (31), angiotensin-converting enzyme inhibition or angiotensin receptor blockade, salt loading (32), methylprednisolone therapy (33), or hypocholesterolemic therapy (34). Substitution of soy protein in the diet of Han:SPRD-cy rats results in slower disease progression and dramatic reduction in interstitial inflammation and fibrosis and is associated with decreased production of long chain n-6 PUFA (9,35).

Based upon the beneficial results in the renal ablation model described in the previous section (3), we undertook a study to test the hypothesis that dietary flaxseed and derivatives would modify the clinical course, renal pathology, and renal biochemistry of male Han:SPRD-*cy* rats. Our laboratory uses computerized image analysis to provide objective measurements of pathologic change, based upon optical characteristics of tissue sections. In our first study, experimental animals received a diet supplemented with 10% by weight ground whole flaxseed (11). Affected Han:SPRD-*cy* animals fed both flaxseed-supplemented and control diets for 8 wk thrived, with mean weights at sacrifice of 381 ± 17 g and 369 ± 18 g, respectively (ns), demonstrating that any benefit derived from flaxseed was not due to protein or energy deprivation.

Flaxseed had no significant effect on total renal volume in Han:SPRD-*cy* rats. Littermates with normal kidneys also did not show any difference in renal volume in response to diet. Histologic studies revealed that flaxseed feeding was associated with a modest reduction in cystic change. Markers of tubular remodeling, including frequency of both apoptotic nuclei and nuclei expressing proliferating cell nuclear antigen (PCNA), a marker of recent passage through the cell cycle, did not differ between the two treatment groups. Renal fibrous volume, measured by methyl blue staining that corresponded to interstitial type III collagen (11), however, was reduced in flaxseed-supplemented animals. A parallel decrease was seen in the number of macrophages infiltrating the kidney. Serum creatinine was significantly lower in flaxseed fed Han:SPRD-*cy* animals (69 μM/L *vs.* 81 μM/L; $P = 0.02$). Serum creatinine in normal animals was uninfluenced by diet. The diet did not have any hypocholesterolemic effect on Han:SPRD-*cy* animals.

Analysis for fatty acid content from affected kidneys revealed that flaxseed supplementation was associated with a small reduction in renal palmitate, a modest but significant reduction in arachidonic acid (AA), and significant increases in n-3 PUFA precursors to prostanoid synthesis compared with controls. The ratio of n-6/n-3 PUFA was significantly reduced in renal tissue from animals fed the flaxseed diet. The n-3 PUFA inhibitor of macrophage activity, docosahexaenoic acid (DHA), was unexpectedly reduced in animals fed a flaxseed diet.

We then applied nuclear magnetic resonance spectroscopic techniques to study biochemical changes that occurred in renal tissue in association with flaxseed feeding to determine if there were any factors independent of lipid effects that might be associated with a reduction in renal injury (36). We had previously shown that expression of disease in rats on a standard lab chow diet was associated with increased excretion and tissue loss of anions involved in the citric acid cycle (35). There was also significant reduction in tissue content of osmolytes, particularly betaine, critical to cell volume regulation in the hypertonic regions of the renal medulla. Amelioration of the disease by soy protein feeding was associated with an increase in renal betaine and succinate content above the levels normally seen in healthy animals (11).

Affected animals on either diet demonstrated increased excretion of citric acid cycle metabolites and reduced excretion of creatinine consistent with declining renal function, which we have described in previous studies. However, despite somewhat milder disease, animals on the flaxseed diet demonstrated significantly greater citrate excretion ($P = 0.0004$). This result was not associated with a significant change in urinary ammonium excretion, suggesting that this was not a response to relative alkalosis induced by the dietary modification. ^{1}H-NMR analysis of tissue from affected animals revealed that flaxseed feeding was associated with a significantly higher content of glutamate, succinate, and betaine, reminiscent of the effects seen with a soy protein based diet.

To test the hypothesis that dietary flaxseed amelioration of this disease is related to changes in PUFA composition, we undertook a study to determine the effects of flaxseed oil on renal injury in Han:SPRD-*cy* rat PKD (37). This study examined biochemical and histologic markers of renal progression in animals fed synthetic AIN 93 type diet with casein as the protein source. Renal disease progresses more quickly in animals fed pure animal-based protein (9,11). Serum creatinine was elevated in Han:SPRD-*cy* heterozygotes with PKD, but was significantly lower in *post hoc* analysis in flaxseed oil-fed heterozygotes compared with corn oil-fed heterozygotes. Both diet and disease significantly influenced serum cholesterol, with flaxseed oil producing a reduction ($P < 0.001$) and disease expression producing an elevation in serum cholesterol on both diets ($P = 0.029$). Serum triglyceride was not different between diet and disease groups. Flaxseed oil supplementation was associated with significant reductions in cystic change (Fig. 16.1), with a decrease in epithelial proliferation (Fig. 16.2) that was proportionate to the observed decrease in cystic change. A comparable reduction was also seen in interstitial fibrosis (Fig. 16.3) and macrophage infiltration (Fig. 16.4). Staining oxidised low density lipoprotein (ox-LDL) detection (Fig. 16.5) was also reduced, demonstrating that the increased incorporation of more unsaturated PUFA was not increasing susceptibility to oxidative injury.

Flaxseed oil-fed animals showed higher renal proportion of α-linolenic acid (ALA), n-3 eicosapentaenoic acid (EPA), and n-3 DHA. Flaxseed oil-fed animals also

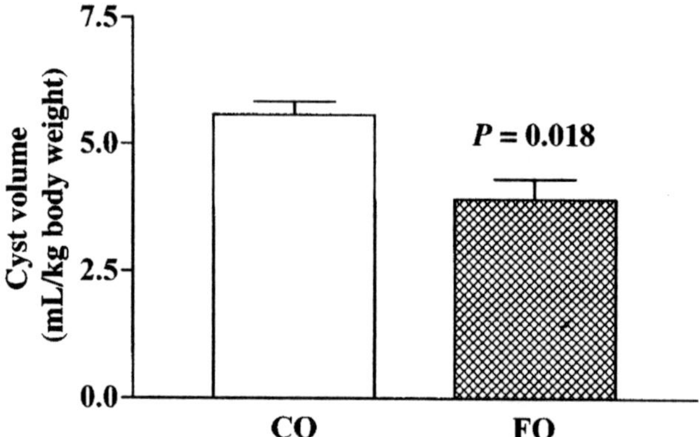

Fig. 16.1. Renal cystic volume in heterozygous Han:SPRD-*cy* rats fed flax oil- or corn oil-based diets. Results are group means with error bars representing SEM. *Abbreviations:* CO, corn oil; FO, flax oil.

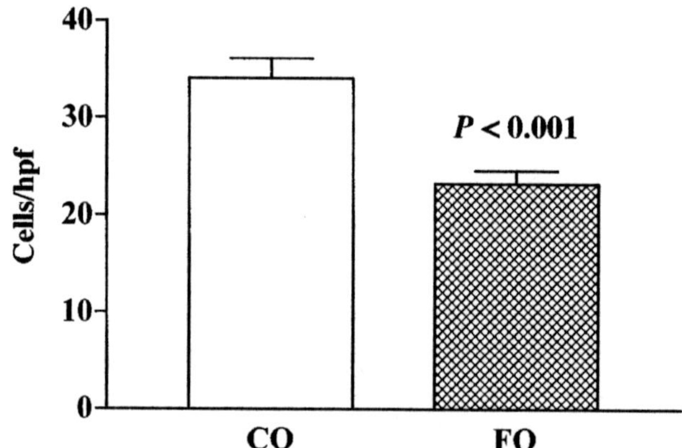

Fig. 16.2. Renal epithelial proliferation as PCNA-positive nuclei in heterozygous Han:SPRD-*cy* rats fed flax oil- or corn oil-based diets. Results are group means with error bars representing SEM. *Abbreviations:* See Figure 16.1; PCNA, proliferating cell nuclear antigen.

had lower renal proportion of n-6 γ-linolenic acid (GLA) and n-6 linoleic acid (LA) only in flaxseed oil-fed animals with PKD. The relative amount of renal n-6 AA in renal lipids did not change with diet.

Increased relative amounts of ALA, EPA, and DHA were seen in liver tissue from all animals, with or without PKD. LA, GLA, AA, and eicosatrienoic acid (ETA, 20:3n-6) in liver were also significantly lower in the animals fed a flaxseed oil diet. Total long chain PUFA was significantly decreased in the liver by flaxseed oil feed-

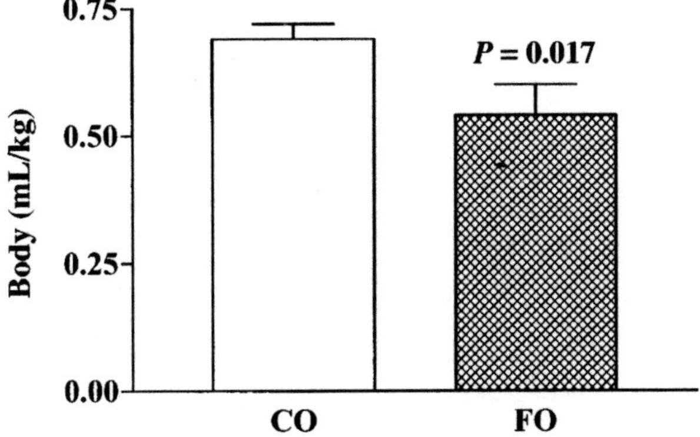

Fig. 16.3. Renal interstitial fibrosis in heterozygous Han:SPRD-*cy* rats fed flax oil- or corn oil-based diets. Results are group means with error bars representing SEM. *Abbreviations:* See Figure 16.1.

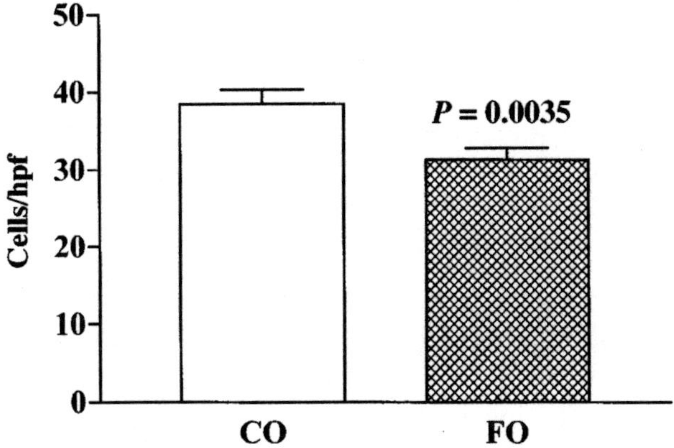

Fig. 16.4. Renal interstitial inflammation, represented as macrophages (cells expressing the rodent equivalent of the CD68 antigen) in heterozygous Han:SPRD-*cy* rats fed flax oil- or corn oil-based diets. Results are group means with error bars representing SEM. *Abbreviations:* See Figure 16.1.

ing, and increased by PKD in the livers of animals on the corn oil diet only. The ratio of LA/AA, although not significantly altered in kidney by diet or disease, was increased in liver by diet, but also decreased in liver in the presence of renal disease. Diet also produced a significant increase in the ratio of ALA/(EPA+DHA) in kidney and liver. The reduced elongation of both n-6 and n-3 PUFA strongly suggested reduced activity of Δ6-desaturase, which has been reported both with flax oil feeding in poultry and with other interventions that reduce n-6/n-3 ratio (38,39). We plan to

M.R. Ogborn

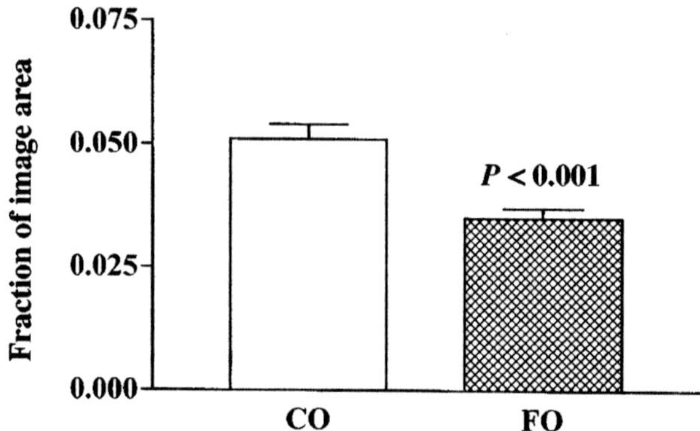

Fig. 16.5. Renal ox-LDL staining in heterozygous Han:SPRD-*cy* rats fed flax oil- or corn oil-based diets. Results are group means with error bars representing SEM. *Abbreviations:* See Figure 16.1; ox-LDL, -low density lipoprotein.

explore the effect of these changes in PUFA content on renal prostanoid release in studies in progress at this time.

We have recently explored the role of SDG alone in a similar experimental design. Animals were fed a synthetic AIN 93 diet with casein as the protein source and corn oil as the lipid source, with or without 20 mg/kg diet of SDG. SDG did not result in a significant difference in serum creatinine among animals with PKD after 8 wk of feeding. The extent of cystic change (Fig. 16.6), epithelial proliferation (Fig. 16.7), interstitial fibrosis (Fig. 16.8), and macrophage infiltration (Fig. 16.9) was sig-

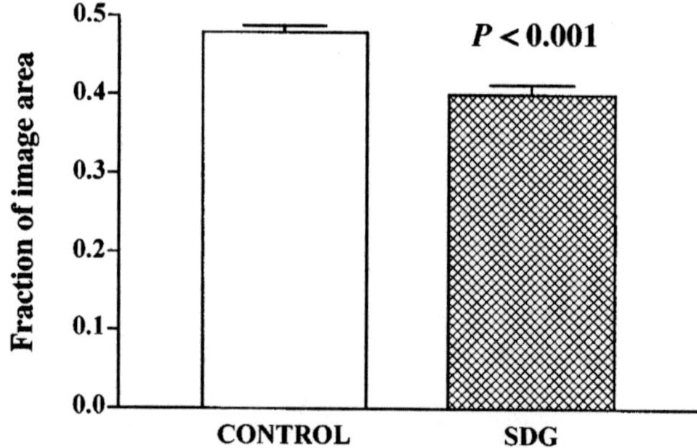

Fig. 16.6. Renal cystic change in heterozygous Han:SPRD-*cy* rats fed control or 20 mg/kg SDG-supplemented diet. Results are group means with error bars representing SEM. *Abbreviation:* SDG, secoisolariciresinol diglycoside.

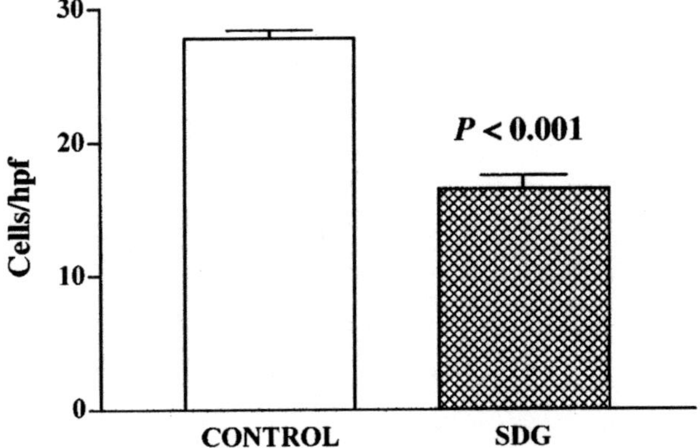

Fig. 16.7. Renal epithelial proliferation as PCNA-positive nuclei in heterozygous Han:SPRD-*cy* rats fed control or 20 mg/kg SDG-supplemented diet. Results are group means with error bars representing SEM. *Abbreviations:* See Figures 16.2, 16.6.

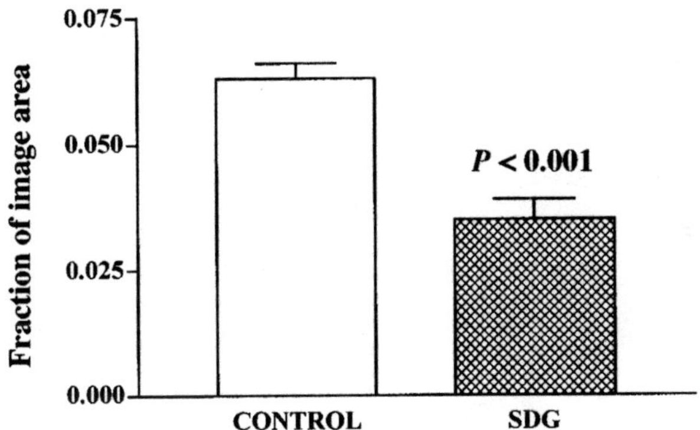

Fig. 16.8. Renal interstitial fibrosis in heterozygous Han:SPRD-*cy* rats fed control or 20 mg/kg SDG-supplemented diet. Results are group means with error bars representing SEM. *Abbreviation:* See Figure 16.6.

nificantly reduced. Oxidant injury as indicated by tissue deposition of oxidized-LDL was also reduced (Fig. 16.10). Studies comparing the effect of combining flax oil and SDG in the absence of other elements of flaxseed are currently in progress. Analysis of tissue fatty acid content did not reveal any indirect effects on tissue fatty acid composition that may explain findings that were similar to those seen with flax oil that has little if any lignan content. Our results with flaxseed derivatives to date suggest that not only can they reproduce the benefits of flaxseed in this model, but the effects may be greater.

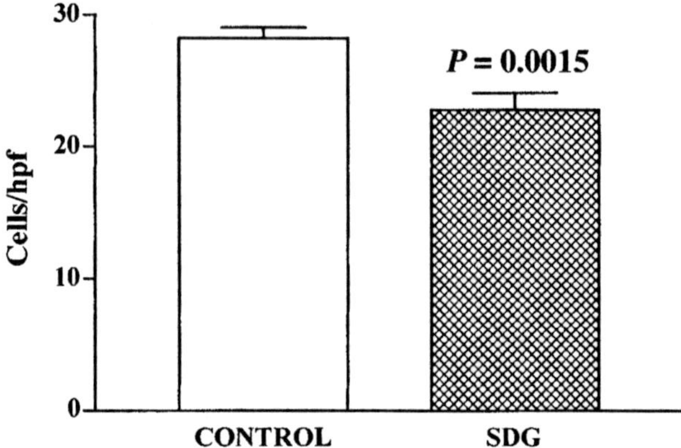

Fig. 16.9. Renal interstitial inflammation, represented as macrophages (cells expressing the rodent equivalent of the CD68 antigen) in heterozygous Han:SPRD-*cy* rats fed control or 20 mg/kg SDG-supplemented diet. Results are group means with error bars representing SEM. *Abbreviation:* See Figure 16.6.

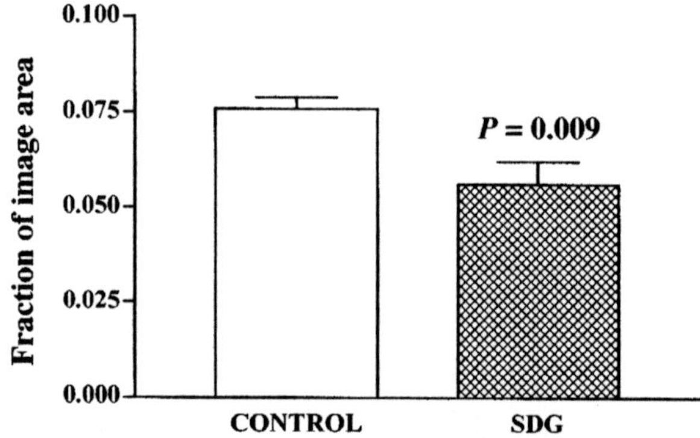

Fig. 16.10. Renal ox-LDL staining in heterozygous Han:SPRD-*cy* rats fed control or 20 mg/kg SDG-supplemented diet. Results are group means with error bars representing SEM. *Abbreviation:* See Figure 16.6.

Flaxseed and Flaxseed Components in Immune Renal Injury

A substantial proportion of chronic renal failure is due to immune-mediated injury to the kidney, particularly in young people. This may take the form of glomerulonephritis, where the kidney is the sole target of the process, or systemic disorders loosely categorized as vasculitis, where a range of immune pathophysiologies produce multiorgan injury through damage largely mediated by inflammation of small blood vessels.

A disease in the latter category, systemic lupus erythematosus (SLE), has been the subject of experimental studies with flax-based treatments. SLE is a complex disorder with manifestations that can affect any system in the body, with a pattern of involvement affecting different organ systems required for diagnosis (40–42). The disease is associated with autoimmune phenomena, particularly the presence of antibodies against double stranded DNA, although multiple autoantibodies are common. There is usually evidence of complement consumption by both classical and alternative pathways. Despite enhanced immune activity, appropriate immune response against infection may be compromised (43). Renal involvement in lupus is common and is a major determinant of morbidity and mortality in the disease (44). The pattern of renal injury varies considerably and may shift between classifications in the same patient over time. PAF is one of the mediators of renal injury through alterations of vascular structure and permeability that are part of the pathophysiologic process of renal injury (45,46). Because the receptor for PAF has been identified as a specific target for SDG, this provides a strong rationale for the application of flaxseed in this illness (47). The conventional treatment of renal SLE involvement invariably uses high-dose glucocorticoids such as prednisone, often in association with antimitotic agents such as cyclophosphamide and azathioprine. Such treatment carries substantial risk from sepsis, malignancy, and steroid-induced problems with obesity, gastrointestinal hemorrhage, and bone mineral loss.

The MRL/lpr mouse is a useful animal model for human SLE (48) and has been the subject of studies using flaxseed supplementation. Hall *et al.* (49) studied the effect of a 15% flaxseed diet compared with standard rodent chow in these animals. Outcomes measures included survival, proteinuria, and glomerular filtration rate. The effects of flaxseed can be summarized as delaying the onset of renal injury, with a later onset of proteinuria in the flaxseed group and better initial preservation of glomerular filtration rate in the flaxseed-fed animals. By 24 wk of age, however, the groups had converged. Despite this, mortality was significantly lower in the flaxseed group at all time points, including 24 wk. In another part of the same study, control and flaxseed fed animals were challenged with PAF. Flaxseed-fed animals demonstrated weaker and reversible platelet aggregation responses, whereas platelet aggregation after PAF was irreversible and lethal in animals fed the control diet. This result is strongly suggestive of a beneficial SDG effect.

Clark *et al.* (50) have tested the hypothesis that SDG may mediate renal injury directly in the MRL/lpr mouse, using pure SDG. Using daily dosing of 0, 600, 1200, or 4800 µg, they demonstrated dose-dependent protection of glomerular filtration rate and a dose-dependent delay in onset of proteinuria similar in extent to that seen with the 15% flaxseed diet in the earlier study (49). They also demonstrated that SDG was the predominant compound found in the circulation of the animals, with much lower levels of the estrogenic compounds enterodiol and enterolactone. Biological effectiveness of the SDG was confirmed by survival of a usually lethal dose of PAF, as was also seen with the 15% flaxseed diet.

Clark *et al.* (51) have also studied the effects of dietary supplementation with flaxseed in human lupus nephritis. In their first study, eight patients with SLE and evidence of nephritis took daily flaxseed supplements at doses of 15, 30, and 45 g for 4-wk periods with 5-wk washout periods in between. The dosing was sufficient to alter platelet aggregation response to PAF and increase n-3 PUFA levels at all doses. The 30 and 45 g doses were associated with a fall in serum creatinine and a rise in creatinine clearance, an approximation of glomerular filtration rate. Urine protein decreased in those taking the 15 and 30 g doses. The lower doses were well tolerated but a laxative effect was sometimes a problem at 45 g. In a subsequent 2-yr crossover study, the same group studied a 12-mon program of 30 g/d flaxseed supplements (52). The study showed a small drop in serum creatinine on the flaxseed diet compared with a small rise in these levels in controls. However, the range of these changes makes the implication of clinical significance to the findings difficult. No difference in proteinuria was detected. The study was seriously hampered by the small numbers willing to complete the study ($n = 15$), and this group possibly comprised the most compliant patients who might have better outcomes than average anyway. The study did however establish that chronic flaxseed ingestion can be achieved and suggests that larger multicenter studies could be done to investigate this treatment more effectively.

A possible future role for flaxseed in kidney disease has been suggested by studies with n-3 rich fish oil in IgA nephropathy. IgA nephropathy is the most common chronic glomerulonephritis in most developed countries, and has a benign prognosis in many patients. The 10–25% of patients who do develop chronic renal failure represent a substantial disease burden (53). The disease is characterized by intraglomerular deposition of IgA class immunoglobulin, with associated renal injury that may range from mild proliferative changes to severe sclerosis (54). Fish oil has been advocated based upon a number of positive results in some studies (55,56), although debate as to the strength of the evidence continues (57). No studies using flax oil in IgA nephropathy have been published as yet, but the possible merits of fish oil in kidney disease have been uncritically trumpeted across the Internet, and clinical nephrologists are finding that their patients are turning to sometimes more easily available flax oil as an n-3 PUFA source. Caughey *et al.* (58) demonstrated that immunological effects, specifically reduction in tumor necrosis factor α and interleukin 1β, occur when EPA levels increase whether that increase comes from the ingestion of EPA-rich fish oil or flaxseed oil. With the uncertainties facing the fishing industries of many countries, a rationale at least exists for a proper trial of flax oil in this setting.

Flaxseed Components as a Treatment for Nonrenal Morbidity in Chronic Renal Failure

In countries that can afford the luxury of dialysis, it is possible to prolong life for years and sometime decades with expensive high technology therapy. Kidney transplantation reduces technology dependence and improves well-being and longevity. In either case, however, normal health is not totally restored. Although the importance of nutrition is perhaps appreciated more in nephrology than in many other disciplines, the potential of specific nutrients to alter progression of renal injury or reduce risk of nonrenal morbidity does not usually influence clinical decision making. Indeed, few dietary products are specifically developed for renal failure patients, despite the size of this patient population. The principal determinants of the composition of such products is the need to deliver calories in small fluid volumes and the need to minimize phosphate content, with little consideration of broader metabolic issues. In this rather speculative section of this review, the case for flax-based dietary supplements for renal failure patients will be made.

Uremic Anorexia and Wasting

Progressive wasting is common in patients with chronic renal failure (59). Difficulty in maintaining adequate energy intake and lean body mass is a major challenge in most patients with chronic renal failure (60), particularly children, where nutritional supplementation with nasogastric or gastrostomy feeding has become the norm for many programs (61). The weight gain with tube feeding may be due more to increased adiposity than to a more useful increase in lean body mass (62). Renal failure is recognized as a proinflammatory state, with evidence of increased exposure to endotoxin and production of tumor necrosis factor α (63), a critical factor in the arrest of new protein synthesis in kwashiorkor states. Flaxseed supplementation has been associated with improvement in lean tissue mass in livestock applications (64). Decreased production of inflammatory cytokines has been demonstrated in patients with rheumatoid arthritis (65). Studies are currently in progress in our laboratory to examine the effect of flax-based diets upon body composition of animals with progressive renal failure.

Accelerated Cardiovascular Disease

Dialysis patients have a substantial increase in age-adjusted cardiovascular morbidity, even allowing for the concomitant hyperlipidemia and hypertension frequently seen in this group (66). Cardiovascular disease inordinately increases the risk of death among renal transplant patients, even those with a functioning kidney, a situation made more tragic when one considers the scarcity of donor organs (67). As a result, dialysis and kidney transplant patients are an easily identifiable and closely followed population at substantially increased risk of myocardial infarction and other atherosclerotic risks. Although a full discussion of flaxseed applications in cardiovascular disease is beyond the mandate of this review, this would seem a good population on which to investigate these effects. Experimental evidence exists of antiatherosclerotic properties of flaxseed and SDG (68–71). Clark *et al.* (51) demonstrated a rapid and significant reduction in total and LDL cholesterol in the specific context of SLE patients. Ingram (3) noted favorable effects of flaxseed and flaxseed oil upon cholesterol in the remnant kidney rat model of renal failure; however, we did not reproduce results of this level in animals on a synthetic, cholesterol free diet (37).

Metabolic Bone Disease

Hyperphosphatemia, depressed production of 1,25 hydroxy-vitamin D_3, and secondary and sometimes tertiary or autonomous hyperparathyroidism are responsible for a large burden of morbidity in renal failure patients, manifest as childhood rickets and growth failure, and adult bone pain and fractures (72). Hyperparathyroidism has been implicated as both an accelerant of progressive renal injury and a potential major factor in accelerated cardiovascular disease (73). We have recently shown that flax oil feeding reduced prostaglandin E_2 (PGE_2) release from bone in 11-wk-old Han:SPRD-*cy* rats with moderate renal dysfunction and hyperparathyroidism, but did not alter markers of bone turnover at this early stage of disease (74). Other studies have suggested a reciprocal relationship between bone PGE_2 release and bone formation rate (75,76); thus, the long-term effects of flax oil on bone remodeling in uremia thus warrant further study.

Conclusions

Flaxseed, flax oil, and flax lignans demonstrate a fascinating range of effects on renal pathology that are of sufficient magnitude in experimental models to warrant the development of human trials. Such trials, however, will need to overcome the barriers created by the frequently late diagnosis of serious renal damage in humans and the slower and less predictable course to end-stage renal disease than is found in rodent models. It is to be hoped that animal studies will develop markers of dietary effect that correlate with beneficial renal outcomes that can be used to monitor human compliance and response.

Knowledge from studies in other aspects of disease makes a strong case for the incorporation of flaxseed products in special renal dietary products, independent of any possible benefit on the progression of renal injury. This is a sustained market of hundreds of thousands of patients in the developed world, and represents a significant opportunity for cooperation between health care providers, food producers and processors, and medical and nutrition scientists.

Acknowledgments

Research into the possible therapeutic applications of flaxseed in our laboratory has been supported by the Flax Council of Canada, the Manitoba Health Research Council, and the Children's Hospital Foundation of Manitoba Inc., Canada.

References

1. Levey, A.S., T. Greene, G.J. Beck, A.W. Caggiula, J.W. Kusek, L.G. Hunsicker, and S. Klahr, Dietary Protein Restriction and the Progression of Chronic Renal Disease: What Have All of the Results of the MDRD Study Shown? Modification of Diet in Renal Disease Study Group, *J. Am. Soc. Nephrol 10*:2426–2439 (1999).
2. Klahr, S., The Modification of Diet in Renal Disease Study, *N. Engl. J. Med. 320*:864–866 (1989).

3. Ingram, A.J., A. Parbtani, W.F. Clark, E. Spanner, M.W. Huff, D.J. Philbrick, and B.J. Holub, Effects of Flaxseed and Flax Oil Diets in a Rat-5/6 Renal Ablation Model, *Am. J. Kidney Dis. 25*:320–329 (1995).

4. Tomobe, K., D.J. Philbrick, M.R. Ogborn, H. Takahashi, and B.J. Holub, Effect of Dietary Soy Protein and Genistein on Disease Progression in Mice with Polycystic Kidney Disease, *Am. J. Kidney Dis. 31*:55–61 (1998).

5. Aukema, H.M., T. Yamaguchi, K. Tomobe, D.J. Philbrick, R.S. Chapkin, H. Takahashi, and B.J. Holub, Diet and Disease Alter Phosphoinositide Composition and Metabolism in Murine Polycystic Kidneys, *J. Nutr. 125*:1183–1191 (1995).

6. Tomobe, K., D. Philbrick, H.M. Aukema, W.F. Clark, M.R. Ogborn, A. Parbtani, H. Takahashi, and B.J. Holub, Early Dietary Protein Restriction Slows Disease Progression and Lengthens Survival in Mice with Polycystic Kidney Disease, *J. Am. Soc. Nephrol. 5*:1355–1360 (1994).

7. Aukema, H.M., M.R. Ogborn, K. Tomobe, H. Takahashi, T. Hibino, and B.J. Holub, Effects of Dietary Protein Restriction and Oil Type on the Early Progression of Murine Polycystic Kidney Disease, *Kidney Int. 42*:837–842 (1992).

8. Jayapalan, S., M.H. Saboorian, J.W. Edmunds, and H.M. Aukema, High Dietary Fat Intake Increases Renal Cyst Disease Progression in Han:SPRD-cy Rats, *J. Nutr. 130*:2356–2360 (2000).

9. Ogborn, M.R., E. Nitschmann, H.A. Weiler, and N. Bankovic-Calic, Modification of Polycystic Kidney Disease and Fatty Acid Status by Soy Protein Diet, *Kidney Int. 57*:159–166 (2000).

10. Bankovic-Calic, N., A. Eddy, S. Sareen, and M.R. Ogborn, Renal Remodelling in Dietary Protein Modified Rat Polycystic Kidney Disease, *Pediatr. Nephrol. 13*:567–570 (1999).

11. Ogborn, M.R., E. Nitschmann, H. Weiler, D. Leswick, and N. Bankovic-Calic, Flaxseed Ameliorates Interstitial Nephritis in Rat Polycystic Kidney Disease, *Kidney Int. 55*:417–423 (1999).

12. Fischbach, H., S. Mackensen, K.E. Grund, A. Kellner, and A. Bohle, Relationship Between Glomerular Lesions, Serum Creatinine and Interstitial Volume in Membrano-Proliferative Glomerulonephritis, *Klin. Wochenschr. 55*:603–608 (1977).

13. Remuzzi, G., P. Ruggenenti, and A. Benigni, Understanding the Nature of Renal Disease Progression, *Kidney Int. 51*:2–15 (1997).

14. Truong, L.D., B. Krishnan, J.T. Cao, R. Barrios, and W.N. Suki, Renal Neoplasm in Acquired Cystic Kidney Disease, *Am. J. Kidney Dis. 26*:1–12 (1995).

15. Klahr, S., Oxygen Radicals and Renal Diseases, *Miner. Electrolyte Metab. 23*:140–143 (1997).

16. Klahr, S., Urinary Tract Obstruction, *Semin. Nephrol. 21*:133–145 (2001).

17. Vitzthum, H., I. Abt, S. Einhellig, and A. Kurtz, Gene Expression of Prostanoid Forming Enzymes Along the Rat Nephron, *Kidney Int. 62*:1570–1581 (2002).

18. Hannedouche, T., P. Chauveau, F. Kalou, G. Albouze, B. Lacour, and P. Jungers, Factors Affecting Progression in Advanced Chronic Renal Failure, *Clin. Nephrol. 39*:312–320 (1993).

19. Modification of Diet in Renal Disease Study Group, Dietary Protein Restriction, Blood Pressure Control, and the Progression of Polycystic Kidney Disease, *J. Am. Soc. Nephrol. 5*:2037–2047 (1995).

20. Davies, M.R., and K.A. Hruska, Pathophysiological Mechanisms of Vascular Calcification in End-Stage Renal Disease, *Kidney Int. 60*:472–479 (2001).

21. Eddy, A.A., Interstitial Fibrosis in Hypercholesterolemic Rats: Role of Oxidation, Matrix Synthesis, and Proteolytic Cascades, *Kidney Int. 53*:1182–1189 (1998).
22. Hostetter, T.H., J.L. Olson, H.G. Rennke, M.A. Venkatachalam, and B.M. Brenner, Hyperfiltration in Remnant Nephrons: a Potentially Adverse Response to Renal Ablation, *Am. J. Physiol. 241*:F85–93 (1981).
23. Nath, K.A., M.K. Hostetter, and T.H. Hostetter, Pathophysiology of Chronic Tubulo-Interstitial Disease in Rats: Interactions of Dietary Acid Load, Ammonia, and Complement Component C3, *J. Clin. Invest. 76*:667–675 (1985).
24. Brenner, B.M., Nephron Adaptation to Renal Injury or Ablation, *Am. J. Physiol. 249*:F324–337 (1985).
25. Ogborn, M.R., Polycystic Kidney Disease—A Truly Pediatric Problem, *Pediatr. Nephrol. 8*:762–767 (1994).
26. Nadasdy, T., Z. Laszik, G. Lajoie, K.E. Blick, D.E. Wheeler, and F.G. Silva, Proliferative Activity of Cyst Epithelium in Human Renal Cystic Diseases, *J. Am. Soc. Nephrol. 5*:1462–1468 (1995).
27. Woo, D., Apoptosis and Loss of Renal Tissue in Polycystic Kidney Diseases, *N. Engl. J. Med. 333*:18–25 (1995).
28. Rankin, C.A., J.J. Grantham, and J.P. Calvet, C-fos Expression is Hypersensitive to Serum-Stimulation in Cultured Cystic Kidney Cells from the C57BL/6J-cpk Mouse, *J. Cell Physiol. 152*:578–586 (1992).
29. Cowley, Jr., B.D., S. Gudapaty, A.L. Kraybill, B.D. Barash, M.A. Harding, J.P. Calvet, and V.H.d. Gattone, Autosomal-Dominant Polycystic Kidney Disease in the Rat, *Kidney Int. 43*:522–534 (1993).
30. Gretz, N., I. Ceccherini, B. Kranzlin, I. Kloting, M. Devoto, P. Rohmeiss, B. Hocher, R. Waldherr, and G. Romeo, Gender-Dependent Disease Severity in Autosomal Polycystic Kidney Disease of Rats, *Kidney Int. 48*:496–500 (1995).
31. Ogborn, M.R., and S. Sareen, Amelioration of Polycystic Kidney Disease by Modification of Dietary Protein Intake in the Rat, *J. Am. Soc. Nephrol. 6*:1649–1654 (1995).
32. Keith, D.S., V.E. Torres, C.M. Johnson, and K.E. Holley, Effect of Sodium Chloride, Enalapril, and Losartan on the Development of Polycystic Kidney Disease in Han:SPRD Rats, *Am. J. Kidney Dis. 24*:491–498 (1994).
33. Gattone II, V.H., B.D. Cowley, Jr., B.D. Barash, S. Nagao, H. Takahashi, T. Yamaguchi, and J.J. Grantham, Methylprednisolone Retards the Progression of Inherited Polycystic Kidney Disease in Rodents, *Am. J. Kidney Dis. 25*:302–313 (1995).
34. Gile, R.D., B.D. Cowley, Jr., V.H. Gattone II, O.D. MP, S.K. Swan, and J.J. Grantham, Effect of Lovastatin on the Development of Polycystic Kidney Disease in the Han:SPRD Rat, *Am. J. Kidney Dis. 26*:501–507 (1995).
35. Ogborn, M.R., N. Bankovic-Calic, C. Shoesmith, R. Buist, and J. Peeling, Soy Protein Modification of Rat Polycystic Kidney Disease, *Am. J. Physiol. 274*:F541–549 (1998).
36. Ogborn, M.R., E. Nitschmann, N. Bankovic-Calic, R. Buist, and J. Peeling, The Effect of Dietary Flaxseed Supplementation on Organic Anion and Osmolyte Content and Excretion in Rat Polycystic Kidney Disease, *Biochem. Cell Biol. 76*:553–559 (1998).
37. Ogborn, M.R., E. Nitschmann, N. Bankovic-Calic, H. Weiler, and H.M. Aukema, Dietary Flax Oil Reduces Renal Injury, Oxidized LDL Content and Tissue w6:w3 Fatty Acid Ratio in Experimental Polycystic Kidney Disease, *Lipids*, in press.
38. Weiler, H.A., and S. Fitzpatrick-Wong, Dietary Long-Chain Polyunsaturated Fatty Acids Minimize Dexamethasone-Induced Reductions in Arachidonic Acid Status but not Bone Mineral Content in Piglets, *Pediatr. Res. 51*:282–289 (2002).

39. Chuang, L.T., J.M. Thurmond, J.W. Liu, S.J. Kirchner, P. Mukerji, T.M. Bray, and Y.S. Huang, Effect of Conjugated Linoleic Acid on Fungal Delta6-Desaturase Activity in a Transformed Yeast System, *Lipids* 36:139–143 (2001).

40. Alegre, V.A., R.K. Winkelmann, and D.A. Gastineau, Cutaneous Thrombosis, Cerebrovascular Thrombosis, and Lupus Anticoagulant—the Sneddon Syndrome. Report of 10 Cases, *Int. J. Dermatol.* 29:45–49 (1990).

41. Balow, J.E., and H.A. Austin III, Renal Disease in Systemic Lupus Erythematosus, *Rheum. Dis. Clin. North Am.* 14:117–133 (1988).

42. Doherty, N.E., and R.J. Siegel, Cardiovascular Manifestations of Systemic Lupus Erythematosus, *Am. Heart J.* 110:1257–1265 (1985).

43. Klippel, J.H., Systemic Lupus Erythematosus: Demographics, Prognosis, and Outcome, *J. Rheumatol. Suppl.* 48:67–71 (1997).

44. Tax, W.J., C. Kramers, M.C. van Bruggen, and J.H. Berden, Apoptosis, Nucleosomes, and Nephritis in Systemic Lupus Erythematosus, *Kidney Int.* 48:666–673 (1995).

45. Baldi, E., S.N. Emancipator, M.O. Hassan, and M.J. Dunn, Platelet Activating Factor Receptor Blockade Ameliorates Murine Systemic Lupus Erythematosus, *Kidney Int.* 38:1030–1038 (1990).

46. Zoja, C., and G. Remuzzi, Role of Platelets in Progressive Glomerular Diseases, *Pediatr. Nephrol.* 9:495–502 (1995).

47. Clark, W.F., and A. Parbtani, Omega-3 Fatty Acid Supplementation in Clinical and Experimental Lupus Nephritis, *Am. J. Kidney Dis.* 23:644–647 (1994).

48. Theofilopoulos, A.N., and F.J. Dixon, Murine Models of Systemic Lupus Erythematosus, *Adv. Immunol.* 37:269–390 (1985).

49. Hall, A.V., A. Parbtani, W.F. Clark, E. Spanner, M. Keeney, I. Chin-Yee, D.J. Philbrick, and B.J. Holub, Abrogation of MRL/lpr Lupus Nephritis by Dietary Flaxseed, *Am. J. Kidney Dis.* 22:326–332 (1993).

50. Clark, W.F., A.D. Muir, N.D. Westcott, and A. Parbtani, A Novel Treatment for Lupus Nephritis: Lignan Precursor Derived from Flax, *Lupus* 9:429–436 (2000).

51. Clark, W.F., A. Parbtani, M.W. Huff, E. Spanner, H. de Salis, I. Chin-Yee, D.J. Philbrick, and B.J. Holub, Flaxseed: A Potential Treatment for Lupus Nephritis, *Kidney Int.* 48:475–480 (1995).

52. Clark, W.F., C. Kortas, A.P. Heidenheim, J. Garland, E. Spanner, and A. Parbtani, Flaxseed in Lupus Nephritis: A Two-Year Nonplacebo-Controlled Crossover Study, *J. Am. Coll. Nutr.* 20:143–148 (2001).

53. Hunley, T.E., and V. Kon, IgA Nephropathy, *Curr. Opin. Pediatr.* 11:152–157 (1999).

54. Rantala, I., J. Mustonen, M. Hurme, J. Syrjanen, and H. Helin, Pathogenetic Aspects of IgA Nephropathy, *Nephron.* 88:193–198 (2001).

55. Donadio, Jr., J.V., J.P. Grande, E.J. Bergstralh, R.A. Dart, T.S. Larson, and D.C. Spencer, The Long-Term Outcome of Patients with IgA Nephropathy Treated with Fish Oil in a Controlled Trial, Mayo Nephrology Collaborative Group, *J. Am. Soc. Nephrol.* 10:1772–1777 (1999).

56. Donadio, Jr., J.V., Use of Fish Oil to Treat Patients with Immunoglobulin A Nephropathy, *Am. J. Clin. Nutr.* 71:373S–375S (2000).

57. Wyatt, R.J., and R.J. Hogg, Evidence-Based Assessment of Treatment Options for Children with IgA Nephropathies, *Pediatr. Nephrol.* 16:156–167 (2001).

58. Caughey, G.E., E. Mantzioris, R.A. Gibson, L.G. Cleland, and M.J. James, The Effect on Human Tumor Necrosis Factor Alpha and Interleukin 1 Beta Production of Diets Enriched in n-3 Fatty Acids from Vegetable Oil or Fish Oil, *Am. J. Clin. Nutr.* 63:116–122 (1996).

59. Monteon, F.J., S.A. Laidlaw, J.K. Shaib, and J.D. Kopple, Energy Expenditure in Patients with Chronic Renal Failure, *Kidney Int. 30*:741–747 (1986).

60. Akner, G., and T. Cederholm, Treatment of Protein-Energy Malnutrition in Chronic Nonmalignant Disorders, *Am. J. Clin. Nutr. 74*:6–24 (2001).

61. Kari, J.A., C. Gonzalez, S.E. Ledermann, V. Shaw, and L. Rees, Outcome and Growth of Infants with Severe Chronic Renal Failure, *Kidney Int. 57*:1681–1687 (2000).

62. Cochat, P., P. Braillon, J. Feber, A. Hadj-Aissa, L. Dubourg, I. Liponski, M.H. Said, C. Glastre, P.J. Meunier, and L. David, Body Composition in Children with Renal Disease: Use of Dual Energy X-Ray Absorptiometry, *Pediatr. Nephrol. 10*:264–268 (1996).

63. Levine, B., J. Kalman, L. Mayer, H.M. Fillit, and M. Packer, Elevated Circulating Levels of Tumor Necrosis Factor in Severe Chronic Heart Failure, *N. Engl. J. Med. 323*:236–241 (1990).

64. Petit, H.V., R.J. Dewhurst, N.D. Scollan, J.G. Proulx, M. Khalid, W. Haresign, H. Twagiramungu, and G.E. Mann, Milk Production and Composition, Ovarian Function, and Prostaglandin Secretion of Dairy Cows Fed Omega-3 Fats, *J. Dairy Sci. 85*:889–899 (2002).

65. Kremer, J.M., n-3 Fatty Acid Supplements in Rheumatoid Arthritis, *Am. J. Clin. Nutr. 71*:349S–351S (2000).

66. Fleischmann, E.H., J.D. Bower, and A.K. Salahudeen, Are Conventional Cardiovascular Risk Factors Predictive of Two-Year Mortality in Hemodialysis Patients? *Clin. Nephrol. 56*:221–230 (2001).

67. Ritz, E., V. Schwenger, M. Wiesel, and M. Zeier, Atherosclerotic Complications after Renal Transplantation, *Transpl. Int. 13*:S14–19 (2000).

68. Bierenbaum, M.L., R. Reichstein, and T.R. Watkins, Reducing Atherogenic Risk in Hyperlipemic Humans with Flax Seed Supplementation: A Preliminary Report, *J. Am. Coll. Nutr. 12*:501–504 (1993).

69. Prasad, K., Dietary Flax Seed in Prevention of Hypercholesterolemic Atherosclerosis, *Atherosclerosis 132*:69–76 (1997).

70. Prasad, K., S.V. Mantha, A.D. Muir, and N.D. Westcott, Reduction of Hypercholesterolemic Atherosclerosis by CDC-Flaxseed with Very Low Alpha-Linolenic Acid, *Atherosclerosis 136*:367–375. (1998)

71. Prasad, K., Reduction of Serum Cholesterol and Hypercholesterolemic Atherosclerosis in Rabbits by Secoisolariciresinol Diglucoside Isolated from Flaxseed, *Circulation 99*:1355–1362 (1999).

72. Weinreich, T., Prevention of Renal Osteodystrophy in Peritoneal Dialysis, *Kidney Int. 54*:2226–2233 (1998).

73. Rostand, S.G., and T.B. Drueke, Parathyroid Hormone, Vitamin D, and Cardiovascular Disease in Chronic Renal Failure, *Kidney Int. 56*:383–392 (1999).

74. Weiler, H., H. Kovacs, E. Nitschmann, S. Fitzpatrick-Wong, N. Bankovic-Calic, and M. Ogborn, Elevated Bone Turnover in Rat Polycystic Kidney Disease Is Not Due to Prostaglandin E2, *Pediatr. Nephrol. 17*:795–799 (2002).

75. Watkins, B.A., C.L. Shen, J.P. McMurtry, H. Xu, S.D. Bain, K.G. Allen, and M.F. Seifert, Dietary Lipids Modulate Bone Prostaglandin E2 Production, Insulin-Like Growth Factor-I Concentration and Formation Rate in Chicks, *J. Nutr. 127*:1084–1091 (1997).

76. Watkins, B.A., C.L. Shen, K.G. Allen, and M.F. Seifert, Dietary (n-3) and (n-6) Polyunsaturates and Acetylsalicylic Acid Alter *ex vivo* PGE2 Bbiosynthesis, Tissue IGF-I Levels, and Bone Morphometry in Chicks, *J. Bone Miner. Res. 11*:1321–1332 (1996).

Chapter 17

Effect of Flaxseed on Bone Metabolism and Menopause

Wendy Elizabeth Ward

Department of Nutritional Sciences, Faculty of Medicine, University of Toronto, Toronto, Ontario M5S 3E2, Canada

Introduction

Osteoporosis and menopausal symptoms such as hot flushes and vaginal dryness are common maladies during aging. Both osteoporosis and menopausal symptoms occur, in part, as a result of the cessation in estrogen production by the ovaries. Because of the aging population, over the next decades there will be a worldwide exponential rise in the numbers of individuals developing osteoporosis and suffering from menopausal symptoms (1). Although hormone replacement therapy (HRT) provides some protection against fractures and has been shown to relieve menopausal symptoms, recent findings from the Women's Health Initiative trial suggesting an increased risk of cardiovascular disease and breast cancer with HRT have raised concerns over who should start or continue to receive HRT (2–4). Moreover, compliance to HRT is generally poor due to side effects and fears over long-term safety (5,6). With respect to osteoporosis, other pharmacological agents such as bisphosphonates have been shown to reduce fracture rates but can result in gastrointestinal problems (7,8). As a result of all these factors, there is increasing interest in natural approaches such as dietary strategies that can maintain bone health and relieve hot flushes and other menopausal symptoms (9,10). It is therefore timely to summarize and evaluate the study findings regarding the effects of flaxseed on bone metabolism and menopausal symptoms.

Bone Metabolism

It is estimated that the incidence of osteoporosis, worldwide, is increasing by 1–3% each year (1). Osteoporosis is a condition in which bones become brittle and susceptible to fragility fracture due to a low bone mineral density and an overall deterioration of the quality of bone. Among the most common fractures are the hip, wrist, and compression fractures in individual vertebra. Estrogen is important for the maintenance of strong, healthy bones throughout the life cycle (11). Based on the fact that the mammalian lignans enterodiol and enterolactone share a chemical structure similar to that of estrogen, it is possible that feeding flaxseed may result in positive effects in bone. Other phytoestrogen-containing foods (e.g., soy foods, a rich source of isoflavones with potential estrogen-like effects) have been shown to prevent or slow bone loss in ovariectomized rodents (12,13) and postmenopausal women (14,15). Moreover, feeding isolated isoflavones slows bone loss in ovariectomized rodents (16–19). By comparison, however, there are considerably fewer published studies reporting the effects of flaxseed or its purified lignan on bone metabolism in either rats (20–23) (Table 17.1) or humans (24–26) (Table 17.2).

Although estrogen is usually considered to be important for bone metabolism in females rather than in males, recent studies using knockout mice have demonstrated that estrogen, as well as testosterone, is important for maintenance of a healthy skeleton in males (27). Mice that lack the gene for aromatase, the enzyme responsible for the conversion of testosterone to estrogen, have impaired bone metabolism (28,29). Moreover, men with aromatase deficiency have a lower bone mass and reduced linear skeletal growth (30).

Flaxseed and Bone Development in Rats

Based on the knowledge that bone development is regulated by hormones, combined with the fact that early exposure to flaxseed or its purified lignan can alter reproductive indices (31–33), two studies were conducted in both female and male rats to determine whether early exposure to flaxseed or its purified lignan was safe with respect to bone metabolism (20,21). These studies were also designed to determine whether early exposure to flaxseed or its purified lignan component conferred any benefit to bone development, particularly among female offspring. In both studies, femur bone mass [bone mineral content (BMC) and bone mineral density (BMD)] was measured by dual energy X-ray absorptiometry, and the biomechanical strength properties of femurs were determined by three-point bending using a materials testing system (20,21). Measurement of BMC indicates the quantity of mineral in a bone, whereas BMD is BMC expressed as a function of bone size and thus is often used for comparisons as it corrects for differences in the size (area) of a bone. Biomechanical strength properties provide functional measures of bone, such as the force that a bone can withstand before it fractures. Other properties include the resilience and toughness of a bone. The biomechanical properties of a bone provide information regarding the contribution of minerals or matrix proteins such as type 1 collagen to bone strength. Also, these properties provide insight into the interaction between mineral and matrix proteins in bone (34,35).

To determine the effect of early and continuous exposure to flaxseed lignan (SDG), lactating rats were fed diets containing purified SDG at a level present in a 5 or 10% flaxseed diet during lactation (20). It was previously shown that lignan present in a mother's diet was transferred to suckling pups via mother's milk (32). At the end of the suckling period, female pups either consumed a control diet, devoid of lignan, or continued on the SDG diets equivalent to lignan in a 5 or 10% flaxseed diet until they reached 50 or 132 d of age. Female pups that received SDG only during suckling had femur BMC, BMD, and biomechanical strength properties similar to those of rats receiving no lignans. In contrast, selected biomechanical strength properties of femurs (ultimate bending stress, Young's modulus) from female pups that received lignans during suckling and through to 50 d of age were greater than those for pups fed con-

TABLE 17.1
Bone Metabolism in Male and Female Rats[a,b]

Intervention	Stage of development, gender and sample size	Dose, route of administration, and duration of treatment	Findings	Reference
Purified SDG	Lactation, females $n = 7–11$/group	Mothers fed SDG at a quantity equivalent to SDG in a 5 or 10% flaxseed diet during lactation	At PND 50 or 132: No differences in femur BMC No differences in femur biomechanical strength	20
Purified SDG	Lactation to adulthood, females $n = 7–11$/group	Mothers fed SDG at a quantity equivalent to SDG in a 5% or 10% flaxseed diet during lactation, and then pups fed mother's diet until PND 50 or 132	At PND 50: No differences in femur BMC Groups fed SDG at 5 or 10% level from birth through PND 50 had greater femur biomechanical strength At PND 132 Groups fed SDG at 10% level from birth–PND 50 or 132 had a greater BMC	20
Flaxseed or Purified SDG	Lactation, Males $n = 6–11$/group	Mothers fed 10% ground flaxseed or equivalent quantity of SDG during lactation	At PND 50 or 132 No differences in femur BMC or BMD No differences in femur biomechanical strength	21
Flaxseed or purified lignan SDG	Lactation to adulthood, males $n = 6–11$/group	Mothers fed 10% ground flaxseed or equivalent quantity of SDG during lactation and then pups were fed mother's diet until PND 50 or 132	At PND 50: No differences in femur BMC or BMD Groups fed 10% flaxseed diet from birth through PND 50 had lower femur biomechanical strength At PND 132 No difference in femur BMC or BMD No difference in femur biomechanical strength	21
Flaxseed or flaxmeal	Females $n = 8–10$/group 3–4 weeks old	Fed 5 or 10% ground flaxseed, or 6.2% flaxmeal[c] for 56 d	No change in plasma levels of calcium or phosphorus Reduced alkaline phosphatase activity in bone among rats fed flaxseed or flaxmeal	22

[a]Femur BMC and BMD were measured by dual energy x-ray absorptiometry.
[b]Femur biomechanical strength was measured by three-point bending using a materials testing system.
[c]Defatted flaxseed equivalent to the 10% ground flaxseed diet.
Abbreviations: BMC, bone mineral content; BMD, bone mineral density; PND, postnatal day; SDG, secoisolariciresinol diglycoside.

TABLE 17.2
Bone Metabolism in Postmenopausal Women

Intervention	Sample Size	Dose and duration of intervention	Findings	Reference
Flaxseed or wheat	$n = 58$ ($n = 29$/treatment)	40 g ground flaxseed or 40 g wheat control	No change in biochemical markers of bone turnover Bone-specific alkaline phosphatase Deoxypyridinoline	25
	Double-blind, parallel study	3 mon	Helical peptide of type 1 collagen Tartrate-resistant acid phosphatase	
Flaxseed or sunflower seed	$n = 38$	38 g flaxseed or 38 g sunflower seed (control)	No change in bone-specific alkaline phosphatase	24
	Crossover study	baked in three muffins and four slices of bread per day	↓ Tartrate-resistant acid phosphatase with flaxseed	
		6 wk/intervention		
Flaxseed, soy, or placebo	$n = 48$ ($n = 16$/treatment)	25 g ground flaxseed or 25 g soy or placebo baked in one muffin per day	No change in biochemical markers of bone turnover Bone-specific alkaline phosphatase Deoxypyridinoline	26
		16 wk		

trol diet. It is interesting that this beneficial effect on biomechanical bone strength was not observed into adulthood, postnatal day (PND) 132. The rise in endogenous estrogen production after puberty, occurring at about 32 d of age, likely diluted the effects of early exposure to lignan (33).

A similar study design was used to determine whether a 10% flaxseed diet or the equivalent quantity of purified lignans affected, positively or negatively, BMC, BMD, and biomechanical bone strength among male offspring (21). Enterolactone had been previously reported to inhibit aromatase activity (36), and suckling is known to be a hormone-sensitive stage of development (37). Exposure to either the 10% flaxseed diet or the equivalent quantity of SDG during suckling did not alter BMC or BMD at young adulthood (PND 50) or adulthood (PND 132) compared with rats not exposed to flaxseed or purified SDG. Moreover, the biomechanical strength of femurs did not differ among any groups at either age. Some but not all measures of femur biomechanical strength (ultimate bending stress and Young's modulus) were lower among males that received lignans, in the form of flaxseed, through mother's milk during suckling and then through diet until PND 50, but femur BMC and BMD were not different among groups. The fact that the femurs of rats receiving purified SDG were not different from the femurs of controls suggests that a nonlignan component in flaxseed affected bone strength. Although flaxseed contains phytates that can bind minerals, the fact that BMC and BMD were unchanged did not support the hypothesis that the phytate content of flaxseed contributed to the reduced biomechanical strength.

Because flaxseed is a rich source of n-3 fatty acids (n-3 PUFA), it is possible that the ratio of n-3 to n-6 PUFA influenced specific regulators of bone metabolism such as insulin-like growth factor-I (IGF-I), specific insulin-like growth factor binding proteins or prostaglandins. The role of fatty acids, particularly n-3 PUFA, in modulating bone cell metabolism has been reviewed in several recent papers (38,39). Despite the lower femur strength at PND 50 among males continuously exposed to the 10% flaxseed diet, this effect did not persist into adulthood. Thus, lifelong exposure to flaxseed and lignan appears to be safe with respect to bone development in male rats as assessed by femur BMC, BMD, and biomechanical strength.

Another study reported that young female rats had a lower bone alkaline phosphatase activity after consuming a flaxseed (5 or 10%) or flaxmeal (6.2%) diet (22). Other biochemical indices such as plasma calcium or phosphorus were not altered with flaxseed or flaxmeal feeding (22). The reduction in alkaline phosphatase activity has been reported in another study in which rats were fed flaxseed (23). Because it is hypothesized that most of the effect on bone alkaline phosphatase activity and femur zinc content is due to the phytate or fiber content of flaxseed or flaxmeal, it does not appear that lignans have a measurable effect on bone metabolism in rats that have endogenous production of sex steroid. Because these studies did not test the effect of purified lignan, it is not possible to delineate the effect of the lignan component in flaxseed or flaxmeal on bone metabolism.

Future studies could use ovariectomized rodent models to test whether flaxseed feeding is protective against the loss of bone mass that occurs after the cessation of endogenous estrogen production. Measurement of biomechanical strength properties of bones would provide specific insight into whether flaxseed or its lignan provides protection against fracture, and whether the matrix and/or mineral component of bone is altered.

Flaxseed and Bone Metabolism in Postmenopausal Women

Two studies have reported associations between bone mass and urinary phytoestrogens, including lignans (40,41). In a group of 75 Korean postmenopausal women (52–65 yr of age), urinary enterolactone was positively correlated with BMD at all three sites measured: lumbar vertebrae (LV2–LV4), femoral neck, and Ward's triangle (41). In contrast, a 10 yr follow-up study of 145 postmenopausal Dutch women reported that urinary enterolactone excretion was significantly higher among women who experienced the greatest bone loss in the radius over the 10 yr study (40). The authors state that this relationship existed even though a number of variables were taken into consideration (i.e., age, years since menopause, body mass index, as well as intakes of calcium, vegetable protein, and dietary fiber). A major difference between these two studies is that the study in Dutch women measured BMC and BMD of the cortical bone of the radius by single photon absorptiometry (40), whereas the study in Korean women measured BMC by dual energy X-ray absorptiometry at multiple sites, none of which included the radius (41). Single photon absorptiometry as opposed to dual energy X-ray absorptiometry provides a measure only of cortical bone, not cortical and trabecular bone. The fact that only cortical and not both trabecular and cortical bone were measured may suggest that lignans have different effects on trabecular bone, more metabolically active bone, versus cortical bone. This requires further investigation.

The effect on bone metabolism of feeding flaxseed has been reported in only a few studies (24–26). These studies have used surrogate measures of bone metabolism such as biochemical markers of bone turnover that are measured in serum and/or urine samples (24–26) (Table 17.2). One study reported that feeding flaxseed in the form of breads or muffins reduced serum tartrate-resistant acid phosphatase, a marker of osteoclastic activity, without altering serum levels of bone-specific alkaline phosphatase activity, a marker of osteoblastic activity (24). A later study, however, reported that feeding flaxseed for 3 mon did not result in any significant changes in biochemical markers of bone turnover. Neither markers of bone formation (indicators of osteoblastic activity) nor markers of bone resorption (indicators of osteoclastic activity) were affected by flaxseed intervention. The study lasted 3 mon, adequate time to observe changes in biochemical markers of bone turnover. Similarly, another study demonstrated that feeding a lower quantity of ground flaxseed (25 g) to postmenopausal women for 4 mon did not alter biochemical markers of bone turnover (26).

No studies have measured changes in bone mass by clinical densitometry (i.e., dual energy X-ray absorptiometry) with flaxseed feeding. A minimum of 6 mon of intervention would be required to detect changes in BMD by dual energy x-ray absorptiometry because the bone-remodeling cycle requires many months. Because BMD and risk of fracture are related, BMD is used clinically as a surrogate measure of risk for fragility fracture (42). A limitation of conducting fracture studies is the long duration that is required to observe a meaningful number of fractures in the cohort and the large sample sizes required to obtain sufficient power for statistical testing. From an ethical perspective, it would be difficult to design a study to evaluate whether flaxseed feeding prevents fractures, because pharmacological agents are available to slow bone loss and reduce the risk of fragility fracture. One would have to study women who had chosen not to take pharmacological agents to preserve bone mass.

No studies have tested the effect of purified lignan precursor (SDG) on bone metabolism. Because n-3 PUFA may also positively modulate bone cell metabolism through modification of specific prostaglandins and cytokines, it is possible that flaxseed contains several components that affect bone cell metabolism (38,39). There is still much to be learned about a potential role of flaxseed in bone health, such as whether there are certain times in the life cycle when consuming flaxseed is particularly beneficial to overall skeletal health.

Menopausal Symptoms

The loss of endogenous estrogen production during and after menopause often results in hot flushes, night sweats, and vaginal dryness that, ultimately, can compromise a woman's quality of life. Although HRT has been proved to prevent or provide relief from menopausal symptoms (4), there is a growing body of evidence that suggests that HRT is associated with an increased risk of specific diseases (as discussed in the Introduction) (2,3). Moreover, some women choose not to take drugs to manage menopausal symptoms and actively seek natural, dietary remedies (9,10,43). Because flaxseed is a rich source of SDG, the precursor of the mammalian lignans (enterodiol and enterolactone) that may mimic the activity of estrogen, determining whether a flaxseed-rich diet attenuates menopausal symptoms is an active area of research interest.

Flaxseed and Menopausal Symptoms

Studies have investigated the effect of feeding flaxseed on the frequency of hot flushes, vaginal cell maturation, vaginal dryness, and/or measurement of specific serum sex hormones (44–48) (Table 17.3). To date, only one of these studies has directly compared flaxseed with the gold standard, HRT, at attenuating menopausal symptoms (44). In this randomized, crossover study, feeding flaxseed for 2 mon attenuated menopausal symptoms as effectively as 2 mon of treatment with HRT (Table 17.3). Flaxseed feeding did not alter hormone levels, whereas HRT resulted in higher sex hormone binding globulin. The sample size was smaller than most of the other reported studies (Table 17.3). Because this study was also designed to assess whether flaxseed intervention improved cardiovascular disease outcomes including blood lipid profile, all of the women studied had an abnormal lipid profile.

TABLE 17.3
Clinical Studies: Menopausal Symptoms

Intervention	Sample Size	Dose and duration of intervention	Findings	Reference
Flaxseed Hormone replacement therapy (HRT)	$n = 25$ Crossover study 2 mon washout	40 g crushed flaxseed (20 g incorporated into 2 slices of bread +20 g crushed flaxseed); or HRT 0.625 mg conjugated equine estrogens (if no uterus) or combined with 100 mg micronized progesterone (if uterus was intact).	↓ Score of menopausal symptoms with both treatments[a] ↑ Sex hormone binding globulin with HRT	44

Phytoestrogen-containing foods	$n = 78$ for phytoestrogen -rich diet
	
	$n = 36$ for omnivorous diet
		Phytoestrogen-rich diet: 25% of daily energy derived from foods containing high levels of phytoestrogens; or omnivorous diet devoid of phytoestrogens
		
		12 wk
			No difference in total score from menopausal symptoms questionnaire but subanalysis showed differences with phytoestrogen-rich diet
			  ↓ Hot flushes
			  ↓ Vaginal dryness
			
			No change in serum sex steroids: estradiol, luteinizing hormone, follicle stimulating hormone
			
			No change in vaginal cytology
			
			With phytoestrogen-rich diet
			  ↑ Sex hormone binding globulin
			  ↑ Urinary enterodiol, enterolactone, daidzein, equol, genistein, O-desmethylangolensin
				45

Flaxseed	$n = 25$	Flaxseed (25 g/d) Soy flour (45 g/d); or	↑ Vaginal cytology maturation index with soy or flaxseed	46
Soy flour	Crossover study	Red clover (10 g dry seed/d)		
Red clover	No washout	No control group	↓ Follicle stimulating hormone after 6 wk	
		2 wk/intervention		
Flaxseed	$n = 52$	Soy (45 g/d) Flaxseed (45 g/d)	No differences in menopausal symptoms when three treatment groups compared, but subanalysis showed	47
	Crossover study	Wheat (45 g/d)	↓ hot flushes with flaxseed or wheat	
		12 wk/intervention	↑ vaginal cytology maturation index with soy ↑ urinary phytoestrogen excretion (flaxseed or soy)	
Flaxseed, soy, or placebo	$n = 99$ ($n = 33$/treatment)	25 g flaxseed or 25 g soy or placebo per day in the form of a muffin	No difference in menopausal symptoms including measures of vasomotor, psychosocial, sexual, or physical symptoms[b]	48
		16 wk	No differences in hot flush symptoms	

[a]Assessed by the Kuppermann Index that is based on the 11 most common menopausal complaints and their severity.
[b]Assessed by a menopause-specific quality of life questionnaire called the MENQOL (49).

Other studies that have not directly tested the effects of flaxseed and HRT report mixed findings with respect to alleviation of menopausal symptoms (45–48). In one study, women were instructed to consume either a phytoestrogen-rich diet that incorporated foods that contained substantial levels of lignans (e.g., flaxseed) that constituted about 25% of their daily caloric intake, or to consume a diet that contained negligible quantities of phytoestrogens (45). Women who consumed the diet rich in phytoestrogens, including flaxseed, for 12 wk experienced no difference in their total score of menopausal symptoms. However, a subanalysis of these data revealed that when the symptoms were assessed individually, there was a reduction in hot flushes and an attenuation of vaginal dryness among women consuming the phytoestrogen-rich diet (45) (Table 17.3). Whether hot flushes and vaginal dryness were alleviated due to flaxseed and/or its lignan component is unclear because other phytoestrogens such as isoflavones were also present in the diet.

In another study, the diets of women were supplemented with either flaxseed, soy flour, or red clover for 2 wk each, for a total supplementation period of 6 wk (46) (Table 17.3). All treatments resulted in a reduction in follicle stimulating hormone at the end of the 6-wk intervention (46). Vaginal cell maturation was increased after a 2-wk supplementation with either flaxseed or soy flour, but not with red clover. A limitation of this study is the fact that there were no washout periods among dietary interventions. Because the incidence of hot flushes was not evaluated, it is uncertain whether the changes in follicle stimulating hormone or vaginal cell maturation altered menopausal symptoms.

Another study reported that women who consumed soy, flaxseed, or wheat (control) for 12 wk experienced no differences in the incidence of hot flushes when groups were compared (47). However, comparison of menopausal scores at baseline and at the end of treatment revealed that flaxseed or wheat reduced hot flushes, whereas soy increased the vaginal cytology maturation index, suggestive of an estrogenic effect. The reason for the reduction in hot flushes while consuming the wheat diet, a diet low in phytoestrogens, is unclear.

The effect of flaxseed on menopausal symptoms is likely dependent on the quantity of flaxseed consumed. For example, studies have shown that feeding 40 g or more of flaxseed a day attenuates menopausal symptoms (44,47), but feeding 25 g of flaxseed does not reduce the incidence of hot flushes or improve the overall menopausal symptom scores (46,48).

The interpretation of the findings from these studies is complex. Unlike the measurement of outcomes such as BMD, biomechanical bone strength, or biochemical markers, assessment of menopausal symptoms is subjective. Some study subjects may report an improvement in menopausal symptoms after commencing a study, regardless of whether they are receiving a placebo or experimental intervention. This may be because menopausal symptoms are attenuated over time and/or it is difficult to conduct a truly randomized, blinded trial due to the fact that subjects are often aware of whether they are consuming a diet rich in flaxseed and/or other phytoestrogens. To date, no studies have tested whether consuming purified lignan precursor (SDG) attenuates menopausal symptoms. Further investigation is required to compare and contrast the effect of providing HRT or of feeding purified lignan and/or flaxseed on reducing menopausal symptoms.

Summary

In summary, flaxseed and its lignan component SDG do not seem to have either a positive or a negative effect on developing bone in male or female rats when provided at levels equivalent to levels attainable in the human diet. It is speculated that any potential hormone-like effect of lignans is likely attenuated when endogenous sex steroids are produced. With respect to attenuation of bone loss, there is currently little evidence to suggest that flaxseed protects against the development of osteoporosis among postmenopausal women. However, additional studies are needed to confirm the effect of flaxseed on bone during aging, after estrogen production has diminished, using measures of BMD. Based on the most recent study (44), menopausal symptoms are effectively attenuated by flaxseed, because the reduction with flaxseed feeding was similar to the reduction observed with HRT. However, further studies are needed to confirm this finding. Although the health benefits of flaxseed are largely attributed to its lignan component, it is important to consider that flaxseed is also a good source of other nutrients that may have biological effects on bone and/or menopausal symptoms. Such nutrients include antioxidants, essential vitamins and minerals, fiber, and polyunsaturated fatty acids, particularly n-3 PUFA.

Acknowledgments

The author gratefully acknowledges the National Institute of Nutrition for the postdoctoral fellowship that she held (1998–2000) while conducting the works cited in this chapter in which she is first author. The studies that were conducted during her postdoctoral fellowship were funded by a grant to L. Thompson from the Natural Sciences and Engineering Research Council of Canada.

References

1. Cummings, S.R., and L.J. Melton, Epidemiology and Outcomes of Osteoporotic Fractures, *Lancet 359*:1761–1767 (2002).
2. Nelson, H.D., L.L. Humphrey, P. Nygren, S.M. Teutsch, and J.D. Allan, Postmenopausal Hormone Replacement Therapy: Scientific Review, *JAMA.* 288, 872-881 (2002).
3. Writing Group for the Women's Health Initiative Investigators (2002). Risks and Benefits of Estrogen Plus Progestin in Healthy Postmenopausal Women, *JAMA 288*:321–333.
4. MacLennan, A., S. Lester, and V. Moore, Oral Oestrogen Replacement Therapy Versus Placebo for Hot Flushes, *Cochrane Database Syst. Rev.*:CD002978 (2001).
5. Mattimoe, D., and W. Newton, Concerns of Women Regarding Hormone Replacement Therapy, *J. Fam. Pract. 48*:999–1000 (1999).
6. Hill, D.A., N.S. Weiss, and A.Z. LaCroix, Adherence to Postmenopausal Hormone Therapy During the Year after the Initial Prescription: A Population-Based Study, *Am. J. Obstet. Gynecol. 182*:270–276 (2000).

7. Hodsman, A.B., D.A. Hanley, and R. Josse, Do Bisphosphonates Reduce the Risk of Osteoporotic Fractures? An Evaluation of the Evidence to Date, *CMAJ. 166*:1426–1430 (2002).

8. Graham, D.Y., What the Gastroenterologist Should Know About the Gastrointestinal Safety Profiles of Bisphosphonates, *Dig. Dis. Sci. 47*:1665–1678 (2002).

9. Israel, D., and E.Q. Youngkin, Herbal Therapies for Perimenopausal and Menopausal Complaints, *Pharmacotherapy 17*:970–984 (1997).

10. Eisenberg, D.M., R.B. Davis, S.L. Ettner, S. Appel, S. Wilkey, M. Van Rompay, and R.C. Kessler, Trends in Alternative Medicine Use in the United States, 1990–1997: Results of a Follow-Up National Survey, *JAMA. 280*:1569–1575 (1998).

11. Riggs, B.L., S. Khosla, and L.J. Melton III, Sex Steroids and the Construction and Conservation of the Adult Skeleton, *Endocr. Rev. 23*:279–302 (2002).

12. Arjmandi, B.H., M.J. Getlinger, N.V. Goyal, L. Alekel, C.M. Hasler, S. Juma, M.L. Drum, B.W. Hollis, and S.C. Kukreja, Role of Soy Protein with Normal or Reduced Isoflavone Content in Reversing Bone Loss Induced by Ovarian Hormone Deficiency in Rats, *Am. J. Clin. Nutr. 68*:1358S–1363S (1998).

13. Arjmandi, B.H., R. Birnbaum, N.V. Goyal, M.J. Getlinger, S. Juma, L. Alekel, C.M. Hasler, M.L. Drum, B.W. Hollis, and S.C. Kukreja, Bone-Sparing Effect of Soy Protein in Ovarian Hormone-Deficient Rats is Related to its Isoflavone Content, *Am. J. Clin. Nutr. 68*:1364S–1368S (1998).

14. Potter, S.M., J.A. Baum, H. Teng, R.J. Stillman, N.F. Shay, and J.W. Erdman, Jr., Soy Protein and Isoflavones: Their Effects on Blood Lipids and Bone Density in Postmenopausal Women, *Am. J. Clin. Nutr. 68*:1375S–1379S (1998).

15. Alekel, D.L., A.S. Germain, C.T. Peterson, K.B. Hanson, J.W. Stewart, and T. Toda, Isoflavone-Rich Soy Protein Isolate Attenuates Bone Loss in the Lumbar Spine of Perimenopausal Women, *Am. J. Clin. Nutr. 72*:844-852 (2000).

16. Draper, C.R., M.J. Edel, I.M. Dick, A.G. Randall, G.B. Martin, and R.L. Prince, Phytoestrogens Reduce Bone Loss and Bone Resorption in Oophorectomized Rats, *J. Nutr. 127*:1795–1799 (1997).

17. Picherit, C., B. Chanteranne, C. Bennetau-Pelissero, M.J. Davicco, P. Lebecque, J.P. Barlet, and V. Coxam, Dose-Dependent Bone-Sparing Effects of Dietary Isoflavones in the Ovariectomised Rat, *Br. J. Nutr. 85*:307–316 (2001).

18. Picherit, C., V. Coxam, C. Bennetau-Pelissero, S. Kati-Coulibaly, M.J. Davicco, P. Lebecque, and J.P. Barlet, Daidzein is More Efficient than Genistein in Preventing Ovariectomy-Induced Bone Loss in Rats, *J. Nutr. 130*:1675–1681 (2000).

19. Fanti, P., M.C. Monier-Faugere, Z. Geng, J. Schmidt, P.E. Morris, D. Cohen, and H.H. Malluche, The Phytoestrogen Genistein Reduces Bone Loss in Short-Term Ovariectomized Rats, *Osteoporos. Int. 8*:274–281 (1998).

20. Ward, W.E., Y.V. Yuan, A.M. Cheung, and L.U. Thompson, Exposure to Purified Lignan from Flaxseed (*Linum usitatissimum*) Alters Bone Development in Female Rats, *Br. J. Nutr. 86*:499–505 (2001).

21. Ward, W.E., Y.V. Yuan, A.M. Cheung, and L.U. Thompson, Exposure to Flaxseed and Its Purified Lignan Reduces Bone Strength in Young but Not Older Male Rats, *J. Toxicol. Environ. Health A. 63*:53–65 (2001).

22. Babu, U.S., G.V. Mitchell, P. Wiesenfeld, M.Y. Jenkins, and H. Gowda, Nutritional and Hematological Impact of Dietary Flaxseed and Defatted Flaxseed Meal in Rats, *Int. J. Food Sci. Nutr. 51*:109–117 (2000).

23. Kaup, S.M., S.C. Hight, S.M. Ahn, J.I. Rader, Flaxseed and Mineral Metabolism in Rats, in *Proceedings of the 55th Flax Institute,* edited by J. Carter, U.S. Flax Institute, Fargo, North Dakota, 1994, pp. 29–38.

24. Arjmandi, B.H., S. Juma, E.A. Lucas, L.L. Wei, S. Venkatesh,, and D.A. Khan, Effects of Flaxseed Supplementation on Bone Metabolism in Postmenopausal Women, in *Proceedings of the 57th U.S. Flax Institute,* edited by J. Carter, U.S. Flax Institute, Fargo, North Dakota, 1998, pp. 65–74.

25. Lucas, E.A., R.D. Wild, L.J. Hammond, D.A. Khalil, S. Juma, B.P. Daggy, B.J. Stoecker, and B.H. Arjmandi, Flaxseed Improves Lipid Profile Without Altering Biomarkers of Bone metabolism in Postmenopausal Women, *J. Clin. Endocrinol. Metab. 87*:1527–1532 (2002).

26. Brooks, J., W.E. Ward, J. Hilditch, J. Lewis, L. Nickell, E. Wong, and L.U. Thompson, Flaxseed, but Not Soy Significantly Altered Urinary Estrogen Metabolite Excretion in Healthy Post-Menopausal Women, *FASEB J. 16*:A1005 (2002).

27. Bilezikian, J.P., The Role of Estrogens in Male Skeletal Development, *Reprod. Fertil. Dev. 13*:253–259 (2001).

28. Miyaura, C., K. Toda, M. Inada, T. Ohshiba, C. Matsumoto, T. Okada, M. Ito, Y. Shizuta, and A. Ito, Sex- and Age-Related Response to Aromatase Deficiency in Bone, *Biochem. Biophys. Res. Commun. 280*:1062–1068 (2001).

29. Oz, O.K., G. Hirasawa, J. Lawson, L. Nanu, A. Constantinescu, P.P. Antich, R.P. Mason, E. Tsyganov, R.W. Parkey, J.E. Zerwekh, and E.R. Simpson, Bone Phenotype of the Aromatase Deficient Mouse, *J. Steroid Biochem. Mol. Biol. 79*:49–59 (2001).

30. Rochira, V., A. Balestrieri, M. Faustini-Fustini, and C. Carani, Role of Estrogen on Bone in the Human Male: Insights from the Natural Models of Congenital Estrogen Deficiency, *Mol. Cell. Endocrinol. 178*:215-220 (2001).

31. Tou, J.C., J. Chen, and L.U. Thompson, Dose, Timing, and Duration of Flaxseed Exposure Affect Reproductive Indices and Sex Hormone Levels in Rats, *J. Toxicol. Environ. Health A. 56*:555–570 (1999).

32. Tou, J.C., J. Chen,, and L.U. Thompson, Flaxseed and Its Lignan Precursor, Secoisolariciresinol Diglycoside, Affect Pregnancy Outcome and Reproductive Development in Rats, *J. Nutr. 128*:1861–1868 (1998).

33. Ward, W.E., J. Chen, and L.U. Thompson, Exposure to Flaxseed or Its Purified Lignan During Suckling Only or Continuously Does Not Alter Reproductive Indices in Male and Female Offspring, *J. Toxicol. Environ. Health A. 64*:567–577 (2001).

34. Turner, C.H., Biomechanics of Bone: Determinants of Skeletal Fragility and Bone Quality, *Osteoporos Int. 13*:97–104 (2002).

35. Turner, C.H., and D.B. Burr, Basic Biomechanical Measurements of Bone: A Tutorial, *Bone 14*:595–608 (1993).

36. Adlercreutz, H., C. Bannwart, K. Wahala, T. Makela, G. Brunow, T. Hase, P.J. Arosemena, J.T. Kellis Jr., and L.E. Vickery, Inhibition of Human Aromatase by Mammalian Lignans and Isoflavonoid Phytoestrogens, *J. Steroid Biochem. Mol. Biol. 44*:147–153 (1993).

37. Santti, R., S. Makela, L. Strauss, J. Korkman, and M.L. Kostian, Phytoestrogens: Potential Endocrine Disruptors in Males, *Toxicol. Ind. Health 14*:223–237 (1998).

38. Watkins, B.A., Y. Li, H.E. Lippman, and M.F. Seifert, Omega-3 Polyunsaturated Fatty Acids and Skeletal Health, *Exp. Biol. Med. (Maywood) 226*:485–497 (2001).

39. Watkins, B.A., Y. Li, and M.F. Seifert, Nutraceutical Fatty Acids as Biochemical and Molecular Modulators of Skeletal Biology, *J. Am. Coll. Nutr. 20*:410S–416S (2001).

40. Kardinaal, A.F., M.S. Morton, I.E. Bruggemann-Rotgans, and E.C. van Beresteijn, Phyto-Oestrogen Excretion and Rate of Bone Loss in Postmenopausal Women, *Eur. J. Clin. Nutr.* 52:850–855 (1998).

41. Kim, M.K., B.C. Chung, V.Y. Yu, J.H. Nam, H.C. Lee, K.B. Huh, and S.K. Lim, Relationships of Urinary Phyto-Oestrogen Excretion to BMD in Postmenopausal Women, *Clin. Endocrinol. (Oxf)*. 56:321–328 (2002).

42. Cummings, S.R., D. Bates, and D.M. Black, Clinical Use of Bone Densitometry: Scientific Review, *JAMA.* 288:1889–1897 (2002).

43. Sidani, M., and J. Campbell, Gynecology: Select Topics, *Prim. Care* 29:297–321, vi (2002).

44. Lemay, A., S. Dodin, N. Kadri, H. Jacques, and J.C. Forest, Flaxseed Dietary Supplement Versus Hormone Replacement Therapy in Hypercholesterolemic Menopausal Women, *Obstet. Gynecol.* 100:495–504 (2002).

45. Brzezinski, A., H. Adlercreutz, R. Shaoul, A. Rosler, A. Shmueli, V. Tanos, and J.G. Schenker, Short-Term Effects of Phytoestrogen-Rich Diet on Postmenopausal Women, *Menopause* 4:89–94 (1997).

46. Wilcox, G., M.L. Wahlqvist, H.G. Burger, and G. Medley, Oestrogenic Effects of Plant Foods in Postmenopausal Women, *Br. Med. J.* 301:905–906 (1990).

47. Dalais, F.S., G.E. Rice, M.L. Wahlqvist, M. Grehan, A.L. Murkies, G. Medley, R. Ayton, and B.J. Strauss, Effects of Dietary Phytoestrogens in Postmenopausal Women, *Climacteric.* 1:124–129 (1998).

48. Lewis, J.E., L. Nickell, L.U. Thompson, J.P. Szalai, E.Y.Y. Wong, and J.R. Hilditch, The Effect of Dietary Soy and Flax on Menopause Quality of Life—a RCT, *Menopause* 9:488 (2002).

49. Hilditch, J.R., J. Lewis, A. Peter, B. van Maris, A. Ross, E. Franssen, G.H. Guyatt, P.G. Norton, and E. Dunne, A Menopause-Specific Quality of Life Questionnaire: Development and Psychometric Properties, *Maturitas* 24:161–175 (1996).

Chapter 18

Effect of Flaxseed and α-Linolenic Acid on Inflammatory Diseases and Immune Function

Les G. Cleland and Michael J. James

Rheumatology Unit, Royal Adelaide Hospital, Adelaide, Australia

Introduction

Standards of Evidence

When considering the influence of nutrients on inflammatory diseases and on protective inflammation, there are a number of generic issues to consider. The first relates to evidence and standards of proof. With regard to treatment for diseases, the highest level of evidence is the randomized, controlled trial. Treatments that consistently show efficacy in this test can be regarded as being effective. Some treatments may have been shown to favorably influence events that contribute to a disease process but may not have been tested in trials of sufficient power to assess efficacy. Such treatments can be considered candidate interventions only.

However, it must be recognized that preventive effects are often inferred based on the findings of epidemiological or observational studies rather than randomized controlled trials. This evidence has a certain currency, particularly when free of obvious confounders and when coupled with studies of mechanisms that provided a strong rationale for biological plausibility. However, it does not carry the same authority as a randomized trial, in which subjects are matched for all important variables other than the test interventions.

Prevention vs. Cure

It needs to be recognized that nutritional factors, by their nature, are generally more suited to delivering preventive than therapeutic benefit. The basis for this assertion is that overt diseases generally involve gross alterations in physiology that materials, with sufficiently modest pharmacological actions to be used as foods, are unlikely to redress alone. By contrast, certain food components may have a preventive effect, demonstrable at the level of a population sample as reducing the frequency of emergence of a disease.

This effect can best be considered in terms of causality, with a cause being defined as a component of conditions that influence the likelihood that an index event will occur. (A disease can properly be regarded as an event, or class of events, because it formally fulfills criteria for an event in having location within afflicted individuals and duration within their biography.) A component whose presence makes an event more likely (positively causal) is called a risk factor, and a component whose presence makes the event less likely (negatively causal), a preventive factor. Obversely, the removal of the risk factor will be preventive (negatively causal), and removal of the preventive factor will increase risk (positively causal). Such factors in themselves generally are neither sufficient to cause the disease, nor necessary for its occurrence.

Although these considerations may seem abstruse, they are fundamental to understanding how factors such as food components can influence disease. It is important to note that they lead to the implication that causality is bidirectional and that any candidate factor may increase or decrease risk for a particular event or disease. Furthermore, factors that increase risk for one event or disease may decrease risk for another and *vice versa*.

Dietary Change: Market Forces vs. Expert Opinion

In light of the above considerations, it can be argued that failure to implement interventions that deliver overall benefit increases risk unnecessarily. This is especially pertinent in nutritional matters where unguided market-driven changes in diet, particularly with regard to processed food, can by displacement and omission increase health risks. Such instances can in principle be remedied by informed guidance or edict. Paradoxically, substantial market-driven changes in Western diets have occurred either through ignorance or indifference to nutrition considerations. These have led to dietary changes that have been defined largely in retrospect when they have been accepted as the *status quo* that conservative scientific opinion has displayed reluctance to redress.

Importance of Context

A further matter that needs to be considered is the influence of a factor within a particular context. For example, a nutrient as a treatment for a disease may not have sufficient potency to yield worthwhile benefit alone, but within the context of other treatments may improve efficacy, reduce toxicity, reduce drug requirements, or otherwise improve outcomes.

Nontrivial Constituents in Recommended Food

Another matter to be considered in attributing therapeutic properties to nutrients is the presence of candidate compounds in foods of interest, at levels that exceed substantially those present in mundane and staple food sources. Because foods are likely to be used prophylactically as well as being used as an adjunct to chemotherapy, the issue of possible collateral health effects also needs to be considered.

This review of the putative health benefits of the salient candidate anti-inflammatory components of flaxseed, α-linolenic acid (18:3n-3; ALA) and secoisolariciresinol diglycoside (SDG) has been prepared with the above considerations in mind.

α-Linolenic Acid

Abundance

ALA is the dominant fatty acid in flaxseed and comprises up to 55% of oil w/w in standard flax strains. Considering the generally low levels of dietary n-3 fats in Western diets, ingestion of foods containing flaxseed or flaxseed oil can have a substantial effect on dietary n-3 fatty acid intake. It should be noted that not all flaxseed cultivars contain a high level of ALA. For example, the chemically mutated variant Linola, which is the principle constituent of a product marketed in Australia as "linseed cereal," contains <2% ALA.

Conversion of ALA to Eicosapentaenoic Acid (EPA) and Docosahexaenoic Acid (DHA)

The evidence for anti-inflammatory properties of dietary n-3 fatty acids centers on clinical trials that have used fish oils rich in EPA (20:5n-3) and DHA (22:6n-3) as a source of n-3 fats (1). Although ALA is a metabolic precursor of EPA and DHA, most dietary ALA is oxidized and, especially within the context of typical Western diets that are rich in linoleic acid (18:2n-6, LA), little is converted to EPA or DHA (2). It is thus pertinent to consider conditions under which conversion of ALA to EPA can be increased and possible advantages of fortifying the diet with both ALA and EPA.

We have shown that replacing LA with oleic acid (18:1n-9) by using olive oil-based products to provide visible fats (spreads, cooking oil, dressings) increased the incorporation of EPA from fish oil into blood cells (3). Relative to an n-6 rich comparator oil, using flaxseed oil-based products increased cellular EPA directly and increased EPA levels further when a fish oil supplement was given concurrently. Notwithstanding, the levels of EPA achieved with the various treatments established the fish oil supplement as the prime determinant of EPA levels achieved (4).

Anti-Inflammatory Effects of Dietary n-3 Fats

Dietary fish oil supplements have been shown to reduce symptoms of rheumatoid arthritis (RA) when given as an adjunct to usual therapy, to reduce progress to renal failure in IgA nephropathy, and to reduce relapse rates substantially in Crohn's disease. Some forms of psoriasis also appear to improve with fish oil supplements. These findings have been reviewed elsewhere (5). To date, there are no published studies of flaxseed or flaxseed oil as a treatment for the above diseases. Based on rates of conversion of ALA to EPA observed in studies of healthy human volunteers, one would not anticipate an effect if this were solely mediated through EPA. However, we have shown that a flaxseed-rich diet caused a modest reduction in the inflammatory cytokines interleukin-1 and tumor necrosis factor-α (6).

As discussed above, the preventive effects of nutrients may be more important than their therapeutic effects. This seems to be the case with RA, where a modest increase in dietary n-3 fats as fish has been associated with a reduction of RA by more than 50% in a case-control study (7). The magnitude of the preventive effect relative to the dietary differences suggests that a similar effect might be seen with diets substantially enriched with ALA through use of flaxseed-based products. Appropriate studies are required to test this hypothesis.

Collateral Health Benefits

Risk for cardiovascular events is increased in RA. The cardioprotective effects of dietary n-3 fats, including ALA, are thus potentially of special benefit to sufferers from this disease (5).

Secoisolariciresinol Diglycoside

Abundance

SDG is the dominant lignan precursor present in flaxseed (8). Although variability in content between flaxseed strains of up to threefold has been observed (9), this is modest compared with differences between flaxseed and other commonly consumed cereals (a factor of more than 100-fold w/w higher in flaxseed and flaxseed meal) (10) and other plant foods (11). The lignan precursor present in processed foods that contain flaxseed is broadly proportional to their flaxseed content (12). As regards dietary lignan precursor intake, an average daily intake of 20 g fresh ground flaxseed delivers 12 mg daily (11), which contrasts with the average daily intake of postmenopausal women in the United States of 0.58 mg (13).

Metabolism and Biological Plausibility

SDG is a lignan phytoestrogen, which is deglycosylated to secoisolariciresinol in the large intestine by bacterial hydrolases. Secoisolariciresinol is, in turn, dehydroxylated and demethylated by enteric bacteria to form enterodiol, which is subject to oxidation to enterolactone (14). Enterolactone, enterodiol, and secoisolariciresinol account for 75–80% of the urinary metabolites of tritiated SDG (15). Enterolactone, can also be formed by bacterial action on the lignan precursor matairesinol, although this is a comparatively minor constituent of flaxseed (8). Enterolactone and enterodiol are metabolized in the liver to glucuronides (16), and in common with steroid hormones, they undergo an enterohepatic circulation (17).

The "mammalian lignans" enterodiol and enterolactone are members of the broader class known as phytoestrogens, which also includes the flavonoids and isoflavonoids. The phytoestrogens are biphenols, which have broad three-dimensional structural similarity with estrogen. The lignans and other phytoestrogens have been shown to exert agonist effects on cells expressing estrogen receptors *in vitro* at physiologically relevant concentrations (18) and to exert effects on estrogen biochemistry *in vivo* (19). Competitive binding to estrogen receptors has been demonstrated directly for some phytoestrogens (20,21). Binding affinities are generally weak relative to estradiol and remain to be defined for the mammalian lignans. On the other hand, plasma lignan levels are high relative to those of steroid hormones, especially in vegetarians (22). Dietary flaxseed has been shown to influence endogenous hormone metabolism in postmenopausal women by decreasing serum 17 β-estradiol and estrone sulfate and increasing serum prolactin concentrations (23).

Considering the influence estrogens appear to exert on the emergence and expression of autoimmune/inflammatory diseases, it is appropriate to consider the case for candidacy of lignans as preventives or adjuncts to treatment of inflammatory diseases.

Effect of Estrogens on Autoimmune and Inflammatory Diseases

The effects of steroid hormones on autoimmune diseases are intrinisically complex because they involve not simply the influence of particular hormones but interactions between different classes of steroids and other effects mediated by gender. *In vitro* studies and animal models of inflammatory diseases have been informative in defining mechanisms of action. For example, estrogens have been shown to promote B-cell responses and B-cell-dependent immune-complex-mediated disease (e.g., lupus nephritis in mice) but to suppress T-cell responses and T-cell-dependent pathology (e.g., collagen-induced arthritis in mice and rats) (24,25). As shown in MLR autoimmune mice, immune deviation induced by exogenous estrogens can aggravate some aspects of the disease (antibody-mediated glomerulonephritis), while ameliorating others (lymphocytic infiltration of salivary glands, T-cell responses to antigen and mitogen) (26,27). In terms of human disease, it is notable that the case-control study of Bengtsson and coworkers (28) did not establish exogenous estrogen as a risk for systemic lupus. In addition, a small double-blind crossover trial of the estrogen antagonist tamoxifen in lupus revealed no evidence of an ameliorative effect on clinical or laboratory indices of activity (29).

Inhibition of Platelet Activating Factor (PAF)

Synthetic compounds containing a lignan moiety have been shown to inhibit PAF, an action that has clear relevance to inflammatory diseases (30). Although the activity of lignan derivatives as PAF inhibitors has been established, the activity of naturally occurring lignans in this regard remains to be established.

Effects on Lymphocyte Proliferation

Plant lignans, although not those derived from flaxseed specifically, have been shown to inhibit lymphocyte proliferation to mitogen *in vitro* (31).

Cardiovascular Effects

Flaxseed feeding has been shown to reduce cardiac dysfunction and oxidative stress following experimentally induced endotoxic shock in dogs (32). SDG has been shown to be antiatherogenic and to have low density lipoprotein-cholesterol lowering and antioxidant effects in rabbits fed an atherogenic diet (33). It is notable that systemic inflammatory diseases, such as RA and lupus, confer increased special risk for cardio-vascular diseases (34), and accordingly the putative cardiovascular benefits of lignans and n-3 fatty acids may be especially important in patients with these diseases.

Protection Against Malignancies

Lignans have been reported to have preventive effects against sex hormone influenced cancers, thereby illustrating a potential collateral health benefit for those choosing to ingest these nutrients as a preventive or adjunct to treatment for inflammatory diseases (35). Differential risk reduction with dietary lignans seen in breast cancer (36) high-lights the potential for genetic factors to condition responses to nutrients in relation to a disease of interest. The suggestion that men may receive benefit from ingestion of lignan precursors was not supported by a recent longitudinal case-control study which failed to show a correlation between serum enterolactone and prostate cancer (37).

Possible Interactions Between Dietary Lignans and Antirheumatic Drugs

The use of long acting antirheumatic drugs that have antimicrobial properties in the treatment of RA (e.g., sulphasalazine, minocycline, hydroxychloroquine, methotrex-ate) could, to varying degrees, influence the conversion of SDG to mammalian lig-nans, which is mediated by enteric bacterial flora (38).

Clinical Studies of Lignan Precursors in Flax Meal in Arthritis

There are no published studies that allow an evaluation of the possible benefit of lig-nan precursors present in flaxmeal on inflammatory diseases.

Summary

Flaxseed contains two prime candidate components for benefit in relation to inflam-matory diseases, especially with regard to possible prevention. It is conceivable that these factors, which act via different mechanisms, may have additive or even syner-gistic effects. In this regard fresh ground flaxseed or flaxseed meal in processed foods may deliver more benefit than flaxseed oil alone taken either as a food component or a supplement. More studies of appropriate design are needed to address this speculation.

References

1. James, M.J., and L.G. Cleland, Dietary n-3 Fatty Acids and Therapy for Rheumatoid Arthritis, *Semin. Arthritis Rheum.* 27:85–97 (1997).
2. Brenna, J.T., Efficiency of Conversion of Alpha-Linolenic Acid to Long Chain n-3 Fatty Acids in Man, *Curr. Opin. Clin. Nutr. Metab. Care* 5:127–132 (2002).
3. Cleland, L.G., M.J. James, M.A. Neumann, M. D'Angelo, and R.A. Gibson, Linoleate Inhibits EPA Incorporation from Dietary Fish-Oil Supplements in Human Subjects, *Am. J. Clin. Nutr.* 55:395–399 (1992).

4. Mantzioris, E., M.J. James, R.A. Gibson, and L.G. Cleland, Dietary Substitution with an Alpha-Linolenic Acid-Rich Vegetable Oil Increases Eicosapentaenoic Acid Concentrations in Tissues, *Am. J. Clin. Nutr. 59*:1304–1309 (1994).

5. James, M.J., and L.G. Cleland, Fats and Oils: The Facts, Available at www.goldncanola.com.au, Accessed December 2002.

6. Caughey, G.E., E. Mantzioris, R.A. Gibson, L.G. Cleland, and M.J. James, The Effect on Human Tumor Necrosis Factor Alpha and Interleukin 1 Beta Production of Diets Enriched in n-3 Fatty Acids from Vegetable Oil or Fish Oil, *Am. J. Clin. Nutr. 63*:116–122 (1996).

7. Shapiro, J.A., T.D. Koepsell, L.F. Voigt, C.E. Dugowson, M. Kestin, and J.L. Nelson, Diet and Rheumatoid Arthritis in Women: A Possible Protective Effect of Fish Consumption, *Epidemiology 7*:256–263 (1996).

8. Meagher, L.P., G.R. Beecher, V.P. Flanagan, and B.W. Li, Isolation and Characterization of the Lignans, Isolariciresinol and Pinoresinol, in Flaxseed Meal, *J. Agric. Food Chem. 47*:3173–3180 (1999).

9. Thompson, L.U., S.E. Rickard, F. Cheung, E.O. Kenaschuk, and W.R. Obermeyer, Variability in Anticancer Lignan Levels in Flaxseed, *Nutr. Cancer 27*:26–30 (1997).

10. Axelson, M., J. Sjovall, B.E. Gustafsson, and K.D. Setchell, Origin of Lignans in Mammals and Identification of a Precursor from Plants, *Nature 298*:659–660 (1982).

11. Thompson, L.U., P. Robb, M. Serraino, and F. Cheung, Mammalian Lignan Production from Various Foods, *Nutr. Cancer 16*:43–52 (1991).

12. Nesbitt, P.D., and L.U. Thompson, Lignans in Homemade and Commercial Products Containing Flaxseed. *Nutr. Cancer 29*:222–227 (1997).

13. de Kleijn, M.J., Y.T. van der Schouw, P.W.Wilson, H. Adlercreutz, W. Mazur, D.E. Grobbee, and P.F. Jacques, Intake of Dietary Phytoestrogens Is Low in Postmenopausal Women in the United States: The Framingham Study (1–4), *J. Nutr. 131*:1826–1832 (2001).

14. Borriello, S.P., K.D.R. Setchell, M. Axelson, and A.M. Lawson, Production and Metabolism of Lignans by the Human Faecal Flora, *J. Appl. Bacteriol. 58*:37–43 (1985).

15. Rickard, S.E., and L.U. Thompson, Lignans in Homemade and Commercial Products Containing Flaxseed, *J. Nutr. 130*:2299–2305 (2000).

16. Brunner, G., F. Tegtmeier, D.N. Kirk, S. Wynn, and K.D. Setchell, Enzymatic Synthesis and Chromatographic Purification of Lignan Glucuronides, *Biomed. Chromatog. 1*:89–92 (1986).

17. Axelson, M., and K.D. Setchell, The Excretion of Lignans in Rats—Evidence for an Intestinal Bacterial Source for This New Group of Compounds, *FEBS Lett. 123*:337–342 (1981).

18. Wang, C., and M.S. Kurzer, Phytoestrogen Concentration Determines Effects on DNA Synthesis in Human Breast Cancer Cells, *Nutr. Cancer 28*:236–247 (1997).

19. Adlercreutz, H., Y. Mousavi, J. Clark, K. Hockerstedt, E. Hamalainen, K. Wahala, T. Makela, and T. Hase, Dietary Phytoestrogens and Cancer: *in vitro* and *in vivo* Studies, *J. Steroid Biochem. Mol. Biol. 41*:331–337 (1992).

20. Schmitt, E., W. Dekant, and H. Stopper, Assaying the Estrogenicity of Phytoestrogens in Cells of Different Estrogen Sensitive Tissues, *Toxicol. in vitro 15*:433–439 (2001).

21. Bowers, J.L., V.V. Tyulmenkov, S.C. Jernigan, and C.M. Klinge, Resveratrol Acts as a Mixed Agonist/Antagonist for Estrogen Receptors Alpha and Beta, *Endocrinology 141*:3657–3667 (2000).

22. Adlercreutz, H., T. Fotsis, J. Lampe, K. Wahala, T. Makela, G. Brunow, and T. Hase, Quantitative Determination of Lignans and Isoflavonoids in Plasma of Omnivorous and

Vegetarian Women by Isotope Dilution Gas Chromatography-Mass Spectrometry, *Scand. J. Clin. Lab. Invest. (Suppl) 215*:5–18 (1993).

23. Hutchins, A.M., M.C. Martini, B.A. Olson, W. Thomas, and J.L. Slavin, Flaxseed Consumption Influences Endogenous Hormone Concentrations in Postmenopausal Women, *Nutr. Cancer 39*:58–65 (2001).

24. Wilder, R.L., Hormones and Autoimmunity: Animal Models of Arthritis, *Baillieres Clin. Rheumatol. 10*:259–271 (1996).

25. Holmdahl, R., H. Carlsten, L. Jansson, and P. Larsson, Oestrogen Is a Potent Immunomodulator of Murine Experimental Rheumatoid Disease, *Br. J. Rheumatol. 28*:54–58 (1989).

26. Carlsten, H., N. Nilsson, R. Jonsson, and A. Tarkowski, Differential Effects of Oestrogen in Murine Lupus: Acceleration of Glomerulonephritis and Amelioration of T Cell-Mediated Lesions, *J. Autoimmun. 4*:845–856 (1991).

27. Carlsten, H., A. Tarkowski, R. Holmdahl, and L.A. Nilsson, Oestrogen Is a Potent Disease Accelerator in SLE-Prone MRL lpr/lpr Mice, *Clin. Exp. Immunol. 80*:467–473 (1990).

28. Bengtsson, A.A., L. Rylander, L. Hagmar, O. Nived, and G. Sturfelt, Risk Factors for Developing Systemic Lupus Erythematosus: A Case-Control Study in Southern Sweden, *Rheumatology 41*:563–571 (2002).

29. Sturgess, A.D., D.T. Evans, I.R. Mackay, and A. Riglar, Effects of the Oestrogen Antagonist Tamoxifen on Disease Indices in Systemic Lupus Erythematosus, *J. Clin. Lab. Immunol. 13*:11–14 (1984).

30. Shen, T.Y., Chemical and Biochemical Characterization of Lignan Analogs as Novel PAF Receptor Antagonists, *Lipids 26*:1154–1156 (1991).

31. Hirano, T., A. Wakasugi, M. Oohara, K. Oka, and Y. Sashida, Suppression of Mitogen-Induced Proliferation of Human Peripheral Blood Lymphocytes by Plant Lignans, *Planta Medica 57*:331–334 (1991).

32. Pattanaik, U., and K. Prasad, Reactive Oxygen Species and Endotoxic Shock: Effect of Dimethylthiourea, *J. Cardiovasc. Pharmacol. Ther. 3*:305–318 (1998).

33. Prasad, K., Reduction of Serum Cholesterol and Hypercholesterolemic Atherosclerosis in Rabbits by Secoisolariciresinol Diglucoside Isolated from Flaxseed, *Circulation 99*:1355–1362 (1999).

34. Riboldi, P., M. Gerosa, C. Luzzana, and L. Catelli, Cardiac Involvement in Systemic Autoimmune Diseases, *Clin. Rev. Allergy Immunol. 23*:247–261 (2002).

35. Adlercreutz, H., Phyto-Oestrogens and Cancer, *Lancet Oncol. 3*:364–373 (2002).

36. McCann, S.E., K.B. Moysich, J.L. Freudenheim, C.B. Ambrosone, and P.G. Shields, The Risk of Breast Cancer Associated with Dietary Lignans Differs by CYP17 Genotype in Women, *J. Nutr. 132*:3036–3041 (2002).

37. Stattin, P., H. Adlercreutz, L. Tenkanen, E. Jellum, S. Lumme, G. Hallmans, S. Harvei, L. Teppo, K. Stumpf, T. Luostarinen, M. Lehtinen, J. Dillner, and M. Hakama, Circulating Enterolactone and Prostate Cancer Risk: A Nordic Nested Case-Control Study, *Int. J. Cancer 99*:124–129 (2002).

38. Kilkkinen, A., P. Pietinen, T. Klaukka, J. Virtamo, P. Korhonen, and H. Adlercreutz, Use of Oral Antimicrobials Decreases Serum Enterolactone Concentration, *Am. J. Epidemiol. 155*:472–477 (2002).

Chapter 19

Effect of Flaxseed Consumption on Male and Female Reproductive Function and Fetal Development

Robert L. Sprando, Thomas F.X. Collins, and Paddy W. Wiesenfeld

Food and Drug Administration, Center for Food Safety and Applied Nutrition, 8301 Muirkirk Road, Laurel, Maryland 20708, USA

Introduction

Seed from the flax plant (*Linum usitatissimum*), also known as linseed, has been used for food in Europe and Asia since at least 5000 B.C.E. The fiber from the stem has been used for linen, cloth, and paper, and linseed oil meal and the flour remaining after the removal of oil from the seed have been used in animal feed. Neither the whole seed nor the meal has been considered a significant component of human food until recently. In the United States, flaxseed (FS) and flaxseed meal (FLM, partially defatted FS) have found market acceptability as a component in some cereals, specialty breads, cookies, and salad dressings (1,2). Its growing popularity is due to food components that may provide health benefits beyond basic nutrition (3–7). Among the reported potential health benefits associated with FS and/or FLM are decreased risk of cardiovascular disease (1,6,8–13), decreased risk of cancer, particularly breast and prostate (6,14–19), anti-inflammatory activity (20–24), laxative effect (10), and alleviation of menopausal symptoms and osteoporosis (4).

Despite the beneficial effects reported for FS, some controversy still exists regarding the effect of FS consumption on the function and development of the reproductive system. This is due, in part, to the complex nature of that system. FS contains nutrient, non-nutrient (including phytoestrogens), and antinutrient components. Nutrients and non-nutrients can have beneficial and/or adverse effects depending upon dose, timing, and length of exposure. Among the bioactive, bioavailable nutrients in FS is the n-3 fatty acid, α-linolenic acid (ALA, C18:3), a polyunsaturated fatty acid (PUFA), which can alter fatty acid composition and function (25; Wiesenfeld *et al.*, unpublished data). Multiple double bonds in PUFA make these fatty acids subject to the abstraction of hydrogen from PUFA that initiates the process of lipid peroxidation. If FS diets increase oxidative stress, (e.g., due to increased PUFA oxidation), the result could be a decrease in antioxidant capacity in key tissues such as serum, liver, and heart. Ingestion of PUFA or increase in PUFA by fatty acid metabolism enhances the susceptibility of tissues to lipid peroxidation, and increases the requirement for vitamin E (27). Vitamin E disrupts the formation of reactive oxygen species by capturing the radical. The radical scavengers thus terminate the chain reaction of free radical damage.

Data from several studies support the hypothesis that the increased consumption of FS can lead to an increase in lipid peroxidation. For example, rats fed diets containing linseed oil (FS oil) were reported to increase lipid peroxidation as assessed by urinary, heart, and liver tribarbituric acid reacting substances (TBAR) (28). Additionally, Ratnayake *et al.* (8) reported that rats fed 40% FS for 90 d had lower tissue vitamin E levels than rats fed FS-free diets. A significant decrease in liver and heart vitamin E has been reported in 40% FS-fed rats compared with 20% FS- and 26% FLM-fed rats, suggesting an FS dose-and-time-dependent depletion of stored vitamin E in liver and heart (26). A reduction in vitamin A, another major antioxidant, in both the liver and heart of rats fed either 26% FLM diets or 40% FS diets has been reported (26). Human studies using hyperlipidemic subjects fed FS or FLM diets reported a significant reduction in serum protein thiol groups compared with subjects on a self-selected National Cholesterol Education Program II diet (13), suggesting oxidative stress. Results from human studies suggested that FS plasma antioxidants and lipid hydroperoxides were unchanged (10). In contrast, Babu *et al.* (29) reported an increase in liver vitamin E in rats fed 10% FS. Taken together the above studies suggest that eating a high FS diet for an extended period of time may result in an increase in oxidative stress possibly resulting in a decrease in antioxidant vitamins, and an extra demand on superoxide dismutase and glutathione peroxidase.

Antinutrient components of FS include linatine, phytic acid, and cyanogenic glycosides (CG). Because linatine binds to vitamin B_6, eating diets rich in FS could cause a B_6 deficiency (3), leading to increased homocysteine and renal insufficiency (30). Phytic acid is known to bind positively charged minerals such as zinc and calcium (3,31) which could result in a deficiency in minerals which could then affect bone development.

Other important antinutrients are CG, and these compounds can be found in different parts of plants (e.g., roots, seeds, and leaves) (32,33) and can function as feeding deterrents or feeding stimulants in various insect species. CG are widely distributed in the plant kingdom, vary widely in the amount of HCN produced (32), and are the products of secondary metabolism of plants. The types and amounts of CG found in FS are dependent on the variety of FS, the growing season, the environmental conditions, and the analytical method used (34). The major CG present in FS are linustatin, neolinustatin, and methyl ethyl ketone cyanohydrin (34,35). Vetter reported the CG content of FS to range between 16–63 mg/kg fresh weight (33). The presence of CG in FS is of concern, because they may release cyanide upon hydrolysis (36). CG ingested by humans is easily detoxified if a well-balanced diet is maintained with adequate protein levels, especially sulphur-containing amino acids such as methionine and cysteine (32). Humans fed FS excrete thiocyanate. Urinary thiocyanate levels increased twofold in healthy female volunteers fed 50 g FS p/d (37) for 4 wk. Adverse health effects could result if large amounts of raw or unprocessed FS were eaten for an extended period of time by individuals with a low protein diet (34,35). Dietary exposure from CG in insufficiently processed seeds (e.g., cassava) is reported to be a contributing factor in child growth retardation (38). CG (linamarin, linustatin, neolinustatin) are undetectable in baked products, suggesting that if FS or other cyanogenic foods are processed and/or baked, the possibility of an adverse health effect from CG would be greatly reduced (37). Further studies that address these types of antinutritional effects of FS are needed.

Lignans are phytoestrogenic compounds possessing a 2,3-dibenzylbutane structure and can be found in whole grains, legumes, vegetables, and seeds, especially FS (39). The phytoestrogenic component of FS is found in the seed coat which contains 75–800 times more lignans than any plant food source studied to date (39). FS lignans appear to have weak antioxidant and estrogenic properties. The weak antioxidant properties (3) may counteract some deleterious oxidative effects of ALA and its long chain PUFA metabolites. *In vitro* data have shown that FS lignans have structure-specific antioxidant activity (40). In at least one study, where subjects were fed full-fat FS, there was no significant reduction in TBAR despite a high concentration of lignans as potential phenolics (10,37). Future studies will have to determine whether lignans in FS provide adequate oxidant protection.

The predominant plant lignan found in FS is secoisolariciresinol diglycoside (SDG); however, other lignans such as matairesinol have also been detected but at lower concentrations. Intestinal microflora convert both secoisolariciresinol and matairesinol to the mammalian lignans enterodiol (ED) and enterolactone (EL) (41,42) which enter into the hepatic circulation and are subsequently excreted as glucuronide or sulphate conjugates (43,44). The level of SDG in FS is variable (between 0.6–1.8 g/100 g) (45). Numerous reports have suggested that SDG has antioxidant properties (40,46,47). Natural antioxidants, such as SDG, may function as reducing agents, free radical scavengers, inhibitors of singlet oxygen formation, or deactivators of pro-oxidant metals (46). SDG has been shown to prevent lipid peroxidation of liver homogenates in a dose-dependent manner by scavenging exogenously generated hydroxyl radicals (47). At a 100 µM concentration, SDG, ED, and EL all lowered lipid peroxidation in both lipid emulsions and aqueous media, and ED and EL were most efficient in reducing deoxyribose oxidation and DNA strand breakage (40). Rabbits fed 15 mg SDG/kg b.w. reduced hypercholesterolemic atherosclerosis by reducing lipid peroxidation product and increasing antioxidant reserves (45). Future studies are needed in which animals are first fed SDG and then challenged with oxidative stress to determine if SDG prevents oxidative damage *in vivo*.

Phytoestrogens, Estrogen Receptors, and Fetal Development

The consumption of dietary substances capable of altering hormone levels is of particular concern not only during pregnancy but also during postnatal development. Pregnancy is a hormone-sensitive period for mother and offspring. Mammalian lignans, because of their chemical similarity to estrogens, exhibit hormonelike properties that mimic endogenous estrogens, and may act as weak estrogens or antiestrogens. Functioning as antiestrogens, mammalian lignans compete for estrogen receptor sites and may thus restrict access to pathways that produce potentially adverse pathogenesis (17,18,48) or to pathways required for the normal maintenance and/or development of the male and/or female reproductive system. High levels of maternal estrogen are needed during pregnancy to allow implantation, enhance uterine growth as the fetus grows, and initiate parturition (49). Sexual differentiation of the fetal genitalia and the central nervous system is also hormone sensitive. Alteration of estrogen balance, through inadequate or excessive estrogen, can disrupt pregnancy.

Recent animal experiments with estrogen receptor knock-out mice have indicated that estrogens are important not only for female sexual differentiation but, in addition to androgens and Muellerian inhibiting hormone, play an essential role in normal male sexual differentiation. In the absence of the estrogen, testis weight, sperm counts, and fertility are reduced (50). Animal experiments have also confirmed that estrogens administered during the critical periods of development result in long-term effects on the male reproductive tract, including adenocarcinoma of the rat testis (51–53), epididymal cysts and epididymal inflammation (54,55), persistent Muellerian duct remnants (56), increased incidence of undescended testes, and testicular cancer (54). These studies have suggested that the effects of estrogens on the male sex accessory glands are time- and organ-dependent. For example, perinatal and neonatal exposure to estrogens (estradiol-17β, estradiol benzoate, ethinyl estradiol, or estradiol valerate) produces alterations in the male genital tract, including undescended testes, testicular hypotrophy, testicular teratoma, hyperplasia of Leydig cells, and hypotrophic prostate, seminal vesicles, and other sex accessory glands (57–64). Additionally, neonatal estradiol benzoate administration decreased organ weights and DNA content of the seminal vesicles, ventral and dorsal prostates, epididymides, and coagulating glands of treated animals. Similar, but less severe, effects were observed in animals exposed to estrogens at the time of puberty. Neonatal estrogenization had its greatest effect on ventral prostate and seminal vesicles by reducing their secretory function as well as their secretory protein level (58).

Two possible mechanisms explaining these effects have been proposed: (i) Exogenous estrogens may disturb male hormone secretion and metabolism, exerting their effect by acting directly on the hypothalamic-pituitary-testis axis to suppress the secretion of gonadotropin and testosterone or by affecting the metabolism of androgens in the male reproductive tract (58). (ii) Estrogens may interact with the estrogen receptors or affect the distribution and expression of the androgen receptor or gonadotropin receptors. Two major estrogen receptors with biological activity have been identified, and each is located on a different chromosome. Estrogen receptor alpha (ER-alpha) has been localized on chromosome 6, and estrogen receptor beta (ER-beta) has been localized on chromosome 14 (65). The relative amount of each receptor varies among tissues. ER-beta is believed to be more generally expressed and plays an important role in the physiology of several tissues including the cardiovascular system, central nervous system, immune system, urogenital system (including the kidney), gastrointestinal tract, and lungs (30,65–70). In contrast, ER-alpha is the predominant receptor in a few tissues and appears to be involved in reproductive events. Although ER-beta is present, ER-alpha appears to be the predominant receptor in the uterus and mammary tissue and may play a role in preventing cellular proliferation. It has been suggested that ER-beta is involved in the control of antioxidant-regulated genes, the products of which are known to control the concentrations of free radicals and reactive oxygen in the cell. Phytoestrogens, in general, appear to have a greater affinity for ER-beta than for ER-alpha (71) and therefore may have the potential to adversely affect the development of both the male and female urogenital systems and normal reproductive function in both the male and female.

Flaxseed, Reproductive Function, and Fetal Development

In the first study by Tou *et al.* (72), pregnant female rats were randomly assigned to the basal diet or the basal diet supplemented with 10% FS, 5% FS, or a daily gavage of 1.5 mg SDG in 1 mL distilled water. The SDG level was estimated to be approximately equivalent to that given to rats on the 5% FS diet. Each group contained seven pregnant females. Each animal was monitored daily. At the time of littering, litter size, birth weight, and live birth index were measured. After birth, pup survival to postnatal day (PND) 21 and percentage of females were measured. After birth, the rat dams continued to consume their respective diets. Maternal weight, food intake, and offspring weight were measured every other day during PND 0–21. At PND 21, all dams and two offspring per litter (one of each sex) were euthanized. Reproductive and major organs were removed and weighed. All remaining offspring were weaned and fed the basal diet so that the offspring were exposed only during gestation and lactation. At PND 50 and 132, subgroups of male and female offspring were euthanized and the effects of FS or SDG exposure were measured on reproductive and major organs. Each subgroup consisted of 2–3 offspring (1–2 of each sex) per litter.

To determine whether mammalian lignans could be transferred to the offspring through the dam's milk, Tou *et al.* (72) gavaged three lactating dams from the SDG group with labeled SDG, and three lactating dams also from the SDG group with unlabeled SDG on PND 20. The dams and their offspring were euthanized 24 h later. Mean total body radioactivity of offspring given labeled SDG was significantly greater than that of the other groups of offspring, indicating lignan transfer *via* milk.

Offspring weight gain was measured every other day. Male and female offspring were distinguished by the anogenital distance (AGD). Hormone imbalance in male or female offspring due to *in utero* FS consumption was determined by the difference in the AGD at PND 3 when compared with the controls. Effect on pup genitalia during lactation was determined from the AGD difference between PND 21 and 3. AGD measurements were adjusted for body size by calculating relative AGD (PND 21 AGD – PND 3 AGD)/body weight gain from PND 3 to 21. Puberty onset was determined in female offspring by visible opening of the vaginal aperture. Estrous cycling was determined by vaginal smears on PND 40–50 and PND 100–132.

FS diet (5%) and SDG produced no significant differences in maternal feed intake, weight gain, gestation length, or pregnancy outcomes (litter size, live birth index, postnatal pup survival, or sex distribution among offspring). FS diet (10%) caused significantly lowered mean birth weight of the offspring, but other measured parameters were similar to the basal ration (control).

At PND 3, 10% FS caused shortening of AGD in female offspring. When calculated as relative AGD, the effect of 10% FS persisted only in the female offspring. From PND 3–21, 10% FS also decreased body weight gain and decreased AGD. FS (10%) resulted in female puberty onset at a significantly earlier age and lighter body weight, whereas FS (5%) resulted in puberty onset at an older age but at the same body weight as the basal diet group.

All female offspring exhibited estrous cycling at PND 50, but the female offspring of the dams fed 10% FS had lengthened estrous cycles due to prolonged time in the estrus phase. By PND 132, 20% (2/10) of the 10%-treated offspring were acyclic due to persistent estrus, but 14.3% (1/7) of the SDG-treated rats and 16.7% (1/6) of the 5%-treated rats were acyclic as a result of persistent diestrus. All offspring (7/7) of the control rats were cycling (5.4 d) at PND 132.

Only the reproductive organs differed in weight at PND 21, 50, and 132. At PND 21, the relative uterine weight was significantly greater in all treated groups than in the control group, but the difference disappeared by PND 50. At PND 50 and 132, relative ovarian weight was significantly greater in the 10% FS group than in the other groups. The authors concluded that the shortened AGD, greater uterine and ovarian relative weights, earlier age and lighter body weight at puberty, lengthened estrous cycle, and persistent estrus suggested estrogenic effects at 10% FS. At 5% FS, the reduced immature ovarian relative weight, delayed puberty, and lengthened diestrus suggested an antiestrogenic effect. SDG produced effects similar to those of 5% FS, and that suggested that lignans were responsible for the observed effects. These results caused the authors to caution against the consumption of FS during pregnancy and lactation.

Lower body weights were observed in male offspring treated *in utero* with 10% FS but were not observed in those animals treated with the basal diet, 5% FS diet, or 1.5 mg SDG diets. Effects were not observed on the male AGD. A decrease in body weight gain and AGD was observed in male rats exposed to 10% FS during lactation. Effects were not observed in male offspring exposed to the other diets. Exposure to FS or SDG had no effect on male accessory sex glands or testis weights at PND 21 or 50; however, a greater accessory sex gland and prostate relative weight was observed at PND 132 in male offspring treated with 10% FS. In contrast, exposure of male rats during pregnancy and lactation to 5% FS resulted in a mild inhibition of prostate growth at PND 132. Effects were not observed at PND 132 in male offspring treated with 1.5 mg SDG diets. Although the prostate is composed of a dorsal lobe, dorso-lateral lobe, a ventral lobe, and coagulating gland, effects were observed only in the ventral prostate. These effects included extensive cell proliferation and increased amounts of secretory epithelial cells forming an increased number of papillary infoldings and secondary projections in the 10% FS-treated offspring. In contrast, the epithelium of those animals treated with 5% FS appeared cuboidal or flattened with fewer infoldings. Some areas contained an increased stroma. The epithelium of those rats exposed to 1.5 mg SDG was indistinguishable from that of rats on the basal diet.

In summary, feeding rat dams 5 or 10% FS or SDG levels comparable to that present in 5% FS had no effect on pregnancy except for lower birth rates in offspring from the 10% FS treatment group. In contrast, dose-dependent hormonal-related effects were observed in the offspring of dams fed 5 or 10% FS or SDG levels comparable to that present in 5% FS during gestation and lactation. The authors suggested that these effects were produced by the mammalian lignans derived from SDG and transferred from the rat dams to their offspring through their milk and not from the ALA-rich oil.

A study of the effects of FS and SDG on estrous cycling in rats by Orcheson *et al.* (73) showed that supplementation of a diet with 2.5, 5, or 10% FS or SDG (0.75, 1.5, or 3.0 mg/d) for 4 wk produced a dose-related cessation, irregularity, or lengthening of the cycle in rats.

In a second FS reproduction study, Tou *et al.* (74) investigated whether the timing of treatment affected sex hormone levels and reproductive indices. Pregnant female rats (5 p/group) were fed a basal (control) diet, 5, or 10% dietary FS from gestation through PND 21. At PND 21, one male and one female per litter were euthanized to determine the effect on sex organs (ovaries and uterus in females, serum estradiol and/or testosterone in males). All remaining offspring were weaned. The offspring of the basal diet (control) animals were continued on basal diet, or were fed 5 or 10% FS from the time of weaning to maturation at PND 50 or 132 (lifetime). At PND 50 and 132, subgroups of male rats (5–7) and female rats (5–6) from all the rat dams were euthanized. All female offspring were killed in the estrus phase of the estrous cycle. Reproductive and other organs were removed and weighed to determine the effect of dose and timing of FS exposure.

In female rats, the age and weight at puberty onset were noted until PND 50 to determine the first full estrous cycle. Vaginal smears were also taken on PND 100–132 to determine estrous cycling. Rat estrous cycle length was determined as the number of days needed to complete the four phases of the estrous cycle. A normal cycle was defined as being 4–5 d and having 1–2 d of estrus. Cycles were considered prolonged if rats remained in one phase for more than 3 d and acyclic if they remained in one phase for more than 10 d.

Exposure to FS after weaning had no significant effect on age or weight at puberty onset, and no effect on the length of the first full estrous cycle. At PND 132, female rats treated with FS after weaning had normal estrous cycle lengths and all animals remained cycling. These results indicated that exposure to FS (5 or 10%) after weaning caused no notable reproductive effects.

Lifetime exposure to 10% FS resulted in puberty onset at an earlier age and lighter body weight, whereas lifetime exposure to 5% FS resulted in delayed puberty onset and puberty onset at a heavier body weight. Lifetime exposure to 10% FS lengthened the first estrous cycle due to prolonged estrus, whereas lifetime exposure to 5% flaxseed had no effect on the length of the first estrous cycle. At PND 132, one animal in the 10% FS group (1/6) had persistent estrus, and 5/6 animals were still cycling, but the cycles were significantly longer than in basal diet animals (7.8 *vs.* 5.4 d).

At PND 21, free serum estradiol levels were significantly lower in the 5% flaxseed than in the basal diet or 10% flaxseed offspring. At PND 50 and 132, female offspring of the 10% FS-fed groups had higher estradiol levels than other treatment groups, and there was a positive relationship between serum estradiol and relative ovarian weight.

Of all organs examined, only sex organs showed weight differences among groups. Relative immature uterine weight was significantly higher than in the basal diet group in female rats treated with 5 or 10% FS from gestation to PND 21, but lifetime 5 or 10% exposure did not result in significant differences in relative uterine weight at PND 50 and 132. Relative immature ovarian weight was markedly lower in rats exposed to 5% FS from gestation to PND 21 than in those on the basal diet or 10% FS diet, but lifetime exposure to 10% FS produced higher ovarian weight in female offspring than in the control at PND 50 and 132.

The authors of the second study (74) concluded that the effects of FS on reproductive indices and sex hormone levels depend on the dose, timing, and duration of exposure. In female rats, lifetime exposure to 10% FS produced higher relative ovarian weight, higher serum estradiol, earlier puberty onset, and lengthened estrous cycle compared with the basal diet controls. When exposure took place after weaning, no changes were observed, indicating that the gestation and lactation stages are a critical period for reproductive effects.

In the male offspring there was no effect on serum testosterone or serum estradiol at weaning (PND 21) or puberty (PND 51). In those male offspring exposed to 10% FS until PND 132, relative sex organ weights, seminal vesicle weights, prostate and testis weights, as well as testosterone and serum estradiol levels were higher compared with animals treated with the basal diet. In contrast, a lifetime exposure to 5% FS resulted in a reduced prostate weight and cell proliferation, but serum hormone levels (testosterone and estradiol) were not affected. An increase in cellular proliferation was observed at PND 132 in the prostate of male offspring from the 10% FS-treated group. The results from this study suggest that dose, timing, and duration of FS exposure have important implications on reproduction.

In the third study, Tou and Thompson (75) studied the effects of exposure to dietary FS (5 or 10%), FS oil, or SDG during gestation and lactation, gestation to PND 50 (final day), or after weaning to maturity (PND 21–50) on the development of mammary gland structures. Pregnant female rats (28 total) were randomly assigned to basal diet or basal diet supplemented with FS, FS oil, or SDG. At PND

21, the female offspring were weaned from the dams and housed individually. Three periods of treatment were investigated: throughout gestation to lactation at PND 21, after weaning (PND 21–50), or throughout gestation to PND 50. At PND 50, female offspring were euthanized, the ovaries were removed and weighed, and the abdominal gland and mammary glands were dissected, processed, and stained. The numbers of terminal end buds (TEB), alveolar buds (AB), and lobule structures were counted. Vaginal smears were taken to PND 50 to determine the length of the first full estrous cycle. At PND 50, blood was also collected from females for estradiol determination.

TEB density was reduced and AB density was enhanced, compared with controls, by exposure to 10% FS during gestation and lactation or during gestation to PND 50. TEB density was also reduced by exposure to 10% FS after weaning. Exposure to 5% FS during gestation and lactation and during gestation to PND 50 reduced the TEB density but had no significant effect on AB or lobule density. Mammary gland structures were not affected by exposure to 5% FS after weaning. TEB and AB density was reduced by exposure to SDG during gestation and lactation. FS oil had no effect.

Exposure to 10% FS during gestation and lactation (to PND 21) or during gestation to PND 50 resulted in earlier puberty onset than in animals fed basal diet, but exposure after weaning had no effect on puberty onset. Estrous cycles were lengthened after vaginal opening due to prolonged time in the estrus phase compared with the control group, but had no significant effect on the number of estrous cycles from vaginal opening to PND 50. A delay in the age of puberty onset and a reduction in the number of estrous cycles, compared with the control group, from vaginal opening to PND 50 resulted from exposure to 5% FS or SDG during gestation and lactation or gestation to PND 50 but had no significant effect on the length of estrous cycles. Exposure to 5% FS after weaning had no effect on puberty onset or estrous cyclicity.

Exposure to 10% FS during gestation and lactation or during gestation to PND 50 resulted in higher relative ovarian weight and serum estradiol levels than in the control animals. After weaning, an exposure to 10% FS had no significant effect on relative ovarian weight or serum estradiol. No significant effect on relative ovarian weight or serum estradiol levels occurred as a result of exposure to 5% FS, SDG, or FS oil.

Tou and Thompson (75) concluded that flaxseed can influence mammary gland structures depending on the dose and timing of exposure. Flaxseed has components that may alter mammary gland structures. It is the richest source of the mammalian lignan precursor SDG. Mammalian lignans have been shown to exert weak estrogenic or antiestrogenic activity on cancer cells *in vitro* (76,77). It is known that the development and differentiation of mammary gland structures are hormone dependent. The role of estradiol and the ER-alpha receptor in mammary gland development and/or tumor formation has been confirmed in studies on mice where the gene for the ER-alpha has been knocked out (78). In the absence of ER-alpha, the mammary gland cannot be induced to develop with estradiol. Because both ER-alpha and ER-beta are found in the mammary gland, it has been suggested that ER-beta might interact with the ER-alpha receptor and by some unknown mechanism modulate the actions of the ER-alpha receptor, thereby regulating growth (79). In the female, FS and its lignan SDG have been implicated in reducing mammary tumor development or mammary tumor size in a number of animal studies (48,73,75,80–84).

FS is also rich in ALA (85). High dietary ALA inhibits metabolism of linoleic acid (LA) (86). This has important implications because exposure to high dietary LA during the gestation and early neonatal stages has been reported to reduce differentiation of TEB to AB and lobules (87,88). To determine which component was responsible for the observed effects, female offspring were exposed to FS oil or SDG at the level found in 5% FS during gestation and lactation. SDG but not FS oil reduced TEB and AB. This suggested that the lignans were responsible for the observed effects on mammary gland structures.

In the Tou and Thompson (75) study, the number of estrous cycles from vaginal opening until PND 50 in the treated groups was not significantly different from the control group due to the lengthening of estrous cycles. Prolonged estrous cycles due to estrus are associated with elevated estradiol levels (89).

Ward *et al.* (90) expanded on the results obtained by Tou and Thompson (75). Pregnant female rats (28 total) were obtained on gestation day 2 and fed basal diet for the remainder of the pregnancy. At the time of delivery, dams were randomized to one of three diet groups: basal diet (control), 10% FS, or SDG equivalent to the amount found in 10% FS (17.7 mg/100 g diet). At PND 21, female offspring were weaned and continued on their mother's diet or switched to basal diet. The offspring (7–8/group) were euthanized at PND 50. The abdominal gland and mammary glands were dissected and processed. Mammary gland structures (TEB, AB, and lobules) were counted.

Feed intake and body weight did not differ among groups at PND 2, 21, or 50. Exposure to 10% FS or SDG equivalent during lactation or continuously significantly reduced the density of TEB. The reduction of TEB was attributed to the differentiation to ABs which were more numerous in animals fed 10% FS or SDG equivalent than in animals fed basal diet. Only SDG equivalent increased lobule density. Based on the results of the study, that 10% FS and SDG equivalent reduced the density of TEB to a similar extent, the authors concluded that the lignans in FS mediate the maturation of the mammary glands. Corresponding elevations were observed in the density of AB, indicating that the reduction of TEB was due to the enhanced differentiation of TEB to AB. Because the TEB are considered most susceptible to carcinogenesis (91), changes induced by FS may prevent carcinogenesis.

In a further study of the estrogenization effects of dietary FS on *in utero* development of offspring, Ward *et al.* (92) studied lactation as the critical period for estrogenization of the offspring. The offspring were exposed to FS (10%) or purified SDG during suckling only (birth to PND 21) or continuously (birth to PND 50 or PND 132). Flaxseed (10%) or the lignan component (SDG) at the level present in a 10% FS diet was fed to rats during lactation. Timed-pregnant females (total 20) were fed basal diet until the pups were born. At PND 2, the dams were continued on the basal diet or were switched to FS or to SDG diets until weaning (PND 21). At weaning, the pups were continued on the mother's diet or were switched to basal diet until PND 50 or 132 (5 test groups in all). At PND 132, each group contained 7–11 males and 7–10 females.

Feed consumption, maternal body weight, and ovarian body weight were similar in all groups at the end of lactation. Among male and female offspring, weight gain and AGD did not differ among the groups through PND 21. Age and body weights at the onset of puberty were similar in female offspring. All but two rats were cycling (one from the basal diet group, and one from the SDG group) during PND 100–132. Of the cycling rats, there were no significant differences in the length of the estrous cycle among groups. At PND 50 or 132, body weight and prostate and testes weights in males, and body weight and uterine and ovarian weights in females did not differ among groups. The authors concluded that a diet of 10% FS or the equivalent amount of lignan (SDG) does not alter reproductive indices in male and female offspring when fed during lactation only or continuously to maturity (PND 132). In humans, the achievement of peak bone mass, which begins *in utero* and continues until the late teens or early twenties (93), is dependent upon a complex interaction of hormones and other factors including growth factors, nutrition, genetics, and physical activity (94–98).

In males, a number of reproductive indices including AGD, and age and body weight at puberty were measured, and a histopathological analysis of the prostate was done at PND 132. No effects were observed in any of the reproductive indices examined. The results from these studies suggested that, in the rat, maternal exposure to FS during gestation could affect the development of the male fetus; however, maternal feeding of FS during lactation did not appear to adversely affect postnatal development.

Two studies have been done to determine the effects of FS and its purified lignan on bone development (bone mineral content and bone strength) in male and female rats (31,99). In the first study, male rats were exposed to one of three diets *via* mother's milk (basal diet, 10% dietary FS, or SDG equivalent to that present in 10% FS diet), during lactation only (birth to PND 21) from birth to PND 50 or from birth to PND 132. At PND 50, bone strength was reduced in male offspring of rats fed 10% FS, but it was not affected in the offspring from rats fed SDG equivalent. At PND 132, bone strength did not differ among groups. The authors suggested that the differences in bone strength observed at PND 50 and PND 132 may be due to the developmental changes in circulating testosterone that is essential for bone development in male rats (100,101). In the second study, female offspring were exposed to basal diet or to one of two doses of SDG equivalent to the amount found in 5 or 10% FS diet during lactation only (birth to PND 21) or from birth to PND 50 or PND 132. At PND 50, female rats exposed to SDG equivalent to 5 or 10% dietary FS had stronger femurs but no change in bone mineral content. At PND 132, there were no differences in femur strength in the treated offspring, and bone mineral content was less than in the offspring fed basal diet. The authors concluded that lignan does not have negative effects on bone strength in female rats, that the female rat bone is more sensitive to the estrogen-like action of lignans during early life when endogenous levels of sex hormones are low, but that the improved bone strength does not persist into adulthood.

In the most recent study, Collins *et al.* (102) evaluated the effects of high levels of FS (20 or 40%) or defatted FLM (13 or 26%) on reproduction, development, and metabolic parameters in rats. They attempted to separate the reproductive effects of the lignan and oil components by giving the pregnant animals diets that were identical in the amount of lignan but differed in the amount and type of oil. The effects of the diet were evaluated during gestation only, during gestation and lactation (to PND 21), or during gestation to maturation (to PND 90). Each group contained 18–22 pregnant females. AGD was measured, and anogenital index (AGI) was determined for offspring from birth to PND 21. At the termination of gestation, developmental toxicity parameters were measured.

Neither FS nor FLM affected fertility, body weight gain of dam, litter size, individual fetal weight, AGD, or skeletal or soft-tissue development of fetuses. These results differ from those found in earlier studies described above and are attributed to the difference in the number of animals tested per group. Small numbers of animals per test group can provide more scattered results than those found in groups with larger numbers.

FS did not affect gestation length, but 26% FLM shortened gestation length. Neither FS nor FLM affected pup weight or viability at birth, and neither FS nor FLM affected the survival of the offspring during the period of lactation to PND 21. FS (40%) decreased the body weight of female offspring at PND 21, but males were not affected. FLM did not affect the weight of either males or females. FS increased the AGI of F1 females at PND 21, but FLM did not. The AGI of F1 males was not affected by either FS or FLM.

During development of the offspring from PND 21 to maturation (PND 90), F1 females fed 20% FS, 13% FLM, or 26% FLM gained more weight than the controls. FLM (13 and 26%) delayed puberty in F1 males, but FS did not. Neither FS nor FLM affected the age or weight at the onset of puberty in females. FS and FLM caused dose-related increases in the number of F1 females with irregular estrous cycles. In conclusion, flaxseed did not affect fetal development but did affect indices of postnatal reproductive development such as the estrous cycle.

Few studies have specifically addressed the effect of maternal FS exposure on spermatogenesis and endocrine function in the adult. The consumption of phytoestrogen-rich foods, especially FS, has increased because of its reported cancer protective effects especially against prostate and breast cancers. Consumption of FS may increase in the United States population by both direct (metabolites in food) and indirect means. Because FS is fed to poultry and livestock, there is the possibility of additional amounts being consumed indirectly. At the present time, FS does not have generally regarded as safe (GRAS) status, but it is approved by FDA to be used, up to 12% of the dietary intake. Consumption of FS by women of childbearing age could result in consumption during pregnancy. The effects of FS and its lignans on the developing fetus, particularly the male fetuses, and subsequent postnatal effects are not known, and thus the potential for toxicity and/or adverse effects has not been adequately investigated.

Studies specifically addressing the effect of maternal FS exposure on the development and maturation of the male reproductive system have recently been conducted (103,104). The male animals utilized in these studies were obtained from the Collins *et al.* (102) study. In these studies the effects of high levels of FS (20 or 40%) or FLM (13 or 26%) on spermatogenesis and endocrine function in the male rat were examined. These studies attempted to separate the reproductive effects of the lignan and oil components by giving the pregnant animals diets that were identical in the amount of lignan but differed in the amount and type of oil. The effects of the diet were evaluated from gestation to maturation (to PND 90). In these studies, pregnant Sprague-Dawley rats were exposed to FS (20 or 40%), FLM (13 or 26%), or standard NIH AIN-93 (0% FS control) diet throughout gestation and lactation. Postlactation F1 generation males received their mother's diet for an additional 70 d at which time tissues were collected and various reproductive indices were examined.

A qualitative and quantitative analysis of testicular tissues indicated that testicular structure was not adversely affected by exposure to either FS or FLM. The seminiferous tubules and interstitial spaces from the FS- and FLM-treated animals were indistinguishable from those of the control group, and deleterious effects on spermatogenesis were not observed (103,104). No effects were observed on testicular spermatid numbers from the FS and FLM treatment groups compared with the control group, suggesting that spermatogenesis was not affected by exposure to FS or FLM. Cauda epididymal sperm numbers in the FS and FLM groups were significantly higher than in the control group, suggesting that FS/FLM exposure may have altered the ability of the epididymis to eliminate excess spermatozoa because testicular spermatid numbers were not significantly elevated in the treated groups. Sperm are typically eliminated from the cauda epididymis by two mechanisms. These methods include (i) a steady movement of small quantities of sperm from the testis through the reproductive tract (105), or (ii) a spontaneous seminal emission which occurs daily (106,107), during the light period (108,109), and with increased frequency in unmated male rats (106,107,110). It is possible that exposure to FS/FLM may have had a subtle effect on the hormonal differentiation of the central nervous system which in the rat is completed by PND 10 and establishes male *vs.* female patterns of hormone secretion and differences in gender behavior (111). It has previously been demonstrated that rats lactationally exposed to coumestrol during the first 10 PND demonstrated an increase in the latency to mount and ejaculate and a reduced mount and ejaculation rate. These effects were observed in the absence of perturbations in serum testosterone levels and therefore not attributed to defects in male gonadal function (112). Unfortunately, the FS/FLM-exposed rats utilized by Sprando *et al.* (103) were not mated. Further studies are required to investigate the effect of phytoestrogens on fetal development and behavior.

Sperm head or tail abnormalities were not observed in animals exposed to either FS or FLM. Sperm membranes are rich in PUFA, which makes them very susceptible to oxygen-induced damage mediated by lipid peroxidation. Sperm plasma membranes damaged by peroxidation can lead to male infertility (113,114). This increase in oxidative damage may be due to their high content of PUFA and diminished superoxide disumutase and glutathione peroxidase activities (115,116). An increase in induced oxidative stress has a damaging effect on sperm PUFA as shown by an increase in double bond index, and increased thiobarbituric acid reactive substance (TBAR) (113).

Significant increases in serum luteinizing hormone (LH) levels were observed in the FS groups and the high FLM dose group (113). Under certain physiologic states, low dose exposure to estrogens can sensitize the pituitary and hypothalamus and enhance LH release (117,118). The results obtained by Sprando *et al.* (103) suggested that exposure to FS/FLM, neonatally or postnatally, resulted in a slight augmentation of basal LH secretion in male rats. A statistically significant increase in serum testosterone was observed only in the 26% FLM group.

Statistically significant differences were not observed in testicular weight when animals from the FS/FLM-treated groups were compared with the control group; however, a statistically significant decrease in prostate weight was observed in the FLM treatment groups (13% and 26%) and the low FS treatment group (20%). A nonstatistically significant decrease in prostate weight was observed in the 40% FS treatment group, perhaps suggesting that, in the rat, exposure to FS and/or FLM may regulate prostate size.

Epidemiological data indicate that the incidence of prostate cancer in the United States is greater than in Japan (119). A close evaluation of Asian diets, and in general vegetarian diets, reveals that the Asian and vegetarian diets not only contain both low-fat and high-fiber components but also are rich in various phytoestrogens (i.e., isoflavonoids, flavonoids, and lignans). It appears that phytoestrogens, especially the lignans, may protect the prostate in several ways. For example, the mammalian lignans EL and ED may prevent the production of dihydrotestosterone, known to promote prostate enlargement. Numerous studies have suggested that prostate cancer may result from free radical damage to healthy prostate cells, converting them to cancer cells. The prostate, among other urogenital organs, expresses significant amounts of ER-beta mRNA to a much greater extent than it expresses ER-alpha (120,121). As mentioned above, ER-beta data from various studies suggest that ER-beta may play a role in preventing cellular proliferation. It has been suggested that ER-beta is involved in the control of antioxidant-regulated genes, the products of which are known to control the concentrations of free radicals and reactive oxygen in the cell. Free radicals play the role of mediator in carcinogenesis (i.e., in the stages of initiation and promotion). It is possible that phytoestrogens exert their cancer protective effect by interacting with the ER-beta receptor to activate antioxidant-regulated genes, thus decreasing the concentration of free radicals in the prostate and thereby preventing cellular proliferation. Data from other studies have suggested that phytoestrogens may affect prostate growth. Makela *et al.* (122) demonstrated that prenatal exposure to a diet containing soy (or diethylstilbestrol) resulted in a persistent decrease of the enzyme activity and growth of the prostate in rats. Hempstock *et al.* (123) showed that several phytoestrogens (biochanin A, daidzein, genistein, genistin, and nordihydroguaiaretic acid) inhibited growth and metabolism in four different prostate cell lines (PNT-1/A, PNT-2, PC-3, and DU145) *in vitro*. Further studies that specifically address these findings are needed.

Conclusion

In conclusion, a review of the available literature has suggested that the consumption of FS can affect various reproductive indices in both the male and female rat. The reviewed animal studies have suggested that the effect of flaxseed on various reproductive indices and sex hormone levels depends on the dose, timing, and duration of exposure. In the female, effects included a lengthened estrous cycle, changes in the AGD, extended or shortened onset of puberty, ovarian weight changes, and effects on the maturation of the mammary gland. In the male, effects included changes in serum hormone levels and secondary sex organ weight, especially of the prostate. Effects were not observed in the male fetus. Neither testis structure, the process of spermatogenesis, sperm production, or sperm morphology were affected by FS exposure during gestation or postnatal development. Epididymal effects (i.e., extended sperm storage times) have been reported; however, further studies are required to corroborate or dispute this finding. At this time, the relevance of these findings with respect to the human population is unknown and further research is required.

References

1. Carter, J.F., Potential of Flaxseed and Flaxseed Oil in Baked Goods and Other Products in Human Nutrition, *Cereal Foods World 38*:753–759 (1993).
2. Nesbitt, P.D., and L.U. Thompson, Lignans in Homemade and Commercial Products Containing Flaxseed, *Nutr. Cancer 29*:222–227 (1997).
3. Thompson, L.U., Potential Health Benefits and Problems Associated with Antinutrients in Foods, *Food Res. Int. 26*:131–149 (1993).
4. Kurzer, M.S., and X. Xu, Dietary Phytoestrogens, *Ann. Rev. Nutr. 17*:353–381 (1997).
5. Brezinski, A., and A. Debi, Phytoestrogens: The "Natural" Selective Estrogen Receptor Modulators? *Eur. J. Obstet. Gynecol. Reproduc. Biol. 85*:47–51 (1999).
6. Craig, W.J., Health-Promoting Properties of Common Herbs, *Am. J. Clin. Nutr. 70 (Suppl.)*:491S–499S (1999).
7. Hasler, C.M., S. Kundrat, and D. Wool, Functional Foods and Cardiovascular Disease, *Current Atheroscler. Rep. 2*:467–475 (2000).

8. Ratnayake, W.M.N., W.A. Behrens, P.W.F. Fischer, M.R.L. L'Abbe, P. Mongeau, and J.L. Beare-Rogers, Chemical and Nutritional Studies of Flaxseed (Variety Linott) in Rats, *J. Nutr. Biochem. 3*:232–240 (1992).

9. Bierenbaum, M.L., R. Reichstein, and T.R. Watkins, Reducing Atherogenic Risk in Hyperlipidemic Humans with Flaxseed Supplementation: A Preliminary Report, *J. Am. Coll. Nutr. 12*:501–504 (1993).

10. Cunnane, S.C., M.J. Hamedeh, A.C. Liede, L.U. Thompson, T.M.S. Wolever, and D.J.A. Jenkins, Nutritional Attributes of Traditional Flaxseed in Healthy Young Adults, *Am. J. Clin Nutr. 61*:62–68 (1995).

11. Harris, W.S., n-3 Fatty Acids and Serum Lipoproteins: Human Studies, *Am J. Clin Nutr. 65(Suppl.)*:1645S–1654S (1997).

12. Nestel, P.J., S.E. Pomeroy, T. Sasahara, T. Yamashita, T.Y. Liang, A.M. Dart, G.L. Jenning, M. Abbery, and J.D. Cameron, Arterial Compliance in Obese Subjects Is Improved with Dietary Plant n-3 Fatty Acid from Flaxseed Oil Despite Increased LDL Oxidation, *Arterioscler. Thromb. Vasc. Biol. 17*:1163–1170 (1997).

13. Jenkins, D.J.A., C.W. Kendall, E. Vidgen, S. Agarwal, A.V. Rao, R.S. Rosenberg, E.P. Diamandis, R. Novokmet, C.C. Mehling, T. Perera, L.C. Griffin, and S.C. Cunnane, Health Aspects of Partially Defatted Flaxseed, Including Effects on Serum Lipids, Oxidative Measures, and *ex vivo* Androgen and Progestin Activity: A Controlled Crossover Trial, *Am. J. Clin Nutr. 69*:395–402 (1999).

14. Tominaga, S., Cancer Incidence in Japanese in Japan, Hawaii and Western United States, *Monog. Natl. Cancer Inst. 69*:83–92 (1985).

15. Don, A.S., and C.S. Muir, Prostate Cancer: Some Epidemiological Factors, *Bull. Cancer 72*:381–390 (1985).

16. Lee, H.P., L. Gourley, S.W. Duffy, J. Esteve, F. Lee, and N.E. Duy, Dietary Effects on Breast Cancer Risk in Singapore, *Lancet 2*:1197–1200 (1991).

17. Serraino, M.R., and L.U. Thompson, The Effect of Flaxseed Supplementation on Early Risk Markers for Mammary Carcinogenesis, *Cancer Lett. 60*:125–142 (1991).

18. Jenab, M., and L.U. Thompson, The Influence of Flaxseed and Lignans on Colon Carcinogenesis and β-Glucuronidase Activity, *Carcinogenesis 17*:1343–1348 (1996).

19. Thompson, L.U., Experimental Studies on Lignans and Cancer, *Baillieres Clin. Endocrinol. Metab. 12*:691–705 (1998).

20. Ingram, A.J., A. Parbtani, W.F. Clark, E. Spanner, M.W. Huff, D.J. Philbrick, and B.J. Holub, Effects of Flaxseed and Flax Oil Diets in a Rat 5/6 Renal Ablation Model, *Am. J. Kidney Dis. 25*:320–329 (1995).

21. Clark, W.F., A.D. Muir, N.D. Westcott, and A. Parbtani, A Novel Treatment for Lupus Nephritis: Lignan Derived from Flax, *Lupus 9*:429–436 (2000).

22. Clark, W.F., C. Kortas, A.P. Heidenheim, E. Spanner, and A. Parbtani, Flaxseed in Lupus Nephritis: A Two-Year Nonplacebo-Controlled Crossover Study, *J. Am. Coll. Nutr. 20*:143–148 (2001).

23. Ogborn, M.R., E. Nitschmann, H. Weiler, D. Leswick, and N. Bankovic-Calic, Flaxseed Ameliorates Interstitial Nephritis in Rat Polycystic Kidney Disease, *Kidney Int. 55*:417–423 (1999).

24. Ranich, T., S.J. Bhathena, and M.T. Velasquez, Protective Effects of Dietary Phytoestrogens in Chronic Rrenal Disease, *J. Renal Nutr. 11*:183–193 (2001).

25. Babu, U.S., V.K. Bunning, P.W. Wiesenfeld, R.B. Raybourne, and M. O'Donnell, Effect of Dietary Flaxseed on Fatty Acid Composition, Superoxide, Nitric Oxide

Generation and Antilisterial Activity of Peritoneal Macrophages from Female Sprague-Dawley Rats, *Life Sci. 60:*545–554 (1997).

26. Wiesenfeld, P.W., U.S. Babu, T.F.X. Collins, R. Sprando, M.W. O'Donnell, T.J. Flynn, T. Black, and N. Olejnik, Flaxseed Increased α-Linolenic and Eicosapen-taenoic Acid and Decreased Arachidonic Acid in Serum and Tissues of Rat Dams and Offspring, Submitted to *Food Chem. Toxicol.* (in press).

27. Cheon, S-H., M-H. Huh, Y-B. Lee, J-S. Park, H-S. Sohn, and C-W. Chung, Effect of Dietary Linoleate/Alpha-Linolenate Balance on Brain Lipid Composition, Reproductive Outcome and Behavior of Rats During Their Prenatal and Postnatal Development, *Biosci. Biotech. Biochem. 64:*2290–2297 (2000).

28. L'Abbe, M.R., K.D. Trick, and J.L. Beare-Rogers, Dietary (n-3) Fatty Acids Affect Rat Heart, Liver and Aorta Protective Enzyme Activities and Lipid Peroxidation, *J. Nutr. 121:*1331–1340 (1991).

29. Babu, U.S., G.V. Mitchell, P.W. Wiesenfeld, M.Y. Jenkins, and N. Gowda, Nutritional and Hematological Impact of Dietary and Defatted Flaxseed Meal in Rats, *Int. J. Food Sci. Nutr. 51:*109–117 (2000).

30. Lindner, A., D.D. Bankson, C. Stehman-Breen, J.D. Mahuren, and S.P. Coburn, Vitamin B6 Metabolism and Homocysteine in End-Stage Renal Disease and Chronic Renal Insufficiency, *Am J. Kidney Dis. 39:*134–145 (2002).

31. Ward, W.E., Y.V. Yuan, A.M. Cheung, and L.U. Thompson, Exposure to Purified Lignan from Flaxseed (*Linum usitatissimum*) Alters Bone Development in Female Rats, *Brit. J. Nutr. 86:*499–505 (2001).

32. Jones, D.A., Why Are So Many Food Plants Cyanogenic? *Phytochemistry 47:*155–162 (1998).

33. Vetter, M., Plant Cyanogenic Glycosides, *Toxicol. 38:*11–36 (2000).

34. Oomah, B.D., G. Mazza, and E.O. Kenaschuk, Cyanogenic Compounds in Flaxseed, *J. Agric. Food Chem. 40:*1346–1348 (1992).

35. Fan, T.W., and E.E. Conn, Isolation and Characterization of Two Cyanogenic Beta-Glucosides from Flax Seeds, *Arch. Biochem. Biophys. 243:*361–373 (1985).

36. Wanasundara, P.K., and F. Shahidi, Process-Induced Compositional Changes of Flaxseed, *Adv. Exper. Med. Biol. 434:*307–325 (1998).

37. Cunnane, S.C., S. Ganguli, C. Menard, A.C. Liede, M.J. Hanadeh, Z.Y. Chen, T.M. Wolever, and D.J. Jenkins, High Alpha-Linolenic Acid Flaxseed (*Linum usitatissimum*): Some Nutritional Properties in Humans, *Br. J. Nutr. 69:*443–453 (1993).

38. Banea-Mayambu, J.P., T. Tylleskar, K. Tylleskar, M. Gebre-Medhin, and H. Rosling, Dietary Cyanide from Insufficiently Processed Cassava and Growth Retardation in Children in the Democratic Republic of Congo (Formerly Zaire), *Ann. Trop. Paediatr. 20:*34–40 (2000).

39. Thompson, L.U., P. Robb, M. Serraino, and F. Cheung, Mammalian Lignan Production from Various Foods, *Nutr. Cancer 16:*43–52 (1991).

40. Kitts, D.D., Y.V. Yuan, A.N. Wijewickreme, and L.U. Thompson, Antioxidant Activity of the Flaxseed Lignan Secoisolariciresinol Diglycoside and Its Mammalian Lignan Metabolites Enterodiol and Enterolactone, *Mol. Cell Biochem. 202:*91–100 (1999).

41. Axelson, M., J. Sjovall, B.E. Gutafsson, and K.D.R. Setchell, Origin of Lignans in Mammals and Identification of a Precursor from Plants, *Nature 298:*659–660 (1982).

42. Boriello, S.P., K.D.R. Setchell, M. Axelson, and A.M. Lawson, Production and Metabolism of Lignans by the Human Fecal Flora, *J. Appl. Bacteriol. 58:*37–43 (1985).

43. Axelson, M., and K. Setchell, The Excretion of Lignans in Rats: Evidence for an Intestinal Bacteria Source for This New Group of Compounds, *FEBS Lett. 123*:237–342 (1981).

44. Setchell, K.D.R., and H. Adlercreutz, Mammalian Lignans and Phytoestrogens: Recent Studies on Their Formation, Metabolism, and Biological Role in Health and Disease, in *Role of the Gut Flora in Toxicity and Cancer*, edited by R. Rowland, Academic Press, London, 1988, pp. 315–345.

45. Prasad, K., Reduction of Serum Cholesterol and Hypercholesterolemic Atherosclerosis in Rabbits by Secoisolariciresinol Diglucoside Isolated from Flaxseed, *Circulation 99*:1355–1362 (1999)

46. Amarowicz, R., U. Wanasundara, J. Wanasundara, and F. Shahidi, Antioxidant Activity of Ethanolic Extracts of Flaxseed in a β-Carotene-Linoleate Model System, *J. Food Lipids 1*:111–117 (1993).

47. Prasad, K., Hydroxyl Radical-Scavenging Property of Secoisolariciresinol Diglucoside (SDG) Isolated from Flaxseed, *Mol. Cell. Biochem. 168*:117–123 (1997).

48. Thompson, L.U., S.E. Rickard, L.J. Orcheson, and M.M. Seidl, Flaxseed and Its Lignan and Oil Components Reduce Mammary Tumor Growth at a Late Stage of Carcinogenesis, *Carcinogenesis 17*:1373–1376 (1996).

49. Pasqualini, J.R., F.A. Kind, and C. Sumida, The Binding of Hormones in Maternal and Fetal-Biological Fluids, in *Hormones and the Fetus*, edited by J.R. Pasqualini, Pergamon Press, New York, 1985, pp. 1–50.

50. Lubhann, D.B., J.S. Moyer, T.S. Golding, J.F. Couse, K.S. Korach, and O. Smithies, Alteration of Reproductive Function but Not Prenatal Sexual Development After Insertional Disruption of the Mouse Estrogen Receptor Gene, *Proc. Natl. Acad. Sci. USA 90*:11162–11166 (1993).

51. Bellido, C., F. Gaytan, R. Aguilar, L. Pinilla, and E. Aguilar, Prepubertal Reproductive Defects in Neo-Natal Estrogenized Male Rats, *Biol. Reprod. 33*:381–387 (1985).

52. Newbold, R.R., B.C. Bullock, and J.A. McLachlan, Lesions of the Rete Testis in Mice Exposed Prenatally to Diethylstilbestrol, *Cancer Res. 45*:5145–5150 (1985).

53. Newbold, R.R., B.C. Bullock, and J.A. McLachlan, Adenocarcinoma of the Rete Testis: Diethylstilbestrol-Induced Lesions of the Mouse Rete Testis, *Am. J. Pathol. 123*:625–628 (1986).

54. Bullock, B.C., R.R. Newbold, and J.A. McLachlan, Lesions of Testis and Epididymis Associated with Prenatal Diethylstilbestrol Exposure, *Environ. Health Perspect. 77*:29–31 (1988).

55. Newbold, R.R., and J.A. McLachlan, Neoplastic and Non-Neoplastic Lesions in Male Reproductive Organs Following Perinatal Exposure to Hormones and Related Substances, in *Toxicology of Hormones in Perinatal Life*, edited by T. More and H. Nagasawa, CRC Press Inc., Boca Raton, Florida, 1988, pp. 89–109.

56. Newbold, R.R., Y. Suzuki, and J.A. McLachlan, Mullerian Duct Maintenance in Heterotypic Organ Culture After *in vivo* Exposure to Diethylstilbestrol, *Endocrinology 115*:1863–1868 (1984).

57. Fenci, M., and C.A. Villee, Effect of Estradiol on the Metabolism of Testosterone by Rat Prostate, *Steroids 21*:537–552 (1973).

58. Higgins, S.T., D.E. Brooks, F.M. Fuller, P.J. Jackson, and S.E. Smith, Functional Development of Sex Accessory Organs of the Male Rat, *Biochem. J. 194*:895–905 (1981).

59. Terakawa, N., R.A. Huseby, S.M. Fang, and L.T. Samuels, Quantitative Changes in Estrogen Receptor Produced by Chronic DES Treatment of Two Mouse Strains Differing in Susceptibility to Leydig Cell Tumor Induction, *J. Steroid Biochem. 16:*643–652 (1982).

60. Yasuda, Y., T. Kihara, and T. Tanimura, Effects of Ethinyl Estradiol on the Differentiation of Mouse Fetal Testis, *Teratology 32:*113–118 (1985).

61. Gayton, F., C. Bellido, R. Aguilar, and M.C. Lucena, Morphometric Analysis of the Rat Ventral Prostate and Seminal Vesicles During Prepubertal Development: Effects of Neonatal Treatment with Estrogen, *Biol. Reprod. 35:*219–225 (1986).

62. Yasuda, Y., T. Kihara, and T. Tanimura, Leydig Cell Hyperplasia in Fetal Mice Treated Transplacentally with Ethinyl Estradiol, *Teratology 33:*281–288 (1986).

63. Cooke, P.S., and V.P. Eroschenko, Inhibitory Effects of Technical Grade Methoxychlor on Development of the Neonatal Male Mouse Reproductive Organs, *Biol. Reprod. 42:*585–596 (1990).

64. Walker, A.H., Bernstein, L., Warren, D.W. Warren, N.E., Zheng, X., and Henderson, B.E. (1990). The Effect of *in utero* Ethinyl Estradiol Exposure on the Risk of Cryptorchid Testis and Testicular Teratoma in Mice, *Br. J. Cancer 62:*599–602 (1990).

65. Enmark, E., M. Pelto-Huikko, K. Grandien, S. Lagercrantz, J. Lagercrantz, G. Fried, M. Nordenskjold, and J-A. Gustafsson, Human Estrogen Receptor β-Gene Structure, Chromosomal Localization, Expression Pattern, *J. Clin. Endocrinol. Metab. 82:*4258–4265 (1997).

66. Arts, J., G.G.J.M. Kuiper, J.M.M.F. Janssen, J-A. Gustafsson, C.W.G.M. Lowik, H.A.P. Pols, and J.P.T.M. van Leeuwen, Differential Expression of Estrogen Receptor α and β mRNA During Differentiation of the Human Osteoblast, SV-HFO Cells, *Endocrinology 138:*5067–5070 (1997).

67. Kuiper, G.G.J.M., B. Carlsson, K. Grandien, E. Enmark, J. Haggbald, S. Nilsson, and J-A. Gustafsson, Comparison of the Ligand Binding Specificity and Transcript Tissue Distribution of Estrogen Receptors α and β, *Endocrinology 3:*863–870 (1997).

68. Kuiper, G.G.J.M., M. Carlquist, and J-A. Gustafsson, Estrogen is a Male and Female Hormone, *Sci. Med. 5:*36-45 (1998).

69. Kuiper, G.G.J.M., P.J. Shughrue, M. Aelto-Huikko, I. Merchenthaler, and J-A. Gustafsson, The Estrogen Receptor β Subtype: A Novel Mediator of Estrogen Action in the Neuroendocrine System, *Frontiers Neuroendocrinology 19:*253–286 (1998).

70. Osterlund, M., G.G.J.M. Kuiper, J-A. Gustafsson, and Y.L. Hurd, Differential Distribution and Regulation of Estrogen Receptor α and β nRNA Within the Female Rat Brain, *Molec. Brain Res. 54:*175–180 (1998).

71. Kuiper, G.G.J.M., J.G. Lemmen, B. Carlsson, J.G. Corton, S.H. Safe, P.T. van der Saag, B. van der Burg, and J-A. Gustaffson, Interaction of Estrogenic Chemicals and Phytoestrogens with Estrogen Receptor β, *Endocrinology 139:*4252–4263 (1998).

72. Tou, J.C., J. Chen, and L.U. Thompson, Flaxseed and Its Lignan Precursor, Secoisolariciresinol Diglycoside, Affect Pregnancy Outcome and Reproductive Development in Rats, *J. Nutr. 128:*1861–1868 (1998).

73. Orcheson, L.J., S.E. Rickard, M.M. Seidl, and L.U. Thompson, Flaxseed and Its Mammalian Lignan Precursor Cause a Lengthening or Cessation of Estrous Cycling in Rats, *Cancer Lett. 125:*69–76 (1998).

74. Tou, J.C., J. Chen, and L.U. Thompson, Dose, Timing and Duration of Flaxseed Exposure Affect Reproductive Indices and Sex Hormone Levels in Rats, *J. Toxicol. Environ. Health 56:*555–570 (1999).

75. Tou, J.C., and L.U. Thompson, Exposure to Flaxseed or Its Lignan Component During Different Developmental Stages Influences Rat Mammary Gland Structures, *Carcinogenesis 20:*1831–1835 (1999).

76. Hirano, T., K. Fukuoka, K. Oka, K. Hosaka, H. Mitsuhashi, and Y. Matumoto, Antiproliferative Activity of Mammalian Lignan Derivatives Against the Human Breast Carcinoma Cell Line, ZR-75-1, *Cancer Invest. 8:*595–602 (1990).

77. Mousavi, Y., and H. Adlercreutz, Enterolactone and Estradiol Inhibit Each Other's Proliferative Effect on MCF-7 Breast Cancer Cells in Culture, *J. Steroid Biochem. Mol. Biol. 41:*615–619 (1992).

78. Bocchinfuso, W.P., and K.S. Korack, Mammary Gland Development and Tumorigenesis in Estrogen Receptor Knock-Out Mice, *J. Mammary Gland Biol. Neoplasia 2:*323–334 (1997).

79. Hall, J.M., and D.P. McDonnel, The Estrogen Receptor Beta-Isoform (ERbeta) of the Human Estrogen Receptor Modulates ERalpha Transcriptional Activity and Is a Key Regulator of the Cellular Response to Estrogens and Antiestrogens, *Endocrinology 140:*5566–5578 (1999).

80. Zava, D.T., and G. Duwe, Estrogenic and Antiproliferative Properties and Other Flavonoids in Human Breast Cancer Cells *in vivo*, *Nutr. Cancer 27:*31–40 (1997).

81. Sung, M.K., M. Lautens, and L.U. Thompson, Mammalian Lignans Inhibit the Growth of Estrogen-Independent Human Colon Tumor Cells, *Anticancer Res. 18:*(3A) 1405–1408 (1998).

82. Yan, L., J.A.Yee, D. Li, M.H. McGuire, and L.U. Thompson, Dietary Flaxseed Supplementation and Experimental Metastasis of Melanoma Cells in Mice, *Cancer Lett. 124:*181–186 (1998).

83. Rickard, S.E., Y.V. Yuan, and L.U. Thompson, Dose Effects of Flaxseed and Its Lignan on *N*-Methyl-*N*-Nitrosourea-Induced Mammary Tumorigenesis in Rats, *Nutr. Cancer 35:*50–57 (1999).

84. Rickard, S.E., Y.V. Yuan, and L.U. Thompson, Plasma Insulin-Like Growth Factor I Levels in Rats Are Reduced by Dietary Supplementation of Flaxseed or Its Lignan Secoisolariciresinol Diglycoside, *Cancer Lett. 161:*47–55 (2000).

85. Johnston, P.V., Flaxseed Oil and Cancer: α-Linolenic Acid and Carcinogenesis, in *Flaxseed in Human Nutrition*, edited by S.C. Cunnane and L.U. Thompson, AOCS Press, Champaign, Illinois, 1995, pp. 207–218.

86. Drevon, C.A., Marine Oils and Their Effects, *Nutr. Rev. 50:*38–45 (1992).

87. Hilakivi-Clarke, L., R. Clarke, I. Onojafe, M. Raygada, E. Cho, and M.E. Lippman, Breast Cancer Risk in Rats Fed a Diet High in n-6 Polyunsaturated Fatty Acids During Pregnancy, *J. Natl. Cancer Inst. 88:*1821–1827 (1996).

88. Hilakivi-Clarke, L., A. Stoica, M. Raygada, and M.B. Martin, Consumption of a High-Fat Diet Alters Estrogen Receptor Content, Protein Kinase C Activity and Mammary Gland Morphology in Virgin and Pregnant Mice and Female Offspring, *Cancer Res. 58:*634–660 (1998).

89. Anderson, M.E., H.J. Clewell III, J. Gearhart, B.C. Allen, and H.A.Barton, Pharmacodynamic Model of the Rat Estrous Cycle in Relation to Endocrine Disruptors, *J. Toxicol. Environ. Health 52:*189–209 (1997).

90. Ward, W.E., F.O. Jiang, and L.U. Thompson, Exposure to Flaxseed or Purified Lignan During Lactation Influences Rat Mammary Gland Structures, *Nutr. Cancer 37:*187–192 (2000).

91. Russo, I.H., and J. Russo, Developmental Stage of the Rat Mammary Gland as Determinant of Its Susceptibility to 7,12-Dimethylbenz[a]anthracene, *J. Natl. Cancer Inst. 61:*1439–1449 (1978).

92. Ward, W.E., F.O. Jiang, and L.U. Thompson, Exposure to Flaxseed or Its Purified Lignan During Suckling Only or Continuously Does Not Alter Reproductive Indices in Male and Female Offspring, *J. Toxicol. Environ. Health, Part A 64:*567–577 (2001).

93. Teegarden, D., W.R. Proulx, B.R. Martin, J. Zhao, G.P. McCabe, R.M. Lyle, M. Peacock, C. Slemenda, C.C. Johnston, and C.M.Weaver, Peak Bone Mass in Young Women, *J. Bone Miner. Res. 10:*711–715 (1995).

94. Slemenda, C.W., M.Z. Miller, S.L. Hui, T.K. Reister, and C.C. Johnston, Role of Physical Activity in the Development of Skeletal Mass in Children, *J. Bone Miner. Res. 6:*1227–1233 (1991).

95. Slemenda, C.W., T.K. Reister, S.L. Hui, J.Z. Miller, J.C. Christian, and C.C. Johnston, Influences on Skeletal Mineralization in Children and Adolescents, *J. Pediatr. 125:*201–207 (1994).

96. Johnston, C.J., J.Z. Miller, C.W. Slemenda, T.K. Reister, S. Hui, J.C. Christian, and M. Peacock, Calcium Supplementation and Increases in Bone Mineral Density in Children, *New Engl. J. Med. 327:*82–87 (1992).

97. Lonzer, M.D., R. Imrie, D. Rogers, D. Worley, A. Licata, and M. Secic, Effects of Heredity, Age, Weight, Puberty, Activity, and Calcium Intake on Bone Mineral Density in Children, *Clin. Pediatrics 35:*185–189 (1996).

98. Molgaard, C., B.T. Thomsen, and K.E. Michaelsen, Whole Body Bone Mineral Accretion in Healthy Children and Adults, *Arch. Dis. Child 81:*10–15 (1999).

99. Ward, W.E., Y.V. Yuan, A.M. Cheung, and L.U. Thompson, Exposure to Flaxseed or Its Purified Lignan Reduces Bone Strength in Young but Not Older Male Rats, *J. Toxicol. Environ. Health, Part A 63:*53–65 (2001).

100. Vanderschueren, D., E. Van Herck, R. De Coster, and R. Bouillon, Aromatization of Androgens Is Important for Skeletal Maintenance of Aged Male Rats, *Calcif. Tissue Int. 59:*179–183 (1996).

101. Vanderschueren, D., E. Van Herck, J. Nijs, A.G.H. Ederveen, R. De Coster, and R. Bouillon, Aromatase Inhibition Impairs Skeletal Modeling and Decreases Bone Mineral Density in Growing Male Rats, *Endocrinology 138:*2301–2307 (1997).

102. Collins, T.F.X., R.L. Sprando, T.N. Black, N. Olejnik, P.W. Wiesenfeld, U.S. Babu, M. Bryant, T.J. Flynn, and D.I. Ruggles, Effect of Flaxseed and Defatted Flaxseed Meal on Reproduction and Development in Rats, *Food Chem. Toxicol.*, in press.

103. Sprando, R.L., T.F.X. Collins, T.N. Black, N. Olejnik, J.I. Rorie, P. Wiesenfeld, U.S. Babu, and M. O'Donnell, The Effect of Maternal Exposure to Flaxseed on Spermatogenesis in F(1) Generation rats, *Food Chem. Toxicol. 38:*325–334 (2000).

104. Sprando, R.L., T.F.X. Collins, P. Wiesenfeld, U.S. Babu, C. Rees, T. Black, N. Olejnik, and J. Rorie, Testing the Potential of Flaxseed to Affect Spermatogenesis: Morphometry, *Food Chem. Toxicol. 38:*887–892 (2000).

105. Beach, F.A., Variables Affecting "Spontaneous" Seminal Emission in Rats, *Physiol. Behav. 15:*91–95 (1975).

106. Orbach, J., Spontaneous Ejaculation in the Rat, *Science 134:*1072–1073 (1961).

107. Beach, F.A., W.A. Westbrook, and L.G. Clemens, Comparison of the Ejaculatory Response in Men and Animals, *Psychosom. Med. 28:*749–763 (1966).

108. Kihlstrom, J.E., Diurnal Variation in the Spontaneous Ejaculation of Male Albino Rats, *Nature 209:*513–514 (1966).

109. Stefanick, M.L., The Circadian Patterns of Spontaneous Seminal Emission, Sexual Activity and Penile Reflexes in the Rat, *Physiol. Behav. 31:*737–743 (1983).

110. Orbach, J., M. Miller, A. Billimoria, and N. Sohlkhan, Spontaneous Seminal Ejaculation and Genital Grooming in Rats, *Brain Res. 5:*520–523 (1967).

111. Manson, J.M., and Y.J. Kang, Test Methods for Assessing Female Reproductive and Developmental Toxicology, in *Principles and Methods of Toxicology*, 2nd edn., edited by A.W. Hayes, Raven Press, New York, 1989, pp. 311–359.

112. Whitten, P.L., C. Lewis, E. Russell, and F. Naftolin, Potential Adverse Effects of Phytoestrogens, *J. Nutr. 125:*771S–776S (1995).

113. Zalata, A.A., A.B. Christophe, C.E. Depuydt, F. Schoonjans, and F.H. Comhaire, White Blood Cells Cause Oxidative Damage to the Fatty Acid Composition of Phospholipids of Human Spermatozoa, *Internat. J. Androl. 21:*154–162 (1998).

114. Sikka, S.C., Relative Impact of Oxidative Stress on Male Reproductive Function, *Current Med. Chem. 8:*851–862 (2001).

115. Alvarez, J.G., and B.T. Storcy, Role of Glutathionine Peroxidase in Protecting Mammalian Spermatozoa from Loss of Motility Caused by Spontaneous Lipid Peroxidation, *Gamete Res. 23:*77–90 (1989).

116. Alvarez, J.G., and B.T. Storcy, Evidence for Increased Lipid Peroxidative Damage and Loss of Superoxide Dismutase Activity as a Mode of Sublethal Cryodamage to Human Sperm During Cryopreservation, *J. Androl. 13:*232–241 (1992).

117. Cooper, K.J., C.P. Fawcett, and S.M. McCann, Inhibitory and Facilitatory Effects of Estradiol-17β on Pituitary Responsiveness to a Luteinizing Hormone-Follicle Stimulat-ing Hormone Releasing Factor (LHRF/FSHRF) Preparation in the Ovariectomized Rat, *Proc. Soc. Exp. Biol. Med. 145:*1422–1426 (1974).

118. Libertum, C., R. Orias, and S.M. McCann, Biphasic Effect of Estrogen on the Sensitivity of the Pituitary to Luteinizing Hormone-Releasing Factor (LHRF), *Endocrinology 94:* 1094–1100 (1974).

119. Adlercreutz, H., and W. Mazur, Phytoestrogens and Western Disease, *The Finnish Medical Society Duodecim. Ann. Med. 29:*95–120 (1997).

120. Sharpe, R.M., Regulation of Spermatogenesis, in *The Physiology of Reproduction*, edited by E. Knobil and J.D. Neill, Raven Press, New York, 1994, pp. 1–72.

121. Hess, R.A., D.H. Gist, D. Dunick, D.B. Lubahn, A. Farrell, J. Bahr, P.S. Cooke, and G.L. Greene, Estrogen Receptor (α and β) Expression in the Excurrent Ducts of the Adult Male Rat Reproductive Tract, *J. Androl. 18:*602–611 (1997).

122. Makela, S., R. Santti, P. Martikainen, W. Niensted, and J. Pranko, The Influence of Steroidal and Nonsteroidal Estrogens on the 5α-Reduction of Testosterone by the Ventral Prostate of the Rat, *J. Steroid Biochem. 35:*249–256 (1990).

123. Hempstock, J., J.P. Kavanagh, and N.J. George, Growth Inhibition of Prostate Cell Lines *in vitro* by Phytoestrogens, *Br. J. Urol. 82:*560–563 (1998).

Chapter 20

Processing of Flaxseed Fiber, Oil, Protein, and Lignan

B. Dave Oomah

National Bioproducts and Bioprocesses Program, Pacific Agri-Food Research Centre, Agriculture and Agri-Food Canada, Summerland, British Columbia V0H 1Z0, Canada

Introduction

Flaxseed is undoubtedly the nutraceutical food of the twenty-first century given its unlimited potential in preventing and/or reducing the risk of several major diseases, including diabetes, lupus nephritis, atherosclerosis, and hormonally dependent cancers. It is a leading source of the n-3 fatty acid, α-linolenic acid (ALA) (52% of the total fatty acids), and of phenolic compounds commonly known as lignans (>500 µg/g, as is basis), in addition to containing hydrocolloidal gum, also referred to as mucilage (about 8% of seed weight), and good quality protein (1). These components, although strategically located in different structural parts of the seeds, interact with each other, thereby providing significant challenges during their extraction and processing. The challenges highlight the limited commercial availability of ingredients and value-added products from flaxseed in spite of its consumption for over 5,000 y as a food ingredient and because of its medicinal properties (2). The recent unveiling of the first product, LinumLife™, by Acatris (Netherlands) (3) and the announcement by Sentex Systems Ltd. (Oshawa, ON, Canada) to invest in the construction of a flaxseed dehulling plant to produce kernels and hulls from flaxseed (4) are encouraging trends.

Until recently, the only flax ingredients available to manufacturers were flax oil and whole or milled flaxseed. The difficulties in processing flaxseed lie in the inherent seed structure—flat, oval, and pointed at one end (2). Cotyledons form 55%, the seed coat and the endosperm 36%, and the embryo axis 4%, respectively, of the total weight of hand-dissected flaxseed (5). Structurally, the testa, endosperm, cotyledons, and mucilage represent 10, 21, 57, and 11%, respectively, of manually dissected flaxseed. The endosperm and cotyledons contribute 17 and 82%, respectively, of the total seed oil and 16 and 76%, respectively, of the total seed protein. The testa and the mucilage contain 24.6 and 26.3%, respectively, of the total dietary fiber (6). The seed coat, which makes up most of the hull, can theoretically be removed completely, thereby allowing further processing of value-added products from flaxseed. However, this has proven to be more difficult than expected. The location, complexities, and interactions of the various biologically active components in flaxseed illustrate the difficulties and limitations involved in processing these functional food ingredients.

Processing of Flaxseed

The use of flaxseed as a valuable nutritional product dates back to ancient times as long ago as 9000 y (7). Conditioning prior to milling of whole flaxseed has been practiced in the animal feed industry and can easily be applied for human food use. Thus, in order to protect fatty acids from hydrogenation in the rumen, flaxseed is treated with lignosulfonate or formaldehyde, ground and heated to 155°C and steeped for 30 min at beginning and end temperatures of 124 and 118°C, respectively (8). The lignosulfonate binds and precipitates proteins and is used to make rumen by-pass protein. Both lignosulfonate and formaldehyde can protect flaxseed from biohydrogenation in the rumen, and this offers flaxseed a unique opportunity to substantially improve the ALA in milk. Recently, micronization has proven to be an effective method for protecting flaxseed protein from ruminal degradation (9). Micronization is a dry-heat treatment produced by infrared gas generators applied to feedstuff at 110–115°C for a short time. It reduces soluble crude protein and increases neutral detergent-insoluble crude protein of flaxseed by 114 and 71%, respectively. The heat treatment reduces in sacco ruminal degradability of flaxseed protein by increasing the size of the slowly degradable protein fraction, thereby reducing its ruminal degradation rate.

Proper selection of seed is a prerequisite for high-quality full-fat milled flaxseed. This is demonstrated in a patent (10) which claims that full-fat milled flaxseed obtained from seeds containing 5% or less of visually distinguishable darker color has good stability and shelf life measured as lipid peroxides (5 meq/kg oil), free fatty acids (2% oleic acid/sample), malonaldehyde (30 nM/mL) of sample, and alkenals (300 nM/mL) of sample. Furthermore, primary and secondary oxidation in freshly milled flaxseed evaluated with Saftest™ and MAPP™ systems reveals a highly significant increase in levels of undesirable oxidation products with an increase in percentage of dark seed content (11).

Milled full-fat flaxseed has good stability even when stored over 128 d at ambient temperature (23 ± 2°C) in triple-wrapped paper bags with plastic liners (12). The presence of endogenous antioxidants capable of preventing oxidation of the unsaturated fatty acids and the corresponding development of off-flavor is credited for such storage stability. The shelf life of whole ground flaxseed can be extended to 6 mon by grinding the flaxseed to a particle size of about (0.26–0.51 mm) in diameter in the presence of vitamin B_6 and zinc sulphate at 50–200 and 150–300 ppm, respectively, at (15.5°C) or room temperature (13). The flaxseed are packaged in a tin-foil-oxygen barrier bag with oxygen removed by flushing with 3–10 volumes of carbon dioxide or nitrogen.

Processing of Oil and Meal

Oil, which constitutes the largest fraction of flaxseed (35–43% dwt. basis), has been the primary focus of commercial flaxseed processing for industrial utilization. The meal, still considered a low-value secondary by-product of oilseed processing, is generally underutilized as an animal feed. Commercial processing of flaxseed for oil and meal is similar to that of other oilseeds and includes seed cleaning, flaking, cooking, pressing, solvent extraction, and solvent removal steps to yield oil and a meal. Processing of flaxseed for oil and meal has recently been reviewed (2,14), and new investigations that have appeared since are discussed here.

Cold-pressed flaxseed oil is one of the most recognized functional foods by health-conscious consumers. Continuous expellers or screw presses are used for the mechanical extraction of oil. Several expellers from the hydraulic press to the more sophisticated continuous screw press are commercially available for cold pressing flaxseed. The KOMET oil expeller, for example, features a special cold pressing system with a single conveying screw to squeeze the oil from the seeds. Expellers achieve the pressure needed to express oil by means of an auger that turns inside a closed barrel with an opening through which the oil drains. Small-scale oil presses or expellers are popular in many countries for batch processing of oil-bearing seeds including flaxseed. This process reduces the oil content of flaxseed from about 34 to about 5%. Screw pressing also compresses the meal into more dense cake, thereby facilitating further oil extraction. Industrial cold pressing of flaxseed, started in 1887 by Vaneputte in Belgium, still continues to this day, although the processing has been modernized to guarantee a very pure first-grade oil. This first press is followed by a second warm pressing to extract a maximum amount of oil (hot-pressed linseed oil) and produce nice flaxseed expellers. The flaxseed oil from mechanical extraction or expellers is sold to consumers at a premium as "First Press" oil, or encapsulated and marketed in specialty and health food stores.

New alternative advanced technology utilizing supercritical fluid extraction with liquid CO_2 has also been applied for the extraction of flaxseed oil. Flaxseed oil has lower solubility in CO_2 than other vegetable oils (corn, soybean, canola) due to its high unsaturated fatty acid content. Therefore, supercritical CO_2 extraction recovery of flaxseed oil is only 55–66% of that obtained by conventional extraction method. However, the recovery of ALA with supercritical CO_2 extraction is high (15). Oil recovery is increased when ethanol is used as a modifier because ethanol increases oil solubility and facilitates extraction of polar material. Optimum oil recovery with supercritical CO_2 can be obtained by a triple extraction involving two consecutive 30-min extractions without a modifier followed by a 30-min extraction with 15% ethanol (16). This triple extraction produces a 10% decrease in acyl lipid and an increase in phospholipids in the extracted oil. The fatty acid composition of flaxseed oil differs in accordance with the supercritical CO_2 method used for its extraction (Table 20.1). Supercritical CO_2 has also been used to extract oil (about 15%) in the press cake after cold pressing whole flaxseed (17).

Flaxseed oil has been extracted using isopropanol, instead of hexane, as an alternative solvent. Press cake from dried dehulled seed is defatted using azeotropic isopropanol (91% vol/vol), and the oil is separated from isopropanol by chilling the miscella (6). Oil yield with isopropanol is dependent on the particle size of dehulled seed and extraction time with optimum yield (13.6%) achieved at long extraction

TABLE 20.1

Fatty Acid Composition (%) of Flaxseed Oils Obtained with Supercritical Fluid Extraction Method[a]

Fatty acids	Method A	Method B	Soxhlet (hexane)	FOSFA (Petroleum ether)
C18:1	14.1–14.3	18.0–18.9	16.0	18.4
C18:2	12.8–13.4	14.5–14.9	14.0	14.7
C18:3	60.5–61.0	55.7–56.9	56.7	57.0

[a]Data from Bozan and Temelli (15) for Method A and Soxhlet, and from Barthet and Daun (16) for Method B and Federation of Oils, Seeds and Fats Associations (FOSFA) method.

time (7 h) and small particle size (1400 μm). Yield, composition, and quality of oil extracted from commercial flaxseed meal depend on the solvent/s used for extraction (18). For instance, oil extracted with methanol as primary solvent followed by petroleum ether or hexane contains substantial amounts of phosphatidylcholine, a major component of brain structural lipids. Similarly, the physicochemical characteristics of the solvent-extracted meal changes in accordance with the solvent system used and depends greatly on the polarity of the primary extracting solvent (19). Ethanol is the alternative solvent of choice for oil extraction because of its greater selectivity for specific lipid fractions and nonlipid extracts. In order to respond to the rising organic market, an extraction plant has been designed and operated in New Holland (England) to handle 10 different oilseeds, including flaxseed, using either hexane or ethanol. This plant is equipped with fluted and smooth rolls and immersion instead of a percolation extraction for maximum flexibility in both full- and pre-pressing during cold pressing of oilseeds (20).

Attempts at aqueous extraction process, touted for its beneficial environmental impact, improved process safety, and simultaneous production of oil and protein isolate or concentrate have been unsuccessful as a viable alternative for flaxseed oil extraction (6). Aqueous extraction of oil from dehulled flaxseed results in the formation of a highly stable emulsion, high residual oil, and dark color of the meal. However, high-quality flaxseed oil has also been extracted by the FRIOLEX® process based on water whereby a water soluble agent is used as a processing aid. In this process (Fig. 20.1) ground flaxseed is mixed in a chamber with water and the adjuvant alcohol, the suspension is reground, mixed thoroughly in a retention tank at 80°C for 30 min, and the oil is separated from the suspension in a decanter centrifuge. Alcohol and water are recovered and recirculated (21). The yield of flaxseed oil by this process is about 20%.

The oil and antinutritional compounds from flaxseed can be simultaneously extracted by a two-phase extraction system (22). This process involves (i) mixing flaxseed (10 kg) with a polar phase solution (0.08% NaOH 10% water, 90% methanol, 20 L) and grinding the slurry twice in a 100 mm i.d. Szego mill, and (ii) diluting the slurry to a solvent/seed ratio of 6.7:1 (vol/wt) with additional polar solvent before the pilot-plant scale extraction in the Karr column where the miscella and raffinate are separated and oil is recovered from the miscella.

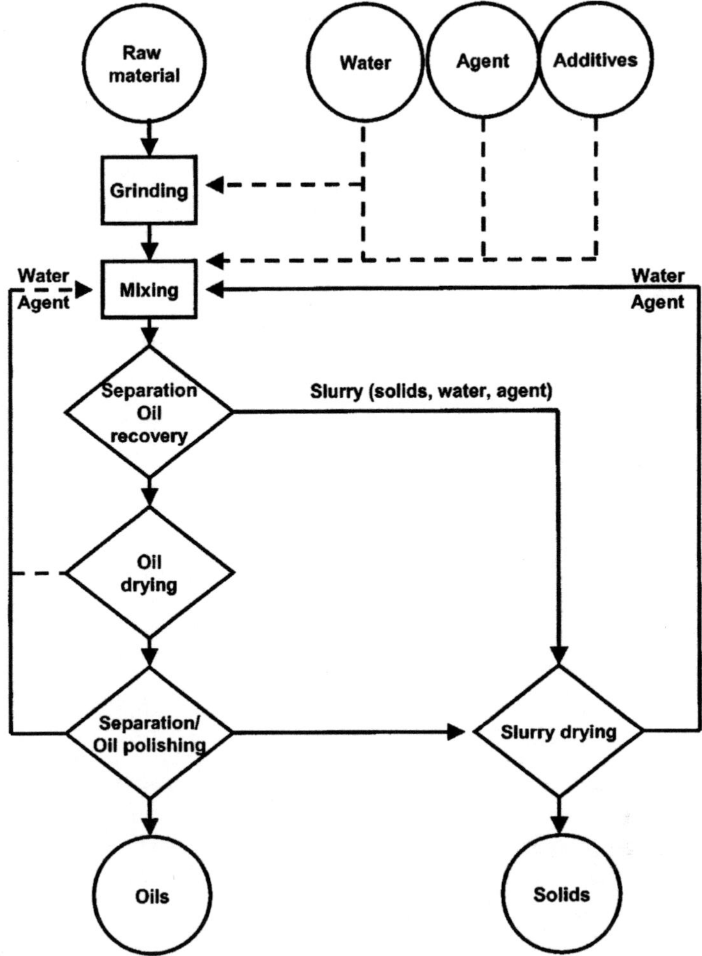

Fig. 20.1. The FRIOLEX® process for oil extraction. *Source:* Reprinted with permission from Hruschka and Frische (21).

Flaxseed oil can be protected from oxidation with improved consumer acceptance by the OmegaDry® process, the new alternative to microencapsulation using γ-cyclodextrin (23). γ-Cyclodextrin and omega oils form very stable dispersion in water that can easily be converted into white powders using conventional drying techniques. These complexes show significant increase in stability even at highly elevated temperatures of 100°C and against shear forces, and are compressible and adjustable in particle size, an advantage over microencapsulation techniques. γ-Cyclodextrin formulated flaxseed oil may be used to fortify various foods such as beverages, cereals, baked goods, spreads, dressings/sauces, and dairy products. It can also be used in nutraceutical formulations, such as tablets, powder drinks, and health bars.

Phosphatides

Flaxseed phosphatides together with flaxseed oil are claimed to reduce platelet adhesiveness, thereby inhibiting or reducing the incidence of coronary thrombosis in humans (24). This formulation has been proposed as a dietary anticoagulant prophylactic therapy. Flaxseed phosphatides are extracted from cold-pressed or expeller-pressed flaxseed cake with hexane, and the mixture is distilled to obtain a hexane-free oil that is then separated by centrifugation into a refined oil fraction and a gum fraction enriched in phosphatides. The resulting phosphatide contains 12% saturated fatty acids, 50% linolenic acid, 15% linoleic acid, 22% oleic acid, and 0.5% crude phosphatide.

Phosphatides may also be separated from triglycerides by membrane filtration because their theoretical molecular weights are approximately 800 and 900, respectively (25). Filtration, according to this process, is carried out at 50–60°C at recirculation flow rate of 0.3–0.4 m^3/h and 1–6 bars pressure. Isopropanol is used for pretreatment and cleaning of the membrane. Two flat-sheet ultrananofiltration membranes are generally used for this process.

Processing of Dietary Fiber

Flaxseed is well recognized for its high dietary fiber content consisting of soluble and insoluble fibers (9 and 20%, respectively) (2,26). The soluble fiber, commonly referred to as flaxseed gum or mucilage, has recently gained prominence as a functional food and nutraceutical ingredient due to its mitigating effect on diabetes, coronary heart disease, obesity, and colon and rectal cancer (27,28). In animal nutrition, the dietary fiber and especially the mucilage limits the metabolizable energy of the seed, influences satiety (29), exhibits laxative effects, and dilutes the beneficial effects of ALA and protein. The mucilage occurs mainly in the epidermis layer of the seed coat (30) which in turn makes up most of the hull. Because the hull and mucilage constitute about 40% of the total seed (31), their removal and recovery have been a primary concern. Extraction of flaxseed gums dates back to 1932 and has constantly been revisited and extensively reviewed (1,2). Several new developments have occurred since the recently published reviews (2,26) and are briefly discussed here.

The traditional aqueous extraction of flaxseed having been optimized (32) can now be spray dried optimally (33). The highest yield of spray-dried flaxseed gum is obtained at a water/seed ratio of 18L/kg and 62 and 92°C feed and outlet air temperatures, respectively. The flaxseed gum powder produced by this process has rheological characteristics that may be beneficial in certain applications where low viscosity is required such as in new-age health drinks and beverages, coatings, and encapsulations. Mucilage of high viscosity and elasticity can be obtained by separating seed surface mucilage produced by swelling flaxseed in water followed by fractional extraction (34). In this procedure water (1.2–5 times wt/wt of seed) is added to flaxseed, allowed to swell, and the mucilage separated by exerting physical energy such as vibration, impact, or shear on the seed. The mucilage is eluted by water addition and agitation followed by drying. An aqueous solution (1%) of this mucilage is claimed to have a viscosity of 100–1000 cps (25°C, 30 rpm, β-type viscometer) that turns into a semi-jel (viscosity of 500–20,000 cps) when heated to 50°C followed by cooling.

Many of the dry dehulling technologies as a process for recovering hulls, mucilage, and lignans from flaxseed have their foundation based on recent studies (35–37) (Fig. 20.2). Earlier reports claim to completely dehull flaxseed in a dehulling device at a minimal rotor speed of 4000 rpm with 450 kg/h throughput in a single pass through the mill and screening devices (38). The dehulling efficiency for flaxseed was not provided, although it was 90 and 95% for sunflower and cottonseed, respectively. The device operates on the principle of bouncing (impacting) the seed repeatedly between a rotor and a baffle plate until the impact cracks the hull and releases it from the kernel.

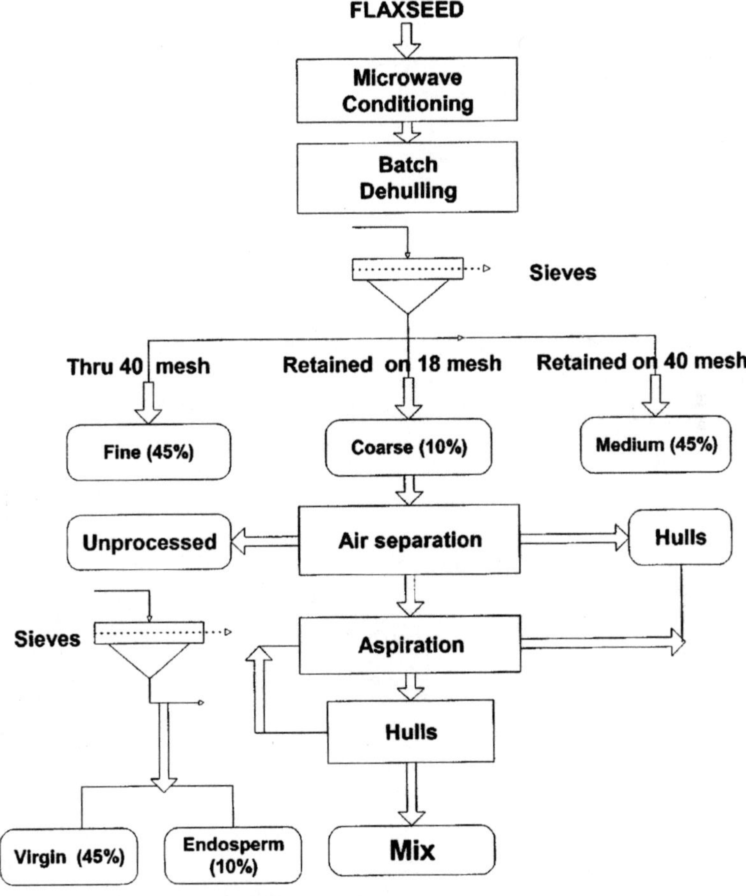

Fig. 20.2. Flaxseed batch fractionation.

A dry mechanical process, primarily based on the concept of abrasive dehulling (37) for obtaining a hull-rich fraction from flaxseed as a source of secoisolariciresinol diglycoside (SDG) has been evaluated (39). In this process, flaxseed is milled in an Urschel Comitrol Processor fitted with a 5-blade impeller rotating at 5120 rpm at 5kg/min, followed by sieving with a Ro-Tap sieve shaker using 16-, 20-, 30-, and 50-mesh U.S. standard sieves, and separated into a lighter hull-rich material and heavier, embryo-rich material with an air aspirator. The Comitrol Processor disintegrates the seed primarily by cutting and impact forces, resulting in the production of a broad spectrum of particle sizes thereby reducing the recovery of the hull fraction. Comparison can be made if a maximum possible effectiveness value of 42% for a dry mechanical process is assumed based on data for hand dissected flaxseed (39). The effectiveness of the mechanical processes (35,39,40) thus calculated is 32-36%, 27% and, 10-12%, respectively. Thus, the roller milling/ sieving process of Smith *et al.* (40) and the abrasive milling process of Oomah *et al.* (35) have better overall hull separation from the embryo compared with the process of Madhusudhan *et al.* (39).

Further improvements have taken place with the development of a bench-scale continuous dehulling process seeking to produce a high yield SDG concentrate from flaxseed (41). According to this process (Fig. 20.3), flaxseed is pearled in a Strong-Scott barley pearler modified for continuous flow, followed by sifting over 20-mesh screen to obtain a fine and a coarse fraction that is reprocessed through the pearler. The pearled, sifted flaxseed is then separated on a gravity table into whole seed, broken seed, and hulls with virtually no cotyledons. The yield of the hull and its oil content (246 g/kg and 30% oil, respectively) is similar to that obtained using an abrasive type dehuller (37).

A method of removing a hull layer from flaxseed by abrasion, pioneered earlier (35–37), has recently been patented (42). The hull is further divided into an outermost layer rich in water-dispersible mucilage and an inner layer particularly rich in fiber and lignan. The abrasion equipment consisting of millstones, such as a modified rice polishing device or device designed for oat abrasion, can be used to obtain a high lignan fraction (14.8 g/kg) at 4% yield after 1 min of abrasion. The protein and fat contents of such a product are 12 and 18% (db), respectively (42).

Seed conditioning has also been implemented prior to dehulling to generally improve the process and yield of hulls (35,37). Thus, flaxseed dried to 2.7% moisture content in an oven (50°C, 72 h) produces comparatively higher yields of hulls (296 *vs.* 246 g/kg for nontreated seeds) with the Strong-Scott barley pearler (41). The drying method may consist of oven drying, microwave drying, or fluid-bed drying, although the latter is apparently most suitable for the dehulling process (43). Optimization of the flaxseed dehulling process with the fluid-bed dryer, milling by barley pearler, and hull recovery by air separation produces hull yield of 22.6% with the following composition: carbohydrates, 48.3%; proteins, 16.8%; crude oil, 26.5%; moisture, 5%; and ash, 3.5% (43). Although this process is considered optimized, the resulting yield and composition of hulls are similar to those reported earlier (41), suggesting that very little improvement can be made with such a process.

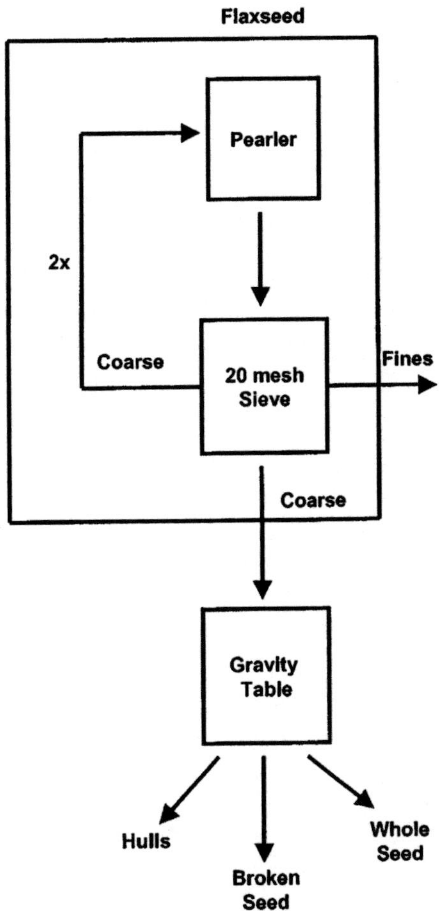

Fig. 20.3. Fractionation process to obtain hulls from flaxseed. *Source:* Reprinted with permission from Tostenson *et al.* (41).

Modification of Dietary Fiber

Attempts have been made to modify flaxseed hulls in order to improve its sensory characteristics and increase the amount of dietary fiber (gums or SDG) added to foodstuffs. Modifications include bleaching, deflavoring, and degreasing of hull obtained by abrasive dehulling (42). Bleaching is achieved with a water-alcohol treatment that may include lipase or protease enzymes and 2–7% hydrogen peroxide. Prior to bleaching, thermal treatment at 40–80°C for 0.5–2 h in a vacuum inactivates enzyme activity and removes low molecular weight compounds affecting flavor and taste. Apparently, the best sensory characteristics result from thermally treating flaxseed for 1 h in a vacuum at 60°C. Defatting of the hulls is generally achieved by hexane extraction.

Processing of Lignans

Flaxseed hull, in addition to being a rich source of fiber, also contains a very high concentration of lignans. The numerous health benefits of lignans, including bone and prostate health, renal protection, alleviation of menopausal symptoms, and protection against elevated blood pressure, coronary heart disease, and cancer (2,44), have stimulated technological innovations in its extraction and processing. Notable in this regard are two commercial endeavors interested in extraction of lignan from flaxseed: Archer Daniels Midlands has recently signed an exclusive licensing agreement for flaxseed lignan technology (45,46), and LinumLife™ claims a proprietary manufacturing process for producing a lignan extract 10–30 times higher than that in whole flaxseed or other flax ingredients (3). The most recent commercial interest in lignan is Natunola Health Inc., a subsidiary of Sentex Systems Ltd., which intends to produce lignans from its flaxseed dehulling and processing plant soon to be in production (47). Natunola processing is based on the abrasive dehulling process for separating the hulls from flaxseed established earlier (35–37) and patented recently (43).

The extraction, hydrolysis, and methods of lignan analysis have recently been reviewed (48,49). SDG, the flaxseed lignan of major interest, concentrates about fourfold in dried ethanol extracts of flaxseed initially defatted by cold pressing and supercritical fluid extraction (CO_2). Ethanol (80%) extraction of defatted flaxseed meal yields an SDG-rich fraction that can be incorporated in baked goods (17). SDG and cinnamic acid derivatives can be extracted from defatted flaxseed meal with aliphatic alcohol solvent (65–75% alcohol, at 4–30:1 alcohol to meal ratio for 1–4 h at room temperature), separating residual solids from the phenolic-rich alcohol solvent from which lignans and cinnamic acid derivatives are liberated by base-catalyzed hydrolysis (0.06–2.5% wt/vol sodium or barium methoxide or 1.25% wt/vol triethylamine or 1N potassium hydroxide 3–7% wt/vol for 4–24 h) (Fig. 20.4). The hydrolyzed extract can be further separated into SDG and cinnamic acid derivative fractions and purified to above 95% by anion exchange chromatography (45). The yield of SDG from this process is about 20 mg/g of defatted flaxseed, a 6000-fold increase over previously known techniques. High-speed counter current chromatography (HSCCC) has also been used to isolate SDG from flaxseed. The solvent system for the separation of SDG by HSCCC consists of tert-butyl methyl ether (TBME)-*n*-butanol-acetonitrile-water (1:3:1:5) at 3.0 mL/min flow rate (50).

Lignans can be readily obtained from an aqueous ethanol (85% ethanol) extract of flaxseed meal (Fig. 20.5). The aqueous fraction is ultrafiltered (5000 Mwt, cut-off), and the permeate is fractionated by absorbent resins (XAD-4) to enrich the SDG concentration from 1.9–13.4 mg/g phenolic acid. The SDG prepared by this method combined with isoflavones extracted from flaxseed has been claimed to treat neurological and immunological symptoms, migraine headaches, inflammations, and dementia (51,52). Abrasive dehulling is increasingly being recognized as an economically feasible method for processing flaxseed to obtain products containing high levels of lignans. Thus, a process involving milling, sieving, and aspiration yields a fraction enriched in SDG at 30.9 mg/g of defatted flour and 15 mg/g of oil (39). Improvements

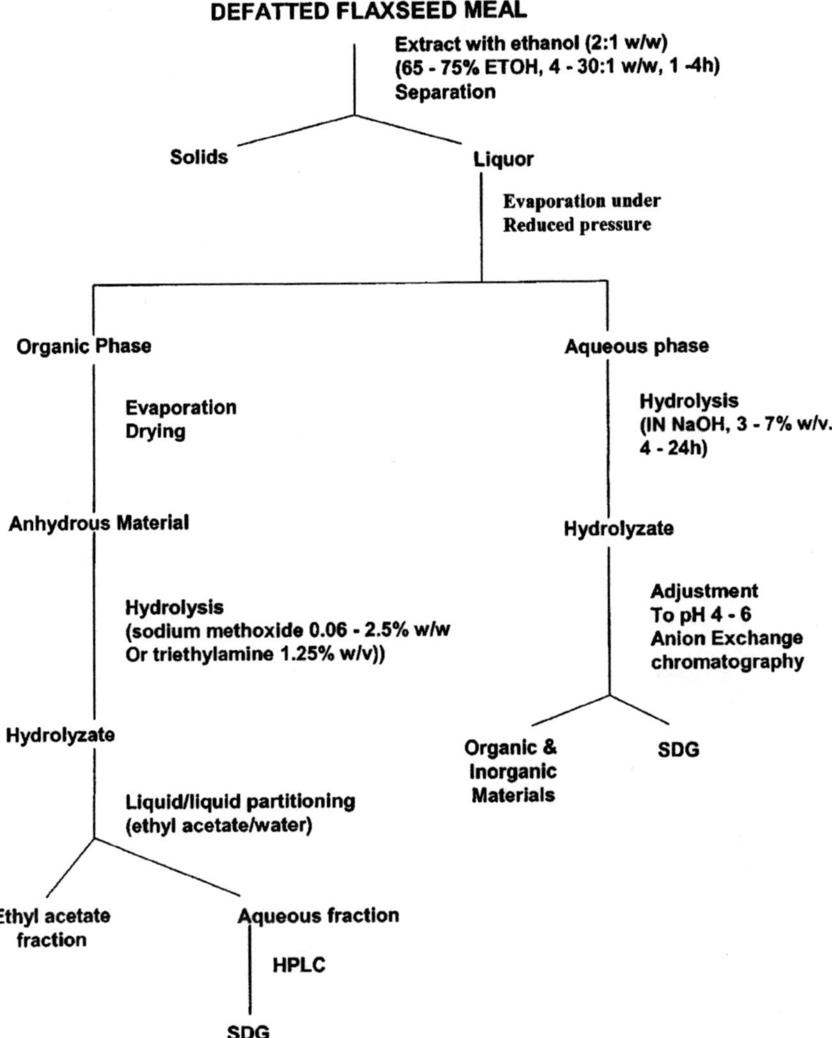

Fig. 20.4. Procedure for extraction of secoisolariciresinal diglycoside (SDG) from flaxseed meal. *Source:* Adapted from Westcott and Muir (45).

of this mechanical flaxseed fractionation intended for continuous processing produces a hull fraction rich in SDG at 21.7–24.2 mg/g of defatted flour (db) at a yield of about 108 mg/g of hull (41). A product obtained by abrasion using a rice polisher produces a powder from 40–180 mg/g yield of flaxseed with lignan content ranging from 10.7–14.8 mg/g, (db) (42).

In additon to SDG, the lignans isolariciresinol, pinoresinol, matairesinol, and anhydrosecoisolariciresinol also known as shonanin or ANHSEC, have been isolated

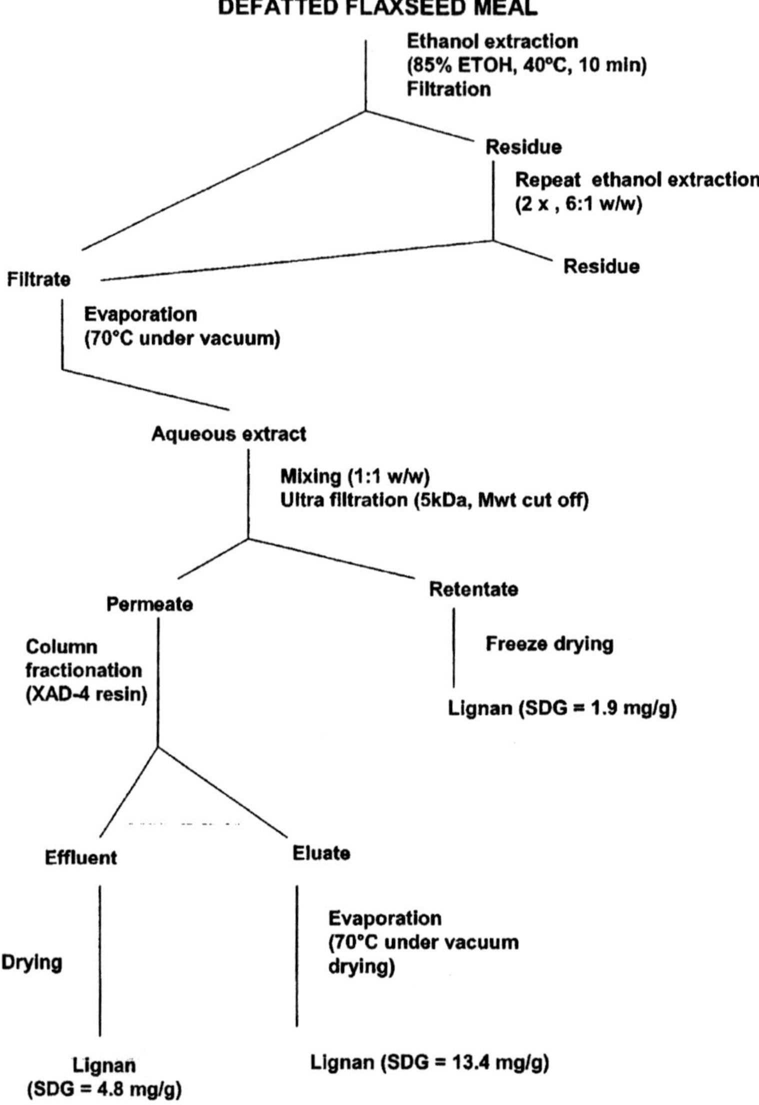

Fig. 20.5. Procedure for extraction of lignan. *Abbreviation:* See Figure 20.4. *Source:* Adapted from Empie and Gugger (51).

and characterized from flaxseed meal (53). Shonanin possesses phytoestrogen activity and is the metabolite of the mammalian lignan enterofuran that may play a role in the prevention of prostate cancer (54). It can be purified by semipreparative HPLC from a hydrolyzed hydroxymethanolic flaxseed extract at 4 mg/100 g yield of flaxseed powder (Fig. 20.6). Flaxseed is defatted with hexane for 1 h at 60°C in a shaker water bath, the supernatant obtained by centrifugation is acid-hydrolyzed at 100°C under reflux, extracted twice with ethyl acetate in hexane, and the organic fraction is evaporated to obtain the dry powder (55). Another compound, hydroxy-methyl glutaric acid (HMGA), considered to have hypocholesteremic effects, can be recovered as a complex containing 50% SDG, 25% total cinnamic acid glucosides and up to 10% HMGA from an alcohol extract of flaxseed meal by ultrafiltration in the 30–50 kDa range (56).

Processing of Protein

Attempts to obtain protein from flaxseed date back to the classical extraction of Osborne (57) (Table 20.2). Flaxseed protein can be isolated by one of the three commonly used technologies: (i) removal of seed coat that hinders the separation of proteins by dehulling, (ii) isoelectric precipitation of aqueous extract, or (iii) reduction of mucilage content by chemical means (i.e., soaking in 1% HCl for 16–24 h) or by enzymes. The difficulties involved in the recovery of protein from flaxseed mainly due to the presence and considerable interference of mucilage have been recognized and explicitly described over half a century ago (58). The same processes, aqueous extraction or mechanical treatments of flaxseed, are still being sought to remove the hull containing almost all the mucilage prior to protein extraction. According to this process, whole flaxseed is mixed with dilute acid (0.3N HCl) or alkali (0.3N NaOH) for 4–16 h in a solid/solvent ratio of 1:12 at 25°C, the residue separated by centrifugation or filtration, delipidated with hexane, and the protein from the mucilage- and oil-free meal extracted with 0.1N NaOH and precipitated at pH 3.8. About 90–95% of the protein content of flaxseed is recovered by this procedure.

Flaxseed mucilage can also be removed successfully by enzymatic treatment. Protein recovery is high (65–75%) from $NaHCO_3$ (0.10 M, 12 h) and viscozyme® (22.5 mg protein/100g, 3 h) treated seed meals at normal pH range (6–7) compared with a maximum of 53% for untreated flaxseed meal (59). Removal of seed coat mucilage has a major effect in enhancing protein yield of flaxseed meal via a solubilization process. The meal obtained from mucilage-reduced flaxseed has been used for the preparation of protein isolates through hexametaphosphate complexation (60). Under optimum sodium hexametaphosphate concentration and meal/solvent ratio, a protein isolate containing 78% of the total meal nitrogen can be obtained.

Protein enrichment can be accomplished by a two-phase solvent extraction system consisting of alkanol-ammonia-water and hexane (61). Flaxseed meal can be fractionated by a liquid cyclone process to obtain hull-free flour (62) from which protein can be recovered (59). The liquid cyclone process produces a flax flour at 62% yield of the meal and contains over 45% protein (N × 6.25) (63). The liquid cyclone process (Fig. 20.7), involves the formation of a defatted meal suspension in a nonaqueous liquid, preferably a mixture of hexane and alcohol, moisture adjustment of the meal (with a particle size of less than 55 μm), in the suspension to below 10% by weight, and subjecting the meal suspension to centrifugal gravity in a liquid cyclone or decanter centrifuge to obtain overflow and underflow streams of flour and hull, respectively.

B.D. Oomah

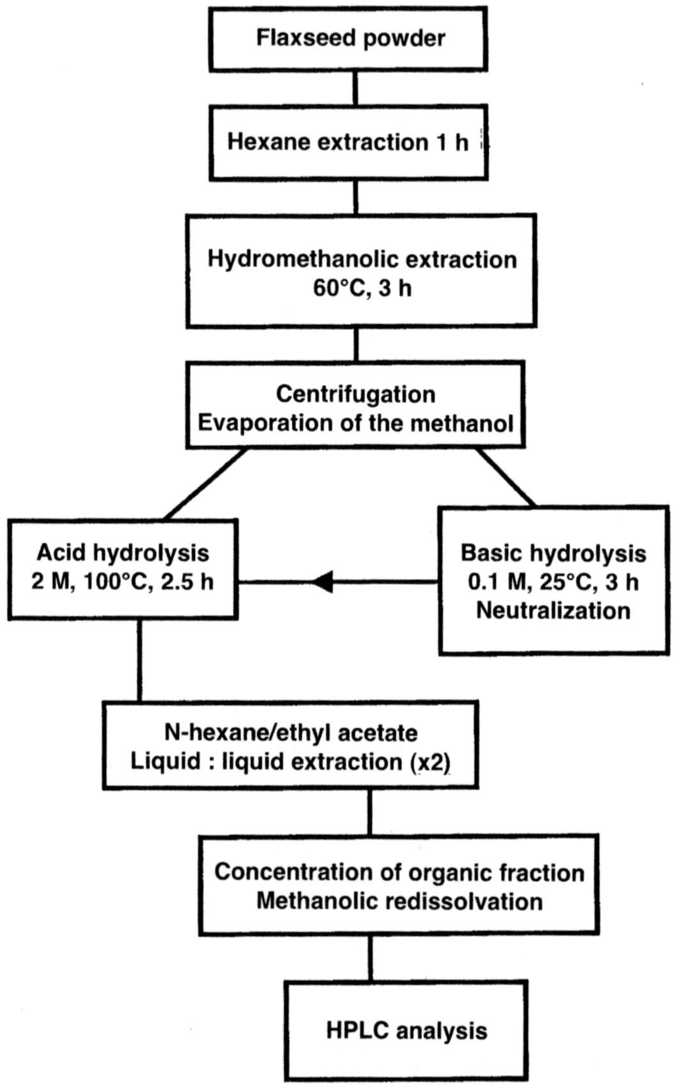

Fig. 20.6. Procedure for extraction and purification of shonanian from flaxseed. *Source:* Reprinted from Charlet *et al.* (55).

Flaxseed protein concentrate can be prepared by fluidization using ethyl alcohol in the presence of live steam (15–26 Psi at 105°C) for at least 30 min (64). The molecular structure of the protein is altered such that the protein digestibility index is reduced by this process, thereby enabling the protein to be more readily attacked by enzymes and microorganisms of the digestive system. The ruminal degradation rate

TABLE 20.2
Milestones in the History of Flaxseed Protein

Date	Achievement	Reference
1882	Classical extraction of flaxseed protein	57
1945	Isolation of major flaxseed protein 12S (linin) and 2S (colinin)	86
1946	Flaxseed protein separated by alkali dispersion and acid precipitation	40
1951	Patent production of flaxseed protein	58
1969	Isoelectric precipitation of flaxseed protein	87
1981	Extraction of low molecular weight of flaxseed protein	88
1983	Fractionation of 12S and 2S flaxseed protein by ammonium sulfate precipitation	89
1986	Characterization of the functional properties of flaxseed protein by isolectric precipitation	30
1994	Optimization of protein extraction with response surface methodology	90
1996	Patent on flax protein preparation	91
2001	Proposed functionality of flaxseed protein in human physiology	1
2002	Preparation of protein isolate by micellization	67
2002	Charcterization of protein complexes in flaxseed	68

per hour and the extent of degradation over 24 h are also considerably reduced by this process. This reduction of degradation is important because it reduces ruminal microdigestion which in turn increases the amount of available protein for subsequent absorption.

Proteins can also be prepared and purified from flaxseed and flaxseed meal (65,66). This process (Fig. 20.8) comprises extraction of the meal with an aqueous food-grade salt solution (NaCl, ionic strength of 0.2, pH 5–6.8 at 15–75°C), separation of the protein by defatting, concentration of the protein solution followed by micellization, separation, and drying to produce a dried protein powder with 90% protein and less than 1% fat. Defatting is accomplished by chilling the protein solution below 10°C without agitation, by decanting, centrifugation, and/or fine filtration. Concentration is achieved by membrane filtration, ultrafiltration or diafiltration.

Flaxseed protein isolates prepared by micellization and isoelectric precipitation consist mainly of 11S globulin also known as linin (67). Isolates prepared by micellization have low contents of phytic acid and pentosans, known to exert marked effects on functional properties (emulsifying, foaming, water- and oil-binding properties) of the protein, and therefore they have significant benefits. Phytic acid and pentosans are removed by the ultrafiltration step during processing. The two procedures for flaxseed protein isolation, isoelectric precipitation and micellization, influence both the amount of nonprotein components and the conformation of the isolate. Micellization preserves the native state of the protein while removing non protein components. Isoelectric precipitation, however, risks partial denaturation and irreversible protein aggregation (67).

Preparation of protein extracts for subsequent fractionation of flaxseed protein requires an effective approach such as dehulling for removal of mucilage (68). Soaking, acid, alkali, and enzyme pretreatments of seeds, recommended to overcome

                                        B.D. Oomah

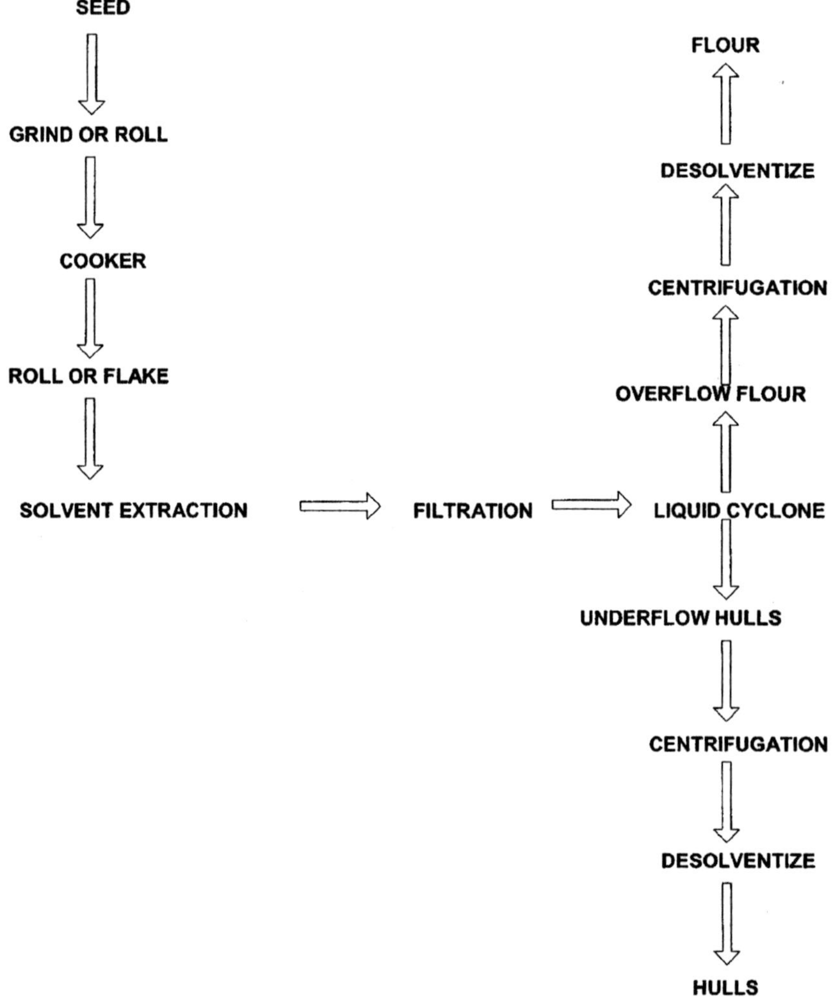

Fig. 20.7. Liquid cyclone fractionation. *Source:* Redrawn from Sosulski and Zadernowski (62).

the interference of mucilage for protein extraction (59), reduce viscosity insufficiently to permit gel chromatography. Protein extracts from dehulled seeds, unlike those from other treatments, have low viscosity that enables fractionation to occur.

Product Development

The processing of flaxseed has sparked interest in the development and proliferation of numerous new products in addition to those described earlier (2), especially for the functional foods and nutraceutical markets. These include the use of flaxseed and/or

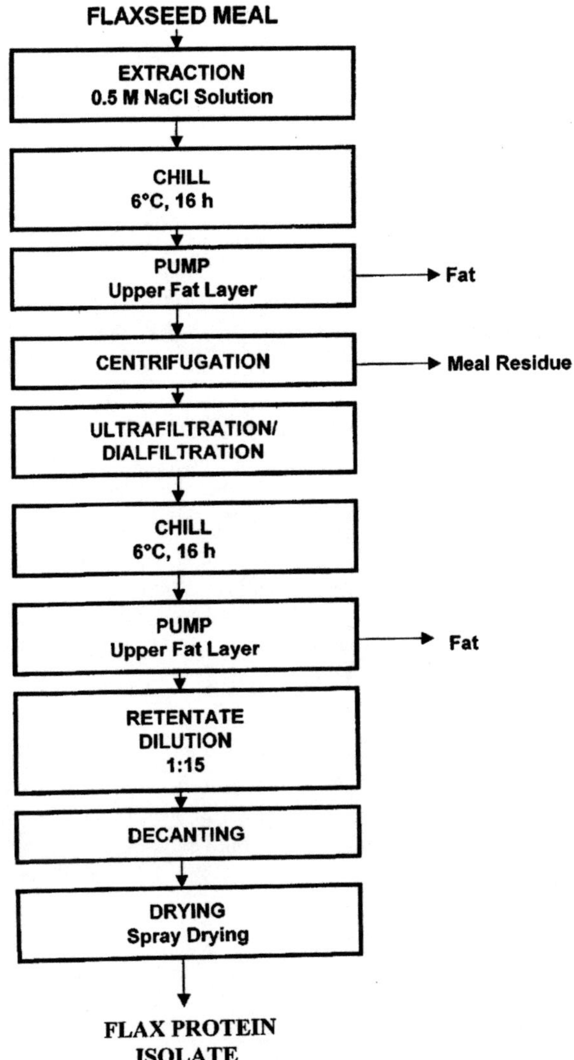

Fig. 20.8. Process for production of flax protein isolate. *Source:* Adapted from Murray (65).

flaxseed oil as a source of n-3 fatty acids (PUFA) in mayonnaise, salad dressing, and bread formulated just for men, and in a range of animal foods and feeds (69). It is interesting to note that the introduction of men's bread at the end of 2001 by French Meadow Bakery, Minneapolis, Minnesota, precedes clinical evidence of flaxseed protection against prostate cancer (70). A standardized lignan extract, LinumLife™, purported to have 30 times the levels of lignans found in other flax ingredients by Acatris (Netherlands) is also aimed at the men's health care market (3).

Flaxseed and flaxseed ingredients are being rediscovered for use in various cereals and bakery products. For example, Sounas (U.K.), recently introduced milled flaxseed bread high in n-3 PUFA and low in cholesterol. Flax Plus New Nature's Path organic boxed cereal (Nature's Path Food, Canada) claims 500 mg of n-3 PUFA per serving. Uncle Sam Cereal (U.S. Mills, Needham, MA), which contains flaxseed as one of its principal ingredients, is one of the oldest cereals being embraced by fitness and weight loss experts. Flaxseed in Uncle Sam Cereal is believed to suppress the appetite resulting in weight loss. New encapsulation technology (71) offering controlled release, shelf-life enhancement, and stability, increased and consistent potency and improved taste and texture, allows flaxseed n-3 PUFA to be used for cereal fortification and baked goods without moisture uptake and migration.

The recent proliferation of breads, cereals, snack foods and other novel foods, containing flaxseed and soy in Australia has been the basis for the pilot study indicating that regular inclusion of these foods in the diet may improve plasma lipids in hypercholesterolemic individuals (72). This study demonstrates the feasibility of supplementing the diet with flaxseed- and soy-containing foods to achieve clinically useful reductions of total, LDL- and non-HDL cholesterol. Cold-pressed flaxseed meal alone is known to be effective in lowering LDL cholesterol (73). Flaxseed, in addition to being as effective as oral estrogen-progesterone in improving mild menopausal symptoms, lowers glucose and insulin levels (74). Recent animal studies suggest that dietary flaxseed and soy may also have a positive effect on obesity and diabetes in humans (75). These two nutritional disorders have become major public health concerns due to their cardiovascular risk factors responsible for excess morbidity and mortality. Products containing both flaxseed and soy already exist, especially in the nutritional bar category, a growing sector of the marketplace. Zoe Foods (Newton, MA), for example, with its combination of flax and soy products, initially targeting women's health, now produces products with health benefits for people of all ages (76). Mais Foods (Boxboro, MA) combines milled flaxseed with soy and a low glycemic sweetener into a health snack, "caramel soy-flax chews," to provide hormone replacement therapy (HRT) benefits without the risk (77). Manufacturers in Australia promote the use of commercially produced bread, Burgen® Soy-Lin™ (Chatswood, NSW, Australia), that combines soy and flaxseed and claims to contain 220 mg of isoflavones and lignans per four slices to meet the phytoestrogenic needs of women seeking alternatives to HRT (48). Products from flaxseed is also being endorsed by the Centre for Advancement in Cancer Education, primarily due to the numerous health benefits related to n-3 PUFA fatty acids and lignans (78).

Flaxseed ingredients may be used in a variety of applications such as bakery products, cereals, snacks, nutrition bars, nutritional beverages, supplements, and pet foods. Flax also has potential value in ice cream and frozen desserts (79). The functionality of flaxseed has been studied in frozen desserts, particularly its use as a partial milk fat replacer, and prototypes using flaxseed oil as a source of n-3 PUFA have been developed. Vanilla frozen desserts with flaxseed oil replacing 10–25% of the milk fat using a bench scale freezer have been produced. Vanilla and chocolate frozen desserts with flaxseed oil replacing 25% of the milk fat have also been produced using a household freezer. Replacement of milk fat by 10% flaxseed oil results in a product with a texture similar to that of controls containing 12% milk fat. Further evaluation is pending on optimizing the level of milk fat replacement by flaxseed oil and its effect on flavor and oxidative stability as well as the functionality of a high-lignan extract on frozen dessert quality (79).

Flaxseed has been specifically designed for smooth nutritional and energy beverage applications. The fine-milled product BevGrad™ blends smoothly into beverages prior to pasteurization or hot-fill processing. The product can boost the energy and nutritional value of vegetable and fruit drinks, smoothies, teas, and soy milks (80).

Products from flaxseed are also being designed for the dietary supplement and pharmaceutical markets. For example, Bioflax®, a mucilage-enriched partially defatted flaxseed meal preparation, has been introduced in the Polish pharmaceutical market (81). Both maceratio and infusium Bioflax® preparations exert gastroprotective effects and are proposed for use in the treatment of peptic ulcers. Orion-Pharma (Finland) has introduced a flax-based dietary supplement (Linus dietary supplement) for stomach well-being (82). Another flaxseed mucilage product, salinum, is recommended as a saliva substitute (83). Sensiline (Silab, Brive, France), active ingredient obtained from flaxseed, rich in polysaccharides (beta-glucan, uronic acids) and in peptidoglycan, is considered to be an important pharmaceutical product. Sensiline limits skin irritation by reducing the release of interleukin, the key molecule in the inflammatory cascade.

Conclusions

Like other grains, flaxseed is subjected to processing to obtain desirable products by optimizing consumer-friendly attributes, such as flavor, color, texture, and appearance, as well as to provide shelf-stable products. Flaxseed may be extruded, puffed, flaked, or altered to improve product quality. These processes condition the grain for digestion and absorption by the body and may also enhance the desirability and therefore the consumption of flaxseed. However, the level of flaxseed that can be added to foods without affecting product quality is limited. In bakery products, for example, milled flaxseed can be used at 6–8% of dry ingredients. This limitation has been a determining factor in the development of processing technologies leading to the fractionation of components, such as oil, fiber, lignans, and proteins, from flaxseed. Nevertheless, the proliferation of products containing flaxseed or flaxseed ingredients for their health benefits has urged major food companies to adopt processing technologies for the introduction of innovative products. Unilever (Unilever, UK), for example, in a radical move, is adding n-3 oils, known for their cardiovascular benefits, to its entire range of polyunsaturated spreads sold under the Flora brand in the U.K. Unilever is using ALA from flaxseed for heart health improvements to the Flora brand and will be adding it to its highly successful cholesterol-lowering spread in

2003 (84). Some of the processing technologies commonly used in the food industry, such as extrusion, fermentation, microwave, and nonthermal processes, have not been adequately explored in the processing of flaxseed. Flaxseed is already one of the fastest growing top 10 supplements in terms of appreciable dollar sales (48% sales growth *vs.* 1 y ago) (85). These and other innovative processing technologies will provide an excellent opportunity for market expansion in the area of functional foods and nutraceuticals.

References

1. Oomah, B.D., Flaxseed as a Functional Food Source, *J. Sci. Food Agric. 81:*889–894 (2001).
2. Oomah, B.D., and G. Mazza, Flaxseed Products for Disease Prevention, in *Functional Foods: Biochemical and Processing Aspects*, edited by G. Mazza, CRC Press, Boca Raton, Florida, 1998, Vol. 1, pp. 91–138.
3. Anonymous, Newly-Formed Acatris Unveils Lignan Product, *Nutraceuticals Intern. 7(6):*16 (2002).
4. Anonymous, Omega-3 Specialist Finds New Owner and Invests in Production Facilities, *New Nutr. Business*, April 19, 2002.
5. Dorrell, D.G., Distribution of Fatty Acids Within the Seed of Flax, *Can. J. Plant Sci. 50:* 71–75 (1970).
6. Kadivar, M., Studies on Integrated Processes for the Recovery of Mucilage, Hull, Oil and Protein from Solin (Low Linolenic Acid Flax), *Ph.D. Thesis*, University of Saskatchewan, Saskatoon, Saskatchewan, Canada, SK, 200, pp. 177.
7. Tolkachev, O.N., and A.A. Zhuchenko, Jr., Biologically Active Substances of Flax: Medicinal and Nutritional Properties (A Review), *Pharm. Chem. J. 34:*360–367 (2000).
8. Goodridge, J.M., Use of Formaldehyde and Lignosulfonate Treated Flaxseed and Linola to Alter Fatty Acid Composition and Increase Omega-3-Fatty Acids in Milk Fat of Dairy Cows, *MSc. Thesis*, University of Manitoba, Canada, 1998, pp. 1–24.
9. Mustafa, A.F., J.J. McKinnon, D.A. Christensen, and T. He, Effects of Micronization of Flaxseed on Nutrient Disappearance in the Gastrointestinal Tract of Steers, *Animal Feed Sci. Technol. 95:*123–132 (2002).
10. Pizzey, G.R., U.S. Patent 6,368,650 (2002).
11. Pizzey, G., and T. Luba, Effect of Seed Selection and Processing on Stability of milled Flaxseed, in *Abstracts 93rd AOCS Annual Meeting & Expo*, Montréal, Québec, Canada, 2002, p. S143.
12. Malcolmson, L.J., R. Przybylski, and J.K. Daun, Storage Stability of Milled Flaxseed, *J. Am. Oil Chem. Soc. 77:*235–238 (2000).
13. Stitt, P.A., U.S. Patent 4,857,326 (1989).
14. Kolodziejczyk, P.P., and P. Fedec, Processing Flaxseed for Human Consumption, in *Flaxseed in Human Nutrition*, edited by S.C. Cunnane and L.U. Thompson, AOCS Press, Champaign, Illinois, 1995, pp. 261–280.
15. Bozan, B., and F. Temelli, Supercritical CO_2 Extraction of Flaxseed, *J. Am. Oil Chem. Soc. 79:*231–235 (2002).
16. Barthet, V.J., and J.K. Daun, An Evaluation of Supercritical Fluid Extraction as an Analytical Tool to Determine Fat in Canola, Flax, Solin, and Mustard, *J. Am. Oil Chem. Soc. 79:*245–251 (2002).

17. Jenkins, D.J.A., Incorporation of Flaxseed or Flaxseed Components into Cereal Foods, in *Flaxseed in Human Nutrition*, edited by S.C. Cunnane and L.U. Thompson, AOCS Press, Champaign, Illinois, 1995, pp. 281–294.

18. Oomah, B.D., G. Mazza, and R. Przybylski, Comparison of Flaxseed Meal Lipids Extracted with Different Solvents, *Lebensm.Wiss.u.-Technol. 29:*654–658 (1996).

19. Oomah, B.D., and G. Mazza, Processing of Flaxseed Meal: Effect of Solvent Extraction on Physicochemical Characteristics, *Lebensm.Wiss.u.-Technol. 26:*312–317 (1993).

20. Anonymous, Multi-Oilseed, Dual-Solvent Plant Opens, *Intern. News Fats, Oils Related Mat. 13:*572 (2002).

21. Hruschka, S., and R. Frische, A New Oil Extraction Process: FRIOLEX®, *Oléagineux, Corps Gras, Lipides 5:*356–360 (1998).

22. Varga, T.K., and L.L. Diosady, Simultaneous Extraction of Oil and Antinutritional Compounds from Flaxseed, *J. Am. Oil Chem. Soc. 71:*603–607 (1994).

23. Reuscher, H., OmegaDry®—The New Alternative to Microencapsulation of Omega Oils Using Gamma Cyclodextrin, in *Abstracts 93rd AOCS Annual Meeting & Expo.*, Montréal, Québec, Canada, 2002, pp. S41–42.

24. Martin, W., U.S. Patent 4,061,738 (1977).

25. Koris, A., G. Vatai, and Z. Kemény, Experimental Investigation and Modeling of Dry Membrane Degumming, in *Abstracts 93rd AOCS Annual Meeting & Expo.*, Montréal, Québec, Canada, 2002, pp. S43.

26. Cui, S.W., Flaxseed Gum, in *Polysaccharide Gums from Agricultural Products Processing, Structures & Functionality*, edited by S.W. Cui, Technomic Publishing Co., Lancaster, Pennsylvania, 2001, pp. 59–101.

27. Oomah, B.D., and G. Mazza, Bioactive Components of Flaxseed: Occurrence and Health Benefits, in *Phytochemicals and Phytopharmaceuticals*, edited by F. Shahidi and C-T. Ho, AOCS Press, Champaign, Illinois, 2001, pp. 106–121 (2000).

28. Jenkins, D.J.A., T.M.S. Wolever, J. Kalmusky, S. Guidici, C. Giordani, G.S. Wong, J.N. Bird, R. Patten, M. Hail, G. Buckley, and J.A. Little, Low Glycemic Index Carbohydrate Foods in the Management of Hyperlipidemia, *Am. J. Clin. Nutr. 42:*604–617 (1985).

29. Novak, C., and S.E. Scheideler, Long-Term Effects of Feeding Flaxseed-Based Diets. 1. Egg Production Parameters, Components, and Eggshell Quality in Two Strains of Laying Hens, *Poult. Sci. 80:*1480–1489 (2001).

30. Dev, D.K., and E. Quensel, Preparation and Functional Properties of Linseed Protein Products Containing Differing Levels of Mucilage, *J. Food Sci. 53:*1834–1837, 1857 (1988).

31. Mandokot, V.M., and N. Singh, Studies on Linseed (*Linum usitatissimum*) as a Protein Source for Poultry. I. Process of Demucilaging and Dehulling of Linseed and Evaluation of Processed Materials by Chemical Analysis and with Rats and Chicks, *J. Food Sci. Technol. Mysore 16:*25–31 (1979).

32. Cui, W., G. Mazza, B.D. Oomah, and C.G. Biliaderis, Optimization of an Aqueous Extraction Process for Flaxseed Gum by Response Surface Methodology, *Lebensm.Wiss.u.-Technol. 27:*363–369 (1994).

33. Oomah, B.D., and G. Mazza, Optimization of a Spray Drying Process for Flaxseed Gum, *Intern. J. Food Sci. Technol. 36:*135–143 (2001).

34. Hanaoka, J., Japanese Patent JP55165902 (1980).

35. Oomah, B.D., G. Mazza, and E.O. Kenaschuk, Dehulling Characteristics of Flaxseed, *Lebensm.Wiss.u.-Technol. 29:*245–250 (1996).

36. Oomah, B.D., and G. Mazza, Effect of Dehulling on Chemical Composition and Physical Properties of Flaxseed, *Lebensm.Wiss.u.-Technol. 30:*135–140 (1997).
37. Oomah, B.D., and G. Mazza, Fractionation of Flaxseed with a Batch Dehuller, *Industrial Crops Products 9:*19–27 (1998).
38. Ashes, J.R., and N.J. Peck, A Simple Device for Dehulling Seeds and Grain, *Animal Feed Sci. Technol. 3:*109–116 (1978).
39. Madhusudhan, B., D. Wiesenborn, J. Schwarz, K. Tostenson, and J. Gillespie, A Dry Mechanical Method for Concentrating the Lignan Secoisolariciresinol Diglucoside in Flaxseed, *Lebensm. Wiss.u.-Technol. 33:*268–275 (2000).
40. Smith, A.K., V.L. Johnsen, and A.C. Beckel, Linseed Proteins: Alkali Dispersion and Acid Precipitation, *Industrial Engineer. Chem. 38:*353–356 (1946).
41. Tostenson, K., D. Wiesenborn, X. Zhang, N. Kangas, and J. Schwarz, Evaluation of Continuous Process for Mechanical Flaxseed Fractionation, in *Proceedings of the 58th Flax Institute of the United States*, Fargo, North Dakota, pp. 17–22 (2000).
42. Myllymäki, O., U.S. Patent 6,440,479 (2002).
43. Cui, W., and G. Mazza, Canadian Patent 2,167,951 (2002).
44. Prasad, K., Canadian Patent 2,311,606 (2000).
45. Westcott, N.D., and A.D. Muir, U.S. Patent 5,705,618 (1998).
46. Anonymous., ADM Licensing Agreement for Flax Lignan Technology, available at http://www.preparedfoods.com/newsletter/articles/0701/0107000_2c.htm, July 17, 2001.
47. Guly, C., Finessing the Flax, The Ottawa Citizen, Oct. 9, 2002, p. F9.
48. Oomah, B.D., Phytoestrogens, in *Methods of Analysis for Functional Foods and Nutraceuticals*, edited by W.J. Hurst, CRC Press, Boca Raton, Florida, 2002, pp. 1–63.
49. Westcott, N.D., and A.D. Muir, Medicinal Lignans from Flaxseed: Isolation and Purification, in *Phytochemicals and Phytopharmaceuticals*, edited by F. Shahidi and C-T. Ho, AOCS Press, Champaign, Illinois, 2000, pp. 122–131.
50. Degenhardt, A., S. Habben, and P. Winterhalter, Isolation of the Lignan Secoisolariciresinol Diglucoside from Flaxseed (*Linum usitatissimum* L.) by High-Speed Counter-Current Chromatography, *J. Chromatogr. A 943:*299–302 (2002).
51. Empie, M , and E. Gugger, U.S. Patent 6,391,310 (2002).
52. Empie, M , and E. Gugger, U.S. Patent 6,399,072 (2002).
53. Meagher, L.P., G.R. Beecher, V.P. Flanagan, and B.W. Li, Isolation and Characterization of the Lignans, Isolariciresinol and Pinoresinol in Flaxseed Meal, *J. Agric. Food Chem. 47:*3173–3180 (1999).
54. Liggins, J., R. Grimwood, and S.A. Bingham, Extraction and Quantification of Lignan Phytoestrogens in Food and Human Samples, *Anal. Biochem. 287:*102–109 (2000).
55. Charlet, S., L. Bensaddek, S. Raynaud, F. Gillet, F. Mesnard, and M-A. Fliniaux, An HPLC Procedure for the Quantification of Anhydrosecoisolariciresinol: Application to the Evaluation of Flax Lignan Content, *Plant Physiol. Biochem. 40:*225–229 (2002).
56. Westcott, N.D., and D. Paton, U.S. Patent 6,264,853 (2001).
57. Osborne, T.M., Proteins of the Flaxseed, *J. Am. Chem. Soc. 14:*629–661 (1892).
58. Vassel, B., U.S. Patent 2,573,072 (1951).
59. Wanasundara, P.K.J.P.D., and F. Shahidi, Removal of Flaxseed Mucilage by Chemical and Enzymatic Treatments, *Food Chem. 59:*47–55 (1997).
60. Wanasundara, P.K.J.P.D., and F. Shahidi, Optimization of Hexametaphosphate-Assisted Extraction of Flaxseed Proteins Using Response Surface Methodology, *J. Food Sci. 61:*604–607 (1996).

61. Wanasundara, J.P.D., and F. Shahidi, Alkanol-Ammonia-Water/Hexane Extraction of Flaxseed, *Food Chem. 49*:39–44 (1994).

62. Sosulski, F., and R. Zadernowski, Fractionation of Rapeseed Meal into Flour and Hull Components, *J. Am. Oil Chem. Soc. 58*:96–98 (1981).

63. Bhatty, R.S., and P. Cherdkiatgumchai, Compositional Analysis of Laboratory-Prepared and Commercial Samples of Linseed Meal and of Hull Isolated from Flax, *J. Am. Oil Chem. Soc. 67*:79–84 (1990).

64. Stahel, N.G., U.S. Patent 4,543,264 (1985).

65. Murray, E.D., U.S. Patent 5,844,086 (1998).

66. Murray, E.D., New Plant Protein Isolate, *Food Sci. Technol. 15(2)*:44–47 (2001).

67. Krause, J-P., M. Schultz, and S. Dudek, Effect of Extraction Conditions on Composition, Surface Activity and Rheological Properties of Protein Isolates from Flaxseed (*Linum usitatissimum* L), *J. Sci. Food Agric. 82*:970–976 (2002).

68. Li-Chan, E.C.Y, F. Sultanbawa, J.N. Losso, B.D. Oomah, and G. Mazza, Characterization of Phytochelatin-Like Complexes from Flax (*Linum usitatissimum*) Seed, *J. Food Biochem. 26*:271–294 (2002).

69. O'Donnell, C.D., Ingredients in Use: Omega-3 Fatty Acids, *Prepared Foods 171*:43–46 (2002).

70. Demark-Wahnefried, W., D.T. Price, T.J. Polascik, C.N. Robertson, E.E. Anderson, D.F. Paulson, P.J. Walther, M. Gannon, and R.T. Vollmer, Pilot Study of Dietary Fat Restriction and Flaxseed Supplementation in Men with Prostate Cancer Before Surgery: Exploring the Effects on Hormonal Levels, Prostate-Specific Antigen, and Histopathological Features, *Urology 58*:47–52 (2001).

71. Wu, W-H., W.S. Roe, V.G. Gimino, V. Seriburi, D.E. Martin, and S.E. Knapp, U.S. Patent 6,153,236 (2000).

72. Ridges, L., R. Sunderland, K. Moerman, B. Meyer, L. Astheimer, and P. Howe, Cholesterol Lowering Benefits of Soy and Linseed Enriched Foods, *Asia Pacific J. Clin. Nutr. 10*:204–211 (2001).

73. Jenkins, D.J.A., C.W.C. Kendall, E. Vidgen, S. Agarwal, A.V. Rao, R.S. Rosenberg, E.P. Diamandis, R. Novokmet, C.C. Mehling, T. Perera, L.G. Griffin, and S.C. Cunnane, Health Aspects of Partially Defatted Flaxseed, Including Effects on Serum Lipids, Oxidative Measures, and *ex vivo* Androgen and Progestin Activity: A Controlled Crossover Trial, *Am. J. Clin. Nutr. 69*:395–402 (1999).

74. Lemay, A., S. Dodin, N. Kadri, H. Jacques, and J-C. Forest, Flaxseed Dietary Supplement Versus Hormone Replacement Therapy in Hypercholesterolemic Menopausal Women, *Obstetrics Gynecol. 100*:495–504 (2002).

75. Anonymous, Dietary Soy and Flaxseed Help Obesity and Diabetes, *Intern. News Fats, Oils Related Mat. 13*:578–579 (2002).

76. Anonymous, Extended Distribution for Flax and Soy Brands, available at http://www.nutraingredients.com/news/news.asp?id=4784, 19/06/02, June 19, 2002.

77. Anonymous, Age-Mediated Cancers Stroke Flaxseed Interest, *Prepared Foods 171(10)*:81 (2002).

78. Anonymous, Health Endorsement for Heintzman Farms, available at http://www.nutraingredients.com/news/news.asp?id=5419 , 17/09/02, September 17, 2002.

79. Pszczola, D.E., 31 Ingredient Developments for Frozen Desserts, *Food Technol. 56(10)*:46–65 (2002).

80. Anonymous, An Instant Flax Back, *Prepared Foods 171(8)*:88 (2002).

81. Kozlowska, J., K. Szczawinska, T. Bobkiewicz, and R. Forjasz-Grus, The Results of Pharmacological Investigations of Bioflax® Nutrition and Healthy Product Obtained from Flaxseed, in *Proceedings of the 58th Flax Institute of the United States*, Fargo, North Dakota, 2000, pp. 100–107 (2000).

82. Anonymous, New Products, *New Nutr. Business 7:*31 (2002).

83. Attstrom, R., P.O. Glantz, H. Hakansson, and K. Larsson, U.S. Patent 5,260,282 (1993).

84. Anonymous, Unilever Decides Omega-3s Have Come of Age, available at http://www.new-nutrition.com/newspage/061202a.htm, 06/12/02, June 12, 2002.

85. Marra, J., The State of Dietary Supplements, *Nutraceuticals World 5 (11):*32–40 (2002).

86. Vassel, B., and L.L. Nesbitt, The Nitrogenous Constituents of Flaxseed.II. The Isolation of a Purified Protein Fraction, *J. Biol. Chem. 159:* 571–584 (1945).

87. Sosulski, F., and A. Bakal, Isolated Proteins from Rapeseed, Flax, and Sunflower Meals, *Canad. Inst. Food Technol. J. 2:*28-32 (1969).

88. Youle, R.J., and A.H.C. Huang, Occurrences of Low Molecular Weight and High Cysteine Containing Albumin Storage Proteins in Oilseeds of Diverse Species, *Am. J. Bot. 68:*44-48 (1981).

89. Madhusudhan, K.T., and N. Singh, Studies on Linseed Proteins, *J. Agric. Food Chem. 31:*959-963 (1983).

90. Oomah, B.D., G. Mazza, and W. Cui, Optimization of Protein Extraction from Flaxseed Meal, *Food Res. Intern. 27:*355–361 (1994).

91. Kankaanpää-Antilla, B., and M. Antilla, International Patent PCT/F196/00042 (1996).

Chapter 21

Flaxseed Proteins: Potential Food Applications and Process-Induced Changes

P.K.J.P.D. Wanasundara[a], and F. Shahidi[b]

[a]Saskatchewan Food Product Innovation Program, Department of Applied Microbiology and Food Science, University of Saskatchewan, Saskatoon, Saskatchewan, S7N 5A8, Canada; and [b]Department of Biochemistry, Memorial University of Newfoundland, St. John's, Newfoundland, A1B 3X9, Canada

Introduction

The seeds of flax or linseed (*Linum usitatissimum* L.) have been used for human consumption since ancient times. Lipids, polysaccharides, and proteins are the major components of this oleaginous seed. Extensive scientific scrutiny carried out on flaxseed during the past two decades has provided a great deal of information related to the chemical, nutritional, and therapeutic value of its lipids, lignans, and polysaccharides. Several food-related applications of flaxseed oil and seed coat polysaccharides have been suggested and adapted in order to increase their dietary intake. However, the protein fraction of flax has not yet attracted as much interest as the other seed macrocomponents. This may partly be attributed to the use of whole seeds in foods without recognizing them as a source of vegetable protein. The inability of flax protein to compete with the versatile physicochemical (e.g., functional) properties that soybean protein ingredients provide may be the most important major drawback in developing flax-based protein products. However, as more information on flaxseed protein becomes available, specific uses and applications may be considered.

Flaxseed protein has been examined nutritionally for its amino acid composition in comparison with other plant proteins and in order to meet human dietary requirements in relation to the presence of antinutritional components. As a food ingredient, concerns have been raised about the bioavailability as well as consistency and stability of the physicochemical properties of proteins during processing and storage (1). Increasing scientific evidence demonstrates that food proteins provide more important physiological and metabolic functions than what is dictated by their essential amino acids; desirable structural and sensory properties all depend on the molecular specifics of proteins. Therefore, the amino acid composition, its sequence and spatial structure should be considered together when characterizing different applications of a protein. Table 21.1 summarizes how the protein molecular organization may relate to the nutritional, physiological, and food-related functional properties. This chapter provides information on the protein fraction

TABLE 21.1

Relationship of Molecular Structure of Food Proteins with Their Physicochemical Characteristics, Functionality, and Nutritional/Physiological Properties

Structural level or processing treatment	Physicochemical property and functionality		Nutritional/physiological property
Structural Level			
Primary structure	Amino acid composition		
	Level of essential amino acods	Curd formation, Distribution in	Nutritional value, N-digestibility
	Isoelectric point	aqueous and liquid phase	Rate of gastric emptying, Rate of
	Hydrophobic/hydrophilic		proteolysis by gastric, pancreatic, and
	interactions		intestinal enzymes, Specificity of
			hydrolysis by proteases, Stimulation of
			gastrointestinal hormones, Rate of
	Amino acid sequence		absorption
	Molecular weight	Antigenicity, Bioactivity, Solubility	
	Site of attack by proteases	Peptide size and amino acid sequence	
Secondary and tertiary structure	Native state	Bioactivity, Solubility, Surface	Site of absorption, Extraction by the gut,
	Denatured state	active properties, Rate of	Rate and appearance in the portal vein,
		enzymatic digestion	Bioactive peptide release in the gut,
			Development of allergenicity/tolero-
			genicity, hormones and growth factors
Processing treatment			
Environment	Food matrix	Accessibility by enzymes, Interactions	
Processing (thermal	Processing treatment affects	with other molecules/nutrients,	
high pressure, etc.)	primary, secondary or	Loss of nutrients	
	tertiary structure		

Source: Adapted from Finot (1).

of flaxseed with respect to its molecular organization, nutritional, therapeutic, and food functional aspects.

Botanical and Chemical Aspects of Flaxseed Proteins

The seed of flax serves as a storage organ for several chemical components synthesized by the plant. The predominant components in the seed endosperm and embryo are lipid and protein; the polysaccharides (mucilage or gum) are concentrated in the testa or seed coat (2,3). Proteins are the main nitrogen reserve for germination and postgermination growth of the seedlings. These storage proteins are deposited in the membrane-bounded intracellular compartments known as protein bodies.

Oomah and Mazza (4) reported that the protein content (nitrogen % × 5.41) of flaxseed varies from 20.9–48.1% in the collection of world accessions. For the period from 1991–2002, protein content of the Canadian-grown flaxseed ranged from 21.8–24.1% (nitrogen % × 6.25 on a dry weight basis [5]). Similar to other oilseeds, the protein content of flaxseed shows a negative correlation with its oil content. As a result, flax grown in cooler climates tends to contain a lower protein content than plants grown in warmer climates (e.g., flax grown in the United States has a higher protein content and a lower oil content than the Canadian-grown seeds). The effect of agronomic, genetic, and environmental factors on the total seed protein content is described elsewhere in this book. It should, however, be noted that flaxseed and its meal contain high levels of nonprotein nitrogen components (6,7) compared with other oilseeds. Therefore, protein calculation data for flax based on the total nitrogen content should be carefully interpreted.

Major Proteins

Similar to other seed proteins, flaxseed protein has an organized molecular structure that is not well understood. Major proteins of flaxseed are 11-12S globulin- and 2S albumin-type.

Globulins. The high molecular weight (HMW) fraction of flaxseed protein is composed of 11-12S globulins, which have properties similar to globulins of other oilseeds such as canola. Vassel and Nesbitt (8) have described the isolation of HMW protein "linin." Purified linin was found to be homogeneous, with an isoelectric point of 4.75, containing 17% nitrogen, 0.6% sulfur, and 0.54% carbohydrate. According to Madhusudhan and Singh (9), the globulin fraction of flax comprised 70–85% of total seed proteins. Marcone *et al.* (10) have estimated that globulins comprise 73.4% of total meal proteins. Globulins can be recovered from the salt-soluble protein fraction (1 M NaCl, at pH 7.5) by salting out (9,10) or cryoprecipitation (11) and further chromatographic purification (e.g., size exclusion and ion exchange). The recent estimation provided by Marcone and coworkers (10) showed that flaxseed globulin has a molecular mass of 320 kDa. Aggregated polymerized protein of molecular mass 459 kDa was also present in the globulin isolate.

The globulin protein of flax is composed of five subunits having molecular mass of 50.9, 35.3, 30.0, 24.6, and 14.4 kDa (10,12,13). Two of these subunits (35.2 and 30.0 kDa) are acidic and one (24.6 kDa) is basic in nature, contrary to what had been reported (1 acidic and 3 basic subunits) by Youle and Huang (14). Weak secondary attractive forces, such as hydrogen, and hydrophobic and electrostatic, rather than disulfide linkages, hold these subunits together (10). High aspartic and glutamic acids of these proteins may also pose repulsive forces between individual subunits and could result in subunit separation in the alkaline pH range. The presence of high amounts of amide (glutamic acids-glutamine and aspartic acid-asparagine and argi-nine) confers a storage role to these proteins (Table 21.2). This regulatory role of globulins can be further confirmed by the observed depletion of the globulin fraction during germination of flaxseed (15). The globulin fraction of flax has a low number of disulfide bonds and is also low in S-containing amino acids compared with albu-mins (Table 21.2). The secondary structure of flax globulin is similar to other dicot globulins by having a low level of α helix (4%) and a high level of α sheet (62.8%) (16–18). A relatively high thermal stability and temperature of denaturation (90 and 114.7°C, respectively) have been reported for 12S globulin of flaxseed (19).

Albumins. The low molecular weight (LMW) protein fraction of flaxseed is mainly composed of albumins (1.6-2S). Vassel and Nesbitt (8) provided the first report on these proteins and termed them "colinin." Youle and Huang (14) reported that 2S proteins of flaxseed account for 42% of the total seed proteins (58% of 11S). However, according to Marcone *et al.* (10), albumins constitute about 26.5% of total proteins. The 2S proteins are soluble in water (about 93%) or dilute salt solutions (0.05 M NaCl, 99% soluble) and can be recovered from the salt-soluble protein frac-tion of seeds after separation of globulins (20). The 2S protein is made up of a single polypeptide chain that has a molecular weight between 16–18 kDa with N-terminal alanine and C-terminal lysine. This protein is rich in lysine, cysteine, glutamic acid, and glycine (20; Table 21.2). It is known that 2S proteins of most dicotyledon seeds have more disulfide linkages than do 11S proteins (21). Albumins of flaxseed exhib-it basic characteristics similar to LMW proteins of castor bean, sunflower, and rape-seed (22). In contrast to the 11S globulin, the albumins of flaxseed are stable over a wide range of pH and temperature (20).

Specific Proteins

Oleosins. Oleosins are proteins associated with the oil bodies of oleaginous seeds. These proteins are embedded with phospholipids (PL) and form the outer surface of oil bodies where triacylglycerols (TAG) form the inner core. Oleosins are highly lipophilic and composed of amphipathic N- and C-terminal domains. The central hydrophobic portion of the protein molecule penetrates the PL layer and into the middle TAG matrix of the oil body. The amphipathic portion resides on the PL layer protruding to the exterior (cytoplasm), thus providing a structural support to the organelle (23). Oleosins of flaxseed have an estimated molecular mass of 16–24 kDa

TABLE 21.2
Amino Acid Composition of Total Protein, Globulins, and Albumins of Flaxseed

| Amino acid | Total protein | | Globulins | | Albumins |
	g/100 g protein of flour[a]	g/16 g N of meal[b]	g/16 g N of purified protein[c]	g/16 g N of crude protein[d]	g/16 g N[d]
Alanine	4.3	4.8	7.9	4.8	1.9
Arginine	10.4	11.5	9.7	12.5	13.1
Aspartic acid + asparagine	8.3	9.2	12.0	11.3	5.5
Cysteine	NR	3.3	1.0	1.4	3.5
Glutamic acid + glutamine	22.8	16.7	21.2	19.8	35.0
Glycine	4.9	6.4	9.2	4.8	8.3
Histidine	5.9	2.7	2.0	2.5	1.6
Isoleucine	4.6	4.8	5.5	4.6	2.8
Leucine	6.5	6.7	5.8	5.8	5.4
Lysine	6.0	4.4	2.7	3.1	4.9
Methionine	3.0	1.5	1.1	1.7	0.8
Phenylalanine	6.5	5.1	4.4	5.9	2.4
Proline	3.0	3.6	NR	4.5	3.0
Serine	4.1	4.9	6.9	5.1	3.9
Threonine	3.1	3.4	3.8	3.9	2.1
Tryptophan	NR	0.5	NR	1.3	2.0
Tyrosine	4.6	2.2	1.7	2.3	1.4
Valine	4.9	5.7	5.6	5.6	2.6

[a]Flax flour (Ref. 14). [b]Flax meall (7). [c](17). [d](18).
Abbreviation: NR, not reported.

and constitute approximately 7.2% (oleosin) of the total seed protein (24). Because of the tight association with the oil bodies, oleosins can be prepared in large quantities by low speed centrifugation of an aqueous extract of the seeds, followed by washing to remove nonoleosin proteins (25). These oleosins have been regarded as suitable sites for production of recombinant proteins in plants of industrial and pharmaceutical importance. It has been shown that proteins of molecular mass up to 68 kDa (e.g., β-glucouronidase) can be expressed as translational fusion at the C-terminus of oleosins without interfering with the oleosin targeting or oil body preparation. Flaxseed oleosins may also have the added advantage in this process, which is based on using the oil as a part of the recovery process (26).

Cadmium-Binding Proteins. Flaxseed is reported to accumulate significant amounts of Cd that exceed the dietary critical value or a maximum level of 0.3 ppm. The Cd content of 109 accessions of flaxseed from the world collection ranged from 0.075–2.775 ppm (27) and in the North American grown flax from 0.14–1.37 ppm (28). In flaxseed, a high concentration of Cd is known to accumulate even at low soil Cd levels. According to Lei *et al.* (29), protein fraction of flax contains Cd-binding components that are rich in SH groups and comprise about 7% of proteins extracted from defatted-dehulled meal with buffered 0.1 M NaCl (pH 8.6). It is estimated that most of the Cd is concentrated in a peptide-rich fraction with a molecular weight of less than 1500 Da. Because the predominant protein fraction of flaxseed has a very low content of Cd, these authors indicated that it was possible to separate the major storage protein of flax with very low Cd levels.

Antifungal Proteins. Proteins possessing antifungal activity have been isolated from flaxseed. Antifungal proteins are part of the plant defense mechanism against invasion by fungal pathogens, and they can be utilized for pharmaceutical purposes if properly isolated. Vigers *et al.* (30) and Borgmeyer *et al.* (31) showed that the protein-containing extract of flaxseed strongly inhibited the growth of the agronomically important pathogen *Alternaria solani* and the human pathogen *Candida albicans*. The purified protein had an estimated molecular weight of 25 kDa and a high degree of homology to other reported pathogenesis-related antifungal proteins (31). The name "linusitin" was suggested for this protein (32).

Amino Acids of Flaxseed Proteins

Several research groups have reported amino acid profiles of flaxseed proteins. Generally, the total flaxseed protein (albumin and globulin together) has an amino acid pattern comparable to soybean protein, and has a relatively high level of aspartic acid (+aspargine), glutamic acid (+glutamine), and arginine (Table 21.2), indicating the high content of amides which serve the storage role in the seeds.

Nutritionally, the limiting amino acids of flaxseed proteins are lysine, threonine, and tyrosine, but the sulfur-containing amino acids are sufficiently available (7,33,34). The percentage ratio of the essential to total amino acids of flaxseed protein was well above the 36% value that is reported for ideal proteins as recommended by FAO/WHO in 1973 (35). Oomah (36) has calculated indicators for flaxseed proteins based on the amino acid composition data available in the literature. Flaxseed protein is high in the branched chain amino acids (BCAA), isoleucine, leucine, and valine. The content of the aromatic amino acids (AAA) phenylalanine and tyrosine is comparatively low. The Fischer ratio (BCCA/AAA) is also high and comparable to that of soybean. Proteins with a high Fischer ratio (high levels of BCCA and low levels of AAA) are suitable for developing physiological functional foods for special needs, such as impaired nutrition associated with burns, cancer, liver failure, and trauma, and for nutritional support of children with chronic or acute diarrhea or milk protein allergies (37). Protein hydrolysates having a high Fischer ratio have been obtained from sunflower globulin and corn zein (38), and those of flax protein show an excellent starting material for such a purpose. Concerning atherogenic effects of proteins, a low value for the lysine/arginine ratio is an indicator of low atherogenicity (39). According to Oomah (36), flax protein has a low lysine/arginine ratio compared with soybean, thus suggesting that it may pose a less lipidemic and atherogenic effect than soy when consumed.

Separation and Isolation of Total Proteins

Isolated or concentrated proteins from flaxseed are not commercially available. The whole seed and full-fat flour (milled flaxseed) are the least refined sources of flaxseed proteins, because they include all the other components of the seed. In order for producers to take advantage of the collective benefits of lipids, polysaccharides, and lignans, flaxseed in the form of whole, ground, or flour offers the best way for their inclusion in food formulations. Prime-quality flaxseed meal (defatted, ground seed) must contain a minimum of 30% protein on a dry weight basis; the available commercial grade meals contain 30–40% protein (40). Table 21.3 illustrates quantitative protein change when flaxseed is processed into its constituent oil and meal.

Influence of Polysaccharides

Isolation of flaxseed proteins in bulk was first reported by Smith *et al.* (41) *via* alkali extraction of oil-free meal followed by their acid precipitation (pH 4.2). This method was a direct adaptation of the protein isolation process from soybean meal; however, it was found unsuitable for flaxseed as the seed coat polysaccharides interfered with the isolation and settling of proteins. It appears that the coextraction of seed coat polysaccharides restricts the production of a high purity protein product from flaxseed. Typical flaxseed processing to obtain oil is by prepress solvent extraction, similar to that of canola crushing. The oil intended for human consumption is produced by cold pressing of the seeds; however, as a second step, hot pressing may be carried out to yield additional oil to be used for industrial purposes (42). Currently existing crushing methods do not remove the seed coat; therefore, polysaccharides remain as an integral part of the seed meal.

Several research groups have suggested different methods to enhance protein recovery without polysaccharide contamination. To separate polysaccharides for the purpose of improving meal digestibility, numerous procedures have been tested since 1945. Removal of seed coat (hull) using graded sieves and air separation for increasing protein recovery was attempted by Smith *et al.* (41). Bolly and McCormack (43)

TABLE 21.3
Quantitative Change of Protein, Oil, and Carbohydrate (percentage on dry weight basis) During Seed Processing

Product	Crude protein[a]	Oil	Carbohydrate
Seed	20.35	43.79	20.35
Flake	21.34	39.31	21.34
Cake	30.42	17.26	30.42
Meal	34.47	5.72	34.47

[a]N% × 5.41.
Source: Adapted from Oomah and Mazza (40).

have patented a method to extract the hull fraction of solvent-extracted meal by air classification. This process afforded a protein-rich kernel fraction and a mucilage-rich hull fraction (<15% protein). A liquid cyclone process that simultaneously extracted oil and separated hulls and meal particles gave 37.5% hulls containing 20.3% protein and 62.4% cotyledon fraction (34,44). A device designed for oat abrasion was successfully utilized to separate seed coat-, protein-, and oil-rich fractions from flaxseed (45). Li-Chan and Ma (19) described an extended procedure as reported by Oomah and Mazza (46) for batch dehulling of flaxseed to obtain a low polysaccharide-containing meal. At first, seeds were mechanically ground using a barley pearler fitted with a 2-mm screen, and then the ground material was air aspirated to remove hulls. Wiesenborn *et al.* (47) have simplified this pearler-based process to separate hulls and embryo of flaxseed which could be scaled up. All these processes show that removal of the seed coat (which is about 40% of seed weight) during milling successfully reduced the polysaccharides of the resultant meal.

A wet process for demucilaging flaxseed has also been attempted with both intact and disintegrated seeds. Soaking the seeds in an acidic solution (1% HCl, wt/vol) followed by washing to remove polysaccharides has been described by Mandokhot and Singh (48). Passing the wet seed mass through a fruit pulper was effective in removing the seed coat (9,49). Use of polysaccharide-degrading enzymes (fungal origin, commercial grade) and pH-elevated water to improve removal of seed coat polysaccharides by solubilization was described by Wanasundara and Shahidi (50). All these methods removed polysaccharides by hydration and solubilization and are known as wet processes. Subsequent protein extraction from the resultant seed fraction may afford a high protein recovery.

Protein Products with Polysaccharides

Dev and Quensel (51) showed that coextraction of polysaccharides may be employed to prepare protein products with varying polysaccharide content. Kankaanpaa-Antilla and Antilla (52) patented a method to produce a dry product containing 45% protein and 11% polysaccharide from cold or hot pressed flaxseed meal. Because the polysaccharides (gum or mucilage) of flaxseed possess beneficial physicochemical properties, they can be used in different food applications based on their performance. In addition, this could be an issue to consider because the polysaccharides of flax have proven physiological benefits (e.g, lowering of glycemic response, 53,54). Flaxseed proteins influence insulin secretion and enhance the effect of associated polysaccharides (55). In addition, association of flaxseed protein with the polysaccharides may have a significant role in the reduction of colon luminal ammonia, thus protecting against the known tumor-promoting effects of ammonia (56). These significant physiological properties of flaxseed polysaccharides and proteins indicate that products that contain both molecular groups may provide a unique combination to develop important ingredients for functional foods.

Protein-Rich Products

Approximately 85% of total proteins can be extracted into 1 M NaCl at pH 7.0 when degummed, defatted, and dehulled meals were used (9). Oomah and coworkers (57) extracted 97% of flaxseed meal proteins using 0.8 M NaCl at pH 8.0. Wanasundara and Shahidi (58) used sodium hexametaphosphate to complex flaxseed proteins under alkaline pH to achieve a high degree of protein recovery (78% of total meal nitrogen); both polysaccharide- and lipid-removed flaxseeds were employed. This protein was mainly composed of globulins and albumins, low in associated phenolic compounds and free of cyanogenic glycosides (12). Krause *et al.* (13) reported preparation of micelle protein isolate from dehulled and defatted flaxseed meal by extraction with 0.5 M NaCl, ultrafiltration, and dilution which afforded a 93% protein isolate. However, how much of the total flaxseed protein could be recovered by the micellization process was unclear. The micelle-isolated product was low in phytate content compared with the protein product obtained by hexametaphosphate complexation. Table 21.4 summarizes the protein content of different protein products reported in the literature.

Functional Properties of Flaxseed Proteins and Processing Effects

As a bulk protein ingredient, the physicochemical behavior of flaxseed proteins (functional properties) is as important as its nutritional value. Vegetable proteins are widely utilized as bulk protein ingredients due to their superior ability to impart and maintain the functional properties that govern food acceptability. In terms of the trend for protein ingredients, plant or vegetable proteins find an increasing demand in the food industry (59). This increased demand is attributed to the negative safety issues of animal-derived protein ingredients, the benefits associated with health-promoting compounds of plant proteins, and the generally driven trend toward vegetarian foods. With regard to all these concerns, flaxseed proteins offer a good potential as a bulk protein ingredient.

To incorporate plant protein ingredients into existing foods or to develop new foods, the physicochemical behavior and functionality of proteins in a complex chemical environment serve as determining factors. According to existing knowledge, food-related functionality of flaxseed protein ingredients may not surpass soy protein, but they have unique properties that can be employed in specific applications or to render certain desired functions.

Several research groups around the world have studied functional properties of flaxseed protein preparations that are important in food products. In general, it may be concluded that the isoelectrically precipitated flaxseed proteins exhibit a more lipophilic character compared with soy which is the most commonly utilized vegetable protein. It has been hypothesized that the coextracted phytic acid and polysaccharides in isoelectrically precipitated proteins also distinctly influence functional properties of proteins (13).

TABLE 21.4

Protein Content of Different Flaxseed Protein Products

Product	Protein (N% 6.25)	Lipid	Ash
Whole flaxseed[a]	23.5	43.8	4.81
Defatted flaxseed meal[b]	43.9	NR	6.40
Commercial flaxseed meal[b]	37.5	NR	7.10
Flax protein concentrate[c]	67.9	2.4	NR
Flax protein isolate			
Extracted at pH 10.0 and precipitated at pH 4.2[d]	83.5	NR	NR
0.8M NaCl, precipitated at pH 4.5[e]			
Extracted at pH 8.9 with 2.75% SHMP and precipitated at pH 3.5[f]	78.1	7.6	9.5
Extracted at pH 8.5 and precipitated at pH 4.0	89.0	NR	NR
Micelle isolate (0.5 M NaCl extracted)[g]	93.0	NR	NR

[a](7). [b](34).

[c]Recalculated on moisture-free basis from (59,60).

[d](61).

[e](57).

[f](58).

[g](13).

Abbreviation: See Table 21.2; SHMP, sodium hexametaphosphate.

The protein isolate prepared by alkali extraction (pH 8.5–10.0) and isoelectric precipitation (pH 3.5–4.2) showed pH-solubility profile, water- and oil-absorption, and emulsifying properties (capacity and stability) comparable to soy protein isolate, but not the foaming properties. Presence of sodium chloride in the system adversely affected the emulsifying activities, but improved its foaming properties (60). The protein products prepared in the same manner, but with different polysaccharide contents (protein isolate with 5% pentosan and protein concentrate with 6 or 9% pentosan), showed that the presence of polysaccharides in the protein preparations improved water and oil absorption, emulsifying capacity, emulsion stability, and foaming capacity, but reduced their solubility (61)). This improved physicochemical behavior of flaxseed protein products reduced cooking loss and improved firmness in cooked meat emulsions and exerted a beneficial stabilizing effect in ice cream preparations and canned fish sauces (62). The isoelectrically precipitated flax protein isolate (very low polysaccharide content) showed poor emulsifying and foaming properties (13,63) within the pH range of 3.0–8.0. The lipophilic properties of flaxseed protein isolates provided a high degree of lipid absorption as a bulk protein ingredient (63). A comparison of gelling behavior of flaxseed protein products showed that isoelectric precipitation produced firmer gels within the pH range of 3.0–8.0 than did the micella-isolated products, which afforded weak gels (13).

Acylation

Attaching acetic or succinyl groups to the reactive side chain-NH_2 groups increases the net negative charge of the protein molecules and markedly changes their conformation. Both acetylation and succinylation significantly improved the pH-dependent solubility of alkali-extracted and isoelectrically precipitated flax proteins (63). Actually, 57% succinylation of flaxseed proteins showed a pronounced increase in protein solubility between pH 3–6. Acylation reduced the lipid-binding capacity of the protein and also the surface hydrophobicity, which relates to its lipophilic nature. This study showed that acylation can be employed to reduce the lipophilic properties of flax proteins, thus improving certain food-related functionalities of the protein.

Detoxification Treatments

Presence of cyanogenic glycosides in flaxseed is well known, and their potential as HCN-releasing molecules remains a concern when seed or meal is used in food (64,65). Researchers have examined several methods to reduce cyanogenic glycoside levels or inactivate them in the seed or meal. Madhusudhan and Singh (66) described an aqueous treatment at high temperature (100°C, 15 min) to detoxify flaxseed meal; however, their process gave a meal with diminished functionalities. The resultant meal had a lower nitrogen solubility and fat absorption capacity, and unfavorable emulsifying and foaming properties, partly owing to the thermal denaturation and dissociation of seed proteins (66,67). Wanasundara *et al.* (65) reported that prior to solvent (hexane) extraction for oil recovery, alkanol (methanol or isopropanol)/ammonia/water [95% (vol/vol)/10% (w/w)/5% (vol/vol)] treatment of crushed seeds reduced the cyanogenic glycosides content without adversely affecting any of the functional properties of the resultant meal (7).

Germination

Quantitative change in the nitrogen-containing components of flaxseed during germination has been reported (15). Germination resulted in a small decrease in the total nitrogen content (from 3.9–3.7% of total dry matter) of the seed. An increase in the water-soluble protein content (LMW proteins) at the expense of the salt-soluble protein fraction was observed. The nonprotein nitrogen content (as percentage of total nitrogen) rose from 9–33.5% during an 8-d germination period (15). Total contents of free amino acids and polyamines were also increased in parallel with the increase in nonprotein nitrogen content (Table 21.5). The amount of free amino acids was significantly increased over the 8-d germination period, and the highest change was seen in the glutamine content (Table 21.5).

The total amount of polyamines (agmatine, putrecine, and spermidine) generated during the germination period was by nearly two orders of magnitude lower than the amount found in beer, cheese, chocolate, fish, sauerkraut, or wine. Polyamines at high levels are known to potentiate tumor growth; hence, their presence at low concentrations in foods is preferred (68). Rapidly growing tissues such as sprouting seeds are considered to have high levels of polyamines; however, germinating flaxseeds were low in polyamines.

Germination appears to offer a natural way for lowering cyanogenic glycoside content of flaxseed. Studies carried out by Wanasundara and others (18) have shown

TABLE 21.5
Contents of Free Amino Acids, Cyanogenic Glycosides, and Polyamines of
Germinating Flaxseed

	Days of germination				
Component	0	2	4	6	8
	Free amino acids (mg/g protein)				
Alanine	0.24 ± 0.01	2.46 ± 0.11	2.43 ± 0.10	3.97 ± 0.20	4.89 ± 0.01
Arginine	1.35 ± 0.09	3.74 ± 0.07	3.16 ± 0.10	4.36 ± 0.13	4.65 ± 0.09
Asparagine	0.31 ± 0.01	0.88 ± 0.01	1.10 ± 0.11	2.00 ± 0.21	3.48 ± 0.05
Aspartic acid	0.43 ± 0.01	0.80 ± 0.03	0.67 ± 0.05	0.92 ± 0.01	1.06 ± 0.04
Cystine	0.06 ± 0.00	0.14 ± 0.10	0.25 ± 0.01	0.33 ± 0.00	0.42 ± 0.03
Glycine	0.19 ± 0.01	0.96 ± 0.08	1.51 ± 0.02	2.58 ± 0.01	6.61 ± 0.10
Glutamine	0.06 ± 0.01	5.11 ± 0.04	8.83 ± 0.12	10.4 ± 0.11	12.6 ± 0.11
Glutamic acid	0.44 ± 0.10	3.17 ± 0.09	3.52 ± 0.11	4.50 ± 0.10	5.83 ± 0.10
Histidine	0.15 ± 0.08	1.24 ± 0.03	1.38 ± 0.08	2.00 ± 0.08	2.47 ± 0.10
Lysine	0.20 ± 0.01	1.14 ± 0.06	1.19 ± 0.02	1.69 ± 0.01	1.95 ± 0.01
Leucine	0.01 ± 0.00	1.84 ± 0.01	1.69 ± 0.10	2.57 ± 0.02	2.63 ± 0.12
Isoleucine	0.04 ± 0.00	1.16 ± 0.10	1.06 ± 0.04	1.32 ± 0.00	1.48 ± 0.00
Methionine	0.01 ± 0.04	0.57 ± 0.02	0.48 ± 0.05	0.55 ± 0.01	0.52 ± 0.00
Phenylalanine	0.13 ± 0.01	1.38 ± 0.06	1.03 ± 0.01	1.57 ± 0.01	1.58 ± 0.01
Proline	0.27 ± 0.02	1.04 ± 0.04	2.35 ± 0.30	4.38 ± 0.10	4.60 ± 0.07
Serine	0.08 ± 0.01	1.92 ± 0.01	1.92 ± 0.01	3.42 ± 0.11	5.08 ± 0.10
Threonine	0.11 ± 0.01	0.99 ± 0.03	1.01 ± 0.00	1.30 ± 0.01	1.62 ± 0.01
Tryptophan	0.37 ± 0.08	0.93 ± 0.05	1.21 ± 0.06	1.79 ± 0.06	1.95 ± 0.08
Tyrosine	0.10 ± 0.00	1.11 ± 0.00	1.14 ± 0.10	1.27 ± 0.00	1.36 ± 0.09
Valine	0.06 ± 0.01	1.43 ± 0.00	1.44 ± 0.01	1.83 ± 0.01	2.32 ± 0.10
Total	4.61 ± 0.47	31.1 ± 0.94	37.4 ± 1.40	52.7 ± 1.19	72.4 ± 1.22
Ammonia	2.27 ± 0.20	5.77 ± 0.50	11.6 ± 0.22	20.1 ± 0.61	32.1 ± 0.35
	Cyanogenic glycosides (mg/g dry matter)				
Linustatin	2.70 ± 0.50	3.50 ± 0.67	3.00 ± 0.55	2.79 ± 0.20	1.60 ± 0.10
Neolinustatin	3.09 ± 0.90	3.60 ± 0.80	2.75 ± 0.18	0.65 ± 0.10	0.28 ± 0.0
	Polyamines (μmol/g dry matter)				
Agmatine	0.08 ± 0.01	1.18 ± 0.10	2.44 ± 0.12	3.32 ± 0.11	4.21 ± 0.44
Putrescine	0.21 ± 0.02	0.09 ± 0.02	0.08 ± 0.01	0.08 ± 0.01	0.07 ± 0.01
Spermidine	0.03 ± 0.01	0.29 ± 0.05	2.54 ± 0.09	4.71 ± 0.23	5.86 ± 0.15

Source: Adapted from Wanasundra *et al.* (15).

that the content of linustatin and neolinustatin of flaxseed was reduced by 40 and
90%, respectively (Table 21.5), thus indicating reduced risk of cyanogenesis.

Hydrolysis

Losso *et al.* (69) have described hydrolysis of flax 11S protein with food-grade
microbial enzymes. High concentrations of free glutamine and glutamic acid were
released compared with the total amino acids. This process may thus serve as a
suitable procedure in developing savory flavorings.

Food Applications and Other Potential Uses

A traditional use of protein-rich meals has generally been for animal feeds. However, the trend is changing as more health benefits of the seed components are recognized. As a component in whole seed and flour, flax proteins are used in a wide variety of bakery, cereal, snack, and health-bar products. Reported applications for isolated bulk protein ingredients (isolated protein) include their use as meat substitutes, emulsifiers, stabilizers, whipping agents, and binders, all of which depend on the higher order (quaternary and tertiary) molecular structure of proteins (16,51,59). More recently, nutraceutical/functional food applications of protein products from plants, especially soy, have increased due to their ability to improve cardiac and bone health (61). It is clear that not only the protein, but also the associated components, such as dietary fiber, phenolic compounds (isoflavones, lignans), glucosinolates, and phytic acid, play an important role in their therapeutic and disease-prevention ability. In addition, development of various physiologically active peptides (e.g., angiotensin-converting enzyme inhibitors, angiogenesis-related enzyme inhibitors, antioxidants, cation binders, opioids) that are released upon *in vivo* digestion of seed proteins has been reported (70,71). This latter area of research for production of novel proteins from flaxseed is still in its infancy, and further developments in this area are warranted.

Summary

Proteins comprise a considerable portion of flaxseed. Nutritionally, flax proteins have properties similar to those of soy proteins. A good understanding of the molecular components of flax proteins and their organization may help in the design of appropriate food applications for protein and protein-containing products. Technological modification(s) that convert proteins into valuable molecules is a new and growing area of innovation in order to use plant proteins, including those of flaxseed.

References

1. Finot, P.A., Physicochemical Characterization of Protein Utilization, in *Proteins, Peptides and Amino Acids in Enteral Nutrition*, edited by P. Furst and V. Young, Nestec, Basel, 2000, pp. 109–119.
2. Dorrel, D.G., Distribution of Fatty Acids Within the Seed of Flax, *Can. J. Plant Sci. 50:*71–75 (1970).
3. Freeman, T.P., Structure of Flaxseed, in *Flaxseed in Human Nutrition*, edited by S. Cunnane and L.U. Thompson, AOCS Press, Champaign, Illinois, 1995, pp. 11–21.
4. Oomah, B.D., and G. Mazza, Flaxseed Proteins—A Review, *Food Chem. 48:*109–114 (1993).
5. Official Web Site of Canadian Grain Commission. Available at: http://www.grain-scand.gc.ca. January 3, 2003.
6. Painter, E.P., and L.L. Nesbitt, Nitrogenous Constituents of Flaxseed—Peptization, *Ind. Eng. Chem. 38:*95–98 (1946).
7. Wanasundara, P.K.J.P.D., and F. Shahidi, Functional Properties and Amino Acid Composition of Solvent-Extracted Flaxseed Meals, *Food Chem. 49:*45–51 (1994).

8. Vassel, B., and L.L. Nesbitt, The Nitrogenous Constituents of Flaxseed. II. The Isolation of a Purified Protein Fraction, *J. Biol. Chem. 159:*571–584 (1945).

9. Madhusudhan, K.T., and N. Singh, Studies on Linseed Proteins, *J. Agric. Food Chem. 31:*959–963 (1983).

10. Marcone, M.F., Y. Kakuda, and R.Y. Yada, Salt-Soluble Seed Globulins of Various Dicotyledonous and Monocotyledonous Plants. I. Isolation/Purification and Characterization, *Food Chem. 62:*27–47 (1998).

11. Dev, D.K., and T. Sienkiewicz, Isolation and Subunit Composition of 11S Globulin of Linseed (*Linum usitatissimum* L.), *Nahrung 31:*167–169 (1987).

12. Wanasundara, P.K.J.P.D., Protein Products and Sprouts from Flax (*Linum usitatissimum* L.) Seed, Ph.D. Thesis, Memorial University of Newfoundland, St. John's, 1995, pp. 169–172.

13. Krause, J-P., M. Schultz, and S. Dudek, Effect of Extraction Conditions on Composition, Surface Activity and Rheological Properties of Protein Isolates from Flaxseed (*Linum usitatissimum* L.), *J. Sci. Food Agric. 82:*970–976 (2002).

14. Youle, R.J., and A.H.C. Huang, Occurrence of Low Molecular Weight and High Cysteine Containing Albumin Storage Proteins in Oilseeds of Diverse Species, *Am. J. Bot. 68:*44–48 (1981).

15. Wanasundara, P.K.J.P.D., F. Shahidi, and M.E. Brosnan, Changes in Flax (*Linum usitatissimum* L.) Seed Nitrogenous Compounds During Germination, *Food Chem. 65:* 289–295 (1999).

16. Dev, D.K., T. Sienkiewicz, E. Quensel, and R. Hansen, Isolation and Partial Characterization of Flaxseed (*Linum usitatissimum* L.) Proteins, *Nahrung 30:*391–393 (1986).

17. Marcone, M.F., Y. Kakuda, and R.Y. Yada, Salt-Soluble Seed Globulins of Dicotyledonous and Monocotyledonous Plants. II. Structural Characterization, *Food Chem. 63:*265–274 (1998).

18. Madhusudhan, K.T., and N. Singh, Effect of Detoxification Treatment on the Physicochemical Properties of Linseed Proteins, *J. Agric. Food Chem. 33:*1219–1222 (1985).

19. Li-Chan, E.C.Y., and C.Y. Ma, Thermal Analysis of Flaxseed (*Linum usitatissimum* L.) Proteins by Differential Scanning Calorimetry, *Food Chem. 77:*495–502 (2002).

20. Madhusudhan, K.T., and N. Singh, Isolation and Characterization of a Small Molecular Weight Protein of Linseed Meal, *Phytochemistry 24:*2507–2509 (1985).

21. Shewry, P.R., Seed Proteins, in *Seed Technology and Its Biological Basis*, edited by M. Black and J.D. Bewley, CRC Press, Boca Raton, Florida, 2000, pp. 42–84.

22. Schwenke, K.D., Studies on Native and Chemically Modified Storage Proteins from Rapeseed (*Brassica napus* L.) and Related Plant Proteins, *Nahrung 34:*225–240 (1990).

23. Tzen, J.T.C., G.C. Lie, and A.H.C. Huang, Characterization of the Charged Components and their Topology on the Surface of Plant Seed Oil Bodies, *J. Biol. Chem. 267:*15626–15634 (1992).

24. Tzen, J.T.C., Y-Z. Cao, P. Laurent, C. Ratnayake, and A.H.C. Huang, Lipids, Proteins, and Structure of Seed Oil Bodies from Diverse Species, *Plant Physiol. 101:*267–276 (1993).

25. Moloney, M.M., Oleosins as Carriers for Foreign Protein in Plant Seeds, in *Engineering Crop Plants for Industrial End-uses*, edited by P.R. Shewry, J.A. Napier, and P.J. Davis, Portland Press. London, 1998, pp. 47–54.

26. Van Rooijen, G.J.H., and M.M. Moloney, Plant Seed Oil-Bodies as Carriers for Foreign Proteins, *Bio/Technol. 13:*72–77 (1995).
27. Oomah, B.D., and E. Kenaschuk, Cultivars and Agronomic Aspects, in *Flaxseed in Human Nutrition*, edited by S. Cunnane and L.U. Thompson, AOCS Press, Champaign, Illinois, 1995, pp. 43–55.
28. Li, Y.M., R.L. Chaney, A.A. Schneiter, J.F. Miller, E.M. Elias, and J.J. Hammond, Screening for Low Grain Cd Phenotypes in Sunflower, Durum Wheat and Flax, *Euphatica 94:*23–30 (1997).
29. Lei, B., E.C.Y. Li-Chan, B.D. Oomah, and G. Mazza, Distribution of Cadmium Binding Components in Flax (*Linum usitatissimum* L.) Seed, *J. Agric. Food Chem. 51:*814–821 (2003).
30. Vigers, A.J., W.K. Roberts, and C.P. Selitrennikoff, A New Family of Plant Antifungal Proteins, *Mol. Plant-Microbe Interactions 4:*315–323 (1991).
31. Borgmeyer, J.R., C.E. Smith, and Q.H. Huynh, Isolation and Characterization of a 25 kDa Antifungal Protein from Flaxseeds, *Biochem. Biophys. Res. Commun. 187:*480–487 (1992).
32. Anžlovar, S., M.D. Serra, M. Dermastia, and G. Menestrina, Membrane Permiabilizing Activity of Pathogenesis-Related Protein Linusitin from Flaxseed, *Mol. Plant-Microbe Interactions 11:* 610–617 (1998).
33. Sosulski, F.W., and G. Sarwar, Amino Acid Composition of Oilseed Meals and Protein Isolates, *Inst. Can. Sci. Technol. J. 6:*1–5 (1973).
34. Bhatty, R.S., and P. Cherdkiatgumchai, Compositional Analysis of Laboratory-Prepared and Commercial Samples of Linseed Meal and of Hull Isolated from Flax, *J. Am. Oil Chem. Soc. 67:*79–84 (1990).
35. Wanasundara, P.K.J.P.D., and F. Shahidi, Process-Induced Compositional Changes of Flaxseed, in *Process-Induced Chemical Changes in Food*, edited by F. Shahidi, C-T. Ho, and N. van Chuyen, Plenum Press, New York, 1998, pp. 307–325.
36. Oomah, B.D., Flaxseed as a Functional Food Source, *J. Sci. Food Agric. 81:*889–894 (2001).
37. Weisdorf, S.A., Nutrition in Liver Disease, in *Textbook of Gastroenterology and Nutrition in Infancy*, 2nd edn., edited by E. Lebenthal, Raven Press, New York, 1998, pp. 665–676.
38. Clemente, A., Enzymatic Hydrolysates in Human Nutrition, *Trends Food Sci. Technol. 11:*254–262 (2000).
39. Czarnecki, S.K., and D. Kritchevsky, Dietary Protein and Atherosclerosis, in *Dietary Proteins: How They Alleviate Diseases and Promote Better Health*, edited by G.U. Liepa, D.C. Bietz, and M.A. Gorman, AOCS Press, Champaign, Illinois, 1992, pp. 42–56.
40. Oomah, D.B., and G. Mazza, Flaxseed Products for Disease Prevention, in *Functional Foods: Biochemical and Processing Aspects*, Technomic, Lancaster, Pennsylvania, 1998, pp. 91–138.
41. Smith, A.K., V.L. Johnsen, and A.C. Beckel, Linseed Proteins, Alkali Dispersion and Peptization, *Ind. Eng. Chem. 38:*353–356 (1946).
42. Grower, D., *Flaxseed, Bi-Weekly Bulletin*, Market Analysis Division, Agriculture and Agri-Food Canada, Winnipeg, Manitoba, Vol. 17, p. 15, 2002.
43. Bolly, D.S., and R.H. McCormack, U.S. Patent 2,593,528 (1952).
44. Sosulski, F.W., and R. Zadernowski, Fractionation of Rapeseed Meal into Flour and Hull Components, *J. Am. Oil Chem. Soc. 58:*96–98 (1981).
45. Myllymaki, O., U.S. Patent 6,440,479 (1998).

46. Oomah, D.B., and G. Mazza, Effect of Dehulling on Chemical Characteristics and Physical Properties of Flaxseed, *Lebensm.-Wiss. u.-Technol. 30:*135–140 (1997).

47. Wiesenborn, D., K. Tostenson, N. Kangas, and C. Osowski, Mechanical Fractionation of Flaxseed for Edible Uses, *Proceedings of the 59th Meeting of Flax Institute of the United States,* Fargo, North Dakota, 2002, pp. 25–29.

48. Mandokhot, V. M., and N. Singh, Studies on Linseed (*Linum usitatissimum*) as a Protein Source for Poultry. I. Process of Demucilaging and Dehulling of Linseed and Evaluation of Processed Materials by Chemical Analysis and with Rats and Chicks, *Lebensm.-Wiss. u.-Technol. 16:*25–31 (1979).

49. Singh, N., Linseed as a Protein Source, *Indian Food Packer 33*(4):54–57 (1979).

50. Wanasundara, P.K.J.P.D., and F. Shahidi, Removal of Flaxseed Mucilage by Chemical and Enzymatic Treatments, *Food Chem. 59:*47–55 (1997).

51. Dev, D.K., and E. Quensel, Preparation and Functional Properties of Linseed Protein Products Containing Differing Levels of Mucilage, *J. Agric. Food Chem. 53:*1834–1857 (1988).

52. Kankaanpaa-Anttila, B., and M. Anttila, U.S. Patent 5,925,401 (1997).

53. Wolever, T.M.S., and Jenkins, D.J.A, *Effect of Dietary Fibre and Foods on Carbohydrate Metabolism*, in *CRC Handbook of Dietary Fibre in Human Nutrition*, edited by G.A. Spiller, CRC Press, Boca Raton, Florida, 1993, pp. 111–152.

54. Wolever, T.M.S., Flaxseed and Glucose Metabolism, in *Flaxseed in Human Nutrition*, edited by S.C. Cunnane and L.U. Thompson, AOCS Press, Champaign, Illinois, 1995, pp. 157–164.

55. Nuttall, F.Q., A.D. Mooradian, M.C. Gannon, C. Billington, C., and P. Krezowski, Effect of Protein Ingestion on the Glucose and Insulin Response to a Standardized Oral Glucose Load, *Diab. Care 7:*465–470 (1984).

56. Clinton, S.K., Dietary Protein and the Origin of Human Cancer, in *Dietary Proteins: How They Alleviate Diseases and Promote Better Health*, edited by G.U. Liepa, D.C. Bietz, and M.A. Gorman, AOCS Press, Champaign, Illinois, 1992, pp. 84–122.

57. Oomah, B.D., G. Mazza, and W. Cui, Optimization of Protein Extraction from Flaxseed Meal, *Food Res. Int. 27:*355–361 (1994).

58. Wanasundara, P.K.J.P.D., and F. Shahidi, Optimization of Hexametaphosphate-Assisted Extraction of Flaxseed Protein Using Response Surface Methodology, *J. Food Sci. 61:*604–607 (1996).

59. Liu, K., Expanding Soybean Food Utilization, *Food Technol. 54* (7):46–58 (2000).

60. Dev, D.K., and E. Quensel, Functional and Micro Structural Characteristics of Linseed (*Linum usitatissimum* L.) Flour and a Protein Isolate, *Lebensm.-Wiss. u.-Technol. 19:*331–337 (1986).

61. Dev, D.K., and E. Quensel, Functional Properties of Linseed Protein Products Containing Different Levels of Mucilage, *J. Food Sci. 53:*1834–1837,1857 (1988).

62. Dev, D.K., and E. Quensel, Functional Properties of Linseed Protein Products Containing Different Levels of Mucilage in Selected Food Systems, *J. Food Sci. 54:*183–186 (1989).

63. Wanasundara, P.K.J.P.D., and F. Shahidi, Functional Properties of Acylated Flax Protein Isolates, *J. Agric. Food Chem. 45:*2431–2441 (1997).

64. Smith, Jr., C.R., D. Weisidler, and R.W. Miller, Linustatin and Neolinustatin: Cyanogenic Glycosides of Linseed Meal that Protect Animals Against Selenium Toxicity, *J. Org. Chem. 45:*507–510 (1980).

65. Wanasundara, P.K.J.P.D., R. Amarowicz, M.T. Kara, and F. Shahidi, Removal of Cyanogenic Glycosides of Flaxseed Meal, *Food Chem. 48:*263–266 (1993).
66. Madhusudhan, K.T., and N. Singh, Effect of Heat Treatment on the Functional Properties of Linseed Meal, *J. Agric. Food Chem. 33:*1222–1226 (1985).
67. Madhusudhan, K.T., and N. Singh, Effect of Detoxification Treatments on the Physicochemical Properties of Linseed Proteins, *J. Agric. Food Chem. 33:*1219–1222 (1985).
68. Bardocz, S., Polyamines in Food and Their Consequence for Food Quality and Human Health, *Trends Food Sci. Technol. 6:*345–346 (1995).
69. Losso, J.N., M. Saito, Y. Horimoto, and S. Nakai, Molecular Weight Distribution of Flaxseed 12S Protein Hydrolyzed with Alcalase® and Flavourzyme® Using Matrix-Assisted Laser Desorption Ionization Time of Flight Mass Spectrometry (MALDI-MS)-53A-6, in *Abstracts of Institute of Food Technologists Annual Meeting and Food Expo*, New Orleans, Louisiana, 1996, p. 118.
70. Ariyoshi, Y., Angiotensin-Converting Enzyme Inhibitors Derived from Food Proteins, *Trends Food Sci. Technol. 4:*139–144 (1993).
71. Losso, J.N., Preventing Degenerative Diseases by Anti-Angiogenic Functional Foods, *Food Technol. 56(6):*78–88 (2002).

Chapter 22

Availability and Labeling of Flaxseed Food Products and Supplements

Diane H. Morris[a] **and Marion Vaisey-Genser**[b]

[a]Mainstream Nutrition, 904-130 Carlton Street, Toronto, Ontario M5A 4K3, Canada; and
[b]University of Manitoba, Human Nutritional Sciences, Winnipeg, Manitoba R3T 2N2, Canada

Introduction

Flaxseed was a recognized food source in prehistoric times in parts of the Middle East, Asia, North Africa, and Europe. It has been said that its use as a food likely preceded the use of flax fibers for cloth (1). At the beginning of the twenty-first century, flaxseed is considered a "functional food" because it satisfies the three requirements for that designation: (i) It is a conventional food, (ii) it is eaten as part of a usual diet, and (iii) it offers physiological benefit or protection against disease (2–4). Of particular interest are the flaxseed components α-linolenic acid (ALA) and the mammalian lignan precursor, secoisolariciresinol diglycoside (SDG), with their potential to decrease the risks of cardiovascular diseases and some forms of cancer (5,6). This chapter describes the various forms of flaxseed available mainly in the North American market or under development; flaxseed's stability to storage and baking; and the labeling of flaxseed food products and supplements in Canada and the United States.

Flaxseed Food Products

Historically, flaxseed has been eaten as a cereal and valued for its medicinal properties. Pioneers in the early 1800s, for example, treated cuts and burns by fashioning a poultice of flaxseed (7). Today, North American consumers are more likely to ingest flaxseed oil and ground flaxseed for a variety of health benefits.

Flaxseed Oil

Flaxseed oil is presently a minor component of the North American vegetable oil market. This situation contrasts with some regions of China, where flaxseed oil is the major oil for human consumption. People in the Gansu region, for example, favor flaxseed oil over mustard and rapeseed oils for frying (8).

Nutritional Value. Flaxseed oil is relatively low in saturated fatty acids, which make up 9% of total fatty acids; it is moderate in monounsaturated fatty acids

(18%) and high in polyunsaturated fatty acids (73%). Flaxseed oil is a rich source of ALA, providing 8 g ALA/14-g portion (Table 22.1). ALA constitutes 57% of the total fatty acids in flaxseed. Linoleic acid makes up 16% of the total fatty acids in flaxseed (9). A 2.5-g serving (1/2 tsp) of flaxseed oil provides about 1.4 g of ALA and is sufficient to meet the Adequate Intake (AI) of 1.1 g ALA/d for women and, when combined with other foods in the diet, the AI of 1.6 g ALA/d for men. AI is a Dietary Reference Intake set by the U.S. Institute of Medicine. For ALA, the AI represents an intake that individuals and groups can aim for and is based on the highest median intake of ALA by adults in the United States. An AI for ALA was also set for children, adolescents, adults, and pregnant and lactating women (10).

Types of Oil. Cold-pressed flaxseed oil is available in low- and high-lignan forms in the North American market. Standard flaxseed oil is naturally low in lignans, because the lignans, which are associated with the fiber fraction of the seed, are removed during the oil-refining process. Standard flaxseed oil contains about 17 μg lignan precursors/g of oil or about 0.2 mg/14-g portion (1 Tbsp), compared with 800 μg lignan precursors/g of flaxseed meal or about 6.4 mg lignan precursors/8-g portion (1 Tbsp) (11).

High-lignan flaxseed oil is manufactured by adding a lignan-containing flax particulate to the standard oil product. The amount of particulate added to the oil is typically 19–20% by weight. Manufacturers report concentrations of 6–11 mg "lignans"/serving (1 Tbsp) (12,13) or 20–55 mg of the lignan precursor SDG/Tbsp (14) in their high-lignan oils. The actual amount of lignan precursors obtained from these oils depends on the degree to which the flax particulate goes into solution and how well it is shaken or stirred prior to consumption (Beutler, J., personal communication). The differences in lignan content among these oils and between the high-lignan oils and ground flaxseed may also be due to the extraction and purification method used by the manufacturer. Muir *et al.* (15) reported significant differences in the concentrations of SDG and the aglycone secoisolariciresinol (SECO) in samples of flaxseed and flax meal, depending on the extraction system. In the case of SECO concentrations in flaxseed, for example, there was a 46-fold difference between values reported for the lowest and highest concentrations. Federal regulations specifying a standard methodology for analyzing SDG and any other lignan precursors in flax products are needed to ensure that label claims are accurate.

Storage and Stability. Until recently, flaxseed oil was shipped refrigerated in opaque bottles to prevent oxidation induced by heat and/or light. Flaxseed oil is commonly shipped today in opaque bottles that do not require refrigeration because the headspace has been flushed with nitrogen to replace any air. Once the bottle has been opened, it must be refrigerated to retard oxidation.

Flaxseed oil, like fish oils and soybean oil, is susceptible to flavor reversion, a term defined as the appearance of an objectionable, hay- or strawlike flavor that

TABLE 22.1

Proximate Composition of Flaxseed Based on Common Measures[a]

Form of Flaxseed	Weight (g)	Common measure	Energy (kcal)	Energy (kJ)	Total fat (g)	ALA (g)	Protein (g)	Total carbohydrate[b] (g)	Total dietary fiber[c] (g)
Proximate analysis	100		450	1890	41	23	20	29	28
Whole seed	180	1 cup	810	3402	74	41	36	52	50
	11	1 Tbsp	50	210	4.5	2.5	2.2	3	3
	4	1 tsp	18	76	1.6	0.9	0.8	1.2	1.1
Ground seed	130	1 cup	585	2457	53	30	26	38	36
	8	1 Tbsp	36	151	3.3	1.8	1.6	2.3	2.2
	2.7	1 tsp	12	50	1.1	0.6	0.5	0.8	0.8
Flaxseed oil	100		884	3712	100	57	—	—	—
	14	1 Tbsp	124	520	14	8.0	—	—	—
	5	1 tsp	44	185	5	2.8	—	—	—

[a]Based on a proximate analysis conducted by the Canadian Grain Commission, May 2001. The fat content was determined using the American Oil Chemists' Society (AOCS) Official Method Am 2-93. The moisture content was 7.7%.

[b]Total carbohydrate includes available carbohydrate (1 g) and total dietary fiber (28 g) per 100 g flaxseed.

[c]Total dietary fiber includes insoluble fiber (19 g) and soluble fiber (9 g) per 100 g flaxseed.

Abbreviation: ALA, α-linolenic acid.

Source: Reprinted with permission of the Flax Council of Canada.

involves oxidation but precedes rancidity (16). Flavor reversion has been associated with a high ALA content, the presence of metallic impurities, and/or traces of lecithin (16–19). The hazard of flavor reversion may have been reduced by the processing and packaging methods used today.

Food Uses and Availability. Flaxseed oil is recommended for "cold" applications like salad dressings, vinaigrettes, and fruit smoothies. It can be used in stir-frying provided a frying temperature less than 150°C is used (20). Its use in frying applications involving higher temperatures is not recommended.

Flaxseed oil can be purchased over the Internet, in health food stores, and in the dietary supplement section of some supermarkets and retail stores. Cold-pressed oil appears to be the most common form. The oil is sold plain; flavored with chili oil or rosemary extract; or blended with other oils such as evening primrose, borage seed, soybean, fish, sesame, and/or sunflower seed oils. Some flaxseed oils contain added nutrients such as conjugated linoleic acid, lecithin, and vitamins A, C, and/or E (12). Many of these products are labeled "organic."

Whole and Ground Flaxseed

The introduction of whole flaxseeds in bakery products likely represents the first large-scale commercial use of flaxseed. Today, both whole seeds and ground flaxseed are added to a growing number of food products.

Nutritional Value. The contribution of whole flaxseeds to the nutrient intake of humans is not known because no clinical studies of their digestibility have been conducted. It is generally accepted that whole seeds tend to pass undigested through the gastrointestinal tract because their resilient seed coat resists the actions of digestive enzymes. Chewing whole seeds shreds the seed coat and exposes nutrients inside the seed to digestive enzymes. Whether humans chew flaxseeds thoroughly enough to achieve a nutritional benefit is not known. Nonetheless, whole flaxseeds provide a crunchy texture in baked goods, cereals, energy bars, and salads and may contribute to the nutrient and soluble fiber intakes of humans, depending on the extent of mastication.

Ground flaxseed is recommended for routine consumption because it is a more digestible form than whole seeds. An 8-g portion of ground flaxseed provides 3.3 g of total fat, 1.8 g of ALA, 1.6 g of protein, and 2.2 g of total dietary fiber (Table 22.1). A daily intake of 8–16 g (1–2 Tbsp) of ground flaxseed, which provides 1.8–3.6 g ALA, meets the AI of ALA for adults (10) and, when consumed regularly, may achieve the spectrum of health benefits reported in clinical studies.

About two-thirds of the total dietary fiber is insoluble and one-third is soluble (9). The total lignan content of raw ground flaxseed has been estimated as 177 µmol/100 g, based on methodology using *in vitro* simulations of colonic fermentation of human fecal samples. Lignan production was significantly greater from cereals containing flaxseed than from cereals containing wheat and rye without flaxseed (21). Adding flaxseed to processed foods increased significantly the production of mammalian lignans.

Types of Whole Seeds. Brown and yellow or "golden" flaxseeds are available in North America. All brown varieties currently purchased by North American consumers are rich in ALA. Of the two varieties of yellow flaxseeds, only the Omega variety is rich in ALA. The other yellow variety, solin, contains less than 5% ALA, and has a higher linoleic acid content of about 70% compared with 15–16% linoleic acid in standard, brown flaxseed (22). In Canada, solin is required to have a yellow seed coat to distinguish it from brown flaxseed (23). Four solin cultivars, Linola™ 947, 989, 1084, and 2047, have been developed by Agricore United. All contain about 1.8–2.2% ALA (Dean, J., personal communication).

Despite claims to the contrary, the nutritional value of the brown and yellow (Omega) flaxseed varieties appears to be similar (Table 22.2). An analysis of a small number of samples showed that the two types have similar moisture and fat contents, but the yellow (Omega) variety contains slightly more protein and monounsaturated fatty acids and slightly less ALA than does the brown flaxseed (Flax Council of Canada, unpublished data). Given the natural variations in oil and fatty acid content due to growing conditions, these differences in nutrient content are negligible.

Brown flaxseeds and the Omega variety are sold directly to consumers through bulk food and health food stores and over the Internet. A few food companies located mainly in Canada and Europe use Linola™ in manufacturing margarine. A small quantity of whole Linola™ seed is sold to the baking industry. Linola™ cultivars are not sold directly to consumers (Dean, J., personal communication).

TABLE 22.2

Comparison of Proximate Analysis (on a Dry Matter Basis) of Brown and Yellow (Omega Variety) Flaxseed[a]

Constituent	Brown flaxseed	Yellow (Omega variety) flaxseed	Brown flaxseed	Yellow (Omega variety) flaxseed
	(g/100 g)		(% of total fatty acids)	
Protein (% nitrogen x 6.25)	22.3	29.2		
Oil/fat	44.4	43.6		
Specific fatty acids				
Saturated fatty acids			8.7	9.0
Monounsaturated fatty acids			18.0	23.5
Polyunsaturated fatty acids				
α-Linolenic acid			58.2	50.9
Linoleic acid			14.6	15.8

[a]Based on a proximate analysis of a small number of samples conducted by the Canadian Grain Commission on the following dates: brown flaxseed, May 11, 2001; yellow flaxseed, July 3, 2001. Moisture content: brown flaxseed, 7.7%; yellow flaxseed, 7.0%.
Source: Reprinted with permission of the Flax Council of Canada.

Stability and Storage. Whether whole or ground, flaxseed appears to be stable to both warm temperatures (autoxidation) and light (photo-oxidation) during common practices for storage and processing. This is surprising, considering that about 40% of the weight of flaxseed is oil of which over half is ALA, a highly unsaturated fatty acid with three double bonds.

Stability of Whole and Ground Seeds. Scientists at Canada's Health Protection Branch and the University of Toronto demonstrated flaxseed stability by measuring the oxygen consumption of 1-g samples that were held in sealed tubes for 280 d at room temperature with 12-h dark/light cycles (24). There was slight use of oxygen and essentially no difference whether the flaxseed was whole or ground (Fig. 22.1).

Two later studies carried out by researchers at the University of Manitoba and the Canadian Grain Commission confirmed the stability of flaxseed when stored at room temperature. In the first of these studies, flaxseed from a mixture of varieties grown in Canada in 1996 was pin-mill ground, packed in 60 lb (27.3 kg) triple-wrapped plastic-lined paper bags, and stored for 128 d at ambient temperatures or about 23°C (25). These conditions simulated those used by commercial bakeries. No significant changes due to storage were found in chemical and physical measurements of peroxide value, free fatty acid content, and conjugated double bond absorbance, or in sensory measurement of the odor intensity of water/flaxseed slurries. Trained panelists examined these slurries at approximately 30-d intervals throughout the study. Furthermore, a panel of 36 consumers was unable to detect a

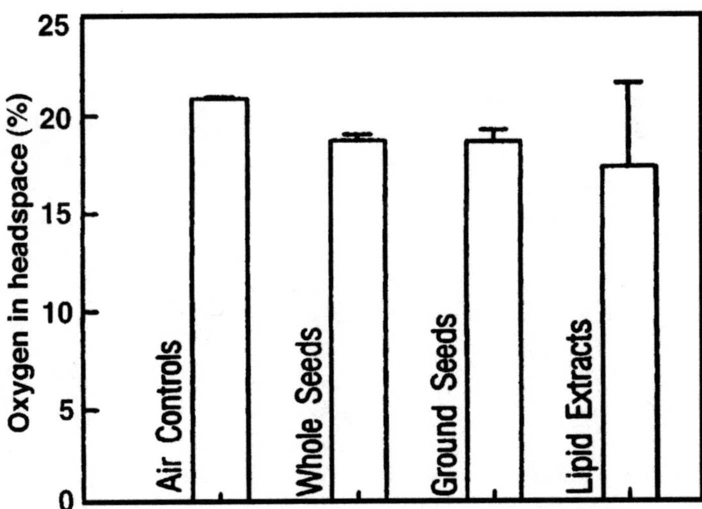

Fig. 22.1 Flaxseed stability to storage. Headspace oxygen concentration in reaction tubes stored for 280 d at room temperature with 12 h dark/light cycles. Values are means ± SD, *n* = 6 tubes. *Source:* Chen *et al.* (24). Used with permission of the American Oil Chemists' Society.

significant flavor difference among slices of yeast bread loaves made with ground flaxseed that had been stored for 0 or 128 d. In both cases flaxseed replaced 10% of the wheat flour (25).

The second study involved ground flaxseed from two different batches that had been warehoused in loosely closed plastic bags and protected from light at ambient temperatures for 11 or 20 mo (26). These stored samples were compared with fresh samples. The ALA content of stored samples was within the normal range of variability, although samples stored 20 mo had a lower ALA content than fresh samples or those stored for 11 mo. Free fatty acids were higher in the stored samples than in fresh ground flax, but peroxide values were similar. This suggested that there was either a low level of oxidation or that the peroxides formed during storage had decomposed. Analyses of the decomposition products of hydroperoxides, a measure of "off-flavor components," showed similar values for the fresh samples and those stored for 20 mo. Tocopherol levels were measured to see if their antioxidant activity was being used to ward off rancidity. However, tocopherol levels for the stored samples were similar to those of fresh, ground flaxseed. An explanation for flaxseed's stability in normal storage conditions remains to be established.

Stability to Baking. Conventional oven temperatures for baking range from about 163–218.5°C. Heat is transferred through air to the food product—that is, the heating is indirect. Despite relatively high oven temperatures, the internal temperature of baking flour mixtures such as breads, muffins, or cakes will seldom exceed the temperature required to gelatinize starch, the major ingredient of wheat flour. In the presence of the amount of sugar used in a cake, starch gelatinization would occur at about 95°C (27).

Considering the internal temperatures of baking products, it is not surprising that studies to date have shown that there was little damage during baking to those flaxseed components that are of major interest to health professionals; these are ALA and the lignans, although gums, fiber, and protein are also important (4). As for ALA, a study published in 1992 (28) showed that heating both whole and ground flaxseed for 60 min at either 100 or 350°C had little if any effect on peroxide values as a measure of oxidation or on the fatty acid composition. Of special note is the fact that gas chromatographic analyses revealed no evidence of the formation of either new *trans* isomers of ALA or of cyclic fatty acids in the flaxseed, even with severe heat treatment.

The stability of the ALA in flaxseed to baking temperatures has been confirmed in follow-up studies. Where samples of a muffin mix had 28.5% of the formula as flaxseed, the ALA content was essentially unchanged after 2 h of baking at 178°C, even though the oxygen consumption during baking was significantly higher than that of a flax-free muffin mix (Fig. 22.2) (24). In a study of the nutritional attributes of flaxseed, the ALA content of muffins containing about 25 g of ground flaxseed again showed no significant reduction due to baking (29). Manthey *et al.* (30) showed that ALA was stable during the processing and cooking of spaghetti fortified with 5, 10, or 15% sieved or unsieved ground flaxseed, and conjugated

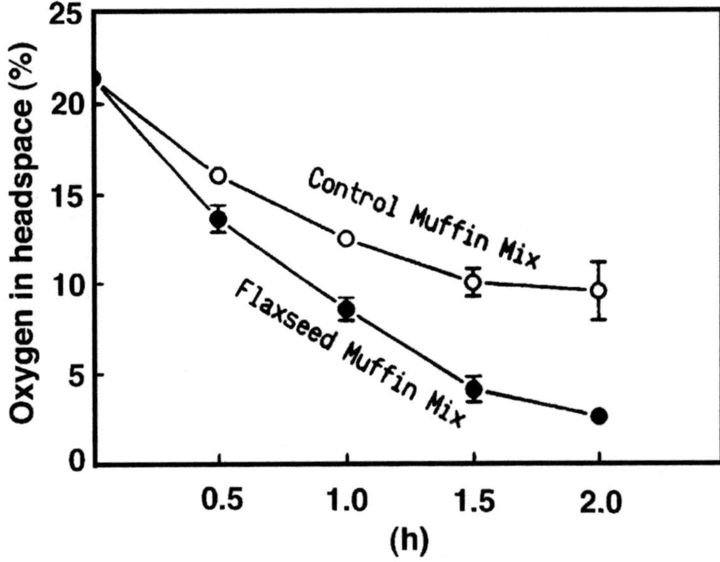

Fig. 22.2. Flaxseed stability to baking at 178°C. Oxygen consumption profiles of flaxseed and plain muffin mixes. Values are means ± SD, *n* = 5 reaction tubes/time point. *Source:* Chen *et al.* (24). Used with permission of the American Oil Chemists' Society.

diene concentrations were lower in the flaxseed-fortified spaghetti than in the unfortified spaghetti.

There is also biological evidence of the baking stability of ALA. No difference was found in the plasma fatty acid profiles of young women consuming 50 g of ground flaxseed daily for 4 wk, whether it was eaten in the raw form added to cereal, juice, soup, or yogurt, and/or as an ingredient of baked bread (31).

Flaxseed is a rich source of the lignan precursor, SDG (32), a phenolic compound that is part of a polar complex associated with the fiber and gums of flaxseed (4,33,34). To date, there has been little information on the storage stability of SDG. However, its baking stability has been confirmed both chemically and biologically. Agriculture and Agri-Food Canada scientists reported no significant difference in the SDG content of the crust and crumb of baked bread, despite the difference in heat exposure between the surface and the internal loaf (35). They later reported that 82% pure SDG added to test loaves was 99.5% recovered, pointing to very good stability to baking temperatures. In contrast, when flax meal was added to bread doughs at three different levels, only 72–75% of the SDG was recovered after baking, regardless of the level of flax meal added (33; Table 22.3). This anomaly remains to be explained. Using an *in vitro* fermentation method, Rickard *et al.* (36) and Nesbitt and Thompson (21) reported that the lignan production from flour mixtures baked at 190°C and pancakes fried at 205°C reflected the

TABLE 22.3

SDG Levels in Flax Breads Prepared with Known Amounts of a Commercially Available Full-Fat Flax Meal[a]

Bread	Actual SDG content[b] (µM)	SDG content of bread by analysis (µM)	Recovery (%)
White flour only	0	0	
White flour + 4% flax meal	154	116 ± 12.3[c]	75.3
White flour + 8% flax meal	313	228 ± 38.6	72.8
White flour + 12% flax meal	463	337 ± 48.6	72.8

[a]Reprinted with permission from Muir, A.D., and N.D. Westcott, Quantitation of the Lignan Secoisolariciresinol Diglucoside in Baked Goods Containing Flax Seed or Flax Meal, *J. Agric. Food Chem. 48:* 4048–4052 (2000). Copyright 2000 American Chemical Society.
[b]The SDG content of the added flax meal was 13.8 ± 0.5 µM/g on a full-fat basis.
[c]± SE, *n* = 3.
Abbreviation: SDG, secoisolariciresinol diglucoside.
Source: Muir and Westcott (33).

amounts of raw flaxseed added to the recipes. Furthermore, they reported that the urinary lignan excretion from nine women was similar whether they ate flaxseed added raw to applesauce or baked in bread or muffins. These biological findings are convincing evidence of the stability of flaxseed SDG to baking temperatures.

Storage at Home. Studies suggest that ground flaxseed can be stored at room temperature for up to 4 mo (25,26). Whole seeds can be stored at room temperature for up to 1 yr. Placing whole or ground flaxseed in the refrigerator or freezer will extend its shelf life.

Food Uses and Availability. Food companies continue to be creative in developing flaxseed ingredients and formulations. Companies perceive that consumers are interested in eating flaxseed as part of a healthy diet that may reduce the risk of chronic diseases and help manage some symptoms of menopause.

Flaxseed Ingredients. Two novel forms of flaxseed have broad applications as ingredients in food products. Encapsulated ground flaxseed was developed by Balchem Corp. as a free-flowing, stable ingredient for use in nutrition/energy bars, breads, sausages, yogurts, ice creams, and other products (37; and Samuels, W., personal communication). The ingredient's capability was showcased at the Institute of Food Technologists' Food Expo in 2002, where participants sampled alligator meat sticks fortified with encapsulated ground flaxseed (38). The other novel ingredient is LinumLife™, a high-lignan flaxseed ingredient developed by the Dutch company Acatris for the men's health market; it will be used in soft gelcaps, tablets, and cereal bars (39). LinumLife™ reportedly provides up to 100 times more lignans than whole flaxseeds or ground flaxseed (40).

Cereals and Porridge. In North America, flaxseed is added to a variety of multigrain cereals like Robin Hood's *Red River Cereal*™\MC, a porridge that was introduced in 1924 and continues to be marketed successfully. However, there is a much longer tradition of using flax as a cereal in East Africa and the Middle East. For example, in Ethiopia, a porridge known as *gumfo* is made using flaxseed that is roasted, crushed, and then cooked, often with salt and red pepper added. Where barley or wheat is available as a porridge base, roasted or stone-ground flax may be added because Ethiopians recognize its health benefits (41).

Baked Goods and Dry Mixes. Ground flaxseed may successfully replace some of the wheat flour in yeast breads (25) and in chemically leavened products such as muffins, tea biscuits, and cookies (42). To date, yeast breads in the form of loaves or bagels have been the most common uses of flaxseed among commercially baked items. Most multigrain breads are marketed to the general public, but one was designed specifically for menopausal women. Burgen® Soy-Lin™ bread was developed by an Australian company, George Weston Foods, based on data from a study conducted at the Monash Medical Center and the Royal Women's Hospital. Four slices of the bread provide 220 mg of soy and flaxseed phytoestrogens (43).

Dry mixes for muffins, pancakes, and waffles are also available in some supermarkets, largely through the efforts of processors such as Pizzey's Milling, which has the largest flax mill in North America. In 2002, Pizzey's Milling shipped about 250 tonnes of flax flour weekly to the food market (44). Several food processors market their flaxseed mixes and waffles over the Internet (45–48).

Sprouts and Sprout Powders. Flax sprouts and powders made from them are relatively new products available in the Canadian market. Flax sprouts are grown to a length of about 1.25–2.5 cm (1/2–1 inch), harvested, mixed with alfalfa or broccoli sprouts, and packaged for sale in plastic containers. Flax sprouts are also dried and powdered. According to company promotional literature, flax sprout powder is lower in dietary fiber than standard ground flax (49). The ALA content of flax sprout powder is reportedly about 15% greater than that of standard ground flax.

Energy Bars. Several flaxseed-containing bars are advertised on the Internet as being ideal for nutritious snacking because of their high-fiber content or their n-3 fatty acid content (50,51). One bar combines flaxseed with rolled oats, brown rice, sesame seeds, raisins, sprouted grains, and blue-green algae (52). A bar designed to help women ease the symptoms of menopause provides 8 g (1 Tbsp) of ground flaxseed per serving (53).

Miscellaneous Products. Flaxseed products marketed over the Internet include a caramel soy-flax chew as a snack food (54), flaxseed-containing granola (46,47), and powdered drink mixes containing ground flax (46). A soy beverage made with flaxseed oil provides 500 mg of n-3 fatty acids/250 mL serving (55), and a whole wheat pasta (spaghetti) made with milled flaxseed provides 500 mg n-3 fatty acids/56-g (2-oz. dry) serving (56). Several companies sell whole seed and/or ground flaxseed over the Internet (48,56,57).

A partially defatted flaxseed meal called Bioflax® was introduced into the Polish pharmaceutical market by the Institute of Natural Fibres. Bioflax® contains about 70% more mucilage than standard, full-fat flaxseed and is recommended by the manufacturer as adjunct therapy in the management of inflammation and ulcers of the stomach and intestines and for the treatment of chronic constipation. Kozlowska *et al.* (58) reported that Bioflax® formulations protected rat gastric mucosa from stress- and ethanol-, but not aspirin-induced injuries.

Products Under Development. Flaxseed oil has been substituted for up to 25% of the milk fat in a frozen dessert resembling ice cream by researchers at North Dakota State University (59). A feasibility study indicated that 60% of members of an untrained sensory panel could not detect the presence of flaxseed oil in frozen dessert samples where 25% of the milk fat was replaced by flaxseed oil. Preliminary data from sensory testing by a trained panel found no objection to a 15% replacement of milk fat with flaxseed oil. A trained sensory panel detected no differences in samples of frozen dessert after 3 mo of storage. Substitution of flaxseed oil for milk fat results in a product containing significantly more ALA than does standard ice cream (Hall, C., personal communication).

Substituting Flaxseed for Wheat in Home Recipes. For successful baked products, some recipe modification is called for when 10–15% of the flour is replaced with whole or ground flaxseed. Payne published a useful review of necessary changes in ingredients, methods of ingredient combination, and baking directions, based on his discussions with commercial bakers (60).

In terms of ingredient quantities, when flaxseed replaces some of the wheat flour, the fat called for in the recipe should be decreased and the water increased. The reason for the fat reduction is that flaxseed itself is high in fat. Oil makes up about 41% of flaxseed weight. Roughly, 240 mL (1-cup) of ground flaxseed will replace 80 mL (1/3-cup) shortening; 360 mL (1-1/2-cups) ground flax can replace 120 mL (1/2 cup) oil. Concurrently, the liquid in the recipe must be increased because flaxseed is hydrophilic. For each 240 mL (1-cup) of flaxseed added, an extra 100 mL (<1/2 cup) of water is recommended. In yeast-leavened doughs containing flax, it has been suggested that the amount of yeast be increased by 25% (60). This is to assure ample loaf volume by compensating for the decrease in the gluten-forming proteins of the wheat that is replaced by flaxseed.

Commercial bakers usually add one or more ingredients to their flaxseed-containing yeast bread formulations to assure good loaf volume. Typical items include "vital gluten" extracted from wheat, ascorbic acid as a flour improver, and a surfactant such as sodium stearoyl lactylate. These ingredients are not readily available on the consumer market. The following list illustrates ingredient amounts in a formulation where 10% of the flour was replaced with ground flaxseed: wheat flour, 855 g; wheat bran, 45 g; ground flaxseed, 100 g; water, 630 mL (630 g); sugar, 40 g; salt, 22.5 g; canola oil, 12 mL (10 g); compressed yeast, 42.5 g; vital wheat gluten, 40 g; sodium stearoyl lactylate, 5 g; and ascorbic acid, 100 ppm (25).

Precautions in the method of combining ingredients for yeast breads have been advised when whole flaxseed is used rather than the ground form. Master bakers recommend conditioning the seed by soaking it in water prior to combining it with the flour. Soaking times vary with the water temperature from as little as 10 min in warm water to 2 h at 21°C. It is further recommended that the soaking water be used as part of the liquid specified in the recipe because it will contain soluble gums that can improve both loaf volume and keeping quality (60).

Dietary Supplements

Supplements containing flaxseed oil are marketed over the Internet and in health food stores. Consumers appear to like the convenience of supplements.

Flaxseed Oil

Most forms of flaxseed oil—low-lignan oil, high-lignan oil, and blended oils (e.g., flaxseed oil and borage oil)—are available in gelatin capsules (12,14). The manufacturers' recommended serving size is 1–6 capsules/d.

Lignan Nutraceuticals

Archer Daniels Midland Company (ADM) announced in 2001 that it had reached an agreement with a Canadian research group called the Flax Consortium to produce and sell a flaxseed lignan complex or purified SDG in dietary supplements, pharmaceuticals, animal feeds, and veterinary products (61). Lignan nutraceuticals will be marketed for the prevention and treatment of chronic diseases.

Labeling of Flaxseed Food Products and Supplements

Labeling regulations in Canada and the United States apply to whole and ground flaxseed, food products containing whole or ground flaxseed, flaxseed oil, products formulated with flaxseed oil, and dietary supplements containing flaxseed oil or purified lignans. This section summarizes briefly the current labeling regulations in Canada and the United States.

Labeling of Foods and Food Products

Canada. The labeling of foods and food products is regulated by the Food and Drugs Act. Two federal departments, Health Canada and the Canadian Food Inspection Agency (CFIA), share responsibility for developing and enforcing food-labeling requirements (62). Health Canada began an extensive review of Canada's nutrition labeling and diet-related health claims in 1998 and released new regulations on nutrition labeling and health claims in December 2002 (63). The Food Directorate of Health Canada's Health Protection Branch is the lead agency in implementing health claims for foods, natural health products, and nutraceuticals and works cooperatively with the CFIA, the Therapeutic Products Programme, and the Office of Natural Health Products (64).

Food Product Labeling. Whole flaxseed, ground flaxseed, or flaxseed oil included in a food formulation must be listed in the ingredient statement on the food product label (65). Linola™ added to margarine formulations may appear in the ingredient declaration as "Linola™ flaxseed oil." This label declaration is intended to help consumers recognize the food with which Linola™ is associated, given that common names are required on food labels and Linola™ is not a common name. However, this label statement is potentially confusing for consumers, who may not realize that Linola™ is a low-ALA variety of flaxseed. The labeling of Linola™ may change under the new regulations (Lee, Y., personal communication).

Nutrition Labeling. Nutrition labeling is the quantitative declaration of energy and of the nutrient content of a food presented in a standard format on the food product label (66). Under the new regulations, nutrition labeling is mandatory on all foods, with some exemptions, such as fresh fruits and vegetables and self-serve bulk foods sold at retail. Therefore, whole flaxseeds sold in bulk to consumers will not require nutrition labeling, but prepackaged whole flaxseeds, ground flaxseed, and flaxseed oil will. Furthermore, when the polyunsaturated fat content of a product is declared on the label, the n-6 and n-3 fat content will also be declared, giving consumers more information about the n-3 fatty acid profile of flaxseed-containing products (67).

Nutrient Content Claims. Nutrient content claims describe, directly or indirectly, the amount of a nutrient in a food (67). The previous regulations permitted label claims and statements related only to linoleic acid and, when authorized by a Temporary Marketing Authorization Letter, claims related to n-3 fatty acids on certain products (68). Under the new regulatory amendments, food manufacturers are allowed to declare the n-3 polyunsaturated fat content of their food products. For example, the label of a food product containing flaxseed may indicate the amount of ALA (e.g., 0.5 g of α-linolenic fatty acid) per serving of stated size. The compositional criterion for an n-3 nutrient content claim is 0.3 g or more n-3 fatty acids per reference amount and per labeled serving (67). Thus, 14 g (1 Tbsp) of flaxseed oil and 8 g (1 Tbsp) of ground flaxseed meet this criterion.

Health Claims. A diet-related health claim is a statement describing the characteristics of a diet that reduce the risk of developing a diet-related disease (67); it is a statement about the health benefits derived from consuming a food product (69). Health claims for n-3 fatty acids are not permitted under current regulations, nor is Health Canada considering an n-3 health claim in its review of generic health claims currently authorized in the United States that may be appropriate for adoption by Canada (70).

Organic Claims. Products such as flaxseed oil or whole or ground flaxseed can be labeled "organic" in keeping with the production, processing, packaging, labeling, storage, and distribution provisions of the National Standard for Organic Agriculture. Certification of products labeled "organic" is voluntary in all provinces except Quebec (71).

United States. Food labeling authority in the United States is shared by two federal agencies, the Food Safety and Inspection Service (FSIS) and the Food and Drug Administration (FDA). FSIS regulates the labeling of meat and poultry products, whereas FDA regulates the labeling of nearly all other food products and the ingredients added to foods. The key legislation governing labeling is the Food, Drug and Cosmetic Act of 1938 and its amendments, including the 1990 Nutrition Labeling and Education Act (72).

Food Product Labeling. Whole flaxseed, ground flaxseed, or flaxseed oil included in a food formulation must be listed in the ingredient statement on the food product label. The FDA has no objection to the use of flaxseed in foods up to a level of 12% flaxseed (9,73).

Nutrition Labeling. The nutrient content of total fat and saturated fat is required on the nutrition label. Declaring the monounsaturated and polyunsaturated content of the food on the nutrition label is optional, unless a claim is made.

Nutrient Content Claims. A nutrient content claim, also called a descriptor, is a label word or phrase used on food packages to describe the level of a nutrient in a serving of food (e.g., "light," "low fat," "high in oat bran"). These claims can be used only if a food meets certain strict definitions specified by government regulations.

Health Claims. The United States defines a health claim as a statement that describes the relationship between a food or food component and a disease or health-related condition. No health claims regarding n-3 fatty acids are currently permitted on food labels (72).

Organic Claims. Products labeled "organic" must conform to regulations of the National Organic Program (NOP). The NOP is directed by the Agricultural Marketing Service of the U.S. Department of Agriculture. A voluntary "USDA Organic" seal can be used on products made from at least 95% organic components (74). Flaxseed products labeled "organic" must comply with NOP regulations.

Labeling of Dietary Supplements

Canada. Dietary supplements are currently regulated as either a food or drug depending on the type of claim made, with the distinguishing characteristic being that dietary supplements (and food products) that purport to cure, treat, mitigate, or prevent illnesses are and will continue to be regulated as drugs (69). For example, some flaxseed oil capsules carry a Drug Identification Number, which is issued by the Therapeutic Products Programme of Health Canada, and they are listed in the Drug Product Database (75). Under the new regulations, gelatin capsules containing flaxseed oil or purified lignans are regulated by Health Canada as natural health products (Lee, Y., personal communication).

United States. In 2000, FDA approved a qualified health claim for n-3 fatty acids and reduced risk of coronary heart disease on the labels of dietary supplements. In reviewing the evidence to support a health claim, FDA concluded that the potential relationship between n-3 fatty acids and coronary heart disease was limited to EPA and DHA (76). ALA was excluded from the definition of n-3 fatty acids in this regulation on the basis that its conversion to EPA and DHA is limited in humans. Therefore, in the United States, manufacturers of dietary supplements containing flaxseed oil cannot use the qualified health claim on their product labels because flaxseed oil contains ALA, not EPA and DHA.

Summary

Consumer awareness of flaxseed as a healthy food ingredient or as a dietary supplement is reflected in the increasing variety of flax products available not only in health food stores but also in supermarkets and over the Internet. These include cereals; baked goods; dry mixes for muffins, pancakes, waffles, and drinks; sprouts and sprout powders; energy bars; pasta; and snack foods. Novel flaxseed ingredients include encapsulated flaxseed and LinumLife™, a lignan-rich ingredient derived from flaxseed. Lignan nutraceuticals are also expected in the North American market. The near future is likely to see the introduction of frozen desserts containing flaxseed oil and an increase in the number of food products formulated with ground flaxseed. Food scientists and nutritionists have bolstered this trend with evidence that ALA and SDG in flaxseed is stable in common storage and baking conditions. Regulations are needed for informative labeling and reliable health claims.

Acknowledgments

The authors thank Barry Hall, Monika Haley, and Paulette Parker at the Flax Council of Canada for their assistance in preparing this manuscript.

References

1. Haggerty, W.J., Flax—Ancient Herb and Modern Medicine, *Herbal Gram 45:*51–57 (1999).
2. Health Canada, Therapeutic Products Programme and the Food Directorate from the Health Protection Branch, Policy Paper: Nutraceuticals/Functional Foods and Health Claims on Foods. Available at www.hc-sc.gc.ca/food-aliment/ns-sc/ne-en/health_claims-allegations_sante/e_nutra-funct_foods.html. Accessed September 30, 2002.
3. Hasler, C.M., Functional Foods: Their Role in Disease Prevention and Health Promotion, *Food Tech. 52:*63–70 (1998).
4. Oomah, B.D., Flaxseed as a Functional Food Source, *J. Sci. Food Agric. 81:*889–894 (2001).
5. Cunnane, S.C., Metabolism and Function of α-Linolenic Acid in Humans, in *Flaxseed in Human Nutrition*, edited by S.C. Cunnane and L.U. Thompson, AOCS Press, Champaign, Illinois, 1995, pp. 99–127.
6. Setchell, K.D.R., Discovery and Potential Clinical Importance of Mammalian Lignans, in *Flaxseed in Human Nutrition*, edited by S.C. Cunnane and L.U. Thompson, AOCS Press, Champaign, Illinois, 1995, pp. 82–98.
7. Buley, R.C., Pioneer Health and Medical Practices in the Old Northwest Prior to 1840, *Mississippi Valley Hist. Rev. 20:*497–520 (1933/1934).

8. Pan, Q., Flax Production, Utilization and Research in China, in *Proceedings of the 53rd Flax Institute of the United States*, Flax Institute, Fargo, North Dakota, 1990, pp. 59–63.

9. Vaisey-Genser, M., and D.H. Morris, *Flaxseed—Health, Nutrition and Functionality*, Flax Council of Canada, Winnipeg, Manitoba, 1997, pp. 9–17.

10. Institute of Medicine, *Dietary Reference Intakes for Energy, Carbohydrate, Fiber, Fat, Fatty Acids, Cholesterol, Protein, and Amino Acids*, The National Academies Press, Washington, D.C., 2002, pp. 8-34–8-39.

11. Axelson, M., J. Sjövall, B.E. Gustafsson, and K.D.R. Setchell, Origin of Lignans in Mammals and Identification of a Precursor from Plants, *Nature 298*:659–660 (1982).

12. Health from the Sun and Other Flaxseed Oils, Marketed on netrition.com's Website. Available at www.netrition.com. Accessed September 23, 2002.

13. Omega Nutrition's Website. Available at www.omeganutrition.com. Accessed August 26, 2002.

14. Barlean's Website. Available at www.barleans.com. Accessed August 26, 2002.

15. Muir, A.D., N.D. Westcott, K. Ballantyne, and S. Northrup, Flax Lignans—Recent Developments in the Analysis of Lignans in Plant and Animal Tissues, in *Proceedings of the 58th Flax Institute of the United States*, Flax Institute, Fargo, North Dakota, 2000, pp. 23–32.

16. Griswold, R.M., *The Experimental Study of Foods*, Houghton Mifflin, Boston, Massachusetts, 1962, p. 270.

17. Armstrong, J.G., and W.D. McFarlane, Flavour Reversion in Linseed Shortening, *Oil Soap 21*:322–327 (1944).

18. Bailey, A.E., Flavor Reversion in Edible Fats, *Oil Soap 23*:55–58 (1946).

19. Smouse, T., Factors Affecting Oil Quality and Stability, in *Methods to Assess Quality and Stability of Oils and Fat-Containing Foods*, edited by K. Warner and N.A.M. Eskin, AOCS Press, Champaign, Illinois, 1994, pp. 17–36.

20. Hadley, M., Stability of Flaxseed Oil Used in Cooking/Stir-Frying, in *Proceedings of the 56th Flax Institute of the United States*, Flax Institute, Fargo, North Dakota, 1996, pp. 55–59.

21. Nesbitt, P.D., and L.U. Thompson, Lignans in Homemade and Commercial Products Containing Flaxseed, *Nutr. Cancer 29*:222–227 (1997).

22. Dean, J.R., Solin: The Newest Crop, in *Proceedings of the Flax Council of Canada Conference: Flax—The Next Decade*, Winnipeg, Manitoba, 1996, pp. 119–138.

23. Flax Council of Canada and the Saskatchewan Flax Development Commission, *Growing Flax: Production, Management and Diagnostic Guide*, 4th edn., Winnipeg, Manitoba, 2002, p. 6.

24. Chen, Z-Y., W.M.N. Ratnayake, and S.C. Cunnane, Oxidative Stability of Flaxseed Lipids During Baking, *J. Am. Oil Chem. Soc. 71*:629–632 (1994).

25. Malcolmson, L.J., R. Przybylski, and J.K. Daun, Storage Stability of Milled Flaxseed, *J. Am. Oil Chem. Soc. 77*:235–238 (2000).

26. Przybylski, R., and J.K. Daun, Additional Data on the Storage Stability of Milled Flaxseed, *J. Am. Oil Chem. Soc. 78*:105–106 (2001).

27. Charley, H., *Food Science*, 2nd edn., John Wiley & Sons, New York, 1992, pp. 33–35, 124–130, 232–245.

28. Ratnayake, W.M.N., W.A. Behrens, P.W.F. Fischer, M.R. L'Abbe, R. Mongeau, and J.L. Beare-Rogers, Chemical and Nutritional Studies of Flaxseed (Variety Linott) in Rats, *J. Nutr. Biochem. 3*:232–240 (1992).

29. Cunnane, S.C., M.J. Hamadeh, A.C. Liede, L.U. Thompson, T.M.S. Wolever, and D.J.A. Jenkins, Nutritional Attributes of Traditional Flaxseed in Healthy Young Adults, *Am. J. Clin. Nutr. 61:*62–68 (1994).

30. Manthey, F.A., R.E. Lee, and C.A. Hall III, Processing and Cooking Effects on Lipid Content and Stability of α-Linolenic Acid in Spaghetti Containing Ground Flaxseed, *J. Agric. Food Chem. 50:*1668–1671 (2002).

31. Cunnane, S.C., S. Ganguli, C. Menard, A.C. Liede, M.J. Hamadeh, Z.-Y. Chen, T.M.S. Wolever, and D.J.A. Jenkins, High α-Linolenic Acid Flaxseed (*Linum usitatissimum*): Some Nutritional Properties in Humans, *Br. J. Nutr. 69:*443–453 (1993).

32. Thompson, L.U., P. Robb, M. Serraino, and F. Cheung, Mammalian Lignan Production from Various Foods, *Nutr. Cancer 16:*43–52 (1991).

33. Muir, A.D., and N.D. Westcott, Quantitation of the Lignan Secoisolariciresinol Diglucoside in Baked Goods Containing Flax Seed or Flax Meal, *J. Agric. Food Chem. 48:*4048–4052 (2000).

34. Harris, R.K., and W.J. Haggerty, Assays for Potentially Anticarcinogenic Phytochemicals in Flaxseed, *Cereal Foods World 38:*147–151 (1993).

35. Muir, A.D., and N.D. Westcott, Quantitation of the Lignan Secoisolariciresinol Diglucoside in Baked Goods Containing Flax Seed or Flax Meal, in *Proceedings of the 56th Flax Institute of the United States*, Flax Institute, Fargo, North Dakota, 1996, pp. 81–85.

36. Rickard, S.E., M. Jenab, J.C.L. Tou, P.D. Nesbitt, and L.U. Thompson, Anticancer Effects and Availability of Flaxseed Lignans, in *Proceedings of the 57th Flax Institute of the United States*, Flax Institute, Fargo, North Dakota, 1998, pp. 8–14.

37. Balchem Corp. Website. Available at at www.balchem.com. Accessed October 4, 2002.

38. Pszczola, D.E., Ingredient Suppliers Define Themselves for Tomorrow, *Food Tech. 56:*70 (2002).

39. Anonymous, Acatris Develops Men's Health Product. Available on the NutraIngredients Website at www.nutraingredients.com. Accessed May 21, 2002.

40. Pszczola, D.E., Products & Technologies, *Food Tech. 56:*85 (2002).

41. Siegenthaler, I.E., The Use of Flaxseed in Ethiopia, in *Proceedings of the 55th Flax Institute of the United States*, Flax Institute, Fargo, North Dakota, 1994, pp. 143–149.

42. Alpers, L., and M.K. Sawyer-Morse, Eating Quality of Banana Nut Muffins and Oatmeal Cookies Made with Ground Flaxseed, *J. Am. Diet. Assoc. 96:*794–796 (1996).

43. Jorgensen, K., D.A.I. Suter, W.K. Thomson, F.S. Dalais, G.E. Rice, and M.L. Wahlqvist, Burgen® Soy-Lin™: Development of an Innovative Functional Staple Food, *Food Aust. 50:*297–299 (1998).

44. Redekop, B., Flaxseed-Bread Operation Rises to the Occasion, *Winnipeg Free Press*, April 13, 2002, A15.

45. Grandma Nunweiler's Website. Available at www.pancakemix.com. Accessed September 10, 2002.

46. Natural Ovens Bakery Website. Available at www.naturalovens.com. Accessed September 10, 2002.

47. Nature's Path Website. Available at www.naturespath.com. Accessed September 10, 2002.

48. Pizzey's Milling Website, Available at www.pizzeys.com. Accessed September 10, 2002.

49. Canadian Organic Sprout Company Website. Available at www.cosc.ca. Accessed September 9, 2002.

50. BumbleBar Website. Available at www.bumblebar.com. Accessed September 10, 2002.

51. Heintzman Farms Website. Available at www.heintzmanfarms.com. Accessed September 10, 2002.

52. Cell Tech Website. Available at www.celltech.com. Accessed September 10, 2002.

53. Zoe Foods Website. Available at www.zoefoods.com. Accessed September 4, 2002.

54. Mai's Foods/Margaret's Kitchen Website. Available at www.maisfoods.com. Accessed September 10, 2002.

55. Soyaworld, Inc. Website. Available at www.sogoodbeverage.com. Accessed September 10, 2002.

56. Hodgson Mill Website. Available at www.hodgsonmill.com. Accessed October 28, 2002.

57. Prairie Flax Products Website. Available at www.flaxproducts.com. Accessed October 29, 2002.

58. Kozlowska, J., K. Szczawinska, T. Bobkiewicz, and R. Forjasz-Grus, The Results of Pharmacological Investigations of Bioflax® Nutrition and Healthy Product Obtained from Flaxseed, in *Proceedings of the 58th Flax Institute of the United States*, Flax Institute, Fargo, North Dakota, 2000, pp. 100–107.

59. Hall, C., and J. Schwarz, Functionality of Flaxseed in Frozen Desserts—Preliminary Report, in *Proceedings of the 59th Flax Institute of the United States*, Flax Institute, Fargo, North Dakota, 2002, pp. 21–24.

60. Payne, T.J., Promoting Health with Flaxseed Bread, *Cereal Foods World 45*:102–104 (2000).

61. Anonymous, ADM to Sell Lignan-Rich Linseed Products, Food Navigator Website. Available at www.foodnavigator.com. Accessed April 9, 2002.

62. Canadian Food Inspection Agency, *Guide to Food Labelling and Advertising*, Section I. Introduction, Section 1.3.1, Canadian Federal Food Labelling Responsibility (amended 09/11/00). Available at www.inspection.gc.ca/english/bureau/labeti/guide/1e.pdf. Accessed September 30, 2002.

63. Health Canada, Regulations Amending the Food and Drug Regulations (Nutrition Labelling, Nutrient Content Claims and Health Claims), December 2002. Available at canadagazette.gc.ca/partII/2003/2003.101/html/sor11-e.html. Accessed March 17, 2003.

64. Health Canada, Product-Specific Authorization of Health Claims for Foods: A Proposed Regulatory Framework, October 2001. Available at www.hc-sc.gc.ca/food-aliment/ns-sc/ne-en/health_claims-allegations_sante/pdf/e_finalproposal.pdf. Accessed September 30, 2002.

65. Canadian Food Inspection Agency, *Guide to Food Labelling and Advertising*, Section II. Basic Labelling Requirements, Section 2.8, List of Ingredients (issued March 25, 1996). Available at www.inspection.gc.ca/english/bureau/labeti/guide/2e.pdf. Accessed September 30, 2002.

66. Canadian Food Inspection Agency, *Guide to Food Labelling and Advertising*, Section V. Nutrition Labelling, Section 5.1.1, What is Nutrition Labelling? (amended January 31, 1997). Available at www.inspection.gc.ca/english/bureau/labeti/guide/5e.pdf. Accessed September 30, 2002.

67. Health Canada, Nutrition Labelling, Nutrient Content Claims and Diet-Related Health Claims: Summary of Proposed Regulatory Amendments (Schedule of Amendments No. 1172), *Canada Gazette*, Part I, June 16, 2001. Available at www.hc-sc.gc.ca/hppb/nutrition/labels/pdf/2001_68e_regsum.pdf. Accessed September 30, 2002.

68. Canadian Food Inspection Agency, *Guide to Food Labelling and Advertising*, Section VI. Nutrient Content Claims, Section 6.2.3.2, Claims for Fatty Acids and Cholesterol (amend-

ed June 10, 1997 and November 9, 2000). Available at www.inspection.gc.ca/english/ bureau/labeti/guide/6e.pdf. Accessed September 30, 2002.

69. Health Canada, Consultation Document—Standards of Evidence for Evaluating Foods with Health Claims: A Proposed Framework, June 2000. Available at www.hc-sc.gc.ca/food-aliment/ns-sc/ne-en/health_claims-allegations_sante/pdf/e_Consultation_doc_en.pdf. Accessed September 30, 2002.

70. Health Canada, Discussion Paper—U.S. Generic Health Claims in Canada: Background and Implementation Issues in the Canadian Context, June 1999. Available at www.hc-sc.gc.ca/food-aliment/ns-sc/ne-en/health_claims-allegations_sante/us-eu/ e_index.html. Accessed September 30, 2002.

71. Canadian Food Inspection Agency, *Guide to Food Labelling and Advertising*, Section IV. Claims as to the Composition, Quality, Quantity and Origin, Section 4.2.9, Organic (Revised 2000). Available at www.inspection.gc.ca/english/bureau/labeti/guide/4e.pdf. Accessed September 30, 2002.

72. Browne, M.B., and the American Dietetic Association, *Label Facts for Healthful Eating*, National Food Processors Association, Washington, D.C., 1993, pp. 1–54.

73. Vanderveen, J.E., Regulation of Flaxseed as a Food Ingredient in the United States, in *Flaxseed in Human Nutrition*, edited by S.C. Cunnane and L.U. Thompson, AOCS Press, Champaign, Illinois, 1995, pp. 363–366.

74. National Organic Program, The National Standards on Organic Agricultural Production and Handling, December 2000. Available at www.ams.usda.gov/nop/index.htm. Accessed September 30, 2002.

75. Health Canada Website, Drug Product Database. Available at www.hc-sc.gc.ca/ hpb/drugs-dpd. Accessed October 9, 2002.

76. Food and Drug Administration, Department of Health and Human Services, Letter to Jonathan W. Emord, Emord & Associates, P.C.: Health Claim: Omega-3 Fatty Acids and Coronary Heart Disease (Docket No. 91N-0103), October 31, 2000.

Chapter 23

Flaxseed in Poultry Diets: Meat and Eggs

Sheila E. Scheideler

University of Nebraska, Department of Animal Science, Lincoln, Nebraska 68583-0908, USA

Introduction

The use of flaxseed in poultry diets as a source of n-3 fatty acids is a relatively new feeding concept in poultry nutrition. However, the use of flaxseed and/or linseed meal as a protein and energy source in poultry diets has been considered since the 1940s (1,2). Despite early work on feeding flaxseed to chickens, until about 1990, the poultry industry did not really investigate the potential of flaxseed as a staple in poultry rations due to several factors including high mucilage content and relative cost as a feed ingredient compared with soybean meal and corn. The interest in utilizing flaxseed in poultry diets as a natural source of n-3 fatty acids flourished in the early 1990s (3–6) and has been further refined in the late 1990s (7–11). A number of management and nutritional factors are of concern when feeding flaxseed. Research at the University of Nebraska resulted in a U.S. Patent on the "Feed to Produce Omega-3 Fatty Acid Enriched Eggs and Method for Producing Such Eggs" (12). A growing number of poultry producers across North America are now feeding flaxseed to commercial laying hens to produce n-3 fatty acid enriched eggs.

Incorporation by poultry of n-3 fatty acids from flaxseed into eggs and meat is actually quite an efficient process. The chicken is able to digest flaxseed in the digestive tract, absorb nutrients from flaxseed in the small intestine, and then further metabolize some of the α-linolenic acid (ALA; C18:3n-3), elongating and desaturating the ALA to eicosapentaenoic acid (EPA;C20:5n-3) and docosahexaenoic acid (DHA;C22:6n-3) for deposition in lipid stores such as egg yolk, and phospholipids in meat products. Flaxseed is a valuable and natural source of n-3 fatty acids for the poultry producer wanting to market an n-3 polyunsaturated fatty acid (PUFA) enriched poultry product.

Flaxseed as a Poultry Feedstuff

Poultry nutrition and diet formulation is a finely tuned scientific process in which the nutrient content of each feed ingredient must be known before a good ration can be formulated. Flaxseed is a good source of ALA and linoleic acid (LA) as well as protein, some essential amino acids, and dietary metabolizable energy. There is some variance in the amount of energy the chicken can derive from flaxseed (13). Very young chicks and chickens are not able to derive as much energy from flaxseed (2086 kcal/kg metabolizable energy) compared with their adult counterparts (3560–4578 kcal/kg metabolizable energy) (13). The young chicken is less able to digest the whole flaxseed due to the high fibrous component of flax. Thus, flaxseed should probably not be fed at high levels (>5% of the ration) to chickens less than 3 wk of age. This is not a problem when creating a poultry egg or meat product enriched with n-3 fatty acids from flaxseed. Broiler chickens (fryers) are reared to 6–8 wk of age before slaughter. If flaxseed is introduced into their diets by 3 wk of age, they will have 3–5 wk to maximize deposition of n-3 fatty acids into the phospholipids of their muscle cell membranes and adipose tissues, which should be an adequate period of time (14). The laying hen will need approximately 21 d of flaxseed consumption to maximize her deposition into the egg and a steady daily intake (10–15% of the diet) thereafter to assure greater than 300 mg n-3 fatty acids/egg (10). Adding up to 30% flaxseed to poultry rations will displace some of the corn, soybean meal, and added fats in the formulation. Dependent on the market price fluctuations of commodities like corn and soybean meal, flaxseed usually increases the cost of the poultry ration. Flaxseed needs to be formulated into a balanced ration, not just added over the top.

Flaxseed oil is not typically utilized in poultry diets as the preferred source of ALA due to its oxidative potential during storage and in mixed feeds. However, some studies have utilized flaxseed oil successfully to alter the n-3 PUFA content of broiler meat (14). If flaxseed oil is utilized, an antioxidant stabilizer such as ethoxyquin needs to be added to the oil during processing and storage to minimize potential free-radical production.

Fatty Acid Metabolism in the Chicken

LA and ALA are required fatty acids in poultry diets. They are provided by a variety of feed ingredients, primarily fat sources such as vegetable oil. They are readily absorbed in the small intestine of the hen as lipid micelles and then reesterified into triglycerides and chylomicrons and transported via the portal vein to the liver. Within the liver, the triglycerides are hydrolyzed by lipoprotein lipase into fatty acid components and then incorporated into triglycerides and phospholipids. Elongation and desaturation of LA and ALA into longer chain fatty acids occurs in the liver (Fig. 23.1). The liver forms a special yolk-targeted very low density lipoprotein (VLDL) which is transported to the ovary and deposited in the developing yolk primarily as lipoprotein. Yolk lipid storage will directly reflect the composition of the dietary fat, hence making yolk lipids a good source of n-3 fatty acids when flaxseed is the major source of dietary fat (15). The elongation and desaturation pathways of LA and ALA are described in Chapter 3. Desaturase and elongase enzymes are essential for these pathways to proceed. The two pathways (n-6 and n-3) compete for enzyme resources, and altering the ratio of dietary n-6/n-3 will alter the end products from each, respectively (15,16). Thus, if dietary ALA is increased by feeding flaxseed, production of EPA and DHA will increase and production of arachidonic acid (AA;C20:4n-6) will decrease. Conversely, overfeeding of LA will impede the production of n-3 PUFA from ALA in the laying hen.

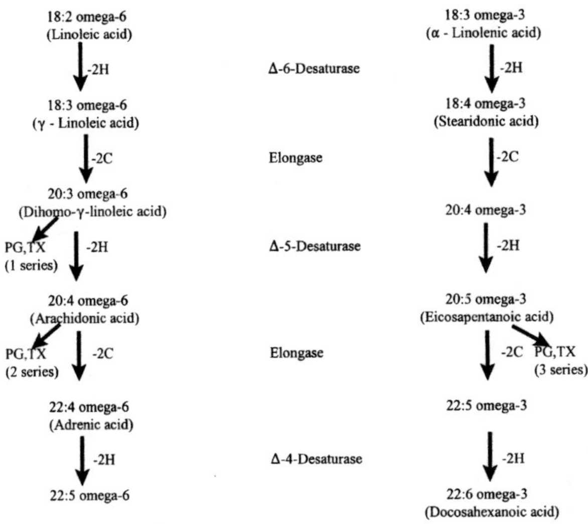

FIG. 23.1. Interactions among polyunsaturated fatty acids (24).

A number of vitamins also play key roles in the activity of elongase and desaturase enzymes in the hen's liver contributing to the efficiency of PUFA metabolism. Pantothenic acid is an important factor in Coenzyme A needed for elongase activity (17). Biotin is also required as a cofactor for elongase and the production of acetyl Co A during carbon chain elongation (17). It is important that chickens being fed flaxseed have an adequate amount of pantothenic acid and biotin supplementation to optimize PUFA synthesis.

Deposition of N-3 Fatty Acids in Poultry Products

Deposition of n-3 PUFA into chicken meat can be done, but this is not necessarily a consumer acceptable means of getting n-3 PUFA into the diet because of the lack of intramuscular fat stores in chicken meat. The broiler chicken deposits n-3 PUFA into phospholipids in the muscle tissue membrance as well as into subcutaneous adipose tissue (18). Most of the fat in poultry meat is subcutaneous, and many consumers do not want to consume that fat and often discard it before consuming the meat. But other consumers obviously like the subcutaneous fat layer associated with "fried" chicken. Nonetheless, chickens will deposit n-3 fatty acids in their lipid stores when fed flaxseed and/or flaxseed oil. Significant accumulation of n-3 PUFA has been found in thigh muscle lipids after 21 d of feeding 1.5–6.0% flaxseed oil in the chickens' ration (14). Born 3 Chicken (19) is an n-3 PUFA enriched chicken product produced from feeding broiler chickens flaxseed (Table 23.1). The product label claims 0.7 g n-3 PUFA/50 g serving size. Little or no taste panel data were available at the time of this review regarding consumer acceptance of n-3 PUFA enriched broiler meat. No research is available regarding n-3 PUFA enrichment of turkey meat at present.

TABLE 23.1
Nutrition Content of a Born 3 Chicken per 50-g Serving[a]

	Born 3 chicken	Regular chicken
Energy (cal/kJ)	75/314	106/444
Protein (g)	9.9	8.7
Fat (g)	3.9	7.9
n-6 polyunsaturates (g)	0.7	0.7
n-3 polyunsaturates (g)	0.7	0.2
Monounsaturates (g)	1.7	4.4
Saturates (g)	0.9	2.6
Cholesterol (mg)	33	39
Carbohydrates (g)	0.1	0.1

Source: Ref. 24.

Deposition of n-3 PUFA into the eggs of laying hens fed flaxseed is a much more feasible source of n-3 PUFA for the consumer than through chicken meat. The egg is a concentrated source of fat (6 g) in the consumers' diet. The deposition of n-3 PUFA from flaxseed into the egg has a direct linear relationship with the amount of flaxseed fed to the laying hen (8,10). The maximum acceptable level of flaxseed in the laying hen diet is 20%. If flaxseed is supplemented above 20% of a complete ration, palatability problems will occur such that the hens will not like to eat the feed. To achieve approximately 350–400 mg n-3 PUFA/egg, hens should be fed between 10 and 20% flaxseed in a complete ration. Consumer tests of taste acceptability of n-3 PUFA enriched eggs have been reported (20). For most of the parameters measured (appearance, texture, flavor, off-flavors) trained panelists could not detect a difference between control *vs.* n-3 PUFA enriched eggs (20).

A number of companies in North America are marketing n-3 PUFA enriched eggs. The University of Nebraska patented process (12) is being produced by several companies and sold as shell eggs in the midwestern part of the United States and as an egg yolk supplement in select baby foods (21; Beechnut First Advantage). Nutrient composition for an Omega egg (12) is given in Table 23.2. The Born 3 company in British Columbia, Canada (19) is also marketing an n-3 PUFA enriched egg. Both companies are reaching their label guarantees for n-3 PUFA by feeding laying hens specialized flaxseed rations.

Human Trial Results with n-3 PUFA Enriched Eggs

A few scientific studies have been reported regarding the health benefits to subjects consuming n-3 PUFA enriched eggs. In one such study, persons with hypercholesterolemia were fed either 12 control or 12 n-3 PUFA enriched eggs/wk for 6 wk in a latin square design including a control no-egg-consumption period (22). The results of the study indicated a 16% reduction in serum triglycerides during the time period of n-3 PUFA enriched egg consumption compared with the no-egg-consumption time period. Other serum parameters such as low density lipoprotein (LDL) and high density

TABLE 23.2
Enriched Egg Composition from Hens Fed a 10% Flax Seed Diet[a]

Egg weight (g)	Average (%)	Average (g/egg)
Yolk	27.5	16.9
Egg yolk lipids	31.6	5.34
Palmitic acid (C16:0)	27.30	1.45
Palmitoleic acid (C16:1)	3.59	0.19
Stearic acid (C18:0)	8.64	0.22
Oleic acid (C18:1)	37.2	1.99
Linoleic acid (C18:2)	14.2	0.760
Linolenic acid (C18:3)	5.61	0.299
Arachidonic acid (C20:4)	0.90	0.048
DPA (C22:5)	0.21	0.011
DHA (C22:6)	1.85	0.099
Total n-6		808 mg
Total n-3		409 mg
n-6/n-3 Ratio		2:1

Source: Scheideler and Froning (8).

lipoprotein (HDL) cholesterol and total cholesterol were not affected by either n-3 PUFA enriched or control egg consumption. Other researchers have failed to find any changes in serum lipid parameters when feeding n-3 PUFA enriched eggs in normal subjects (23). Despite the mixed results of human feeding trials, n-3 PUFA enriched eggs are an economical, concentrated source of n-3 PUFA for the human population.

Summary

Flaxseed is a viable food ingredient in poultry diets to enrich egg yolk lipid or phospholipids in poultry meat and adipose tissue with n-3 PUFA. The chicken will metabolize some of the ALA consumed from flaxseed to EPA and DHA, provided that there is some longer chain PUFA in their lipid stores in addition to abundant ALA. Consumption of n-3 PUFA through enriched eggs could provide the consumer approximately 350 mg n-3 PUFA/egg. Taste and consumer acceptance of n-3 enriched eggs is good to excellent with a number of supermarkets in North America now selling several brands of n-3 enriched eggs.

References

1. Kratzer, F.H., and D.E. Williams, The Improvement of Linseed Oil Meal for Chick Feeding by the Addition of Synthetic Vitamins, *Poultry Sci.* 27:236–238 (1948).

2. Kratzer, F.H., and D.E. Williams, The Relation of Pyridoxine to the Growth of Chicks Fed Rations Containing Linseed Oil Meal, *Poultry Sci.* 27:297–305 (1948).

3. Caston, L., and S. Leeson, Research Note: Dietary Flaxseed and Egg Compostion, *Poultry Sci.* 69:1617–1620 (1990).

4. Jiang, Z., D.U. Ahn, and J.S. Sim, Effect of Feeding Flaxseed and Two Types of Sunflower Seeds on Internal and Sensory Qualities of Eggs, *Poultry Sci.* 70:2467–2475 (1991).

5. Jiang, Z., D.U. Ahn, L. Ladner, and J.S. Sim, Influence of Feeding Full-Fat Flaxseed and Sunflower Seeds on Internal and Sensory Qualities of Eggs, *Poultry Sci.* 71:378–382 (1992).

6. Caston, L.J., E.S. Squires, and S. Leeson, Hen Performance, Egg Quality, and the Sensory Evaluation of Eggs from SCWL Hens Fed Dietary Flax, *Can. J. Anim. Sci.* 74:347–353 (1994).

7. Aymond, W.M., and M.E. Van Elswyk, Yolk Thiobarbituric Acid Reactive Substances and n-3 Fatty Acids in Response to Whole and Ground Flaxseed, *Poultry Sci.* 74:1388–1394 (1996).

8. Scheideler, S.E., and G.W. Froning, The Combined Influence of Dietary Flaxseed Variety, Level, Form, and Storage Conditions on Egg Production and Composition Among Vitamin E Supplemented Hens, *Poultry Sci.* 75:1221–1226 (1996).

9. Scheideler, S.E., G. Froning, and S. Cuppett, Studies of Consumer Acceptance of High Omega-3 Fatty Acid Enriched Eggs, *J. Appl. Poultry Res.* 6:137–146 (1997).

10. Van Elswyk, M.E., Nutritional and Physiological Effects of Flax Seed in Diets for Laying Fowl, *World Poultry Sci. J.* 53:253–264 (1997).

11. Botsoglou, N.A., A.L. Yannakopoulos, D.J. Fletouris, A.S. Tserveni-Goussi, and I.E. Psomas, Yolk Fatty Acid Composition and Cholesterol Content in Response to Level and Form of Dietary Flaxseed, *J. Agric. Food Chem.* 46:4652–4656 (1998).

12. Scheideler, S.E., U.S. Patent 5,897,890 (1999).

13. Gonzalez-Esquerra, R., and S. Leeson, Studies on the Metabolizable Energy Content of Ground Full-Fat Flaxseed Fed in Mash, Pellet, and Crumbled Diets Assayed with Birds of Different Ages, *Poultry Sci.* 79:1603–1607 (2000).

14. Olomu, J.M., and V.E. Baracos, Influence of Dietary Flaxseed Oil on the Performance, Muscle Protein Deposition and Fatty Acid Composition of Broiler Chicks, *Poultry Sci.* 70:1403–1411 (1991).

15. Klasing, K.C., *Comparative Avian Nutrition*, CAB International, Wallingford, Oxon, UK, 1998, pp.171–200.

16. Puthpongsiriporn, U., Effect of Dietary Ratio of Linoleic Acid to Linolenic Acid on Performance and Immune Response in Leghorn Chickens, Ph.D. Thesis, University of Nebraska, Lincoln, 2002, pp. 1–149.

17. McDowell, L.R., *Vitamins in Animal Nutrition*, Academic Press, San Diego, California.

18. Hargis, P.S., and M.E. Van Elswyk, Manipulating the Fatty Acid Composition of Poultry Meat and Eggs for the Health Conscious Consumer, *World Poultry Sci. J.* 49:251–264 (1993).

19. Born 3 Chicken, Available at www.born3.com, Accessed October 4, 2002.

20. Scheideler, S.E., G. Froning, and S. Cuppett, Studies of Consumer Acceptance of High Omega-3 Fatty Acid Enriched Eggs, *J. Appl. Poultry Res.* 6:137–146 (1997).

21. Beechnut First Advantage, Available at www.beechnut.com, Accessed October 4, 2002.

22. Lewis, N.M., K. Schalch, and S.E. Scheideler, Serum Lipid Response to n-3 Fatty Acid Enriched Eggs in Persons with Hypercholesterolemia, *J. Am. Dietetic Assn.* 100:365–367 (2000).

23. Ferrier, L.K., L.J. Caston, S. Leeson, J. Squires, B.J. Weaver, and B.J. Holub, α-Linolenic Acid and Docosahexaenoic Acid Enriched Eggs from Hens Fed Flaxseed: Influence on Blood Lipids and Platelet Phospholipid Fatty Acids in Humans, *Am. J. Clin. Nutr.* 62:81–86 (1995).

24. The Return of ω-3 Fatty Acids into the Food Supply. 1. Land-based Animal Food Products and Their Health Effects, *World Rev Nutr. Diet*, Basel, Karger, 83:166–175.

Chapter 24

Effects of Feeding Flaxseed to Pigs

K.R. Matthews

Meat and Livestock Commission, Winterhill House, Snowdon Drive, Milton Keynes MK6 1AX, UK

Introduction

Flaxseed and the meal that results from extracting the oil from the seed can both provide cost-effective ingredients in pig rations. Whole flaxseed contains, on a percent moisture-free basis: oil, 41; protein, 26; ash, 4; crude fiber, 5; and nitrogen-free extract, 24 (1). Flaxseed meal contains ash, 7; protein, 44; crude fiber, 9; and nitrogen-free extract, 40 percent. However, the first limiting amino acid in flaxseed is lysine, with lower levels than soybean or canola (1). Lysine is often the limiting amino acid in pig diets, and the increased requirement for synthetic lysine can therefore increase ration costs.

In a book on the role of flaxseed in human nutrition, the main interest in feeding whole flaxseed to pigs is the high level of the n-3 fatty acid α-linolenic acid (18:3n-3) at approximately 50% of the fatty acid composition (1). Indeed, in a review carried out for the British Ministry of Agriculture Fisheries and Food in 1997, flaxseed was the land-based plant material identified with the highest level of n-3 fatty acids. Only green forages have similar levels on a percentage (of fatty acids) basis, but the low oil content of these materials makes them a relatively poor source of the fatty acids. Human dietary intakes of n-3 polyunsaturated fatty acids, in particular the longer chain n-3 fatty acids associated with oily fish, are considered too low. It was demonstrated as early as 1926 that dietary polyunsaturated fatty acids (PUFA) are incorporated readily into the pig's fat stores [see the review by Warnants *et al.* (2)]. The pig can therefore provide a useful vehicle for increasing the supply of n-3 fatty acids in the human diet.

Of particular importance for human health is the ability of the pig to use dietary α-linolenic acid as a precursor for the longer chain n-3 PUFA, in particular docosapentaenoic acid (DPA, 22:5n-3) and eicosapentaenoic acid (EPA, 20:5n-3). The desaturase enzyme system is shared by the n-3 and n-6 fatty acids for chain elongation and desaturation. The extent to which elongation of 18:3n-3 occurs can therefore be influenced by the ratio of 18:3n-3 to linoleic acid (18:2n-6) in the diet. This will be discussed.

This chapter reviews the role of flaxseed in feeding pigs with particular emphasis on enhancing the nutritional properties of pork through increased n-3 PUFA levels. Particular attention has been given to work published since 1995.

Animal Performance and Cost of Production

The use of flaxseed in a finishing ration has generally been found to have no adverse effect on production traits (3–7). From the animal husbandry perspective, the main concern related to the use of flaxseed is the possible impact on the cost of the ration. Nevertheless, flaxseed has been successfully incorporated into least-cost formulated rations with minimal impact on the total cost of the diet. The cost implications under any particular circumstance will depend on the relative prices of a wide range of raw materials, and it is beyond the scope of this chapter to discuss this in detail.

Fatty Acid Composition of Pork Products and Role in Human Nutrition

Use of Flaxseed

Over a decade ago, milled flaxseed fed to pigs at 5% of the ration by weight was reported to increase the n-3 fatty acid content of a wide range of pork tissues, but the rise in long chain n-3 PUFA was variable with no effect in brain, fat, skin, skeletal muscle, or kidney but a rise in liver and heart (8). Of particular interest from a human health perspective are the most heavily consumed tissues, muscle and fat. Liver is also of interest because of the high inherent levels of unsaturated fatty acids.

Table 24.1 summarizes the reported effects of feeding flaxseed at different levels on the n-3 PUFA levels of the muscle (*longissimus*), fat (backfat), and liver. Clearly the largest effects are seen in the levels of 18:3n-3. The elongation products are seen to varying degrees in the different tissues.

In the backfat, there are relatively low levels of the long chain n-3 PUFA compared with the high levels of 18:3n-3. There is also an apparent limit to the proportion of 18:3n-3 that can be deposited in the backfat. In our work (5) with probably the longest reported time on the treatment diet, at the highest level (10% flaxseed), the 18:3n-3 in the fat was actually lower than at the intermediate level (5%).

Romans (3) reported on two earlier trials in which four levels of flaxseed (0, 5, 10, and 15%) were fed to pigs for 25 d prior to slaughter, and 15% flaxseed was fed for 0, 7, 14, 21, or 28 d prior to slaughter. They showed an increase in 18:3n-3 in the backfat with flaxseed in the diet and with time. However, flaxseed was substituted for corn and soybean meal in the diet, and the level of dietary oil increased from 3 to 8% as the level of flaxseed went from 0 to 15%. The *de novo* synthesis of fat from starch would therefore be relatively low on the higher flaxseed diet, resulting in a higher proportion of the backfat being from dietary sources. In this work, different backfat layers were also found to respond differently to flaxseed in the diet; the middle/inner layer was more responsive than the outer layer.

Generally, where they have been measured, the longer chain PUFA 22:5n-3 and 20:5n-3 have been increased in backfat through feeding flaxseed, whereas 22:6n-3 is largely unaffected.

As in most areas of meat science, the *longissimus* muscle is the only skeletal muscle that has been measured for fatty acid analysis. This is useful for measuring the impact of the diet on muscle fatty acid levels, but care needs to be taken in interpreting the results. Taugbol and Saarem (9) demonstrated differences between muscles in their fatty acid composition, with the adductor muscles having higher levels of long chain n-3 fatty acids than both *psoas minor* and *longissimus* following supplementation with cod liver oil. On the other hand, Fontanillas *et al.* (6) found no difference in n-3 levels between *longissimus* and *trapezius* muscles. In considering health claims that can be made with pork and products from supplemented pigs, it is important to ensure that the tissues actually used have been analyzed. As with fat tissue, the largest changes observed in the muscle are increases in 18:3n-3 levels. Increases in 20:5n-3 and 22:5n-3 are also usually reported, but 22:6n-3 levels are unchanged.

Enser *et al.* (10) found that the major n-3 PUFA in the liver were all increased, whereas we found that the 22:6n-3 level did not increase (5). The ratio of n-6/n-3 fatty acids was reduced markedly in all tissues in the studies reporting flaxseed feeding.

Warnants *et al.* (2) highlighted the care needed with meat products derived from pigs fed high levels of PUFA. In order to reduce the risk of quality deterioration, it is advisable to reduce the 18:2n-6 level while increasing the 18:3n-3 level. This should maintain a similar total PUFA level while shifting the balance in favor of n-3 fatty acids. Indeed, the most efficient incorporation of 18:3n-3 acid into pig tissues has been achieved by those workers who have avoided increasing the total PUFA of the diet, (e.g., 4). This strategy was also successfully adopted by Enser *et al.* (10). In their work, relatively small changes in the fatty acid composition of the diet were shown to be sufficient to facilitate changes in fatty acid composition.

In terms of the most practical strategies to increase the n-3 PUFA in pig tissues, both long-term and short-term feeding have proved effective. Longer term low-level feeding is generally more effective at achieving higher total levels of elongated n-3 products in porcine tissues. However, over time, proportions of the long chain n-3 PUFA are reduced (Wood, J.D., personal communication). This is probably due to the dilution effect of (saturated) fatty acids deposited as a result of *de novo* synthesis (7). The ratio of precursor fatty acids in diet is important in determining the relative rates of elongation (i.e., 18:2n-6 and 18:3n-3 in the diet). As long ago as 1971, it was proposed that competition between fatty acids of the different series for the desaturase system determined the relative proportions of the elongated products (11).

It is worth observing that genetic differences in the sensitivity to PUFA in the diet have been demonstrated (12). The slightly different effects seen in different studies may, in part, be the result of this difference.

Application Methods Other than Use of Seed

Fontanillas *et al.* (6) compared different supplementary oils at 4% of the diet of finishing pigs for 82 d prior to slaughter and found that there was a more efficient uptake of the n-3 fatty acids using flaxseed oil than is usually reported with use of the seed:

TABLE 24.1

Comparison of Reported Effects of Flaxseed on the Levels of n-3 Fatty Acids in Various Tissues[a]

Flaxseed in diet (%)	Period fed flaxseed[b]	Backfat (outer layer where separated)							
		18:3n-3	20:3n-3/20:4n-3	20:5n-3	22:5n-3	22:6n-3	Total n-3	Ratio n-6/n-3	Reference
0	25 d	4.6	—	0.074	—	0.068	4.74	—	(3)
5	25 d	9.2	—	0.12	—	0.071	9.39	—	(3)
10	25 d	15	—	0.16	—	0.095	15.26	—	(3)
15	25 d	18	—	0.18	—	0.067	18.25	—	(3)
15	0 d	3.4	—	0.023	—	0.028	3.45	—	(3)
15	7 d	6.5	—	0.047	—	0.038	6.59	—	(3)
15	14 d	7.9	—	0.073	—	0.045	8.02	—	(3)
15	21 d	9.3	—	0.069	—	0.036	9.41	—	(3)
15	28 d	12	—	0.11	—	0.053	12.16	—	(3)
1.9	36–107 kg	3.1	—	—	—	—	—	—	(4) (barrows)
3.7	36–107 kg	5.1	—	—	—	—	—	—	(4) (barrows)
5.4	36–107 kg	6.8	—	—	—	—	—	—	(4) (barrows)
		(mg/g fatty acids)							
0	24 d	14	2.1	ND	1.4	0.6	18.1	8.6	(7)
11.4	24 d	46	5.6	0.5	1.8	0.5	54.4	3.2	(7)
0	65 d	13	1.7	0.2	1.5	0.7	17.1	8.4	(7)
1	65 d	20	2.8	0.2	1.8	0.7	25.5	5.4	(7)
2	65 d	29	3.8	0.5	2.2	0.8	36.3	3.9	(7)
3	65 d	38	5.3	0.7	2.5	0.9	47.4	3.2	(7)
0	30 to 77–87 kg	28	—	—	—	—	—	—	(5)
5	30 to 77–87 kg	106	—	—	—	—	—	—	(5)
10	30 to 77–87 kg	88	—	—	—	—	—	—	(5)
To supply 1.5 g $C_{18:3}$/kg diet	25–82 kg	19	0.3	0.5	1.9	0.8	22.5	9.08	(9) (entire males)
To supply 4.5 g $C_{18:3}$/kg diet	25–82 kg	28	0.4	0.7	2.5	1.2	32.8	5.58	(9) (entire males)

M. longissimus (polar lipids)

Flaxseed in diet (%)	Period fed flaxseed[b]	18:3n-3	20:5n-3	22:6n-3	Total n-3	Reference
15	0 d	0.021	0.0067	0.0024	0.03	(3)
15	7 d	0.071	0.012	0.0023	0.09	(3)
15	14 d	0.10	0.017	0.0028	0.12	(3)
15	21 d	0.11	0.025	0.0047	0.14	(3)
15	28 d	0.13	0.027	0.0039	0.16	(3)

M. longissimus (nonpolar lipids)

Flaxseed in diet (%)	Period fed flaxseed[b]	18:3n-3	20:5n-3	22:6n-3	Total n-3	Reference
15	0 d	0.16	0.0048	0.0026	0.16	(3)
15	7 d	0.26	0.0055	0.0021	0.27	(3)
15	14 d	0.36	0.0082	0.0021	0.37	(3)
15	21 d	0.41	0.012	0.0035	0.43	(3)
15	28 d	0.42	0.014	0.0031	0.43	(3)

M. longissimus (total lipids)

Flaxseed in diet (%)	Period fed flaxseed[b]	18:3n-3	20:3n-3/20:4n-3	20:5n-3	22:5n-3	22:6n-3	Total n-3	Ratio n-6/n-3	Reference
0	24 d	0.14	0.02	0.04	0.09	0.03	0.32	8.5	(7)
11.4	24 d	0.43	0.05	0.08	0.11	0.03	0.70	3.8	(7)
0	65 d	0.10	0.01	0.04	0.09	0.04	0.28	8.2	(7)
1	65 d	0.16	0.02	0.06	0.11	0.04	0.39	5.8	(7)
2	65 d	0.23	0.03	0.08	0.13	0.04	0.51	4.7	(7)
3	65 d	0.28	0.03	0.10	0.14	0.04	0.59	3.9	(7)
to supply 1.5 g C$_{18:3}$/kg diet	25 to 82 kg	0.12	—	0.04	0.10	0.04	0.29	9.02	(9) (entire males)
to supply 4.5 g C$_{18:3}$/kg diet	25 to 82 kg	0.17	—	0.08	0.13	0.06	0.44	5.03	(9) (entire males)

(continued)

TABLE 24.1

Comparison of Reported Effects of Flaxseed on the Levels of n-3 Fatty Acids in Various Tissues *(continued)*

Flaxseed in diet (%)	Period fed flaxseed[b]	(mg/g fatty acids)							Reference
0	30 to 77–87 kg	11	—	6	13	8	38	7.2	(5)
5	30 to 77–87 kg	33	—	18	22	8	81	3.5	(5)
10	30 to 77–87 kg	39	—	19	20	7	85	3.9	(5)
1.9	36 to 107 kg	12	—	5	ND	ND	17	—	(16) (barrows)
3.7	36 to 107 kg	19	—	7	ND	ND	26	—	(16) (barrows)
5.4	36 to 107 kg	23	—	8	ND	ND	31	—	(16) (barrows)

Liver

Flaxseed in diet (%)	Period fed flaxseed[b]	18:3n-3	20:5n-3	22:5n-3	22:6n-3	Total n-3	Ratio n-6/n-3	Reference
to supply 1.5 g $C_{18:3}$/kg diet	25 to 82 kg	40.3	26.1	102	73	241	5.39	(9) (entire males)
to supply 4.5 g $C_{18:3}$/kg diet	25 to 82 kg	68.8	82.5	131	112	394	3.06	(9) (entire males)
		(mg/g fatty acids)						
0	30 to 77–87 kg	8	12	38	22	80	4.7	(5)
5	30 to 77–87 kg	28	40	37	17	122	2.9	(5)
10	30 to 77–87 kg	44	59	62	22	187	1.8	(5)

[a]mg fatty acid/g tissue except where stated otherwise.
[b]Period fed flaxseed is given in days fed diet prior to slaughter or the live weight range of pigs on test.
Abbreviations: ND, not detected (as reported by original authors); —, not reported.

18:3n-3 increasing to 9% of the total *longissimus* fatty acids, and 20:5n-3 and 22:5n-3 increasing threefold compared with pomace oil.

The use of calcium salts of flaxseed oil fatty acids has several advantages over the use of the seed or the oil, as described by Barowicz and Pietras (13). The salts can be readily mixed with other fatty acids, making compositional modification easy. They undergo limited oxidation (reducing the need for antioxidants to protect the diet), and the resulting "fodder fat" (mixed salts and additional fatty acids) can be easily mixed with other feed ingredients. These authors have found results with the use of fodder fat to be similar to the use of flaxseed; no effect on production performance or carcass or meat quality but significant improvements in 18:3n-3 and EPA in muscle. These authors also cite a number of other papers reporting experiments supporting these data, but unfortunately most are available only in Polish.

Carcass and Meat Quality

Carcass Quality

In the main, no differences in carcass weight, fat/lean content, or other carcass quality attributes have been associated with higher levels of flaxseed in the diet. Van Oeckel *et al.* (4) found that the highest level of flaxseed in the diet (5.4%) resulted in lower carcass lean content and poorer conformation than the intermediate level (3.7%). Neither was different from the lowest level (1.9%) however.

The major concern related to quality as perceived by the butcher is the effect of PUFA on fat firmness, with higher PUFA levels leading to a risk of soft fat. It would be anticipated that, where total dietary PUFA has not been changed, the fat firmness would be unaffected. In the majority of studies this is the case. Subjective fat firmness scores have even shown a tendency for the higher flaxseed levels to result in more firm fat (4).

Rancidity and Fat Stability

Flaxseed oil has a high propensity to rancidity (fat oxidation) as indicated by its high iodine value (14), and this might be expected to result in a more rapid development of rancidity in the tissues of animals fed flaxseed.

Fat oxidation in animal tissues is usually measured by the use of thiobarbituric acid reactive substances (TBARS). Van Oeckel *et al.* (4) found that TBARS increased with increased flaxseed (2.27–6.19) following 6–18 months of frozen storage in vacuum pack. However, feed levels of vitamin E and selenium were low (only 25 mg/kg and 0.2 mg/kg, respectively). Sheard *et al.* (15) examined the effect of elevated 18:3n-3, 20:5n-3, and 22:5n-3 on the oxidative stability of chops and liver produced from 80 pigs. In these animals, vitamin E was included in the diet at 100 mg/kg. The *longissimus* was removed from one side of each carcass, conditioned for 10 d in vacuum pack, and used to prepare chops for simulated retail display and sensory analysis. Livers were chilled overnight, sliced, packed in modified atmosphere (75% oxygen and 25% carbon dioxide), and placed under simulated retail display. There was no effect of the diet on TBARS in liver displayed for 8 d. In the loin chops displayed for up to 10 d, the test diet (4.5 g 18:3n-3/kg diet) actually gave lower TBARS than the control, although the difference was small, with little effect on color.

Riley *et al.* (7) found that TBARS were increased in loin by feeding at the highest level (3%) for 65 d. All TBARS were below the level likely to be detected by consumers as rancid flavor. Again, there were no color effects.

Rey *et al.* (16), using just 0.5% flaxseed oil, found higher propensity to rancidity associated with relatively small increases in levels of unsaturated fatty acids in muscle. This is likely to be more of a problem in diets containing flaxseed oil rather than the seed, because the seed has endogenous antioxidants that help protect against rancidity development. The authors found that supplementary alpha-tocopherol acetate (vitamin E) in the diet reduced the development of rancidity. It is interesting to note that they also observed that alpha-tocopherol increased the levels of some monounsaturated fatty acids and n-3 PUFA and suggested that this is through an influence on desaturase activity. Supplemental vitamin E may therefore help to protect not only dietary fatty acids against rancidity, but also to increase the meat percentages of the desirable long chain n-3 fatty acids.

Although there are some effects on the development of rancidity in pork products, they are relatively small and can be prevented by sufficient vitamin E included in the diet.

Sensory Quality

No negative effects have been reported on the eating quality (sensory panel scores) of pork loin (17). We (5) reported a small depression of abnormal odor scores at intermediate levels of flaxseed (5%). Riley *et al.* (7) found a slight increase in tenderness and that pork chops had slightly higher juiciness.

Consumer Acceptability

No scientific research has been carried out to evaluate the effects of flaxseed feeding on pork eating quality as assessed by consumers. A few years ago, the Meat and Livestock Commission (U.K.) carried out some focus groups with consumers on their perception of pork products with enhanced levels of n-3 fatty acids. This showed that, although there was a niche group of interested consumers who might select pork labeled in this way, for the vast majority of consumers concerns were raised over the "naturalness" of the process. This may reflect a peculiarly British attitude to animal husbandry, but it is worth considering the consumer reaction to product labeling before investing large sums in marketing pork with enhanced n-3 PUFA.

Further Processed Products

Unsurprisingly, differences in the fatty acid composition of pork tissues have been found to lead to differences in the fatty acid composition of products (3,15). Because the processing conditions often provide for a greater degree of fat oxidation, care needs to be taken in the use of pigs fed flaxseed to avoid rancidity. Bacon was produced from the carcasses resulting from the work reported by Romans (3), and lard from the carcasses was used to make pastry. Flaxseed at more than 5% of the diet resulted in bacon that had flavor differences as detected by a trained taste panel. This was the case regardless of whether the bacon had been subjected to prolonged frozen storage. Loin resulted in a less marked difference in sensory panel scores. Conversely, Sheard *et al.* (15) found that bacon from pigs fed the test diet was slightly less salty and more tender, and that sausages from pigs fed the test diet had a slightly higher overall liking score. The oxidative stability of bacon and English sausages produced from these pigs was tested during simulated retail display following a period of frozen storage. There was no effect of the diet on TBARS in bacon displayed for 9 d or sausages displayed for 9 d. Color measurements indicated little effect of diet in bacon or sausages. In the work of Riley *et al.* (7), in minced pork patties, 2 and 3% flaxseed for 65 d increased TBARS at the start of display but not after 7 d.

Further processed products can, then, be made from meat derived from animals fed flaxseed. Care should be taken to minimize the risk of oxidation, and the use of supplementary vitamin E in the diet of the pigs can help in this regard.

Summary and Conclusions

Flaxseed can be included in the diet of growing and finishing pigs with little effect on the production performance of the animals or sensory qualities of the resulting products. This results in an increase in the n-3 PUFA of the tissues most commonly consumed. Extrapolating from levels of n-3 PUFA achieved through feeding flaxseed, we have concluded that, if the diet of all finishing pigs were modified in this way, the target n-6/n-3 ratio (of 6) for the British population could be achieved. This does not, of course, mean that pork alone would provide all the n-3 fatty acids to the diet, nor that sufficient long chain n-3 fatty acids would be provided, but it does demonstrate the huge potential for flaxseed in contributing to an improved dietary fatty acid balance through the feeding of pigs (5). Flaxseed feeding to pigs can, therefore, contribute to an improved human diet through providing essential fatty acids that are currently consumed in insufficient quantities.

Acknowledgments

The author would like to thank Neil Hudson of the Meat and Livestock Commission library for his help in searching the literature and obtaining references for this review. Thanks are also due to my wife, Jennifer, for allowing me the time to prepare this chapter and for her proofreading service.

References

1. Bhatty, R.S., Nutrient Composition of Whole Flaxseed and Flaxseed Meal, in *Flaxseed in Human Nutrition*, edited by S.C. Cunnane and L.U. Thompson, AOCS Press, Champaign, Illinois, 1995, pp. 22–42.

2. Warnants, N., M.J. Van Oeckel, and M. De Paepe, Fat in Pork: Image, Dietary Modification and Pork Quality, *Pig News and Information 22*:107N-113N (2001).

3. Romans, J.R., Flaxseed and the Composition and Quality of Pork, in *Flaxseed in Human Nutrition*, edited by S.C. Cunnane and L.U. Thompson, AOCS Press, Champaign, Illinois, 1995, pp. 348–362.

4. Van Oeckel, M.J., M. Casteels, N. Warnants, and C.V. Boucque, Omega-3 Fatty Acids in Pig Nutrition: Implications for Zootechnical Performances, Carcass and Fat Quality, *Arch. Anim. Nutr. 50*:31–42 (1997).

5. Matthews, K.R., D.B. Homer, F. Thies, and P.C. Calder, Effect of Whole Linseed in the Diet of Finishing Pigs on Performance, Quality and Fatty Acid Composition of Various Tissues, *Br. J. Nutr. 83*:637–643 (2000).

6. Fontanillas, R., A. Barroeta, M.D. Baucells, and R. Codony, Effect of Feeding Highly *cis*-Monounsaturated, Trans, or n-3 Fats on Lipid Composition of Muscle and Adipose Tissue of Pigs, *J. Agric. Food Chem. 45*:3070–3075 (1997).

7. Riley, P.A., M. Enser, G. Nute, and J.D. Wood, Effects of Dietary Linseed on Nutritional Value and other Quality Aspects of Pig Muscle and Adipose Tissue, *Anim. Sci. 71*:483–500 (2000).

8. Cunnane, S.C., P.A. Stitt, S. Ganguli, and J.K. Armstrong, Raised Omega-3 Fatty Acid Levels in Pigs Fed Flax, *Can. J. Anim. Sci. 70*:251–154 (1990).

9. Taugbol, O., and K. Saarem, Fatty Acid Composition of Porcine Muscle and Adipose Tissue Lipids as Affected by Anatomical Location and Cod Liver Oil Supplementation of the Diet, *Acta Vet. Scand. 36*:93–101 (1995).

10. Enser, M., R.I. Richardson, J.D. Wood, B.P. Gill, and P.R. Sheard, Feeding Linseed to Increase the n-3 PUFA of Pork: Fatty Acid Composition of Muscle, Adipose Tissue, Liver and Sausages, *Meat Sci. 55*:201–212 (2000).

11. Brenner, R.R., The Desaturation Step in the Animal Biosynthesis of Polyunsaturated Fatty Acids, *Lipids 6*:567–575 (1971).

12. Cameron, N.D., J.D. Wood, M. Enser, F.M. Whittington, J.C. Penman, and A.M. Robinson, Sensitivity of Pig Genotypes to Short Term Manipulation of Plasma Fatty Acids by Linseed Feeding, *Meat Sci. 56*:379–386 (2000).

13. Barowicz, T., and M. Pietras, Fattening Performance and Meat Value of Pigs Fed Calcium Salts of Linseed Oil Fatty Acids in their Diet, *Ann. Anim. Sci.-Rocz. Nauk. Zoot. 26*:249–261 (1999).

14. Madsen, A., K. Jakobsen, and H.P. Mortensen, Influence of Dietary Fat on Carcass Fat Quality in Pigs: A Review, *Acta. Agric. Scand. Sect. A, Anim. Sci. 42*:220–225 (1992).

15. Sheard, P.R., M. Enser, J.D. Wood, G.R. Nute, B.P. Gill, and R.I. Richardson, Shelf Life and Quality of Pork and Pork Products with Raised n-3 PUFA, *Meat Sci. 55*:213–221 (2000).

16. Rey, A.Y., J.P. Kerry, P.B. Lynch, C.J. Lopez-Bote, D.J. Buckley, and P.A. Morrissey, Effect of Dietary Oils and Alpha-Tocopherol Acetate Supplementation on Lipid (TBARS) and Cholesterol Oxidation in Cooked Pork, *J. Anim. Sci. 79*:1201–1208 (2001).

17. Van Oeckel, M., M. Casteels, N. Warnants, L. Van Damme, and C.V. Boucque, Omega-3 Fatty Acids in Pig Nutrition: Implications for the Intrinsic and Sensory Quality of the Meat, *Meat Sci. 44*:55–63 (1996).

Index

Atherosclerotic plaques, on the intimal surface
of aortae, 265
Athymic mouse model, effects of flaxseed
and SDG on human breast cancer,
206–208
Atomic absorption spectroscopy (AAS), 19
ATP III. *See* Adult Treatment Panel III
Australian studies, 195
Autoimmune diseases, effects of secoisolariciresinol diglycoside on, 337
Availability of α-linolenic acid to the fetus
and neonate, 176–177
uniquely low α-linolenic acid and linoleic
acid in body fat and brain lipids, 176
Availability of flaxseed food products
and supplements, 404–422
dietary supplements, 415
flaxseed food products, 404–415
labeling of dietary supplements, 417–418
labeling of foods and food products,
415–417
Azoxymethane (AOM) induction, 213, 215

B

β-glucuronidase, 214–215, 391
β-glycosidase, 158
17β-hydroxysteroid dehydrogenase, 137
β-oxidation of common dietary fatty acids, 74
Bacteroides, 101
Baked goods and dry mixes, from flaxseed, 413
Balchem Corp., 412
Base hydrolysis method, 96–97
BCAA. *See* Branched-chain amino acids
Benign prostatic hyperplasia (BPH), 138, 211,
238
BevGrad™, 381
BHA. *See* Butylated hydroxy anisole
BHT. *See* Butylated hydroxy toluene
Bifidobacteria, 101
Bioavailability of lignans, 101–110
absorption and metabolism, 101–102
and diet, 110
disposition of lignans, 102–108
factors affecting, 108–110
and individual variability, 109
and processing, 108–110
Biochemical pathway, for synthesis
of docosahexaenoic acid, 178–179
Bioflax®, 381, 413–414
Biological plausibility, of secoisolariciresinol
diglycoside, 336–337

Biomarkers of α-linolenic acid intake
and breast cancer risk, 234–235
and prostate risk, 237
Biosynthesis of podophyllotoxin, in *Linum
flavum* and *Podophyllum* sp., 50
Biosynthetic pathway
to 5-methoxypodophyllotoxin, proposed, 50
to the lignans, 46
Biosynthetic pathway to secoisolariciresinol
using *Forsythia intermedia* and *Thuja
plicata*
lignans and phenylpropanoids, 45
pinoresinol/lariciresinol reductase, 48–49
secoisolariciresinol dehydrogenase
(SDH), 49
stereoselective coupling and dirigent
proteins, 46–48
Biotechnia Canada, 29
Biotin supplementation, 425
Bipolar disorder, 185
Biscuits, from flaxseed, 413
Bisphenol, 123
Black soil zones, 31
Blood glucose levels, in type-2 diabetes, 282
Blood levels, of lignans, 105–106
Blood pressure effects, of α-linolenic acid,
248–249
Blood triglycerides, ALA increases in, 70
BMC. *See* Bone mineral content
BMD. *See* Bone mineral density
Body tissues, lignans in, 104–105
Bone metabolism, flaxseed effects on, 319–325
Bone mineral content (BMC), 320, 323–324
Bone mineral density (BMD), 320, 323–325,
328
BPH. *See* Benign prostatic hyperplasia
Brain function and infant development,
α-Linolenic acid (ALA) and, 174–193
Branched-chain amino acids (BCAA), 392
Breads, from flaxseed, 26, 108, 150, 341,
379–380, 413
Breast cancer, 199–209, 232–236
biomarkers of α-linolenic acid intake
and breast cancer risk, 234–235
effects of enterolactone and SDG
in the carcinogen-treated rat model,
205–206
effects of flaxseed and SDG in early
exposure and carcinogenesis, 208–209
effects of flaxseed and SDG
in the carcinogen-treated rat model,
201–205

F

T